Chemical Ecology in Aqua

Chemical Ecology in Aquatic Systems

EDITED BY

Christer Brönmark
Lund University, Sweden

and

Lars-Anders Hansson
Lund University, Sweden

OXFORD
UNIVERSITY PRESS

OXFORD
UNIVERSITY PRESS

Great Clarendon Street, Oxford OX2 6DP

Oxford University Press is a department of the University of Oxford.
It furthers the University's objective of excellence in research, scholarship,
and education by publishing worldwide in

Oxford New York

Auckland Cape Town Dar es Salaam Hong Kong Karachi
Kuala Lumpur Madrid Melbourne Mexico City Nairobi
New Delhi Shanghai Taipei Toronto

With offices in

Argentina Austria Brazil Chile Czech Republic France Greece
Guatemala Hungary Italy Japan Poland Portugal Singapore
South Korea Switzerland Thailand Turkey Ukraine Vietnam

Oxford is a registered trade mark of Oxford University Press
in the UK and in certain other countries

Published in the United States
by Oxford University Press Inc., New York

British Library Cataloguing in Publication Data

Data available

Library of Congress Cataloging in Publication Data
Library of Congress Control Number: 2011945452

Typeset by SPI Publisher Services, Pondicherry, India
Printed and bound by
CPI Group (UK) Ltd, Croydon, CR0 4YY

ISBN 978–0–19–958309–6 (Hbk.)
      978–0–19–958310–2 (Pbk.)

1 3 5 7 9 10 8 6 4 2

Contents

List of contributors

Jelle Atema—Boston University Marine Program, 5 Cummington Street, BRB 307, Boston MA 02215, USA

Josh R. Auld—Department of Biology, West Chester University, West Chester, PA 19383, USA

Finn Baumgartner—Department of Marine Ecology, Tjärnö Marine Biological Laboratory, Gothenburg University, 45296 Strömstad, Sweden

Thomas Breithaupt—Department of Biological Sciences, University of Hull, Hull, HU6 7RX, UK

Christer Brönmark—Department of Biology/Aquatic Ecology, Lund University, Ecology Building, SE-223 62 Lund, Sweden

Grant E. Brown—Department of Biology, Concordia University, 7141 Sherbrooke Street West, Montreal, Quebec, Canada, H4B 1R6

Gunnar Cervin—Department of Marine Ecology, Tjärnö Marine Biological Laboratory, Gothenburg University, 45296 Strömstad, Sweden

Douglas P. Chivers—Department of Biology, University of Saskatchewan, 112 Science Place, Saskatoon, Saskatchewan, Canada, S7N 5E2

Charles D. Derby—Neuroscience Institute and Department of Biology, Georgia State University, P.O. Box 5030, Atlanta, GA 30302-5030, USA

Swantje Enge—Department of Marine Ecology, Tjärnö Marine Biological Laboratory, Gothenburg University, 45296 Strömstad, Sweden

Maud C. O. Ferrari—Department of Veterinary Biomedical Sciences, WCVM, University of Saskatchewan, 52 Campus Drive, Saskatoon, Saskatchewan, Canada, S7N 5B4

Gabriele Gerlach—Institute of Biology and Environmental Sciences, Carl von Ossietzky University Oldenburg, 26111 Oldenburg, Germany

Elisabeth M. Gross—Department of Biology, Limnology, PO Box 659, University of Konstanz, 78457 Konstanz, Germany and Laboratoire des Interactions Ecotoxicologie Biodiversité Ecosystèmes (LIEBE) - CNRS UMR 7146, Université de Lorraine (UdL), Campus Bridoux, 57070 Metz, France

Lars-Anders Hansson—Department of Biology/Aquatic Ecology, Lund University, Ecology Building, SE-223 62 Lund, Sweden

Jörg D. Hardege—Department of Biological Sciences, University of Hull, Hull, HU6 7RX, UK

Cornelia Hinz—Institute of Biology and Environmental Sciences, Carl von Ossietzky University Oldenburg, 26111 Oldenburg, Germany

Cynthia Kicklighter—Goucher College, Department of Biological Sciences, 1021 Dulaney Valley Road, Baltimore, MD 21204-2794, USA

Julia Kubanek—School of Biology, Georgia Institute of Technology, 310 Ferst Dr., Atlanta, Georgia 30332, USA

Christian Laforsch—Department of Biology II & GeoBio-Center, Ludwig-Maximilians-University Munich, Grosshaderner Str. 2, 82152 Planegg-Martinsried, Germany

Catherine Legrand—School of Natural Sciences, Linnaeus University, Landgången 3, S-39182, Kalmar, Sweden

Miquel Lürling—Aquatic Ecology & Water Quality Management Group, Wageningen University Droevendaalsesteeg 3a, 6708 PB Wageningen, The Netherlands and Department of Aquatic Ecology, Netherlands Institute of Ecology (NIOO-KNAW), PO Box 50, 6700 AB Wageningen, The Netherlands

Göran M. Nylund—Department of Marine Ecology, Tjärnö Marine Biological Laboratory, Gothenburg University, 45296 Strömstad, Sweden

Henrik Pavia—Department of Marine Ecology, Tjärnö Marine Biological Laboratory, Gothenburg University, 45296 Strömstad, Sweden

Scott D. Peacor—Department of Fisheries and Wildlife, Michigan State University, 10D Natural Resources, East Lansing, MI 48824-1222, USA

Georg Pohnert—Friedrich Schiller University Jena, Institute of Inorganic and Analytical Chemistry, Lessingstraße 8, 07743 Jena, Germany

Karin Rengefors—Department of Biology/Aquatic Ecology, Lund University, Ecology Building, 223 62 Lund, Sweden

Erik Selander—Technical University of Denmark, Charlottenlund Slot, Jægersborg Allé 1, 2920 Charlottenlund, Denmark

Ole B. Stabell—Department of Natural Sciences, Faculty of Engineering and Science, University of Agder, Postboks 422, 4604 Kristiansand, Norway

Ulrich K. Steiner—INSERM U1001, Faculte Cochin, 24 rue Faubourg Saint Jacques, 75014 Paris, France

J. Robin Svensson—Department of Marine Ecology, Tjärnö Marine Biological Laboratory, Gothenburg University, 45296 Strömstad, Sweden

Urban Tillmann—Department of Ecological Chemistry, Alfred Wegener Institute, Am Handelshafen 12, D-27570 Bremerhaven, Germany

Ralph Tollrian—Department of Animal Ecology, Evolution and Biodiversity, Ruhr University Bochum; NDEF 05; Universitätstsstraße 150; D 44780 Bochum, Germany

Gunilla B. Toth—Department of Marine Ecology, Tjärnö Marine Biological Laboratory, Gothenburg University, 45296 Strömstad, Sweden

Andrew M. Turner—Department of Biology, Clarion University, Clarion, Pennsylvania 16214, USA

Eric von Elert—Cologne Biocenter, University of Cologne, Zülpicher Strasse 47b, 50674 Köln, Germany

Linda Weiss—Department of Animal Ecology, Evolution and Biodiversity, Ruhr University Bochum; NDEF 05; Universitätstsstraße 150; D 44780 Bochum, Germany

Marc Weissburg—School of Biology, 310 Ferst Dr, Georgia Institute of Technology, Atlanta, GA, 30332-0230, USA

Richard K. Zimmer—Department of Ecology and Evolutionary Biology, Neurosciences Program, and Brain Research Institute, University of California, Los Angeles, CA 90095-1606, USA

Chemical ecology in aquatic systems—an introduction

Christer Brönmark and Lars-Anders Hansson

Life as we know it started as a chemical reaction, the synthesis of amino acids in the primordial soup, and ever since then chemistry has been involved in all aspects of life on our planet. Ecological interactions can be affected by chemical processes at all scales, from metabolic reactions and signalling systems within cells and individual organisms, chemical communication among individuals, stoichiometric relationships and nutrient dynamics at the ecosystem level, to regional and global element cycles. A plethora of studies have shown that aquatic organisms from many different taxa and functional groups respond to minute concentrations of chemical substances released by other organisms. Although much understanding has been achieved there are still too many questions that lack an answer; a situation that calls for further studies and syntheses of the knowledge already gathered.

In this book we focus on a subset of all the possible chemical–ecological interactions, namely the importance of *chemical defences* and *infochemicals* for ecological processes in aquatic systems at the individual, population, and ecosystem levels, including both marine and freshwater. Chemical defences have been suggested to be particularly well developed in aquatic systems, that is with respect to those chemical substances that are toxic to other organisms and are used as defences against consumers (including both herbivores and predators), weapons against competitors (allelopathy), or to reduce overgrowth (anti-fouling). This is especially true for marine systems because of the high diversity and density of grazers and predators and, further, because many 'prey' organisms are sessile in these environments. Thus, the high grazing/predation

pressure and inability to escape has selected for the evolution of intricate chemical defence systems (Chapter 15 and 16). In addition, chemical signals (infochemicals; Box 0.1) may be particularly advantageous in aquatic, compared to terrestrial, environments since the use of visual signals can be severely restricted due to attenuation of light in deep and/or turbid waters. Further, organisms in aquatic systems virtually live in a solution of chemical signals,

---

Box 0.1 Definitions

Infochemical: a chemical that carries information that mediates an interaction among two individuals and results in an adaptive response in the receiver. Either the sender or the receiver, or both, benefits from the infochemical.

Allelochemical: an infochemical that mediates the interaction between two individuals that belong to different species

Pheromone: an infochemical that mediates an interaction between organisms of the same species.

Kairomone: an infochemical that mediates interactions between individuals of different species and where the information transfer is beneficial for the receiver but not for the individual producing the signal.

Alarm substance (*alarm cue*, '*Schreckstoff*'): chemical cues released by an injured prey that evokes a fright reaction in conspecific individuals.

Allelopathy: an inhibitory or stimulatory effect of one plant on another plant, mediated by the release of a chemical factor.

See also Dicke and Sabelis (1988).

a landscape of smells ('smellscape'), where all organisms, dead or alive, release chemical substances into the water. In recent years there has been an increasing interest in understanding the importance of non-visual signals for communication and information in aquatic systems and many studies have shown that aquatic organisms use chemical cues to locate food, detect the presence of predators, find a partner, and for proper migration and navigation. The use of infochemicals offers many advantages. For example, they can be transmitted over large distances, they form gradients with higher concentrations of the cue close to the source, and cues are also persistent in time (hours or even days). The chemical cues are rich in information and can provide a wealth of reliable information on different attributes of the sender, including for example species identity, sex, social status, and diet. However, the efficiency of the chemical information system is constrained by physical factors connected to odour transport, such as flow velocity, turbulence, and boundary layer dynamics (see Chapters 1 and 7). Detection thresholds, signal-to-noise ratios, and other constraints of the neurobiological mechanisms of the chemosensory system also limit the use of infochemicals (Chapter 11) and, further, anthropogenically driven changes in the environment (pH, organic pollutants, etc.) may reduce the efficiency of the chemical sensory system (Chapter 17).

In its infancy, the science of chemical ecology was typically driven by either chemists that in great detail investigated the chemistry of secondary metabolites (natural products chemistry) in algae, and assumed that there was an ecological function behind them (a reductionist approach), or ecologists that assumed that the patterns and observations they observed were caused by processes involving chemicals (Hay 1996). A huge number of secondary metabolites have been identified by natural products chemists (Chapter 13), searching for compounds that could be used in the pharmaceutical industry. Even today it is suggested that we should concentrate our search for future pharmaceuticals to organisms living in the sea, rather than in rainforests. Ecologists, on the other hand, often use(d) a 'black box' approach, for example by adding water from a tank with a predator to tanks with prey, and assuming that the behavioural or morphological changes in prey are due to a chemical exuded by the predator (e.g. Brönmark and Miner 1992, Hansson 2000). The ecologists often blame the methodological difficulties of analysing infochemicals or question if we really do need to know the exact identity of an infochemical—'It is enough to know that it has an ecological effect'. However, knowing the exact identity of an infochemical would allow us to quantify and evaluate responses in greater detail (e.g. 'Are we working with adequate cue concentrations?'), increase the mechanistic understanding of interactions, and also increase our understanding of how the reception/emission of infochemicals has evolved. Fortunately, as will become evident for the reader of this volume, chemical ecology has matured as a science and we now often see a more holistic approach, possibly because researchers are now trained in both chemistry and ecology or are engaged in active collaboration across disciplines. Further, recent advances in analytical chemistry, as well as in molecular genetics, have resulted in the development of more efficient tools to identify the chemical structure of infochemicals, elucidate the sensory processing mechanisms and the genetics behind them and, importantly, their consequences for natural systems.

Ecology aims to address broad questions, such as the factors that affect biodiversity, the resilience of ecosystems, and how an invasive species, or a species' extinction, will affect the ecosystem. The fitness of individuals and the interactions between species lie at the heart of our conceptualizations and ecological theories constructed to address such problems; to understand ecological problems, we must understand these building blocks. When examining the factors that affect species fitness and species interactions a common theme emerges: chemical ecology persistently plays a large role! That is, the production and/or assessment of chemicals influence almost every aspect of the factors that affect fitness and species interactions. To illustrate the pervasive effect of chemical ecology we can, as suggested by Scott Peacor, one of the contributors to this volume, consider a basic equation (see eg. Fryxell and Lundberg 1998) describing the population growth rate of a consumer (C) feeding on a resource (R) and where there are interactions between consumers:

$$\frac{dC}{dt} = eC\left(\frac{aR}{1 + ahR + awC}\right) - mC \qquad (0.1)$$

where e is the conversion efficiency of the consumed resource into the consumers reproductive potential, a is the attack rate of the consumer on the resource, h is the handling time, w is related to consumer interference and m is the natural mortality of the consumer. The question then is: How are these parameters affected by processes involving chemicals?

First, conversion efficiency (e), which specifies the ability to convert food into energy that could be used for growth, is of course tightly coupled to food quality, and chemical defences should therefore have a considerable effect on this parameter (Fig. 0.1). Both marine and freshwater algae and angiosperms are exposed to an array of grazers and have developed sophisticated chemical defence systems both against the grazers (Chapter 15) and towards competitors (Chapter 14). In most species

the defensive chemicals are produced at all times (constitutive defences), whereas in some species the defence is not produced until attacked by grazers (inducible defences, Chapter 12). Besides defences, some primary producers use chemicals as offences, that is as an efficient weapon against competitors (Fig. 0.1). These allelochemicals reduce the overgrowth of higher plants and macroalgae by epiphytic algae, benefit algae that compete for space, and may affect species composition and succession in phytoplankton communities (Chapter 14). However, even though chemical defences against herbivores may be useful and efficient, they may, curiously enough, also provide an advantage to animals. A range of marine herbivores have evolved resistance to the chemical defences of specific algae and instead selectively feed on these algae, sequester the defence chemicals, and then use them in

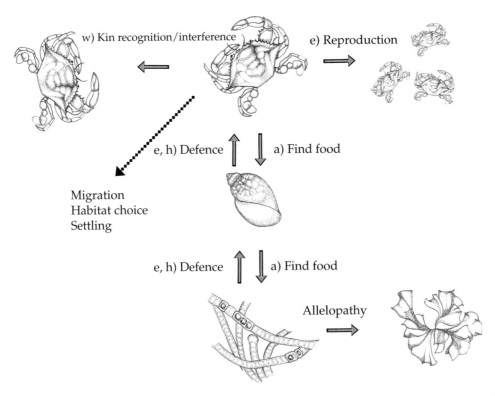

Figure 0.1 Chemical ecology affects most phases in an organism's life, including finding food, defence against enemies, reproduction, migration, habitat choice, kin recognition, and, for plants, also allelopathy. Letters (a, e, h, m, w) refer to Equation 0.1. The alert and skilled reader will recognize that we have mixed marine and freshwater organisms in the figure. This is to emphasize that all these processes are active along the whole salinity gradient, and that despite the fact that there is little communication between freshwater and marine scientists, the chemical ecology of the systems is similar.

their own defence system against their own enemies (Chapter 16). Thus, a chemical that was originally used as a defence against herbivores now serves as a defence against predators and benefits the herbivore!

The next parameter in Equation 0.1 is the attack rate of the consumer (a). The attack rates, or search efficiencies, of consumers is a function of the detectability of their resource, which is affected by foraging efficiency in the consumer and anti-predator traits in the resource. Foraging and defence traits are, in turn, strongly influenced by chemicals produced by both the resource (prey) and the consumer (herbivore/predator) (Fig. 0.1). A number of studies have shown that both herbivores and predators in both benthic and pelagic habitats use infochemicals to locate their prey (see Chapters 2 and 11). In marine systems, it has long been thought that predators use amino acids to locate their prey, but recent studies using ecologically relevant concentrations suggest that amino acids are not as important as previously thought, at least in living prey (Chapter 11). Prey organisms in aquatic habitats, in turn, commonly use infochemicals to assess local predation risk and adaptively respond by changing their behaviour, morphology, life history, or production of toxic substances (Chapters 2, 8, 12, and 15). Changes in behaviour induced by infochemicals include reduced prey activity and a change in habitat use, for example, and although these behaviours may reduce predation rates they also commonly come with a cost of reduced foraging or mating opportunities. The infochemicals that prey respond to may be predator specific and always produced by the predator (kairomones), associated to the diet of the predator, or actively released by prey after an attack by the predator (alarm substances). The evolution of specific cells that release alarm substances in fish has been debated in the literature for a while, but recent studies shed light on this issue; the alarm cells may actually have evolved as part of the immune system (Chapter 9). Most studies on predator cues have focused on one type of infochemical but as there are costs associated with changing behaviour, morphology, and so on prey should integrate different cues to fine-tune their response to the actual predation threat. A recent study on tadpoles indeed suggests that prey do not rely on singular cues, but integrate the information from a whole suite of cues associated with predation (Hettyey *et al.* 2010).

Once the prey has been located and attacked it may still escape the predator by having efficient defences, including morphological (e.g. spines) and chemical (toxic substances) defence systems, and this will affect the handling time of the predator (parameter h above). These defences may be constitutive, that is always present independent on the risk of predation, or they may be inducible, that is they are only expressed in the presence of predators (Chapter 12). The evolution of inducible defences requires that the induced defence is beneficial (reduces predation rate), but also that there is a cost associated with the defence (Tollrian and Harvell 1999). Further, the predation pressure should be variable in space and/or time and prey should have reliable cues of detecting the predator. In aquatic systems the expression of induced defences, both in plants and animals, commonly relies on chemical cues from the consumer (Chapters 2 and 8).

The last parameter in the equation, the predator interference parameter (w), is related to density-dependent effects of interactions between predators that will reduce prey intake rates (Fig. 0.1). This could be competitive interactions among predators for resources other than food or aggressive behaviours among predators. Fights are of course costly in terms of energy, injuries, or even death and should thus be avoided as far as possible. Thus, many organisms use chemical cues to signal dominant status and to form dominance hierarchies (Chapter 3). This information system can be so effective that individual lobsters can identify individuals that they have encountered before, recognize their dominance status, and thereby avoid the costly physical encounters (Chapter 1). Many aquatic animals also live in kin-structured aggregations and benefit from fewer aggressive interactions than in mixed aggregations. This requires that an individual animal is able to recognize its kin and a number of studies have shown that fish and tadpoles, for example, use chemical cues for kin recognition (Chapter 4).

The outline above shows in very general terms how chemicals can affect every aspect of consumer-resource interactions and thus have strong effects on individual fitness, population growth, and food web interactions. A huge amount of literature has shown strong indirect effects in food webs due to the lethal effects imposed by predators (cascading trophic interactions), but as suggested above the mere presence of a predator can alter prey fitness and indirectly affect food web interactions via induced changes in prey traits. It is now becoming increasingly recognized that such trait-mediated indirect effects (TMIEs) are as important, or even more important, than the effects that results of predators consuming prey (Chapter 10). Adaptive changes in predators and prey, such as those resulting from exposure to chemical cues, are now also incorporated into theoretical models, increasing the complexity compared to the traditional food web models and often resulting in counter-intuitive predictions (Abrams 2005).

The processes above mainly focus on how chemicals are used to locate prey and how prey use chemicals to assess predation risk in different defence systems. However, there is more to life than eating and avoiding being eaten. Reproduction for example: aquatic organisms are known to use intricate pheromone signalling systems to attract/locate mates, synchronize reproduction, and determine receptivity (Chapter 3). Experiments where male crabs are attracted to, and even try to copulate with, sponges that have been treated with and subsequently release the female pheromone signal clearly underlines the great importance of chemical cues for reproduction in aquatic systems. Further, chemical cues are also used by migrating organisms to find the way back to the spawning grounds, such as in salmonid fishes that locate their natal stream with chemical cues after spending years in the ocean (Chapter 6) or in larvae and juveniles to find the appropriate habitat to settle in (Chapter 5). Recent studies have shown that minute fish larvae are able to locate and swim long distances to the coral reefs where they hatched using chemical cues!

The picture that emerges from the above, and may be seen even more clearly when reading the chapters of this volume, is that chemicals are involved in all aspects of the life of aquatic organisms and have a tremendous effect at all levels, from the individual to the ecosystem (Fig. 0.1). Unfortunately, human activities often have a considerable, generally negative, impact on ecological interactions and chemical/ecological interactions are no exception. Pollutants, such as heavy metals, pesticides, and hydrocarbons, as well as acidification and eutrophication, affect the chemical environment that aquatic organisms live in. Many chemicals act as endocrine disruptors, that is they disturb the hormone signalling system within an organism, but recent studies also point to an increasing threat of infodisruption, that is where sub-lethal doses of pollutants affect the chemical information system and change individuals' behavioural and morphological responses to infochemicals, resulting in reduced fitness (Chapter 17).

The aim of this book is to summarize the state of the art of chemical interactions in aquatic systems and thereby provide a firm foundation for future studies. Moreover, our intention is to highlight the most important developments and case studies in recent years and to identify the key challenges for the future in order to stimulate discussions and hopefully further the development of the field. As you will see, the book covers a wide range of studies, including both plants and animals, from different geographic regions and habitats—pelagic as well as benthic. Further, over the years we have found that although many of the interactions are similar in freshwater and marine habitats, research focus and tradition seem to have diverged between the two systems. Thus, an important goal of the book is to increase the communication and understanding among researchers working with marine and freshwater organisms, as limitations and challenges are indeed comparable among these systems and should not be affected by historical differences in research tradition.

Finally, we have to conclude that things very seldom work out as planned and despite the fact that we had an original plan and structure of this book, the final product, which you now hold in your hand, has deviated considerably from these intentions. Hence, through the wide wisdom and experience of the many (38!) authors involved in the book, it changed and almost began to live its own life. In retrospect this is very fortunate, as it has resulted in a book that has turned into something we believe is far better than the one we originally planned for. We hope that you as a reader will be inspired and encouraged by the coming chapters which aim at advancing our joint knowledge regarding chemical ecology in aquatic systems. Before finishing this introduction we would like to present the process through which this book was conceived.

Actually, this whole project took off at a restaurant in St John's, Canada, when we had prepared many arguments for writing a book on chemical ecology and approached the Oxford University Press Commissioning Editor Ian Sherman. However, already by the first beer Ian said 'yes, go for it', and our well prepared arguments were of no use...and instead we could concentrate on the European Championship in soccer. Then, after getting the green light from the publisher to start preperation, we contacted our potential first authors and presented them with a preliminary title and a field we felt would fit into a book format. We were a bit surprised when almost all of them immediately accepted the invitation with great enthusiasm! They then had about eight months to write a first draft of their respective chapter. As a suitable deadline for this initial writing phase we had a workshop at the Häckeberga Castle in southern Sweden between 16 and 19 June 2010, where copies of all chapters were stapled together into a first draft of the book which provided a starting point for the discussions. This was indeed stimulating, and the discussions during the workshop were very interesting, creative, and constitute the basis for both this introductory chapter and the final chapter on 'future challenges'. This process also allowed the authors the opportunity to comment on each others' manuscripts and to cross-reference other chapters. After the workshop the authors returned home and finished their manuscripts, which were then sent out for review by two reviewers. Moreover, at this stage we, the editors, also commented on all manuscripts before the final submission to the publisher. Although the road to the finished product has been winding, it has indeed been creative and we have all learnt a lot. Finally, we, as editors, have worked together during the whole production process and are equally responsible for the initiative, workload, as well as the final product; the author/editor order is therefore alphabetic on the cover, as well as on the chapters we have written together.

Acknowledgements

We are very grateful to our authors for taking on the challenge of producing this book. We would also like to especially thank our commissioning editor, Ian Sherman at Oxford University Press, for always being enthusiastic, supportive, and using carrots rather than the whip. We would also like to thank Helen Eaton at OUP for a tremendous job during the production process. Maria Sol Souza has with great enthusiasm made the beautiful cover artwork as well as the line drawings in most of the figures. We also want to express our great appreciation to all the external reviewers of the chapters, but as some of them want to remain anonymous, we will not present their names. We received financial support from the Swedish Research Council (VR), the Institute of Biology/Aquatic Ecology, Lund University (special thanks to Wilhelm Granéli), and from the Center for Animal Migration (CAnMove).

CB and LAH, Lund July 2011

References

Abrams, P. A. (2005) The consequences of predator and prey adaptations for top-down and bottom-up effects. In: Barbosa, P. and Castellanos, I. (eds) *Ecology of Predator-prey Interactions*. Oxford University Press, New York.

Brönmark, C. and Miner, J. (1992) Predator-induced phenotypical change in crucian carp. *Science* **258**: 1348–50.

Dicke, M. and Sabelis, M. W. (1988) How plants obtain predatory mites as bodyguards. *Netherlands Journal of Zoology* **38** (2–4): 148–65.

Fryxell, J. M. and Lundberg, P. (1998) *Individual behaviour and community dynamics*. Chapman and Hall, London.

Hansson, L-A. (2000) Induced pigmentation in zooplankton: a trade-off between threats from predation and UV-damage. *Proc. R. Soc. Lond. B.* **267**: 2327–31.

Hay M. E. (1996) Marine chemical ecology: what's known and what's next? *Journal of Experimental Marine Biology and Ecology* **200**: 103–34.

Hettyey, A., Zsarnoczai, S., Vincze, K., Hoi, H., and Laurila, A. (2010) Interactions between the information content of different chemical cues affect induced defences in tadpoles. *Oikos* **119**: 1814–22.

Tollrian, R. and Harvell, C. D. (Eds.) (1999) *The Ecology and Evolution of Inducible Defences*. Princeton University Press, Princeton, New Jersey.

CHAPTER 1

Aquatic odour dispersal fields: opportunities and limits of detection, communication, and navigation

Jelle Atema

Humans are pretty good at odour discrimination and recognition, but we rely on vision and hearing to map our world and find that good-smelling coffee shop. Then we hear that sharks can sense a drop of blood a kilometre away. Can they really do this? And if they can, will they follow the odour trail to find the source of blood? (See Box 1.1.) In this chapter I will outline some considerations necessary to help evaluate the role of odour in aquatic ecology. It will bring us from the physics of odour dispersal in the ocean to the biology of odour sensing by aquatic animals. We will evaluate limits of sensory detection and the interaction of multiple senses to navigate to an odour source. We will need to know where the odour molecules are going after they leave the source and how many are needed to be detectable; how a shark's nose accesses odour and how a lobster's receptors process information; how animals communicate by odour signals and how they use odour to navigate.

Box 1.1 Can sharks smell a drop of blood a kilometre away?

So, can the shark smell a drop of blood a kilometre away? Let's push the answer in the YES direction realizing that the assumptions are either speculative or unrealistic. The drop travels for a kilometre undiluted to the shark's nose: he can easily smell it but has no direction to follow. Besides . . . no drop can travel that way except in a vial. In perfectly laminar flow the drop is stretched out evenly for a kilometre, a distance of over 1,000,000 drops, diluting the original odour concentration from 10^{-6} to 10^{-12} M (molar), or 10^{12} molecules per litre. This gives us 10^9 molecules per ml, roughly the volume of the shark nose. It approaches tuna detection threshold. However, a fluid drop cannot be stretched this way. In moderately turbulent flow dilution is still fast (Fig. 1.3); measured uniform dilution is 10^3 at 10 cm, maybe 10^9 at 10 m, and 10^{15} at 1 km. Let's simplify the complex blood mixture to a single amino acid with an original concentration of 10^{-6} M. At 1 km it will be diluted to 10^{-21} M or 10^3 molecules per litre which equals 1 molecule in the shark nose. This would not be recognizable as blood and probably not even register as a stimulus. Assuming non-uniform (patchy) dilution and—at 1 km distance—a maximum 100 x peak-to-mean concentration ratio, the shark may be able by chance to pick up one of those very rare patches of 100 molecules per ml floating around in the odour far-field. This would be at the detection threshold if we assume a 20% binding rate and (as in human photon detection) a five-molecule signal-to-noise acceptance rate. Even then this would be detection of a rare patch, not suitable for steering.

Answer: a shark probably CANNOT even *detect* a drop of blood a kilometre away and it can surely not find the blood source on this information alone. Now rephrasing thequestion and the situation: if a lot of blood gets into the water so that a—moderately diluted—drop of it reaches the shark a kilometre away, it can probably locate the source based on encounters with many other such diluted drops and the local flow patterns of the dispersal field.

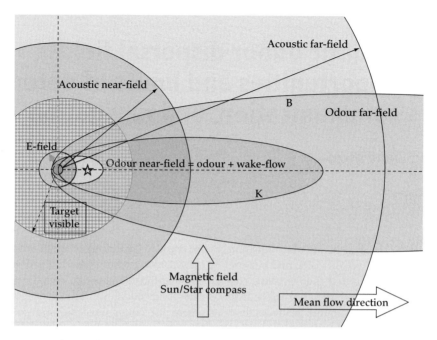

Figure 1.1 Aquatic sensory information fields. From near to far, targets can be detected by bio-electric 'E'-field (short arrow), visibility (checker, broken arrow), acoustic near-field (long arrow), and acoustic far-field (thin, long arrow), and by downstream odour fields: inner near-field (star, small inner ellipse), near-field (middle ellipse), and far-field (large ellipse). In addition, large-scale navigation can be aided by magnetic fields (open vertical arrow), sun/star-compass, wave direction and mean flow direction.

Electro-magnetic fields radiate (biologically instantaneously) through the fluid medium; acoustic near- and far-fields radiate (biologically quickly) as a density-viscosity property of the fluid medium; odour fields travel (biologically slowly) by fluid currents through the medium. In odour dispersal fields we recognize a near-field of odour and flow information extending from the source to the Kolmogorov scale (K) where flow information has ceased and a far-field between the Kolmogorov and Batchelor scale (B) where odour information has ceased; closest to the source is an inner near-field characterized by varying coincidence between odour and flow.

Chemical signals in the environment are ubiquitous since virtually all animate and inanimate objects shed molecules and these molecules are transported by fluid (air and water) currents. Chemical signals are unique: they have no directionality! They are also constantly released and difficult to hide. Organisms can reduce their visual detection by transparency and camouflage, and their acoustic and fluid dynamic detection by immobility. But, like electrical signals, chemicals cannot be easily shielded. Bio-electrical signals are directional but detectable only at small dispersal scales (mm–m) and they are limited to living organisms. Chemical dispersal, however, spans many scales and odours are released from living as well as dead cells and organisms. The different sensory information fields are sketched in Fig. 1.1. This chapter explores the opportunities and limits imposed by physical dispersal of chemical signals, their links with other senses, and their constraints on chemical sensory ecology, including animal behaviour and sensing. To avoid the interesting, but here distracting, debate over smell and taste (Atema 1980) I will use the simple term *odour* to refer to chemical signals. Moreover, most fish, crustacea, and molluscs use their nose or its equivalent for dealing with chemical dispersal. I will ignore a distinction between pheromones and signal molecules for which I refer to Wyatt (2010). Also, substrate-attached odour trails are not explicitly part of this discussion.

Odours provide virtually unlimited possibilities to create uniquely identifying signals, from designer compounds to complex mixtures. This makes them valuable for communication and identification purposes, but they are non-directional *per se* and can only be sensed locally in the nose or its equivalent. While odour concentration *gradients* do have directionality, they are typically difficult to measure, as will be discussed. Odour dispersal results in unpredictable signal intermittency, which poses serious problems for steering and enforces links with directional senses such as vision and flow detection. To understand the biological use of odour we need to ask chemical and physical questions. The chemical questions '*What are the odour molecules and what constitutes a signal?*' are covered in other chapters (see e.g.Chapter 2). In this chapter, we ask the equally important physics question 'Where are the signal molecules?' (Section 1.1 of this chapter), followed by the biological questions 'How to access odour and process its information?' (Section 1.2), 'How to send odour information?' (Section 1.3), and 'How to navigate in odour dispersal fields?' (Section 1.4). Permeating the entire discussion is the close association between odour and flow across a large range of scales, from milliseconds to years and from millimetres to thousands of kilometres.

This chapter is a biologist's view of odour dispersal in water; in other words, an attempt to look at odour dispersal from the perspective of an aquatic animal encountering the chaotic signals of natural dispersal fields. The goal of physics is to advance our understanding by creating mathematical and thus idealized models of this chaotic world. Full understanding benefits from a blend of the two approaches. For a user-friendly treatment of chemical dispersal by a fluid physicist see Webster (2007). An excellent source for biologists remains Vogel's (1996) *Life in Moving Fluids*.

1.1 Odour dispersal: where are the molecules?

1.1.1 A question of scale

After release from a source odours are dispersed by currents. Properties of the source and local currents determine much of where the molecules are at any one time. Odour dispersal takes place at many scales. For global perspective we can briefly consider the largest aquatic dispersal scale on earth: the

'Great Conveyor Belt', a global current transporting water with all its dissolved and suspended material across all oceans. It impacts biology on the century–millennial scale of climate change. Its ecological impact is probably mostly indirect as a driver of ocean-scale currents, such as those comprising the North Atlantic Gyre of which the Gulf Stream is the best-known surface current. The Gulf Stream is a giant jet plume (scale 1,000 km) with its source in the Gulf of Mexico. It transports whole ecosystems, including their odours, originating in the Gulf and arriving on the shores of Cape Cod and elsewhere. The Gulf Stream in turn spins off warm and cold eddies (spatial scale 100 km; temporal scale 1 year). Animals migrating on an ocean-scale could use characteristic odours as markers of particular currents.

The continuity principle states that fluids cannot disappear and thus currents are balanced by vertical and horizontal counter-currents, including gyres. Therefore they can disperse as well as retain organisms and odours. Seen as a current, the Gulf Stream by itself is a dispersing agent (e.g. for eels), but as part of the Gyre it also helps retention within the North Atlantic (e.g. for sea turtles). Also its eddies are retaining agents of entrained odours and organisms. At their periphery, eddies spin off smaller eddies, which do the same in turn. This cascade of ever-smaller eddies then works its way down, eventually to scales of millimetres and seconds (known as the Kolmogorov scale). When the smallest velocity eddies vanish, the initial momentum of the current has been dissipated but small chemical patches still persist; their small size makes them subject to efficient molecular diffusion and they disappear at sub-millimetre scales (Batchelor scale). At that point odours are homogeneously dispersed and no longer detectable by chemoreceptors that typically respond best to brief odour concentration increases, Section 1.2. This eddy cascade is not limited to Gulf Stream cores but applies to all sources of turbulent flow, large and small; the source limits its odour dispersal field and thus the distance over which its odour and eddy turbulence will be detectable.

At a large scale, eddies are also generated when currents encounter solid obstacles such as islands, reefs, and sea mounts where they form lee-side eddies in the horizontal dimension (scale 1–100 km, depending on the size of the obstacle, typically an island). In the vertical dimension, obstacles create upwellings. Continuity dictates that what goes up must come down and the resulting vertical 'eddies' can both retain and disperse. In tidal regimes, lee-side eddies develop alternately on either side of the obstacle. Around the obstacle, these eddies form an odour 'halo', which is refreshed daily and dilutes with distance. This creates a homing target much larger than the object itself (Fig. 1.1 shows a one-sided halo). These retention and homing mechanisms can have important consequences for population structure by limiting dispersal of larvae and maintaining local odour signatures.

It is safe to assume also that most currents have odour signatures. All organisms, dead or alive, are odour sources that release odour into the fluids that surround them; then dispersal starts and proceeds along the local eddy cascade. The scale of odour injection varies with the scale of its source. The floating (low-density) Amazon river plume can be detected at sea hundreds of kilometres from the river mouth; the sinking (denser) Mediterranean Sea water can be identified deep under the Atlantic Ocean surface hundreds of kilometres from its source, the Strait of Gibraltar. Islands and reefs release their particular odour into the surrounding currents and reef fish larvae can discriminate between odours of nearby reefs (Gerlach *et al.* 2007). Oil slicks (Fig. 1.2a) and schools of anchovy may leave odour trails of kilometres that last for days; a lobster urine signal embedded in its gill current (Fig. 1.2b) disperses over a few metres before becoming undetectable; while a copepod leaves a trail of millimetres lasting for seconds (Yen *et al.* 1998). Turbulence cascades are generated constantly at different scales by many different sources so that many scales of turbulence exist simultaneously in each location. The copepod trail can exist within the lobster gill current, which can exist within the tidal current, and each has its characteristic odour. It is not intuitive to humans, but water masses are highly structured to those who can sense it.

Animal size determines how they can use the different turbulence scales. Body *length* (BL) determines whether an eddy (E) transports the animal in

unnoticed drift (E >> BL), rocks the entire animal (E = BL), or impinges on its skin (E << BL). In benthic animals that can sense the bottom (by vision or mechanoreception), the large-scale eddy would represent a mean flow that can be used for navigation. Without sensory contact with the bottom, animals drifting in large-scale currents would need to obtain other information (e.g. magnetic fields) to measure their displacement. Intermediate-scale eddies can be detected by inertial senses, such as the otoliths of vertebrates and statoliths of invertebrates. Small-scale eddies can be sensed by hydrodynamic senses, such as the lateral line of fishes, amphibia, and cephalopods, and the mechanosensory sensilla of arthropods. Body *width* determines bilateral sensor spacing, such as the nostrils of fishes, antennae of arthropods, and many other sensor designs among invertebrates. Sensor spacing is important for detection of bilateral differences in stimulus intensity and arrival time. This applies to sensing odour patches as well as small-scale eddy flow. Weissburg (2000) proposed a non-dimensional scaling factor to link animal sensor spacing to spatial odour information and to predict its navigation behaviour.

In sum, while 'water' to human divers appears rather homogeneous it is clear that in reality all bodies of water are constantly in complex motion and filled with odours dispersing from an endless variety of sources. We will now characterize the most important fluid properties that affect odour detection and navigation before discussing the anatomical, physiological, and behavioural responses they have spurred in the evolution of animals. (The viscous flow regime of algal gametes and microorganisms is not considered here. Absent also is a discussion of odour decomposition, the loss of molecules due to chemical-physical processes such as photolysis or biological processes such as bacterial uptake during dispersal, all affecting odour quality and quantity.)

1.1.2 Odour plumes

Humans can see odour plumes emerge from smoke stacks and deep sea hydrothermal vents (e.g. 'black smokers') because the water vapour and sus-

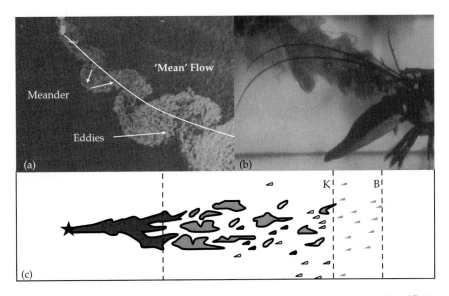

Figure 1.2 Odour dispersal. (a) Surface oil slick at 1 km scale showing three turbulence scales: curvature shows that this 'mean' flow is actually part of a large scale eddy; the flow meanders as a result of smaller eddies, which in turn show eddy fine structure (from 'Album of fluid motion'). (b) Information current. Dye visualization of the lobster gill current, a turbulent jet carrying urine and pheromones up to seven body lengths away from the animal. (c) Sketch of eddy cascade from coherent plume near source (star) to large eddies breaking down to ever smaller eddies up to the Kolmogorov scale (K) followed by odour patches up to the Batchelor scale (B). See also colour Plate 1.

pended particles make the plume visible to us. Soon after leaving the source the plume breaks up into patches that float away and disperse rather randomly (Fig. 1.2c). The plume is the beginning stage of a chemical dispersal field with properties that vary with distance. A physical boundary in the dispersal process exists where the eddy momentum has disappeared while small odour patches still exist. This so-called Kolmogorov scale represents a boundary between a near-field and a far-field—to borrow terminology from the dispersal of sound. The '*odour far-field*', is thus defined as the region between the Kolmogorov and Batchelor scales (Figs 1.1, 1.2c). In acoustics, far-field motion is dominated by sound pressure, near-field motion by hydrodynamic flow. The acoustic analogy holds also in that a sharp dividing line between near and far only exists in mathematical models. In the physics of sound, $L/2\pi$ is a mathematically logical dividing line that can be approximated under carefully controlled laboratory conditions. While practical as an estimate for biological understanding of sound detection, it is hard to measure as a physical

boundary under field conditions. Similarly, in natural odour dispersal fields it will be possible to calculate a rough estimate of where the momentum of the smallest plume eddies has disappeared into the ambient noise, but it will be difficult to measure. As a practical matter, the odour far-field serves the realization that there is an odour region where source-generated flow information is no longer available. Recognizing the different regions of an odour dispersal field will allow estimates of source distance, detectability, and localization, as it does for acoustic and hydrodynamic dispersal fields. The '*odour near-field*' is the region where both odour and momentum can be traced to the source; in the *inner near-field*, close to the source where the plume is still coherent, there can be significant coincidence between odour concentration and flow velocity. Dye tracers provide a qualitative view of the dispersal process. High-resolution quantification is obtained with additional particle tracking in laser light sheets giving superior spatial images of odour flow fields in 2-D (Webster 2007; Fig. 1.3) or 3-D from tomographic reconstruction. However,

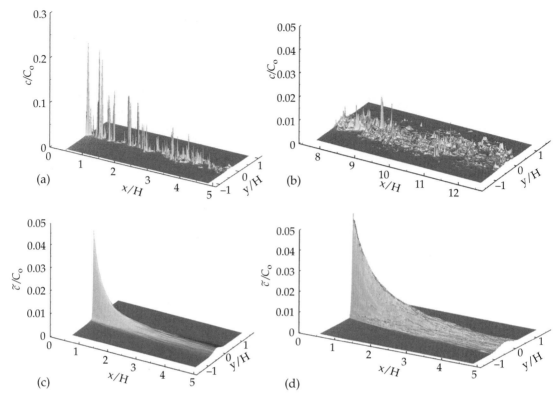

Figure 1.3 Instantaneous and statistical measures of the plume concentration. The measurement plane is at the height of the source nozzle. x is downstream flow coordinate, y is horizontal cross-stream coordinate, H is water depth (here: 200 mm). Concentration is normalized by source concentration, C_0. (a) Instantaneous concentration field near plume source. (b) Instantaneous field farther downstream. (c) Time-averaged concentration field. (d) Standard deviation of the fluctuating concentration. From Webster 2007, used with permission from George Webster.

from this 'superior' human view of 3-D plume dynamics in real time we still want to determine what information animals can and do access. Sharks cannot see the entire plume structure but only its local condition as they swim. They need to respond to local information.

The two most relevant odour source conditions are jets and wakes. Jets result from excess source-generated momentum (essentially velocity), typically released from a nozzle of some sort; wakes result from a momentum deficit at the source, which is typically a solid object. Natural dispersal situations can have both jet and wake properties. The physical properties of jets are best understood. A jet odour source produces a stream of greater velocity than the surrounding water, injecting momentum and odour simultaneously into the environment.

We mentioned deep sea hot vents (100 m scale) and crustacean breathing currents (0.1 m scale, Fig. 1.2b). The initial coincidence of odour concentration and fluid momentum (inner near-field) disappears with distance as eddy energy dissipation is faster than molecular diffusion of odour. The source-generated velocity signal of the jet remains detectable (near-field) until it matches the turbulence of the environment, i.e. the momentum-noise background. This leaves a patchy odour far-field until it diffuses into the chemical noise.

Wakes are generated behind a solid object moving through a fluid (e.g. a fish) or a fluid moving around an object (e.g. a rock). Water velocity slows down along and in the lee of the object thereby creating flow imbalance leading to turbulence that is carried away downstream. Odour released from the object travels down with the wake turbulence and both dissipate at their respective rates as in jets. A variety of animal swimming motions can make their vortex trails specific; when odour is released simultaneously the vortices will be 'flavoured', adding to their specificity. Since chemical compounds are routinely shed from both animate and inanimate objects, it is likely that most wakes and jets are flavoured.

Under carefully controlled conditions one can create 'oozing' odour sources. These are iso-kinetic, meaning that the effluent momentum/velocity matches the magnitude and direction of the mean flow. In this case the source itself does not create significant turbulence, but the turbulence of the environment spreads the plume and creates an odour dispersal field without a *source-generated* turbulence field. While many odour sources ooze (e.g. decaying carcasses), they are typically surrounded by nearby, turbulence-generating objects or they are themselves wake-forming objects. Overall, wakes appear to be the most common odour dispersal fields.

1.1.3   Near-source coincidence

Initially, coincidence between odour source concentration and source momentum is complete whenever a jet plume spews forth an odourous water mass. As soon as the jet leaves the source its momentum begins to decay in shear with the surrounding water. The jet breaks up into odour-laden patches, which lose momentum as they further decay. These patches can briefly maintain their source concentration, but shear and molecular diffusion at the edges quickly cause odour concentration to spread and homogenize (Fig. 1.3). As the patches get smaller, odour concentration and momentum proceed towards their respective noise levels. Since they decay at different rates, their initial coincidence diminishes. This differential decay could make coincidence itself a candidate for estimates of nearby source distance, an important component of plume tracking. The physics of this coincidence have not been well studied, but animals appear well equipped to use this information (Section 1.4).

1.1.4   Boundary layers

Boundary layers have a major ecological impact on odour dispersal and physiological impact on the function of chemoreceptors. As fluids adhere to solid surfaces, a velocity gradient results starting from zero flow at the solid surface and increasing to unimpeded free flow in open water. The fluid layer with reduced flow is the boundary layer around the obstacle. Increased flow velocity reduces boundary layer thickness as well as residence time of suspended and dissolved particles near the solid surface. Thus, faster flow through the nose means more odour passing by, but also less time for oduor diffusion to the receptors. Faster flow past ocean reefs means faster arrival of larvae near the reef but decreased retention time for settlement.

The bottom is a common 'obstacle' to flow, and benthic boundary layers are of particular interest to chemical ecology. As one approaches the bottom, flow velocities diminish creating areas of long odour residence time, which can become oozing sources of chemical information. Benthic obstacles from pebbles to sea mounds create various scales of turbulent upwelling. Turbulence disperses odour and affects the ability of animals to track odour plumes to their source (Moore and Grills 1999). Highly turbulent flow homogenizes odour concentrations so quickly that the source becomes difficult to locate; this can create a refuge from detection by predators (Weissburg and Zimmer-Faust 1993).

Finally, shear layers are found at boundaries between different water masses. Most bodies of water are stratified by small density differences that are caused primarily by temperature and salinity and to a minor degree by other dissolved substances. When bodies of water move relative to one another their interface becomes a shear layer. The turbulence of the shear layer disperses the odours that characterize each body of water with potential for providing information for fish navigation (Doving *et al.* 1985). The water surface presents another boundary, here between fluids of very different densities: water and air. From an aquatic point of view, surface slicks are 2-D dispersal fields (Fig. 1.2a). From an aerial point of view such slicks are bottom boundary layers for evaporating slick components useful to sea birds navigating to fish kill sites (Nevitt 1999; DeBose and Nevitt 2008).

1.2 Signal detection: accessing odour

1.2.1 Detection threshold

A single photon can *trigger* a fully dark-adapted photoreceptor cell. Theoretically, chemical detection should be similar to photon detection: both are quantal events. However, for humans to *recognize* a flash of light, five photons triggering five different (fully dark-adapted) rod receptors within a half second are needed (Hecht *et al*. 1942). As more sensitive receptors are also more instable they occasionally fire randomly in the absence of photons (false positive). Apparently, our brain accepts five different hits in a brief time-window as a non-random event. By analogy, chemoreceptor organs should report detection of a single compound when five molecules bind to five different (fully disadapted) receptor neurons in 0.5 s. For that, and assuming receptor A has a binding efficiency of 1% for molecule A, we would need a flux of 1,000 A-molecules per second through the nose. With a fish nose volume of 1 ml this is a concentration of 10^6/litre, roughly a 10^{-18} M (molar) solution.

One of the lowest reported behavioural thresholds in fish is for tuna responding to tryptophan patches at 10^{-14} M (Atema *et al*. 1980). The discrepancy of four orders of magnitude between the measured and modelled threshold concentration could have many explanations: we lack information on receptor binding efficiency and receptor density, on the tuna brain's estimate of false positives, and on the chemical noise environment during the tests. Fish amino acid response thresholds are of the order of 10^{-6} to 10^{-9} M (Catfish: Caprio 1978; sharks: Meredith and Kajura 2010) which matches ambient concentrations of the particular compound. For example, mean ammonium concentration in coastal water is around 10^{-6} M, which is also a common ammonium detection threshold in lobsters. Increasing the ammonium background raises the lobster response threshold accordingly (Borroni and Atema 1988).

Ecologically more relevant than detection of single compounds is detection of behaviourally significant odour mixtures. We know even less here. Since minor odour components can have a critical role in behavioural responses (Atema *et al*. 1989), it is likely that behavioural response thresholds depend on the least concentrated mixture components.

1.2.2 Signal-to-noise ratios

Most sensory systems are designed to detect contrast. Visual systems adapt to mean ambient light over many orders of magnitude while reporting instantaneous contrast over only a few orders of magnitude. This effect is largely based on lateral inhibition whereby the excitation of one neuron suppresses the sensitivity of its neighbours (Hartline 1969). In chemoreception one sees quick adaptation (<1 second) and slow disadaptation (~25 seconds) by single receptor neurons (Gomez and Atema 1996a, b) (Fig. 1.4). Pulse repetition rates of up to 5Hz can be discriminated by some and not other neurons (Gomez *et al*. 1999). These are the 'flicker-fusion' frequencies of the receptor neurons. As a result of slow disadaptation, chemoreceptor neurons commonly operate in a state of partial adaptation linked to the ambient noise, leaving them insensitive to lower concentration events. All receptor neurons respond best to sudden brief increases of concentration, a characteristic of patchy odour dispersal patterns. In addition, each neuron responds 'best' to a particular compound. It is estimated that there are hundreds of different compounds a fish can detect. As ambient concentrations rise so do the detection thresholds. This occurs both with backgrounds of the 'best' compound and—to a lesser degree—other compounds, a phenomenon known as mixture suppression (Atema *et al*. 1989). Overall, these receptor properties allow animals to detect local chemical contrast, that is resolving sequences of sub-second concentration increases of signal mixtures.

1.2.3 Intermittency: a patch in space is a pulse in time

The receptor properties mentioned above seem well matched to the structure of odour dispersal fields. Odour dispersal creates unpredictable sig-

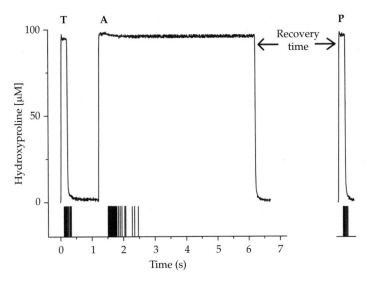

Figure 1.4 Temporal filter properties of lobster olfactory receptor neuron for the amino acid Hydroxyproline. A 0.1 s test pulse (T) produces a clear response; shortly thereafter a 5-s adapting pulse (A) shows the neuron adapting after 0.2 s, reaching zero after 1 s despite the constant stimulus level; some time after the adapting pulse ends a probe pulse (P = T) shows a still decreased response; it takes on average 25 s for full response recovery. Odour pulses measured in the nose with high-resolution electrode (see Figure 1.5a). From Gomez and Atema (1996b), used with permission from George Gomez.

nal intermittency. Once generated at the source, the plume structure moves downstream with continually changing patch shape, orientation, and dilution. (Note: many authors use the term *filament*, i.e. stretched patches, generated typically in boundary layers.) Therefore, as the patches of a plume pass a stationary point (such as an animal nose) the odour concentration signal consists of highly variable pulses with variable gaps of no-odour; the concentration variance exceeds the mean by orders of magnitude (Fig. 1.3; Webster 2007). Local velocity vectors are also highly variable, in this case in magnitude and direction. At any one point, intermittency and variance of both turbulence and odour exist at many temporal scales simultaneously ('from the Gulf Stream to the Kolmogorov and Batchelor scales'). It is therefore important to define the scale of measurement relevant to the biological question. To understand what animals can detect in the stream of odour patches we need to describe odour dispersal fields from their perspective and measure concentration with spatial and temporal resolution that match or exceed the animal's resolution. Micro-electrodes can be made to physically match sensory structures of crustacean olfactory sensilla and their boundary lay-

ers and with sample rates exceeding the flicker-fusion frequencies of chemoreceptor cells using electrochemical detection (Fig. 1.5a). This biomimetic approach was applied to lobster olfactory sensilla (the aesthetasc hairs) and allowed a receptor view of odour dispersal fields (Moore and Atema 1991). Simultaneous animal behaviour measurements gave an indication of central integration of this high-resolution sensory input (Basil and Atema 1994).

Spatial analysis of odour pulse features in a dispersal field showed a gradient pointing to the source (Moore and Atema 1991). Some pulse features (pulse height and onset slope) provided spatial information whereas others did not (pulse width and intermittency). Even different pulse shapes (i.e. concentration–time profiles) could be roughly assigned to different plume areas (Atema 1996). This shows that a spatial gradient of fine-scale concentration information exists near the source of a (jet) plume. Further neurophysiological analysis showed that olfactory receptor cells increase their response to the first 200 ms of an odour square pulse after which they adapt and stop firing (Fig. 1.4; Gomez and Atema 1996a). This brief sensing window corresponds to the duration of onset slopes of

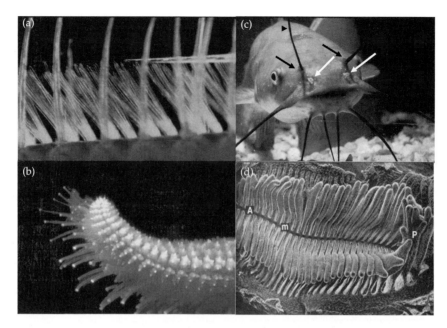

Figure 1.5 Aquatic noses. (a) Lobster olfactory (slanted, 1 mm long) and bimodal (vertical, 2 mm long) sensilla of the antennules. Micro-electrode (horizontal glass probe) matches the 20 μm diameter of olfactory sensilla to mimic their olfactory boundary (diffusion) layer, but its 100 x faster sampling rate allows odour measurement exceeding the scale of a lobster sensillum. (b) Sea star 'nose' of specialized tube feet, arm raised up out of the bottom boundary layer. (c) Catfish (*Ictalurus nebulosus*) with small forward-facing siphons as inflow nares (white arrows) and flat upward-facing outflow nares (black arrows) with valves barely visible behind nasal barbel (arrow head). (d) Olfactory rosette of catfish showing densely packed lamellae; odour receptor cilia located at inner lamellar surfaces: A = anterior, P = posterior, m = midline raphe (from Caprio and Raderman-Little (1978), with permission from Elsevier). See also colour Plate 2.

many odour pulses and may serve to differentiate between slopes. Since slope and height are closely correlated, the first 200 ms of an odour pulse may be a good predictor of its height. This form of *temporal information* extraction to derive *source direction* could be behaviourally important. Slope discrimination by receptor neurons may also be useful in bilateral slope comparisons: steeper slopes are characteristic for 'younger' patches, closer to the source. Steering toward the side with the steeper slope would enhance connection with the plume centre and its source. Local odour patch information—extracted from temporal concentration features—seems useful for steering in patchy dispersal fields. We return to 'patch-by-patch' steering below.

The human—visual—view of odour plumes has led to the assumption that following concentration gradients would lead to the source of dispersal. And indeed we can build such models from a large number of local measurements (Webster 2007). However, the chaotic intermittency and signal con-

centration variance of dispersal make it difficult to determine a source-directed vector from a few local measurements. It has been calculated repeatedly that determining a stable mean concentration in a single sample location takes many seconds and establishing a spatial gradient concentration needs several sample locations (e.g. Elkinton and Carde 1984). A common (human) technique to reduce variability is to integrate. High-resolution receptor information can be integrated over time to reduce *temporal* variability and using a widespread sensor array can reduce *spatial* variability. Sea stars appear to do both: integrating spatial information from the receptors at the tips of each arm and moving slowly to allow for temporal integration (Dale 1999). However, faster moving animals with two bilateral receptor organs could extract such integrated information only at great cost of time. To be efficient these animals need to steer patch-by-patch and/or use additional sensory information, particularly visual and mechanical detection of flow.

1.2.4 Aquatic noses

If we define the nose as the organ that analyses odours in the external fluid medium, we can review a number of interesting aquatic nose designs based on their fluid dynamic properties. It will become apparent that the different anatomy/phylogeny of noses of fish and crustacea have much in common and this commonality leads us to consider what is important for extraction of odour information from the chaotic dispersal fields of nature. Both designs act as physical filters that regulate flow, often in a pulsed manner, to optimize odour access and dynamic contrast.

Fish are solid bodies in a fluid and the motion of fluid over the body creates a boundary layer, which is an impediment to diffusion of odour to the receptors. Fish solve this with a number of adaptations (for a detailed review see Cox 2008). Their noses are bilateral chambers located in front of the eyes. The forward location of the nose is the first adaptation: boundary layers are thinnest in front of a moving object (here motion refers to relative motion). Simple noses are open pits with olfactory epithelium exposed to ambient water flow as the fish moves or faces a current. The epithelial surface is typically enlarged by folds that can form rosettes (Fig. 1.5d). In a small catfish (Fig. 1.5c), the total nasal surface area folded up into its two noses nearly matches its entire body surface (Atema, unpubl.). More elaborate designs develop a bridge creating an inflow and outflow nostril. This allows pressure differences between inflow and outflow that can be increased by a forward facing inflow and an upward facing outflow (Fig. 1.5c), the design of a Pitot tube; a variant operates as a Venturi probe (Vogel 1996). Both are engineering designs for flow measurement. In addition, when the anterior nostril is outfitted with a tube, it breaks through the fish boundary layer for better access to ambient odour (Fig. 1.5c). Also, the wider the posterior nostril the better flow over that nostril sucks fluid out of the nose by viscous entrainment. These designs can all exist simultaneously in different proportions (Cox 2008). Two active mechanisms aid in nose ventilation: cilia and pumping (sniffing). These are particularly important for slow-moving fish. The olfactory epithelium is generally supplied with an abundance of motile cilia that may cause efficient water flow over the epithelium in between the lamellae and provide the proper boundary layer conditions for odour diffusion to the receptors. Their joint action also drives fluid through the nostrils. Pumping is achieved by action of muscles and bones in the buccal cavity involved primarily in mouth and opercular movements. Simply stated: pressure differences in the mouth are transmitted via a soft membrane to the nose cavity. When the nose is given a one-way valve, typically at the posterior (outflow) nostril, the pressure decrease phase of fish inhalation sucks water into the anterior (inflow) nostril while pressure increase blows it out of the posterior nostril. In larval fish the connection may simply be a small soft membrane allowing a sniff with each inhalation cycle (Atema *et al.* 2002) but many adult fish develop more or less extensive accessory sacs serving as bellows that drive nasal flow. In addition to normal breathing movements to drive nasal flow, 'deliberate' mouth snapping in fish such as tuna (Gooding 1963) and flounder (Nevitt 1991) would be an additional form of sniffing.

An interesting case of flow regulation is found in sharks, fish that must swim continuously to provide lift (compensating for the lack of a swim bladder); and unlike bony fishes cannot swim backward. During forward motion a small foil ('wing') between the in- and outflow nostrils creates a pressure differential that continuously sucks water through the nose. To regulate this flow during fast swimming the nose of a hammerhead shark (and probably other sharks) uses a shunt to relieve the excess flow (Abel *et al.* 2010). Benthic sharks that rest on the bottom must use different nose ventilation mechanisms, probably ciliary flow, which may be an additional mechanism in all sharks. The location and anatomy of the shark nose relative to its mouth makes sniffing unlikely and it has not been observed.

Arthropod and fish noses are anatomically totally different. Arthropods use antennae covered with sensilla (small porous tubes filled with receptor cilia) that can form a dense brush (Fig. 1.5a). As each sensillum has a boundary layer, these layers merge to make the brush nearly impenetrable to flow, similar to fish rosettes. Here, too, this seemingly bad design for a nose has the important function of regulating odour access. Crustacea 'flick', the functional equivalent of sniffing. A fast (~100 ms) flick downstroke is followed by a slower return stroke. At this fluid speed, boundary layers become sufficiently reduced to allow odour access only during the downstroke. This odour becomes captured in the boundary layers of the sensilla where it remains somewhat trapped during the slower upstroke. The next down stroke releases the trapped odour while replacing it with a new sample. The maximal 5/s flick rate of lobsters creates a 200 ms odour sample. This time window corresponds to the physiological signal integration properties of lobster olfactory receptor cells, which in turn correspond to the duration of onset slopes of odour pulses, that is, the instantaneous contrast in dispersal fields (review: Atema 1995). Since biological evolution of the nose is enforced by fluid physics (flow and diffusion) it is likely that odour pulses encountered in dispersal fields shaped receptor properties, which in turn constrained nose anatomy and sampling behaviour (flicking). This suggests—so far without experimental evidence—that sniffing noses of fish evolved similarly, which leads us to predict temporal filter properties on the order of 200 ms for fish olfactory receptor neurons.

Flicking and sniffing impose a digitized sample on the natural dispersal field, altering its patchy structure and possibly making patch-by-patch steering more difficult. However, against sufficiently strong currents, when moving speed plus oncoming current match or exceed the speed of the flick downstroke, lobsters no longer flick and natural odour patches can flow through the sensory brush. A similar effect can be expected from the shunting shark nose: odour patches flow through the nose continuously at a near-constant rate because excess flow is shunted past the rosette.

Molluscan noses are not well studied. Slow-moving snails can track odour plumes with great accuracy (Ferner and Weissburg 2005). The benthic marine mud snail, *Illyanassa obsoleta*, sucks water through a single (~1 cm) siphon to wash over the gills and osphradium inside the body cavity. The osphradium looks remarkably like the cilia-covered rosette of a fish nose (Crisp 1973). The siphon not only allows water access while the snail is buried in the mud, but also allows the snail to sample and spa-

tially integrate odour patches over a wider area by swinging it from side to side. Slow crawling speed allows for temporal patch integration. Experimentally reduced siphon length severely impedes their normally well-directed, nearly straight-line (2-D) tracking (Atema *et al.*, unpubl.) suggesting its role in odour sampling. Nautilus employs bilateral noses—and not its many tentacles—for 3-D odour plume tracking. In these noses, water enters through a single pinhole. Temporarily blocking one or both pinholes results in serious tracking deficits (Basil *et al.* 2000). The ventilation mechanism is not known but—given the narrow opening—must be pressure driven.

Echinoderms, such as sea stars, detect odour with specialized tube feet on the tips of their arms (Fig. 1.5b). They typically crawl with arm tips raised out of the viscous bottom boundary layer facilitating access to ambient overhead odour. Disabling chemoreceptors of one or more arm tip with a brief exposure to distilled water increasingly affects odour plume tracking (Dale 1999). It is likely that their (non-centralized) nervous system integrates spatial and temporal odour information to support their highly accurate tracking. Similar to mud snails, their straight-line approach is a good example of tracking a turbulent concentration gradient based on slow motion speed for *temporal* integration, and *spatial* integration over an array of multiple 'noses' or a 'swinging nostril'.

1.3 Odour information currents

As much as *receiving* odour information is dependent on flow, so is the *sending* of odour signals. Animals employ a number of information currents by fanning fins or feathery paddles. The best known examples are in crayfish (Breithaupt 2001; Denissenko *et al.* 2007) and the lobster, *Homarus americanus* (Atema 1985; Fig. 1.2b). The latter uses three different currents to send (and receive) odour. (1) It injects its pheromone-laden urine stream into a gill (breathing) current that extends seven body lengths forward. Individual urine signals are remembered by the loser of a fight (Karavanich and Atema 1998). The release of urine and the direction of the currents are carefully regulated (Breithaupt and Atema 2000). (2) Lobsters can deflect the forward gill current (serving as a sender of odour) sideways and backward by fan organs composed of pairs of exopodites. This results in a return flow directed toward the animal's head where it can be sampled by the antennules (serving a receiver function). (3) Lobsters use a pleopod current to fan odour out of their shelter to attract females. (The same pleopod current propels settlement stage lobster larvae in a remarkable swimming performance and helps adults climb rocks and bulldoze gravel). Similarly, brooding fish are known to fan eggs in their nest. Fanning could be used to generate information currents. Some aggressive and courtship displays may well involve fanning (urine) pheromones toward each other (Rosenthal *et al.* 2011).

While wakes behind a moving animal can serve as information currents, they seem not to be specifically designed to disperse information. However, it would be possible to shape the wake both mechanically and chemically. This may be the case in the small-scale wakes ('footprints') of copepods, a series of odour/velocity patches (flavoured eddies) left behind a moving animal, used by mates as well as predators to locate the source (Yen *et al.* 1998). Since this takes place in a nearly viscous fluid environment, just the mechanical characteristics of the trail may serve a species and sex-specific signal function.

1.4 Navigation in odour fields

Navigation takes place at different spatial scales and different senses can be involved (Fig. 1.1). Gross order of magnitude estimates may be as follows. Large distance navigation (10–1,000 km) is likely compass based (e.g. magnetic). Due to scatter, vision under water has a limited range (<100 m). Extraction of directional information from sound remains poorly understood. The acoustic far-field (pressure) in water can be detected by a swim bladder but thereby loses directional information as the bladder becomes the sound source; the near-field (hydrodynamic flow) is directional but has a limited range depending on source amplitude and frequency, the lower the farther. For example when a 10 Hz source has a near-field detectable out to 25 m, its 150 m wavelength makes localization difficult. Odour fields could span the globe, but in practice they may

be most useful at intermediate and small scales (<10 km). The following laboratory examples are scaled at 1–10 metres and all but one involve other senses, since at this point it is clear that odour alone is rarely sufficient to steer animal behaviour.

Steering by odour alone was tested in a jet plume with a biomimetic robot equipped with nothing but two odour sensors, something not possible with real animals. The robot was patterned after a lobster in terms of speed and turning mobility and its odour sensors were spaced as a lobster antennule pair. As long as the plume was coherent (inner near-field), the robot's algorithm to turn in the direction of highest concentration could lead it to the source. As soon as the plume broke up into patches, the robot would locate the high concentration centre of a patch and move downstream with it. Locating success improved when mean upstream direction information was added (via a gyroscope) together with some turning delay to facilitate integration, but it never came close to real lobster performance (Grasso and Atema 2002). Among many other differences between robots and lobsters, this suggests that adding mean upstream flow direction to the odour field is not sufficient and that local flow and odour together may provide better information; this has not yet been tested.

Human technology (and sea stars) can integrate odour information to derive stable concentration gradients for navigation. However, in food and mate search there is a need for speed. Odour trails can linger after the source stops emitting odour, including the possibility that the source disappears through predation: animals compete for odour targets and the winner gets the prize. Speed may have driven a number of tracking algorithms. But speed must be matched by tracking accuracy, since relocating a lost odour field is costly. Lobsters track at only around 10 cm/s, less than half their maximal walking speed (Moore *et al.* 1991). Sharks, for example *Mustelus canis*, smelling food, speed up from random cruising speeds but remain far below their maximum speed. When they lose the plume they circle back to reconnect with the plume and try again (pers obs), a potentially costly error when competing with another animal that stays on track.

For fast tracking, animals need flow information. *M. canis* uses its lateral line, the water flow detection sense, to swim upstream. It tracks the turbulent wake of an odour source; it then strikes at a *turbulent* odour source in preference to a nearby *pure* odour source, that is without turbulence (Gardiner Atema 2007). Apparently source turbulence adds salience to the signal that is absent from the more concentrated oozing source nearby. Different components of the lateral line may be used to track the mean current upstream and to locate the source of small-scale source turbulence. Tracking is virtually normal in total darkness, but significantly less efficient when its lateral line is disabled. When both vision and lateral line information are unavailable, tracking stops entirely. This shows that for upstream navigation *M. canis* primarily uses its lateral line but can also use the visual flow field. Using vision for plume tracking is seen in flying moths, which steer (patch-by-patch) using the visual flow field (Mafra-Neto and Carde 1994; Vickers and Baker 1994): when encountering an odour patch they immediately steer up the local wind direction and when they lose the odour they start—after a small time delay—to steer across the local wind. (Local wind refers to the flow direction of that odour patch.) Perhaps because they are so small and fast, moths do not appear to use bilateral sensor information. Sharks however, do use bilateral odour information to steer into a patch. In controlled experiments with freely swimming animals carrying head sets, small odour arrival time differences (< 1 s) caused turning toward the side first hit even if the second side was hit with (100 x) higher concentration (Gardiner Atema 2010). In odour fields this would lead to steering into a patch. Since an odour field is a stream of patches it would help the animal to stay in the field. This suggests that sharks, like moths, navigate patch-by-patch, but use bilateral arrival time differences rather than local current for turning direction. This makes sense considering that moths are small relative to the local wind (Eddy E > Body length BL) allowing them to—visually—sense their local drift, but not bilateral odour differences, while sharks are large compared to odour patches (E < BL) preventing them from being moved by small turbulent patches and allowing bilateral sensing. The shark response may be helpful even in the odour farfield. In open water, in the absence of an external reference frame, typically the bottom, animals cannot sense the mean flow field for orientation: they are

drifting in the mean flow (E >> BL) at the same speed and direction as the dispersal field. Only the dispersal *patterns* of odour and flow can provide directional information.

The shark's use of flow information to locate the source of an odour dispersal field can be classified as 'odour-motivated rheotaxis'. This common navigation algorithm is based on odour as the stimulus motivating the animal to follow a current upstream. The odour is the unique signal identifying the source; the current provides direction. Without odour, currents have little meaning; without current the odour source cannot be found. In this usage, current generally seems to refer to the mean current but the scale of the current is rarely specified. As mentioned, in open water the mean current (E >> BL) is difficult to measure, but eddies of animal size (E ≥ BL) can be detected by an otolith system detecting minute body motions. Integrating local current variance may provide a larger-scale flow vector to follow even without an external (visual) reference. Since turbulence exists at many scales simultaneously it is interesting to know which scales different animals use for tracking. Moths are the iconic example of odour-motivated rheotaxis. Their eddy scale is specified and referred to as the local wind, not the mean wind direction. It stands to reason that for efficient navigation the scale of turbulence should remain within a frequency range linked to animal size and speed.

1.4.1 Odour-flow coincidence would signal the close proximity of the source

Eddy chemotaxis is a form of odour-stimulated rheotaxis. It refers specifically to the algorithm of patch-by-patch tracking, where the dispersal of both eddies and odour patches provides dual information including their degree of coincidence (Atema 1996). There is both anatomical and physiological evidence for odour-flow coincidence detection. Lesion studies in lobsters show that their antennules are critical for plume tracking (Devine Atema 1982; Reeder Ache 1980). These antennules have a sensory brush where chemoreceptive and mechanoreceptive sensilla are located next to each other. The spatial co-localization of two different sensor modalities can signal the degree of spatial-temporal coincidence of odour and flow arrival. This external anatomy predicts the presence of odour-flow coincidence detectors in the brain. Indeed, at the next level of the olfactory pathway in the crayfish brain there are two different types of odour-flow coincidence detectors (Mellon 2005). One type responds most vigorously when odour concentration and flow velocity increase simultaneously; the other type stops responding to flow when odour arrives (Fig. 1.6). The two opposing responses now invite the suggestion that there must be a third level integrator—still unknown—receiving input from the first two, one with excitatory and the other with inhibitory synapses. Regardless, anatomy and physiology strongly suggest that detecting degrees of odour-flow coincidence is important for these crustacea. This may be equally so for other animals of similar size and ecology.

Source distance estimates can be important for navigation in order to calibrate the final approach and strike distance. Different animals change their tracking behaviour within one to several body lengths from the source. They typically slow forward motion and may start using additional senses. Lobsters slow down and start grabbing around with their legs (Moore *et al.* 1991), Nautilus slow down and start throwing out their tentacles (Basil *et al.* 2000), sharks and catfish slow down and shake their heads. Since odour concentration and turbulence intensity by themselves do not provide reliable

information for source distance (unless source strength of either is known *a priori*) animals must rely on other plume features. This could be patch encounter frequency, typically greater near the source. Or it could be the—steeply increasing—degree of odour-flow coincidence near the source. It is tempting to think that this could be a critical parameter for estimating source distance. Physical details have not yet been analysed quantitatively.

1.5   Conclusion

There are many variations on the odour-flow tracking algorithm. For narrow odour plumes, animals with widely spaced odour sensors could track concentration patches along the plume edge as shown in blue crabs (Weissburg 2011). A non-dimensional scaling factor to link animal sensor spacing to spatial odour information (Weissburg 2000) was based on concentration detection, but wide sensor spacing also applies to hammerhead sharks that may perhaps additionally steer by arrival time differences (Gardiner and Atema 2010). It could also be expanded to encompass other bilateral sensory information.

For odour-informed navigation on the large scale, odour landscapes should be considered. Animals living in a specific locale can be exposed to predictable odours coming from different

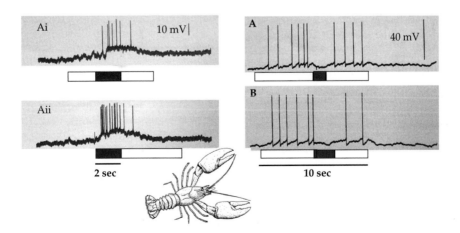

Figure 1.6   Flow–odour coincidence detection in crayfish brain. Left. Type I cells: *simultaneous* arrival of flow (white bar) and odour (black bar) causes faster and stronger response (Aii) than odour arrival preceded by flow onset (Ai). Right. Type II cells: *delayed odor arrival* causes brief increase but then lasting suppression of ongoing response to flow (A); longer odour presence, longer suppression (B). (After Mellon 2005.)

sources in different directions, even far away. They could learn to associate these odours with compass directions provided by their sun or magnetic compass. If they were displaced in any one direction they could use odour to inform them in which direction they had been displaced and return home. This has been proposed for homing pigeons (Wallraff 2000). Open ocean sea birds like albatrosses and shearwaters rely greatly on odour to find food over 100 km distances and then return home to their brood (Nevitt 1999; Bonadonna and Nevitt 2004). There is no reason to believe this could not operate also under water on the scale of ocean currents.

It should be clear that we cannot understand odour-based behaviour and ecology without understanding odour dispersal by currents at scales ranging from 1,000 km down to mm and dispersal by diffusion at smaller scales. It is probably no exaggeration to state that all organisms use chemical signals: life processes are based on an endless variety of highly specific chemical interactions. Among them, the nose of animals is the organ that informs the animal of *what* is around, but not *where* it is.

Acknowledgements

I thank Peter Buston, Gabi Gerlach, George Gomez, Henrik Mouritsen, and Rod O'Connor for valuable comments on the manuscript and the editors for outstanding preparation and memorable hospitality as well as patience at the end. I dedicate this chapter to the memory of my friend, colleague, and musical partner Thomas Eisner, a founder of *Chemical Ecology*, whose inspiration has affected many of us.

References

Abel, R.L., Maclaine, J.S., Cotton, R., Xuan, V.B., Nickels, T.B., Clark, T.H., Wang, Z., Cox, J.P.L. (2010). Functional morphology of the nasal region of a hammerhead shark. *Comp. Biochem. Physiol. A. Comp. Physiol.* **155**: 464–75.

Atema, J. (1980) Smelling and tasting underwater. *Oceanus* **23**: 4–18.

Atema, J. (1985) Chemoreception in the sea: adaptations of chemoreceptors and behaviour to aquatic stimulus conditions. *Symp. Soc. Exp. Biol.* **39**: 386–423.

Atema, J. (1995) Chemical signals in the marine environment: dispersal, detection, and temporal signal analysis. *PNAS* **92**: 62–6.

Atema, J. (1996) Eddy chemotaxis and odor landscapes: exploration of nature with animal sensors. *Biol. Bull.* **191**: 129–38.

Atema, J., Borroni, P., Johnson, B.R., Voigt, R., Handrich, L.S. (1989) Adaptation and mixture interactions in chemoreceptor cells: mechanisms for diversity and contrast enhancement. In: Laing, D.L., Cain, W., McBride, R., Ache B.W. eds., *Perception of Complex Smells and Tastes*. Acad. Press, Sydney, NSW. pp. 83–100.

Atema, J., Holland, K., Ikehara, W. (1980) Olfactory responses of yellowfin tuna (*Thunnus albacares*) to prey odors: Chemical search image. *J. Chem. Ecol.* **6**: 457–65.

Atema, J., Kingsford, M.J., Gerlach, G. (2002) Larval fish could use odour for detection, retention and orientation to reefs. *Mar Ecol. Prog. Ser.* **241**: 251–60.

Basil, J. and Atema, J. (1994) Lobster orientation in turbulent odor plumes: simultaneous measurement of tracking behavior and temporal odor patterns. *Biol. Bull.* **187**:272–3.

Basil, J.A., Hanlon, R.T., Sheikh, S.I., and Atema, J. (2000) Three-dimensional odor tracking by Nautilus pompilius. *J. Exp. Biol.* **203**: 1409–14.

Bonadonna, F., Nevitt, G.A. (2004) Partner-specific odor recognition in an Antarctic seabird. *Science* **396**: 835.

Borroni, P.F., Atema, J. (1988) Adaptation in chemoreceptor cells. I. Self-adapting backgrounds determine threshold and cause parallel shift of response function. *J Comp Physiol A* 164: 67–74.

Breithaupt, T. (2001) The fan organs of crayfish enhance chemical information flow. *Biol. Bull.* 200: 150–4.

Breithaupt, T., Atema, J. (2000) The timing of chemical signalling with urine in dominance fights of male lobsters (*Homarus americanus*). *Behav Ecol Sociobiol* **49**: 67–78.

Caprio, J. (1978) Olfaction and taste in the channel catfish: An electrophysiological study of the responses to amino acids and derivatives. *J Comp Physiol A* **123**: 357–71.

Caprio, J., Raderman-Little, R. (1978) Scanning electron microscopy of the channel catfish olfactory lamellae. *Tissue Cell* **10**: 1–9.

Cox, J.P.L. (2008) Hydrodynamic aspects of fish olfaction. *J. R. Soc. Interface* **5**: 575–93.

Crisp, M. (1973) Fine structure of some prosobranch osphradia. *Marine Biol.* **22**: 231–40.

Dale, J. (1999) Coordination of chemosensory orientation in the starfish *Asterias forbesi*. *Mar. Freshw. Behav. Physiol.* **32**: 57–71.

DeBose, J.L., Nevitt, G.A. (2008) The use of odors at different spatial scales: comparing birds with fish. *J. Chem. Ecol.* **34**: 867–81.

Denissenko, P., Lukaschuk, S., Breithaupt, T. (2007) The flow generated by an active olfactory system of the red swamp crayfish. *J. Exp. Biol.* **210**: 4083–91.

Devine, D.V., Atema, J. (1982) Function of chemoreceptor organs in spatial orientation of the lobster, *Homarus americanus*: differences and overlap. *Biol. Bull.* **163**: 144–53.

Doving, K.B., Westerberg, H., Johnsen, P.B. (1985) Role of olfaction in the behavioral and neuronal responses of Atlantic Salmon, Salmo salar, to hydrographic stratification. *Can. J. Fish. Aquat. Sci.* **42**: 1658–67.

Elkinton, J.S., Carde´, R.T. (1984) Odor dispersion. In *Chemical Ecology of Insects*, W.J. Bell and R.T. Carde´, eds. (London: Chapman and Hall), pp. 73–91.

Ferner, M.C., Weissburg, M.J. (2005) Slow-moving predatory gastropods track prey odors in fast and turbulent flow. *J. Exp. Biol.* **208**: 809–19.

Gardiner, J.M., Atema, J. (2007) Sharks need the lateral line to locate odor sources: rheotaxis and eddy Chemotaxis. *J. Exp. Biol.* **210**: 1925–34.

Gardiner, J.M., Atema, J. (2010) The function of bilateral odor arrival time differences in olfactory orientatation of sharks. *Curr. Biol* **20**: 1187–91.

Gerlach, G., Atema, J., Kingsford, M.J., Black, K.P., Miller-Sims, V. (2007) Smelling home can prevent dispersal of reef fish larvae. *PNAS.* **104**: 858–63.

Gomez, G., Atema, J. (1996a) Temporal resolution in olfaction. I. Stimulus integration time of lobster chemoreceptor cells. *J Exp Biol* **199**: 1771–9.

Gomez, G., Atema, J. (1996b) Temporal resolution in olfaction. II. Time course of recovery from adaptation in lobster chemoreceptor cells. *J Neurophysiol.* **76**: 1340–3.

Gomez, G., Voigt, R., Atema, J. (1999) Temporal resolution in olfaction III: Flicker fusion and concentration-dependent synchronization with stimulus pulse trains of antennular chemoreceptor cells in the American lobster. *J. Comp. Physiol. A.***185**: 427–36.

Gooding, R.M. (1963) The olfactory organ of the skipjack Katsuwonus pelamis. *FAO Fish. Rep.* **3**, 1621–31.

Grasso, F.W., Atema, J. (2002) Integration of flow and chemical sensing for guidance of autonomous marine robots in turbulent flows. *Environmental Fluid Mechanics* **2**: 95–114.

Hartline, H.K. (1969) Visual receptors and retinal interaction. *Science* **164**: 270–8.

Hecht, S., Shlaer, E.L., Pirenne, M.H. (1942) Energy, quanta and vision. *J. Gen. Physiol.* **31**: 459–72.

Karavanich, C., Atema, J. (1998) Olfactory recognition of urine signals in dominance fights between male lobster, *Homarus americanus*. *Behaviour* **135**: 719–30.

Mafra-Neto, A., Carde´, R.T. (1994). Fine-scale structure of phero-mone plumes modulates upwind orientation of flying moths. *Nature* **369**: 142–4.

Mellon, D. Jr. (2005) Integration of hydrodynamic and odorant inputs by local interneurons of the crayfish deutocerebrum. *J Exp Biol* **208**: 3711–20.

Meredith, T.L., Kajura, S.M. (2010) Olfactory morphology and physiology of elasmobranchs. *J Exp Biol.* **213**: 3449–56.

Moore, P.A., Atema, J. (1991) Spatial information contained in three-dimensional fine structure of an aquatic odor plume. *Biol. Bull.* **181**: 408–18.

Moore, P.A., Grills, J.L. (1999) Chemical orientation to food by the crayfish *Orconectes rusticus* influence of hydrodynamics. *Animal Behavior* **58**: 953–63.

Moore, P.A., Scholz, N., Atema, J. (1991) Chemical orientation of lobsters, *Homarus americanus*, in turbulent odor plumes. *J. Chem. Ecol.* **17**: 1293–307.

Mopper, K., Lindroth, P. (1982) Diel and depth variations in dissolved free amino acids and ammonium in the Baltic Sea determined by shipboard HPLC analysis. *Limnol. Oceanogr.* **27**: 336–347.

Nevitt, G.A. (1991) Do fish sniff? A new mechanism of olfactory sampling in pleuronectid flounders. *J Exp Biol.* **157**: 1–18.

Nevitt, G.A. (1999) Olfactory foraging in Antarctic seabirds: a species-specific attraction to krill odors. *Mar Ecol Prog Ser.* **177**: 235–41.

Reeder, P.B., Ache, B.W. (1980) Chemotaxis in the Florida spiny lobster, *Panulirus argus*. *Anim Behav.* **28**: 831–9.

Rosenthal, G.G., Fitzsimmons, J.N., Woods, K.U., Gerlach, G., Fisher, H.S. (2011) Tactical release of a sexually-selected pheromone in a swordtail fish. PloS ONE **6**(2): e16994. doi:10.1371/journal.pone.0016994.

Vickers, N.J., Baker, T.C. (1994) Reiterative responses to single strands of odor promote sustained upwind flight and odor source location by moths. *Proc. Natl. Acad. Sci. USA* **91**, 5756–60.

Vogel, S. (1996) *Life in Moving Fluids: The Physical Biology of Flow* (Second edition), Princeton University Press, New Jersey.

Wallraff, H.G. (2000) Simulated navigation based on observed gradients of atmospheric trace gases (models on Pigeon Homing, Part 3). *J. Theoret. Biol.* **205**: 133–45.

Webster, D.R. (2007) The structure of turbulent chemical plumes, in *Trace Chemical Sensing of Explosives*, edited by Woodfin, R.L., John Wiley & Sons, New York, pp. 109–29.

Webster, D.R., Weissburg, M.J. (2009) The hydrodynamics of chemical cues among aquatic organisms. *Ann. Rev. Fluid Mechanics* **41**: 73–90.

Weissburg, M.J. (2000). The fluid dynamical context of chemosensory behavior. *Biol. Bull.* **198**: 188–202.

Weissburg, M.J. (2011) Waterborne chemical communication: stimulus dispersal dynamics and orientation strategies in crustaceans. In *Chemical Communication in Crustaceans*. Breithaupt T., Thiel M. eds. Springer Verlag, New York, pp. 63–83.

Weissburg, M.J., Zimmer-Faust, R.K. (1993) Life and death in moving fluids: Hydrodynamic effects on chemosensory-mediated predation. *Ecology* **74**: 1428–43.

Wyatt, T.D. (2010) Pheromones and signature mixtures: defining species-wide signals and variable cues for identity in both invertebrates and vertebrates. *J. Comp. Physiol A* **196**: 685–700.

Yen, J., Weissburg, M.J., Doall, M.H. (1998) The fluid physics of signal perception by mate-tracking copepods. *Philos Trans R Soc Lond B Biol Sci.* **353**: 787–804.

CHAPTER 2

Information conveyed by chemical cues

Eric von Elert

Infochemicals are chemical cues mediating inter-actions between organisms, and the study of their ecological functions is the main subject of Chemical Ecology. From an ecological perspec-tive, the same infochemical may be involved in various interactions serving different functions. To account for this, and to consider evolutionary aspects of chemical communication, an infochem-ical terminology based on cost–benefit analysis rather than on the origin of the infochemical has become widely used. However, different func-tions of the same infochemical do still result in different terms, so that a context-specific termi-nology is required in order to avoid ambiguity. Here terms that require cost–benefit analyses will be avoided, and only the terms 'infochemical' or 'pheromone' will be used. This chapter is con-fined to infochemical-mediated interactions, in which the chemical nature of the compounds has been at least elucidated if not identified. The vast majorities of studies that have greatly contributed to our current understanding of the ecological and evolutionary effects of infochemicals (e.g. Brönmark and Miner 1992; Hansson 1996; Hansson 2000; Rundle and Brönmark 2001) have not bothered to identify the chemical structure and have yet come up with major contributions, a situation that emphasizes the urgent need to vali-date these findings on the basis of known concen-trations of defined chemical compounds. Considerations of costs and benefits and of inter-actions with other unknown chemical cues are not covered in this chapter. In line with the struc-ture of the book, the examples presented here are arranged according to biological interactions.

2.1 Habitat and food finding

In mobile animals, evolution should favour the use of chemicals that give reliable information on poten-tial food sources or the quality of a habitat from a distance. This holds too for animals that ontogeneti-cally shift between terrestrial and aquatic habitats. For example, in a number of insect species with aquatic larval stages and terrestrial adult stages, there is strong selection pressure on females to make accurate egg-laying decisions. The oviposition behaviour of mosquitoes is mediated by various cues associated with the aquatic habitat, such as cyanobacterial mats floating in open water, where larvae mature. In the air above such cyanobacterial mats a persistent unidentified volatile, aliphatic C_{15} alcohol with a mass spectrum similar to n-pentadecanol has been found (Rejmankova et al. 2000). Subsequently, n-pentadecanol (Fig 2.1 (1)) was shown to lead to increased egg-laying of the mosquito Anopheles albimanus in experimental water containers with n-pentadecanol, which suggested that n-pentadecanol is involved in localization of the cyanobacterial mats that are known to be preferred sites of oviposition of this mosquito (Rejmankova et al. 2000).

Further, larvae of another mosquito, Aedes aegypti, graze on microbes, and Ponnusamy et al. (2008) have demonstrated that females of this spe-cies preferentially laid eggs in water bodies con-taining tetradecanoic acid and its methyl ester (Fig. 2.1 (2)). These oviposition stimulants were shown to be associated with bacteria (a food source for the mosquito larvae) thus ensuring that growth and development of the offspring is well

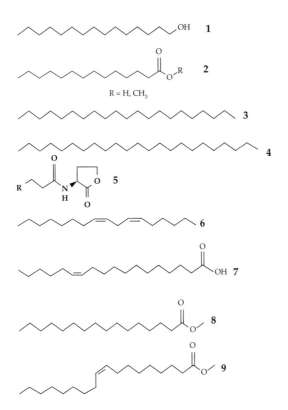

Figure 2.1 Infochemicals mediating the finding of oviposition sites and biofilms exemplified with: Pentadecanol (1), tetradecanoic acid and its methyl ester (2), n-heicosane (3), n-tricosane (4), N-acylhomoserine lactone (AHL, 5), 6(Z),9(Z)-heptadecadiene (6), 12(Z)-octadecenoic acid (7), hexadecanoic acid methyl ester (8), 9(Z)-octadecenoic acid methyl ester (9).

supported in the chosen aquatic habitat. Similarly, the proper assessment of the predation risk in the juvenile habitat is crucial for successful reproduction. Hurst *et al.* (2010) have reported that infochemicals released by the crimson-spotted rainbowfish *Melanotaenia duboulayi* led to reduced oviposition of the freshwater mosquito *Culex annulirostris*, whereas ovipositing *Aedes notoscriptus* was attracted to water with fish. A low persistence of the repellent effect was interpreted as evidence for a volatile cue, though neither was tested for volatility nor were alternative explanations, such as adsorption or degradation, taken into account. Larvae of insects may not only be preyed upon by fish but also by aquatic insects, such as the common backswimmer *Notonecta maculate*. Only recently was it demonstrated that starving *N. maculata* release two hydrocarbons, n-heneicosane

(Fig. 2.1 (3)) and n-tricosane (Fig. 2.1 (4)), which repel ovipositing females of the mosquito *Culiseta longiareolata* (Silberbush *et al.* 2010). In behavioural tests in outdoor mesocosm experiments with environmentally relevant chemical concentrations, the repellent effects of the two compounds were additive. The two hydrocarbons are of moderate to low volatility, which presumably ensures that these infochemicals do not persist longer than backswimmers are present in the system. In this exciting field, identification of further infochemicals released from vertebrate and invertebrate predators would be warranted. As many mosquitoes with aquatic larval stages are vectors for human diseases, such infochemicals might be applied to enhance the efficacy of surveillance and be used in control programmes.

Almost all surfaces in marine and freshwater are colonized by a microbial biofilm that mainly consists of eukaryotic microbes and bacteria. Bacteria in biofilms are known to release so called quorum-sensing (QS) molecules, which are of low molecular weight and are involved in bacterial cell–cell communication. In gram-negative bacteria, one class of quorum-sensing molecules are N-acylhomoserine lactones (AHLs Fig. 2.1 (5), see also Fig. 2.2). Seaweeds and invertebrates attach to these biofilms, and in two cases it has been shown that this selection of surfaces is due to attraction to chemical cues released by the biofilm. Most seaweeds have a motile life stage that is released into the water column before attaching to a surface and developing into a new plant. The zoospores of the green macroalga, *Ulva*, use AHLs released by bacteria in biofilms (Fig. 2.2) to detect a suitable surface for attachment (Joint *et al.* 2007).

This cross-kingdom detection of quorum-sensing molecules is clearly adaptive to the heterotrophic bacteria, which benefit from organic carbon released by the alga. Currently, it can only be speculated that bacteria provide yet unknown compounds that benefit the alga and that could explain the evolution of this specific response in zoospores of *Ulva*. Many invertebrates also release motile larvae into the plankton, where they are dispersed to colonize new surfaces. Settlement of larvae of the marine polychaete *Hydroides elegans* on natural biofilms is induced by the hydrocarbon 6,9-heptadecadiene

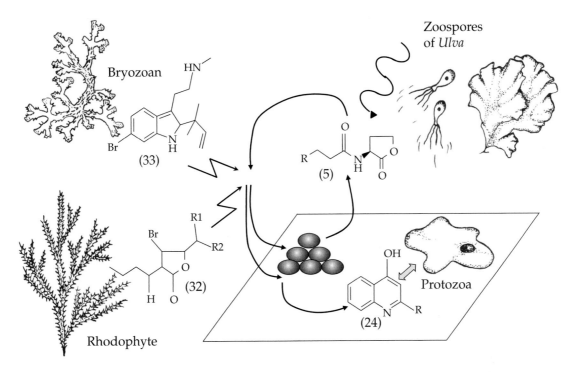

Zoospores
of *Ulva*

Bryozoan

(33)

(5)

R1

Br

R2

(32)

H   O

OH

Protozoa

(24)

Rhodophyte

Figure 2.2 Illustration of the interactions of the bacterial quorum sensing system with eukaryotic organisms. Bacteria and—among other eukaryotes—protozoa are constituents of biofilms on aquatic surfaces. Bacteria release pheromone-like quorum-sensing molecules, e.g. AHLs (5). When bacterial cell densities exceed a certain threshold, as is the case in microcolonies, the increased extracellular AHL-concentration induces the bacterial synthesis of violacein (24), which acts as a chemical defence against predation by protozoa. High AHL-concentrations reduce the swimming speed of zoospores of the green macroalga *Ulva* sp. and induce their attachment, thus promoting settlement of a macroorganism on a bacterial biofilm. Other macroorganisms like the red alga *Delisea pulchra* produce halogenated furanones (32) that interfere with the bacterial AHL-mediated quorum sensing due to similarity in the chemical structure. Bryozoans produce a variety of alkaloids with bromine (33) with the same effect on quorum sensing thus reducing the formation of bacterial biofilms on their surfaces.

(Fig. 2.1 (6)) and the fatty acid 12-octadecenoic acid (Fig. 2.1 (7)), substances that were found to be released by the biofilms (Hung *et al.* 2009).

In laboratory bioassays, adults of the beetle *Bembidion obtusidens* were attracted to volatile metabolites (VOCs) (methyl esters of hexadecanoic (Fig. 2.1 (8)) and 9-octadecenoic acid (Fig. 2.1 (9)) of mat-forming, filamentous cyanobacteria (*Oscillatoria* sp.) growing in the microhabitat of these beetles (Evans 1982). In a more artificial scenario, VOCs from benthic cyanobacteria attracted nematodes (Höckelmann *et al.* 2004). Both studies emphasize the relevance of VOCs for habitat finding in benthic communities and require further investigation into whether the VOC-releasing cyanobacteria constitute a beneficial habitat. The few cues (1–9) identified so far to be involved in finding biofilms and

habitats for oviposition all have long aliphatic chains. This suggests that the cues preferentially accumulate at the surface of waters, so that they can be detected at the surface of waters or biofilms, which thus enables ovipositing females to find a suitable habitat for their offspring.

In the case of ectoparasites, dependence on a suitable habitat requires finding and infecting a host. Despite the great economic importance of fish ectoparasites, the fish-mediated infochemicals that mediate host-finding of the parasite have only been identified in the sea louse *Lepeophtheirus salmonis*. This copepod is a natural marine ectoparasite of wild and farmed salmonids, and the copepodid stage, that is the free-swimming stage, is the infective stage. Using Y-tube behavioural bioassays Bailey *et al.* (2006) investigated the response of copepodid larvae of sea lice to water conditioned by the preferred salmonid host and to an extract thereof. Extraction was performed with C_2-solid phases, thus enriching fairly polar compounds. The biological activity was found to be successfully extracted by the solid phase and to be in the volatile fraction of this extract. Subsequent GC-MS analysis led to the identification of isophorone (Fig. 2.3 (10)) and 6-methyl-5-hepten-2-one (Fig. 2.3 (11)). Unfortunately the non-volatile fraction of the extract was not assayed; hence the presence of other attractants cannot be ruled out. Dose-response studies with isophorone resulted in a bell-shaped curve with a maximized response between 10 and 100 $\mu g \cdot mL^{-1}$. As isophorone is used widely as a solvent of natural and synthetic resins, wax, printing ink, pesticides, and paints, this might represent another case of info-disruption (Chapter 17).

In aquatic systems, studies of animal response to food-finding chemical cues have focused primarily on fish and non-insect invertebrates. In decapod crustaceans the chemical cues used as feeding stimulants are rather unspecific. Carnivorous decapods respond best to small nitrogen-containing compounds, such as amino acids, peptides, amines, and nucleotides, whereas herbivores and omnivores are often sensitive to sugars as well as to amino acids. Similarly, a variety of nitrogenous compounds serve as feeding cues in fish (Derby and Sorensen 2008).

In the case of benthic, slow-moving herbivores, like snails, it might be adaptive to use chemical cues for localization of food patches. Such food-finding can be regarded as a more specific case of habitat finding, in which the adaptive value of the habitat is immediately evident. Evidence for food-finding cues in benthic habitats comes from the release of an odour bouquet that consists of mono- and di-unsaturated alcohols and ketones (primarily with a C_8-skeleton) from the benthic freshwater diatom *Achnanthes biasolettiana* upon cell lysis (Fink *et al.* 2006a). These volatile organic compounds (VOCs) have been proved to attract the freshwater snail *Radix ovata* that actively feeds on *A. biasolettiana*. Attractive VOCs were also released from lysed cells of the benthic freshwater green alga *Ulothrix fimbriata* (Fink *et al.* 2006b). When the major compounds, 2,4-heptadienal (Fig. 2.3 (12)) and the three C_5 compounds 1-penten-3-one (Fig. 2.3 (13)), 1-penten-3-ol

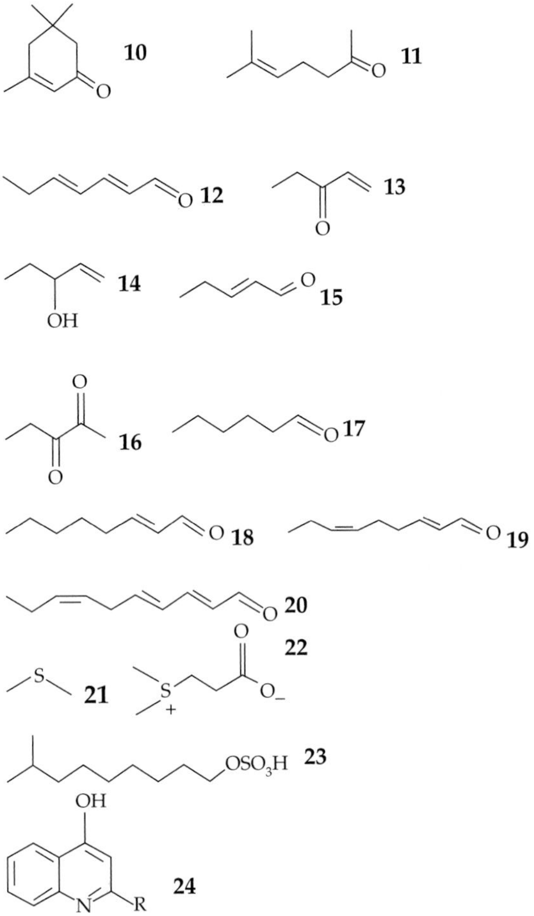

Figure 2.3 Infochemicals involved in finding food and hosts and inducing defences in algae and bacteria exemplified with: Isophorone (10), 6-methyl-5-hepten-2-one (11), 2(E),4(E)-heptadienal (12), 1-penten-3-one (13), 1-penten-3-ol (14), 2(E)-pentenal (15), pentane-2,3-dione (16), hexanal (17), 2(E)-octenal (18), 2(E),6(Z)-nonadienal (19), 2(E),4(E),7(Z) decatrienal (20), dimethylsulfide (21), dimethylsulfoniopropionate (22), 8-methylnonyl sulfate (23), violacein (24).

(Fig. 2.3 (14)), and 2-pentenal (Fig. 2.3 (15)), were assayed, it was revealed that only a mixture of both, C_7 and C_5 compounds was attractive. In a study by Jüttner *et al.* (2010) VOCs released upon cell lysis from the benthic marine diatom *Cocconeis scutellum parva* were tested for behavioural effects against a wide spectrum of naturally associated invertebrates. The diatom released a bouquet of VOCs that consisted of saturated and unsaturated aldehydes and ketones with chain lengths from C_5 to C_{10}. In subsequent behavioural assays 17 animal species associated with *C. scutellum* were tested for their response to an artificial bouquet that contained pentane-2,3-dione (Fig. 2.3 (16)), hexanal (Fig. 2.3 (17)), 2, 4 heptadienal (Fig. 2.3 (12)), 2-octenal (Fig. 2.3 (18)), 2, 6-nonadienal (Fig. 2.3 (19)), and decatrienal (Fig. 2.3 (20)) in relative concentrations similar to those in the natural bouquet. Threshold concentrations and the type of response to the bouquet varied considerably between animal species, which points to a variable but important role of VOCs from benthic primary producers in structuring the associated invertebrate community.

All three studies referred to above have used a behavioural assay, which is a promising approach to reveal effects of infochemicals on spatial community structures. However, all studies have investigated VOCs released by artificial cell lysis, which merely demonstrates a *potential* role of these compounds in natural situations. From the perspective of the benthic algae the release of VOCs that attract herbivores is not adaptive. It is reasonable to assume that actively grazing invertebrates would elicit the release of VOCs that in turn would attract other conspecifics, as has been observed in the mangrove snail *Terebralia palustris* (Fratini *et al.* 2001). Clearly, further studies are needed to investigate in situ mechanisms and rates of VOC-release from benthic algal communities and to identify putatively adaptive functions of these VOCs for the primary producers.

Infochemical-mediated food finding is also reported for pelagic interactions, where the adaptive value can best be explained by the non-homogenous distribution of resources. Steinke *et al.* (2006) observed the herbivorous marine calanoid copepod *Temora longicornis* respond with increased tail-flapping to 'odor plumes' of dimethlysulfide (DMS, Fig. 2.3 (21)). This behaviour might be necessary for the spatial integration of chemical signals and could aid the copepod in tracing patches of DMS-releasing algal prey. Plumes of DMS putatively result from cell lyses of certain phytoplankton species due to pathogens or herbivores. It remains to be seen if increased tail-flapping is correlated with expanded search behaviour and with increased selective grazing, so that patches of infested or actively preyed food would be preferentially grazed upon by *T. longicornis*. Cells of certain phytoplankton species do not only release DMS upon cell lysis, but also

contain dimethylsulfoniopropionate (DMSP, Fig. 2.3 (22)), which serves as a precursor in the production of DMS upon cell rupture. Hence, it is reasonable to assume that upon cell lysis, in addition to DMS, DMSP is also released. Recently Debose *et al.* (2008) have presented evidence that planktivorous fish living in coral reefs are attracted to sites of DMSP release. Given that DMSP release may be caused by herbivores, these results imply that odours linked more closely to feeding activity than to the mere presence of prey alert some planktivorous fish species to potential foraging opportunities. Thus the release of DMSP potentially triggers bi-trophic interactions.

Food finding for herbivores is especially important for specialists. Compared to terrestrial systems, cases of specialized herbivorous insects feeding on macrophytes are relatively rare. The freshwater beetle *Euhrychiopsis lecontei* is a specialist aquatic herbivore that feeds, oviposits, and mates on the Eurasian water milfoil *Myriophyllum spicatum*. In North America, where Eurasian water milfoil is an invasive species, the beetle has been tested as potential means for biological control of *M. spicatum*. A bioassay-driven isolation of infochemicals attracting *E. lecontei* led to the identification of the attractive compounds, such as glycerol and uracil, in the exudates of *M. spicatum* (Marko *et al.* 2005). This is the first example of a freshwater specialist insect being attracted to chemicals released by its host plant. It remains to be seen if other more specific compounds are involved that might explain the particular attraction of the beetle to *M. spicatum*.

2.2 Induced defences in primary producers and bacteria

In planktonic microalgae and cyanobacteria a variety of secondary metabolites are known that are presumably part of an anti-herbivore defence. In a few cases elevated cellular contents of such toxic compounds were observed in response to waterborne cues from herbivores, for example increased paralytic shellfish toxin in the dinoflagellate *Alexandrium minutum* (Selander *et al.* 2006) and elevated levels of microcystin-LR and -RR in the cyanobacterium *Microcystis aeruginosa* in response to cues from the cladocerans *Moina macrocopa* or

Daphnia magna (Jang *et al*. 2003, 2007). The observation that such inductions correlate with increased resistance to grazing by copepods, or decreased somatic growth by *Daphnia*, supports the idea that these metabolites function as defences against herbivores. The inducibility of increased toxin contents suggests that permanently elevated levels are associated with costs for dinoflagellates or cyanobacteria respectively. However, if these costs are due to resource limitations, then resource saturation should result in a constantly high toxin synthesis and no further induction by herbivore cues should be observed.

In many cases herbivores and predators on animals feed size selectively, which does not only affect the size structure of the prey community but also offers a size refuge for prey that has a plastic morphology. In phytoplankton chemical cues from herbivores can induce colony formation in their prey. The bloom forming prymnesiophyte *Phaeocastis globosa* may be preyed upon by large copepods preferring larger sized items or by ciliates that cannot ingest items above a certain size. *P. globosa* can grow in colonies or as single cells, and size increase was shown to be stimulated by actively feeding grazers. Cell-free culture filtrate from the grazing ciliate *Euplotes* sp. led to increased colony formation, whereas filtrates from the actively feeding copepod *Acartia tonsa* resulted in less frequent colony formation in *P. globosa* (Long *et al*. 2007). This finding demonstrates that herbivore-specific, albeit unknown, chemical cues are involved. More information on the chemical nature of the inducing infochemical was obtained in a study using *Phaeocystis antarctica* where the grazer cue passes through a membrane that was permeable for cues less than 12 kDa (Tang *et al*. 2008).

Another example of morphological defence is the transition of various unicellular green algae to grazing resistant colonies. In different green algae this change in morphology is induced by a variety of herbivorous zooplankton taxa (Van Donk *et al*. 1999). The chemical characterization has been restricted to cues released by *Daphnia*, and bioassay-guided approaches have revealed that only actively feeding (and hence digesting), not starved, animals release the infochemical. The cue can neither be extracted from *Daphnia* nor from green alga

(Lampert *et al*. 1994). A subsequent chemical characterization focused on the cue released into the medium and showed that the cue could be concentrated by C_{18}-solid-phase extraction, and an active fraction could be obtained by reversed-phase HPLC (Von Elert and Franck 1999). The cue was found to be of low-molecular weight, and possessed olefinic double bounds essential for biological activity. Binding to anion-exchangers suggested that the cue was negatively charged, however this was not confirmed by van Holthoon *et al*. (2003). In a fundamentally different approach, an extract from 10 kg of frozen *Daphnia* was subjected to bioassay-guided isolation and led to the identification of alkan- and alkensulfates and -sulfamates that were confirmed by chemical synthesis (Yasumoto *et al*. 2005; Yasumoto *et al*. 2008a; Yasumoto *et al*. 2008b). The compounds proved to be active at a concentration of 10 ng/mL. Incubation of *D. magna* in the absence of food and subsequent extraction of anions from the incubation medium led to the confirmation of one of the above mentioned compounds (8-methylnonyl sulfate, Fig. 2.3 (23)) as an exudation product of *D. magna* (Uchida *et al*. 2008); for all the other amphiphilic anions reported from *Daphnia* biomass a functional role as an infochemical remains to be demonstrated.

The reported organosulfates are strong anions and are chromatographed by reverse-phase high pressure liquid chromatography (HPLC); so far these findings confirm earlier findings about the infochemical released by actively feeding *Daphnia*. Discrepancies occur with regard to the essential character of olefinic double bounds for biological activity (Von Elert and Franck 1999) and with the finding that, in the absence of ingestible food, no infochemical is released by *Daphnia* (Lampert *et al*. 1994; Van Donk *et al*. 1999). It remains to be seen if alkensulfates can be detected among the excreted infochemicals when herbivorous zooplankton are actively feeding. The reported compounds are structurally strikingly similar to widely used commercial detergents that are commonly detected in freshwater (Lürling 2006) and which might interfere with the induction of colony size in green algae by grazing zooplankton in nature (see Chapter 17).

Colony formation in response to grazers has also been demonstrated in the cyanobacterium

Microcystis aeruginosa (Ha *et al.* 2004; Jang *et al.* 2003; van Gremberghe *et al.* 2009): cell-free culture medium of the cladocerans *Daphnia magna* and *Moina macrocopa* can induce colony formation in different *Microcystis* strains, which indicates that induction of colonies and toxins by grazer-borne cues are not mutually exclusive. However, high variability between *Microcystis* strains has been shown, and the amphiphilic octylsulfate had no effect (van Gremberghe *et al.* 2009). The latter suggests that the chemical basis of colony induction in *Microcystis* differs from that in green algae.

Size selective predation is also a structuring force in nanoplankton, and bacteria with filamentous and colonial phenotypes have long been known to be relatively protected from protozoan grazing. However, Corno and Jürgens (2006) demonstrated an infochemical-mediated effect on bacterial morphology. When the freshwater bacterial strain, *Flectobacillus* sp., which showed pronounced morphological plasticity, was grown in continuous culture, filament formation was significantly enhanced when a dialysis bag was immersed in the culture. The bag contained the actively feeding and growing bacterivorous flagellate *Ochromonas* sp. Due to the dialysis bags there was only chemical, not direct, contact between bacteria and bacterivores. This suggests that excretory products of the flagellate induced filament formation, which can be regarded as a morphological anti-grazer defence. Recently, the induction of a chemical anti-predator defence against protozoan grazing has been demonstrated in the bacterium *Pseudoalteromonas tunicate* (Matz *et al.* 2008). The chemically-mediated resistance to protozoan grazing is elicited by the purple pigment violacein (Fig. 2.3 (24), see also Fig. 2.2), an L-tryptophan-derived alkaloid. The violacein content of bacterial cells increases with cell density, which explains the high content in bacterial biofilms, and is probably mediated by quorum sensing. Size-selective grazing by protozoa favours the occurrence of bacterial microcolonies with high cell densities in biofilms and thus—indirectly—leads to grazer-mediated induction of violacein-production. It remains to be tested if, in addition to this indirect induction, predator-borne chemical cues are directly capable of inducing violacein-production in bacterial biofilms.

2.3 Induced defences in animal prey

2.3.1 Morphological changes

Size selectivity is also a major parameter of animal predator–prey interactions, and several cases of adaptive changes in morphology due to predator-borne infochemicals are reported. In *Daphnia* sp. changes in morphology can be induced by abiotic factors, but also in response to infochemicals (see Chapter 8). Neckteeth and helmets are induced in response to chemical cues from larvae of *Chaoborus* sp. (Krueger and Dodson 1981). In a bioassay-guided approach with *D. pulex* using incubation water of *Chaoborus flavicans* larvae, Tollrian and Von Elert (1994) enriched the infochemical by lipophilic solid-phase extraction and characterized the cue as a low-molecular weight ($\leq$ 500 Da) anion with a weak basic functional group. pH-dependent interactions with ion-exchangers suggested it was a carboxylic acid, and reversible acetylation and the absence of effects of amino-specific derivations indicated the presence of hydroxyl- rather than amino functional groups as essential for activity. Until the identification of the infochemical it remains speculative whether the predator-borne cue directly interferes with internal signalling cascades or is perceived by external receptors and subsequently integrated into internal signalling in *D. pulex*. The latter is supported by the finding that, in response to infochemicals released by *C. flavicans*, the gene for a juvenile hormone acid methyltransferase (JHAMT) and for a putative juvenile hormone receptor (Met) are upregulated in the postembryonic stage of *D. pulex* (Miyakawa *et al.* 2010).

D. pulex is not the only *Daphnia* species that responds with morphological changes to infochemicals released by *Chaoborus* sp. Morphological changes similar to those observed in *D. pulex* are known in *D. cucullata*, and it remains to be tested if the same chemical cues are causative. The differences in ultrastructure observed in induced *D. ambigua* (Chapter 8) might be due to different chemical cues released by larvae of *Chaoborus* sp. Likewise the adaptive induction of a crest in *D. carinata* by a notonectid predator (Grant and Bayly 1981) and of helmets in *D. lumholtzii* by fish have not been investigated chemically.

Ciliates of the genus *Euplotes* defend themselves by an inducible increase in size against different predators: *Lembadion bullinum* (Ciliophora) (Kusch and Heckmann 1992), the turbellarian *Stenostomum sphagnetorum* (Kusch 1993), and the predatory amoeba *Amoeba proteus* (A-factor) (Kusch 1999). The chemical cues released from these predators proved to be proteins of considerably different molecular mass, which poses the interesting question of the specificity of a putative receptor. Based on the chemical identification of the A-factor it was revealed that this protein functions as a self-recognition cue on the outer surface of the predator (Kusch 1999), which explains why the exploitation of this cue as a predator-mediated infochemical seems to be an evolutionary stable system.

2.3.2 Life history and behavioural changes

Though physiologically fundamentally different, inducible changes in life history are functionally related to inducible morphological changes, as both anti-predator responses provide an adaptive response to a size-selective predator. Machacek (1991) was the first to demonstrate that exudates from planktivorous fish trigger life-history shifts in *Daphnia* that, in general, lead to earlier maturation and a smaller size at first reproduction, which seems adaptive as fish select for larger prey. In a bioassay-guided approach with *D. magna* using incubation water of *Leuciscus idus*, the infochemical could be enriched by lipophilic solid-phase extraction and proved to be an anion (Von Elert and Stibor 2006). Inactivation by acetylation pointed at a hydroxyl- or amino functional group being essential for biological activity. It cannot be excluded that olefinic bonds or ester bonds are part of the infochemical, however, they were shown to be dispensable for biological activity. These chemical features are in accordance with those reported for the infochemical inducing diel vertical migration (DVM) of freshwater fish (see below). However, retention by HPLC differed between cues affecting life history and DVM, which indicates that the infochemicals might be different. Infochemicals from the invertebrate predators *Notonecta glauca* and from larvae of *Chaoborus flavicans* induce different life history shifts in *D. hyalina* (Stibor and Lüning 1994), and chemical

investigations of these infochemicals are certainly warranted.

Predator-mediated induction of DVM constitutes another very efficient defence strategy in zooplankton and is observed in both marine and freshwater zooplankton. Fish-induced DVM results in daytime residence of zooplankton in the dark hypolimnetic refuge. During the night zooplankton ascend to the warmer surface water layers, where developmental time is shorter and algal food is more abundant. In the presence of light DVM is triggered by predator-borne infochemicals. Inverse patterns of DVM as a response to invertebrate predators are well known, but the proximate chemical cues have only been investigated in the case of fish. Infochemicals released from either starving planktivorous or piscivorous freshwater fish species have been shown to induce DVM in *Daphnia magna* (Loose *et al.* 1993, von Elert and Loose 1996, von Elert and Pohnert 2000). The DVM-inducing infochemical released by fish has been suggested to be Trimethylamine (TMA) (Boriss *et al.* 1999), but Pohnert and Von Elert (2000) proved this to be false by subsequently demonstrating that TMA does not induce DVM when tested at ecologically relevant concentrations. Bioassay-guided characterization of the fish-borne infochemical showed that the cue was of low molecular weight (≤ 500 Da) and was thermally and pH stable and resisted digestion with proteinases, and mucus proved to be inactive (Loose *et al.* 1993). The anionic cue could be enriched from water by lipophilic solid-phase extraction, and the activity depended on hydroxylgroups but not on amino-, carboxy-, sulfate-, phosphate groups, nor on olefinic doublebonds (Von Elert and Loose 1996; Von Elert and Pohnert 2000). Digestion with glucuronidase did not decrease activity of the infochemical, which demonstrated that glucuronic acid is not essential for biological activity and suggests that the chemical cue is not a constituent of the mucus of freshwater fish (Von Elert and Pohnert 2000). Identical chemical characteristics and similar retention on HPLC of biological activity released from different planktivorous or piscivorous fish strongly suggested that the infochemicals are chemically similar if not identical, which would point at a generalized rather than a fish species-specific cue that provides an evolutionary stable indicator of predation risk.

The chemistry of DVM-inducing infochemicals from marine fish has only recently been reviewed in detail by Cohen and Forward (2009). For marine zooplankton Forward and Rittschof (1993) demonstrated that chemicals released from planktivorous fish (*Atlantic menhaden* larvae) with a molecular mass of less than 500 Da induced DVM measured as phototactic response in larvae of brine shrimp (*Artemia franziscana*). Their photoresponse was successfully induced with disaccharides with either a sulfamino or acetylamino group on carbon 2 of a hexoseamine. Further, constitutive enzymes from fish mucus generated active chemical cues from heparin and chondroitin sulfate (Forward and Rittschof 1999), which suggested that infochemicals derived from killifish, *Fundulus heteroclitus*, mucus contained sulfated or acetylated amino groups. Together with the findings that fish-incubated water constituted a source of both the glucopolymers for the release of these oligosaccharides and of the hydrolysing enzymes, these results indicated that enzymatic degradation products of sulfated and acetylated fish mucus can serve as fish-mediated infochemicals for the induction of DVM in marine zooplankton. However, fish incubation water was produced at a ratio of 330 g of fish to 1 L water, which may have led to an overestimation of dissolved oligosaccharides and hydrolysing enzymes in fish-incubation water. Similar experiments suggest that mucus-derived cues also are potential infochemicals for the induction of a photoresponse in the marine copepod *Calanopia americana* (Cohen and Forward 2005). Clearly, amino sugars as degradation products of fish mucus at least imitate fish infochemicals in marine zooplankton and may be major constituents of the DVM-inducing infochemicals. It remains to be seen what their exact chemical nature is and if they constitute a major bioactive fraction of fish infochemicals when fish are incubated at the threshold densities required to produce active fish-incubation water.

2.4 Alarm cues in invertebrates

Prey species use various strategies to avoid being detected, captured, and consumed by predators. One strategy is performing anti-predator behaviours in response to cues released by injured or disturbed conspecifics (alarm cues, see also Chapter 9). Despite the overabundance of studies showing alarm responses in aquatic species, very few investigations have identified the responsible molecules and shown their function as alarm cues. The sea anemone *Anthopleura elegantissima* responds with tentacle withdrawal and mouth closing to the quaternary ammonium ion anthopleurine (Fig. 2.4 (25)) that is released by wounded conspecifics (Howe and Sheikh 1975) and functions as an alarm pheromone.

Subsequent investigations revealed that anthopleurine is assimilated by the predatory nudibranch *Aeolidi apapillosa* when it feeds on the sea anemone. The nudibranch then elicits the same anti-predator response in the sea anemone due to release of anthopleurine, so that this alarm pheromone, due to trophic transfer, reliably signals actual and passed

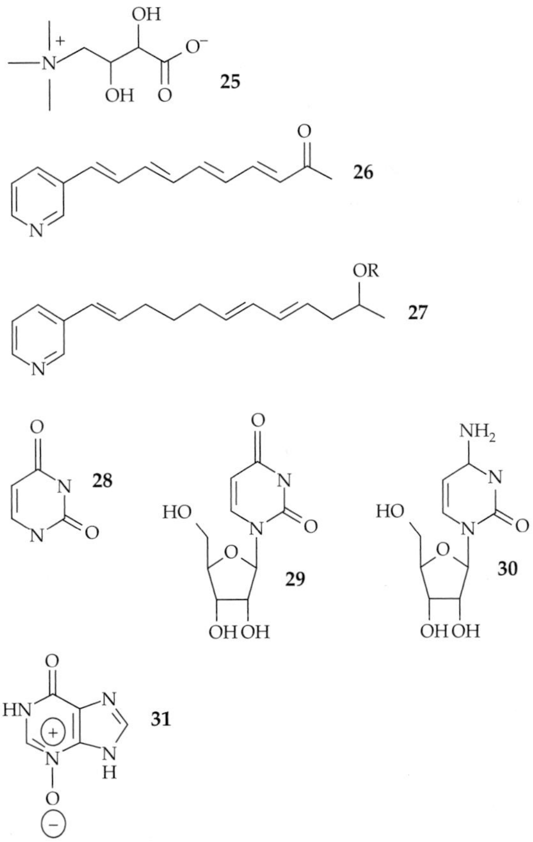

Figure 2.4 Alarm cues in invertebrates and vertebrates, exemplified with: Anthopleurine (25), navenone A (26), haminol-A (R = H) and haminol-B (R = OAc) (27), uracil (28), uridine (29), cytidine (30), hypoxanthine-3(*N*)-oxide (31).

predation risk (Howe and Harris 1978). Alarm pheromone systems in which the cue is actively secreted, that is tissue damage is not a prerequisite for signalling, can be regarded as evolutionarily more advanced. When molested, the opisthobranch molluscs *Navanax inermis* and *Haminoea navicula*, respectively, secrete the alkylpyridines navenones A (Fig 2.4 (26)), B, and C and haminols-A and B (Fig. 2.4 (27)) in their slime trails, which arrest conspecifics following slime trails and cause them to move away from the cues (Sleeper *et al.* 1980; Cimino *et al.* 1991). Active secretion is also found in the sea hare *Aplysia californica*: this mollusc defends itself from predators in many ways (Chapter 16). The defensive secretions, ink and opaline, each from a specific gland, specifically act as alarm cues to nearby conspecifics. Uracil (Fig. 2.4 (28)) and the nucleosides uridine (Fig. 2.4 (29)) and cytidine (Fig. 2.4 (30)) proved to be the active alarming substances in ink, and each of them is sufficient to elicit alarm responses in conspecifics (Kicklighter *et al.* 2007). In opaline there are three alarm cues, all of which are mycosporine amino acids derived from the snail's diet (C. Kicklighter, pers. comm.). For other effects of the ink and opaline secretions see Chapter 16.

Hemolymph has been demonstrated to contain alarm substances in a few other aquatic invertebrate species, in which passive release of hemolymph, for example after injury from predator attack, is the source of the alarm substances. For example, the snail *Nassarius obsoletus* has been shown to respond to molecules > 100 kDa (Atema and Stenzler 1977). Snail alarm responses are widely explored in experimental studies using a variety of species mainly by applying crushed animals, which suggests that release of hemolymph is involved. Though chemical cues inducing alarm responses have been shown to vary between species (e.g. Rahman *et al.* 2000) chemical identification is much needed.

Within decapods, the hermit crab *Clibanarius vittatus* responds to molecules in the hemolymph of conspecifics. These molecules bind to octadecyl silica and show a molecular mass of less than 500 Da (Rittschof *et al.* 1992; Shabani *et al.* 2008). Further, the crayfish *Procambarus clarkii* is alerted by cues of less than 5 kDa from the hemolymph (Acquistapace *et al.* 2005). In the presence of predator-borne info-

chemicals, homogenates from conspecifics have been shown to enhance adaptive behavioural (Pijanowska 1997) and morphological changes (Laforsch *et al.* 2006) in *Daphnia* that earlier had been attributed solely to the predator-borne chemical cue; studies on the chemical structure of alarm substances in cladocerans should be strongly encouraged.

2.5 Alarm cues in vertebrates

Most fish of the superorder Ostariophysi (e.g. minnows, catfish, loaches, suckers) show a characteristic fright reaction when they detect an alarm substance derived from the injured skin of conspecifics or other ostariophysean species (see also Chapter 9). The fright reaction was first discovered by Von Frisch (1941), and it exists only within the Ostariophysi and Gonorhynchiformes, where it is widespread; many tests have shown that interspecific reactions are not only present but are especially strong in taxonomically close species (Pfeiffer 1974). Making use of the interspecific effects of the Schreckstoff, Argentini (1976) elucidated its chemical identity using the skin of the European minnow *Phoxinus phoxinus* in an approach guided by behavioural bioassays with schools of the giant danio (*Danio malabaricus*). Spectroscopical data suggested a purine and were identical to those of hypoxanthine. However, hypoxanthine was chromatographically different from the genuine alarm substance and had no biological activity. Mass spectrometry instead suggested the presence of an N-oxide from hypoxanthine. Chemical synthesis of hypoxanthine-1 (*N*)-oxide and -3(*N*)-oxide yielded identical mass spectra and, further, chromatographic retention of hypoxanthine-3(*N*)-oxide was identical to the genuine alarm substance. In subsequent bioassays hypoxanthine-3(*N*)-oxide proved to be biologically active, whereas hypoxanthine-1(*N*)-oxide was clearly not active. Hence hypoxanthine-3(*N*)-oxide (Fig. 2.4 (31)) is assumed to be a component of the alarm substance or to be identical with this pheromone in this species.

Later studies have shown that hypoxanthine-3(N)-oxide elicits a response in a wide range of fish species, and the compound has thus been suggested to be an active component of the Ostariophysan chemical alarm cue system (Argentini 1976; Brown *et al.* 2000; Pfeiffer *et al.* 1985). By using different types of behavioural response and different fish taxa, Pfeiffer *et al.* (1985) determined a threshold concentration for hypoxanthine-3(N)-oxide of 1.65 nM, whereas Brown *et al.* (2001) found a threshold concentration of 0.4 nM.

An interspecific response to alarm pheromones should be evolutionary adaptive, and indeed behavioural assays suggest at least partial conservation of alarm pheromones within families and orders (Mirza and Chivers 2001). However, the same data showed that salmonids and swordtails (*Xiphophorus helleri*) could distinguish between chemical cues of conspecifics versus heterospecifics and that the intensity of the behavioural response to heterospecific alarm cues decreases as the phylogenetic distance between donor and receiver species increases. These findings question the universal character of hypoxanthine-3(*N*)-oxide as the Ostariophysan chemical alarm cue.

Hypoxanthine-3(*N*)-oxide (Fig. 2.4 (31)) is comprised of a purine with a nitrogen oxide functional group at the three position. When hypoxanthine-3(*N*)-oxide and a suite of seven structurally similar compounds were assayed for induction of the freight response in fathead minnows (*Pimephales promelas*) and finescale dace (*Chrosomus neogaeus*), it was revealed that the purine skeleton is not a functional component of the biological activity (Brown *et al.* 2000). Based on the activity of the functionally similar pyridine-N-oxide, it was concluded that hypoxanthine-3-(N)-oxide may be one of several possible molecules that function as a chemical alarm signal. As data on the release of this compound by fish are lacking, it is still unknown if hypoxanthine-3(*N*)-oxide is the major component of the alarm pheromone or merely one of a suite of potential chemical signals.

Another interesting case of a chemically identified alarm pheromone in aquatic vertebrates is found in larvae of the California newt (*Taricha torosa*) that exhibit strong avoidance behaviour of potentially cannibalistic adults. Tetrodotoxin (Ttx) was found to be passively released from the skin of adults and to elicit the avoidance behaviour of larvae at concentrations of $\geq 10^{-9}$ mol/L by olfactory perception (Zimmer *et al.* 2006). This larval response, and thus the infochemical's function as an alarm

pheromone, is ontogenetically restricted to the young (and thus vulnerable) newt larvae. From this function as a pheromone it is not completely clear that passive release of Ttx from the skin of adults is adaptive, as it leads to predator avoidance of potential prey. However, Ttx in adults serves as anti-predator defence, which might explain the adaptive nature of Ttx production and release in the adult California newt.

2.6 Pheromones and quorum sensing

Among primary producers several algal species have been shown to use pheromones. Female gametes of various genera of brown algae (e.g. *Ectocarpus*, *Fucus*, and *Dictyotus*) produce specific, volatile, cyclic and acyclic hydrocarbons that function as pheromones as they attract sperm at picomolar concentrations (reviewed in Fink 2007). *Volvox carteri* is one of the simplest multicellular green algae with biflagellate somatic cells arranged in a monolayer surface around a hollow sphere with 16 asexual reproductive cells ('gonidia'). When exposed to the sex-inducing pheromone, the gonidia cleave in a different pattern to initiate development of a sexual organism. From the spent medium of a *V. carteri* culture the pheromone was characterized as a 32-kDa glycoprotein that is synthesized by sperm cells. It works at a very low concentration of less than 10^{-16} M, making it one of the most potent infochemicals. This low threshold concentration is due to a special mechanism that amplifies the sexual-pheromone signalling within the spheroid during the sexual-induction process (Sumper *et al.* 1993).

A very interesting type of chemical communication is quorum sensing (QS). This is a pheromone-like cell–cell communication and gene regulatory mechanism that allows bacteria to coordinate swarming, biofilm formation, stress resistance, and production of secondary metabolites in response to threshold concentrations of QS signals that accumulate within the cell's environment. QS is best understood in the AHL-system of gram-negative bacteria (Schaefer *et al.* 2008), where AHLs not only affect gene expression patterns (for more details see Fig. 2.5) but also stimulate the transition of bacteria from a planktonic to a benthic stage and are thus involved in the process of biofilm-formation.

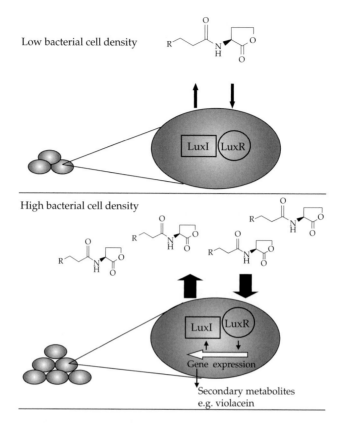

Figure 2.5 Quorum sensing in bacteria. In bacteria the coupling of gene expression to population density is accomplished by quorum sensing. The best understood system is the AHL-system of gram-negative bacteria. In its simplest form an acyl-homoserine-lacton synthase of the LuxI protein family synthesizes an acetylated homoserine lactone (AHL, 5). The acyl-group provides specificity and can vary in size and composition of the acyl chain. The chains range from 4 to 14 carbon atoms, can contain doublebonds, and often contain an oxo or hydroxyl group on the third carbon or consist of p-coumaroly (Schaefer *et al.* 2008). When bacterial cell densities are low, the activity of LuxI results in low intra- and extracellular AHL-concentrations. Increased bacterial cell densities lead to increased extracellular and thus intracellular AHL-concentrations and a binding of AHL to an AHL-receptor belonging to the LuxR protein family. Only the AHL-LuxR complex binds to the promotor region of various genes and thereby increases the expression of LuxI, which constitutes a cell-density dependent positive feedback on the AHL-concentration. AHL-dependent activation of further target genes induces the synthesis of various metabolites, e.g. the anti-predator defence compound violacein.

With the exception of the induction of settlement by biofilms (see below), it should be adaptive for primary producers and animals to suppress bacterial biofilm formation; and two eukaryotes have been shown to do so by chemically interfering with the QS system of gram-negative bacteria. The rhodophyte *Delisea pulchra* produces a number of halogenated furanones (32 in Fig. 2.2), which are structurally similar to the bacterial AHLs and interfere with bacterial processes that involve AHL-driven QS systems (Givskov *et al.* 1996). Similar interference was observed for brominated alkaloids and a diterpene of the bryozoan *Flustra foliacea* (33 in Fig. 2.2; Hung

et al. 2009). These examples demonstrate that eukaryotic chemical interference with the pheromone-like system of bacteria is used to reduce formation of bacterial biofilms on eukaryotic surfaces. Similar mechanisms remain to be demonstrated for QS systems other than AHL, and this type of interference has not been reported for freshwater systems.

Planktonic and benthic invertebrates provide further examples for the use of pheromones. In the marine polychaetes *Platynereis* sp. and *Nereis* sp. fertile individuals assemble at midnight in high densities near the water surface and form swarms of both males and females. Within these swarms, mate recognition is achieved by chemical signals. At the moment of mate recognition, the male discharges an egg-release pheromone. This pheromone stimulates females to swim with high velocity in narrow circles surrounded by swarming males. After an induction period of several seconds, the female spawns, and the discharged fluid includes eggs and the sperm-release pheromone. Then, upon detection of this pheromone, males circling around the egg cloud are stimulated to emit sperm, and fertilization of eggs in the water occurs.

In *Nereis succinea* the volatile compound 5-methyl-3-heptanone (Fig. 2.6 (34)) seems to play a role in mate recognition. Males of this polychaete discharge an egg-release pheromone complex, in which the purine inosine (Fig. 2.6 (35)) was identified as the main component. The sperm-release pheromone of *N. succinea* is identified as L-cysteine-glutathione disulfide (Zeeck *et al.* 1998b). In the related annelid, *Platynereis dumerilii*, the involved pheromones are chemically different from those of *N. succinea*. The males trigger the egg-release in females by liberating L-ovothiol A (Fig. 2.6 (36)) (Rohl *et al.* 1999), and then the females liberate uric acid (Fig. 2.6 (37)), a purine, with their eggs, which triggers the release of sperm (Zeeck *et al.* 1998a). Interestingly purines are involved in both species though with different functions. These annelids represent the first marine invertebrates with identified gamete-release pheromones and provide excellent examples for reproductive synchronization in planktonic invertebrates using infochemicals of low molecular weight.

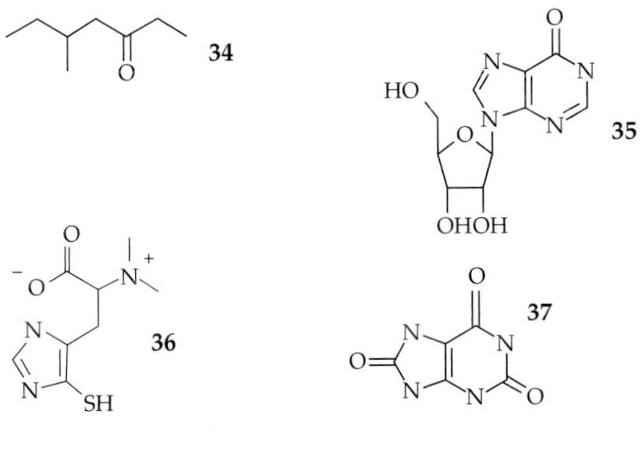

Figure 2.6 Pheromones exemplified with: 5-methyl-3-heptanone (34), inosine (35), L-ovothiol-A (36), uric acid (37).

In contrast, reproduction of the marine mollusc *Aplysia* sp. occurs on recently laid egg cordons and does not involve low molecular weight pheromones. *Aplysia* sp. is a simultaneous hermaphrodite that doesn't normally fertilize its own eggs. During reproduction, aggregations of mating and egg-laying animals in association with recently laid egg cordons are observed, and evidence suggests that egg-laying releases pheromones that establish and maintain the aggregation. From egg cordons of six aplysiid species a family of attractins (58-residue N-glycosylated proteins) has been identified, and these are the first water-borne protein pheromones to be characterized in invertebrates (Painter *et al.* 1998). However, attractins are biologically active only in the presence of another pheromone. Binary blends of attractin with any of three other gland proteins (enticin, temptin, seductin) proved sufficient to attract animals, which suggests that these proteins comprise a bouquet that stimulates mate attraction (Cummins *et al.* 2006). Sex-specific trail-following has been reported in the marine mud snail (*Ilyanassa obsoleta*) (Moomjian *et al.* 2003) and the freshwater snail *Pomacea canaliculata* (Takeichi *et al.* 2007), and it remains to be seen if the infochemicals involved in this surface-associated signalling also are proteinaceous in nature.

Another type of pheromone is the larval release pheromone that is well studied in crustaceans. In decapod crustaceans fertilized eggs are attached to feathery pleopods on the female's abdomen. Chemicals released by the brood prompt grooming and larval release behaviour by the female. The role of the female in larval release is to compress the egg mass by flexing her tail, a so called pumping-reaction. Based on the finding that serine protease degradation products were active egg-release infochemicals, structure–function studies with synthetic peptides showed that neutral–neutral–basic carboxyl terminal tripeptides were most effective and that Arg was more potent than Lys at the carboxyl terminal (Rittschof and Cohen 2004). Conceptually these results suggest that bouquets of rather unspecific degradation products of proteins function with different efficiencies as egg-release pheromones. The infochemicals involved are obviously of low specificity, which might be explained by the fact that the initial events of egg-hatching that trigger the abdominal pumping behaviour of the mother go along with proteolysis-mediated lysis of the eggs and are perceived within the direct vicinity by the mother.

2.7 Dispersal and settlement cues

Many benthic marine invertebrates have a planktonic larval stage, that's primary purpose is dispersal. Unlike larval dispersal, in many species settlement occurs on a relatively small spatial scale and involves larval behaviour, in which biogenic chemical cues have been implicated. These cues might be released from biofilms (see Section 2.2) or from conspecifics and thus function as pheromones. In barnacles, well known for their gregarious settlement, identification of the underlying infochemicals is also of economic interest due to the development of anti-fouling strategies. The most important biogenic cue to gregarious settlement in barnacles is a contact pheromone. Recently the settlement cue in *Balanus amphitrite* was identified as a novel glycoprotein with a sequence similarity to α2-macroglobulin (Dreanno *et al.* 2006). If potential post-translational modifications are ignored, the molecular mass of the glycoprotein can be estimated to be 169 kDa. This protein is produced by the epidermis. In common with barnacles, surface-associated proteins have been implicated in the gregarious settlement of the tubeworm *Phragmatopoma californica* (Jensen and Morse 1990) and the oyster *Crassostrea virginica* (Crisp 1967), although it is far from clear that these are authentic cues. Interestingly, a water-soluble peptide cue is also associated with induction of settlement in both these species. Whether there is a relationship between the surface-associated and waterborne cues is unknown, but it has been suggested that the waterborne cue of *B. amphitrite*, which may also be a peptide, is derived from the surface-associated settlement-inducing protein. The settlement-pheromones of barnacles and oysters are (glyco-)proteins, which probably ensures high specificity of these cues.

2.8 Pheromones

Rotifers and daphnids are well known for their ability to reproduce both asexually and sexually, and the chemically-mediated switches from asexual to sexual modes of reproduction are well known in both taxa. In general, under crowding conditions, reduced food availability, or deterioration of food quality a proportion of asexually-reproducing females within the population switches to the production of sexually-reproducing males and subsequently produces haploid eggs that are to be fertilized. In the rotifer *Brachionus plicatilis* an infochemical that induced sexual reproduction in females was isolated from the spent medium of high density cultures of *B. plicatilis*. It was identified to be a 39 kDa protein with 17 N-terminal amino acids that were 100% similar in sequence to a steroidogenesis-inducing protein from human ovarian follicular fluid (Snell *et al.* 2006). Though crowding is widely used to experimentally induce sexual reproduction in daphnids, the chemical nature of the infochemical is entirely unknown. The finding that high maternal levels of methylfarnesoate, the major terpenoid hormone in crustaceans, induce the production of males in daphnids (*D. magna*, *D. pulex*, *D. pulicaria*) (Leblanc 2007) but not the subsequent production of haploid eggs, suggests that methylfarnesoate is partly involved in hormonally mediating the induction of sexual reproduction in daphnids.

Mate choice is a common feature of sexually reproducing organisms. In the rotifer *Brachionus manjavacas* (formerly known as *B. plicatilis*) males exhibit a distinctive mating behaviour that consists of tight circling around the female while maintaining contact with his corona and penis. Male mate recognition is based upon the mate-recognition pheromone (MRP), which has been described as a 29 kDa surface glycoprotein in *B. manjavacas*. Bioasssay-guided fractionation and N-terminal sequencing of a 29 kDa surface protein yielded enough sequence information for the identification of the mate-recognition pheromone gene from a cDNA library. Subsequent RNAi knockdown confirmed the functional role of the mate-recognition pheromone gene in rotifer mating, demonstrating that this surface-associated protein is critical for *B. manjavacas* mate recognition. In *Daphnia* sp. evidence supports the involvement of surface associated, rather than of dissolved, infochemicals in mate recognition. However, the chemical identity has not been addressed yet, though this would allow for a greater understanding of mechanisms behind speciation and hybridization in daphnids. Male copepods in the genus *Tigriopus* use surface proteins of female copepods to recognize appropriate mates. Using an antibody, a protein was isolated from the homogenate of pooled *T. japonicus*. The resultant 70 KDa protein was partially sequenced revealing sequence similarity to α2-macroglobulin (Ting and Snell 2003).

2.9 Conclusions

Ultimately chemical ecology aims at the identification of benefits and costs associated with the release of infochemicals in order to understand the evolution of a particular infochemical-mediated interaction. As a consequence, compounds are classified by function (e.g. as kairomone or pheromone), which implies that the same compound may hold different functions in different contexts. However, in many cases of chemical communication infochemicals of unknown chemical identity are involved. Despite the lack of knowledge of their chemical identity, the evolutionary aspects of several of these communication systems are investigated. These approaches must be regarded as preliminary, as only knowledge of the chemical nature of an infochemical will allow us to reveal its involvement in other physiological and chemical processes that may add constraints on the potential evolution of particular interactions (as is exemplified elsewhere, Zimmer *et al*. 2006).

On land, long-range chemical communication requires a volatile cue, whereas short-distance communication is generally accomplished by less volatile substances. This distinction does not hold for the aquatic realm, where both volatiles and more polar (and hence less volatile) cues are equally well dispersed (Chapters 4 and 10). In line with this, food finding cues range from volatiles to dissolved amino acids. For generalist herbivores food finding should rely on general rather than food specific chemicals. However, common metabolites as well as mixtures of specific volatiles are found among food finding cues for generalists, which suggests that too few examples have been investigated yet to allow for generalizations. In addition, cases of specialized herbivores have unfortunately not been investigated in detail.

For the induction of anti-predator defences fairly general cues should be favoured by evolution. From the perspective of a *Daphnia* it is not adaptive to differentiate between species of planktivorous fish, as any such species is a visually orientated predator with a strong positive selection for large body size. As a consequence the inducible defences in *Daphnia* (diel vertical migration, smaller size at first reproduction) are the same regardless of fish species. This makes it reasonable to assume that evolution has favoured *Daphnia* to rely on an infochemical signalling the presence of any species of planktivorous fish. On the other hand, predators that select for small body size or who are active during night time require a different inducible defence, and the inducing infochemicals must be chemically different from the infochemicals of planktivorous fish. Data for *Daphnia* showing different inducible defences for vertebrate and invertebrate predators, but responding to chemically very similar, if not identical cues from fish, support this notion.

Attacks by predators may furthermore cause the release of alarm cues. Evolutionary theory predicts that alarm cues should be fairly species-specific, as has been demonstrated for invertebrates releasing pheromones from distinct glands. In many other invertebrates alarm responses are triggered by the release of hemolymph, along with the liberation of general metabolites. It is therefore reasonable to assume that the release of hemolymph is an evolutionarily older mechanism than the release from specific glands and that it induces alarm responses with a low species specificity. In the case of Ostariophysan fish, where distinct cells release the so called 'Schreckstoff' upon rupture, the evidence for one chemically identical compound in a wide range of fish species is under discussion and has led to different hypotheses on the evolution of this alarm substance (Chapter 9).

Intuitively, the highest degree of chemical specificity would be expected within sex pheromones in order to ensure species-specific mating. For mate recognition and attraction, surface-associated proteins ensure a high degree of specificity, not only in benthic animals but also in zooplankton. However, successful reproduction involves many successive steps, and not all of them require species-specific signals. Steps in reproduction that occur after mates have successfully aggregated, such as release of sperm, eggs, and larvae, are often mediated by single compounds only, with some of them being common metabolites. Probably the overall specificity of reproduction is achieved through species-specificity in other steps of reproduction such as mate attraction or sperm fertilization.

Recently, molecular techniques and proteomics have led to major progress in our understanding of surface-associated signalling, mainly with regard to mate recognition and settlement cues. At present it cannot be excluded that, compared to surface-associated cues, the identification of dissolved chemical cues is inherently more difficult because the dissolved cues are to varying degrees subjected to microbial degradation, which might lead to chemically similar but unfortunately not identical infochemicals. Nevertheless, recent improvements in sensitivity and resolution in LC-MS hold strong potential for progress in the identification of low molecular weight infochemicals. Only upon the chemical identification of infochemicals will it be possible to determine rates of release and degradation and the ecological relevance of such compounds. Combining this with behavioural assays holds further promise. In most assays of toxicity or inhibition the receiving organisms are not given the opportunity to avoid the tested compound. For motile organisms, such avoidance behaviour will presumably occur at considerably lower concentrations than effects of intoxication. Therefore accounting for behavioural responses in bioassays holds the potential to reveal the repellant activity of infochemicals at lower, and hence ecologically more relevant, concentrations. It is reasonable to assume that putatively toxic or inhibitory compounds function as repellents and as such have a largely overlooked role in spatially structuring communities.

References

Acquistapace, P., Calamai, L., Hazlett, B.A., and Gherardi, F. (2005). Source of alarm substances in crayfish and their preliminary chemical characterization. *Canadian Journal of Zoology-Revue Canadienne de Zoologie* **83**: 1624–30.

Argentini, M. (1976) Isolierung des Schreckstoffes aus der Haut der Elritze *Phoxinus phoxinus* (L.). Ph.D. thesis, Chemisches Institut der Universität Zürich, University of Zürich, Switzerland.

Atema, J. and Stenzler, D. (1977) Alarm substance of marine mud snail, *Nassarius-obsoletus* - biological characterization and possible evolution. *Journal of Chemical Ecology* **3**: 173–87.

Bailey, R. J. E., Birkett, M.A., Ingvarsdottir, A., Mordue, A. J., Mordue, W., O'Shea, B., Pickett, J.A., and Wadhams, L.J. (2006) The role of semiochemicals in host location and non-host avoidance by salmon louse (*Lepeophtheirus salmonis*) copepodids. *Canadian Journal of Fisheries and Aquatic Sciences* **63**: 448–56.

Boriss, H., Boersma, M., and Wiltshire, K.H. (1999) Trimethyamine induces migration of waterfleas. *Nature* **398**: 382.

Brönmark, C. and Miner, J.G. (1992) Predator-induced phenotypical change in body morphology in crucian carp. *Science* **258**: 1348–50.

Brown, G. E., Adrian, J.C., and Shih, M.L. (2001) Behavioural responses of fathead minnows to hypoxanthine-3-N-oxide at varying concentrations. *Journal of Fish Biology* **58**: 1465–70.

Brown, G. E., Adrian, J.C., Smyth, E., Leet, H., and Brennan, S. (2000) Ostariophysan alarm pheromones: Laboratory and field tests of the functional significance of nitrogen oxides. *Journal of Chemical Ecology* **26**: 139–54.

Cimino, G., Passeggio, A., Sodano, G., Spinella, A., and Villani, G. (1991) Alarm pheromones from the mediterranean opisthobranch *Haminoea navicula*. *Experientia* **47**: 61–3.

Cohen, J. H. and Forward, R. B. (2005) Photobehavior as an inducible defense in the marine copepod *Calanopia americana*. *Limnology and Oceanography* **50**: 1269–77.

Cohen, J. H. and Forward, R.B. (2009) Zooplankton diel vertical migration - a review of proximate control. *Oceanography and Marine Biology: An Annual Review* **47**: 77–109.

Corno, G. and Jürgens, K. (2006) Direct and indirect effects of protist predation on population size structure of a bacterial strain with high phenotypic plasticity. *Applied and Environmental Microbiology* **72**: 78–86.

Crisp, D. J. (1967) Chemical factors inducing settlement in *Crassostrea virginica* (Gmelin). *Journal of Animal Ecology* **36**: 329–35.

Cummins, S. F., Nichols, A. E. , Schein, C. H., and Nagle, G. T. (2006) Newly identified water-borne protein pheromones interact with attractin to stimulate mate attraction in Aplysia. *Peptides*, **27**: 597–606.

Debose, J. L., Lema, S. C., and Nevitt, G. A. (2008) Dimethylsulfoniopropionate as a foraging cue for reef fishes. *Science*, **319**: 1356.

Derby, C. D. and Sorensen, P. W. (2008) Neural processing, perception, and behavioral responses to natural chemical stimuli by fish and crustaceans. *Journal of Chemical Ecology*, **34**: 898–914.

Dreanno, C., Matsumura, K., Dohmae, N., Takio, K., Hirota, H., Kirby, R. R., and Clare, A. S. (2006) An alpha(2)-macroglobulin-like protein is the cue to gregarious settlement of the barnacle *Balanus amphitrite*. *Proceedings of the National Academy of Sciences of the United States of America*, **103**: 14396–401.

Evans, W. G. (1982) *Oscillatoria* sp. (Cyanophyta) mat metabolites implicated in habitat selection in *Bembidion obtusidens* (Coleoptera: Carabidae). *Journal of Chemical Ecology*, **8**: 671–8.

Fink, P. (2007) Ecological functions of volatile organic compounds in aquatic systems. *Marine and Freshwater Behaviour and Physiology*, **40**: 155–68.

Fink, P., Von Elert, E., and Jüttner, F. (2006a) Oxylipins from freshwater diatoms act as attractants for a benthic herbivore. *Archiv für Hydrobiologie*, **167**: 561–74.

Fink, P., Von Elert, E., and Jüttner, F. (2006b) Volatile foraging kairomones in the littoral zone: attraction of an herbivorous freshwater gastropod to algal odors. *Journal of Chemical Ecology*, **32**: 1867–81.

Forward, R. B. and Rittschof, D. (1993) Activation of photoresponses of brine shrimp nauplii involved in diel vertical migration by chemical cues from fish. *J. Plankton Res.*, **15**: 693–701.

Forward, R. B. and Rittschof, D. (1999) Brine shrimp larval photoresponses involved in diel vertical migration: Activation by fish mucus and modified amino sugars. *Limnology and Oceanography*, **44**: 1904–16.

Fratini, S., Cannicci, S., and Vannini, M. (2001) Feeding clusters and olfaction in the mangrove snail *Terebralia palustris* (Linnaeus) (Potamididae: Gastropoda). *Journal of Experimental Marine Biology and Ecology*, **261**: 173–83.

Givskov, M., Denys, R., Manefield, M., Gram, L., Maximilien, R., Eberl, L., Molin, S., Steinberg, P. D., and Kjelleberg, S. (1996) Eukaryotic interference with homoserine lactone-mediated prokaryotic signaling. *Journal of Bacteriology*, **178**: 6618–22.

Grant, J. W. G. and Bayly, I. A. E. (1981) Predator induction of crests in morphs of the *Daphnia-carinata* King Complex. *Limnology and Oceanography*, **26**: 201–18.

Ha, K., Jang, M. H., and Takamura, N. (2004) Colony formation in planktonic algae induced by zooplankton culture media filtrate. *Journal of Freshwater Ecology*, **19**: 9–16.

Hansson, L. A. (1996) Behavioral-response in plants - adjustment in algal recruitment induced by herbivores. *Proceedings of the Royal Society of London Series B- Biological Sciences*, **263**: 1241–4.

Hansson, L. A. (2000) Induced pigmentation in zooplankton: A trade-off between threats from predation and ultraviolet radiation. *Proceedings of the Royal Society of London - Series B: Biological Sciences. [print]*, **267**: 2327–31.

Höckelmann, C., Moens T., and Jüttner, F. (2004) Odor compounds from cyanobacterial biofilms acting as attractans and repellents for free-living nematodes. *Limnology and Oceanography*, **49**: 1809–19.

Howe, N. R. and Harris. L. G. (1978) Transfer of the seaanemone pheromone, Anthopleurine, by the nudibranch *Aeolidia-papillosa*. *Journal of Chemical Ecology*, **4**: 551–61.

Howe, N. R. and Sheikh, Y. M. (1975) Anthopleurine - a sea-anemone alarm pheromone. *Science*, **189**: 386–8.

Hung, O. S., Lee, O. O., Thiyagarajan, V., He, H. P., Xu, Y., Chung, H. C., Qiu, J. W., and Qian, P. Y. (2009) Characterization of cues from natural multi-species biofilms that induce larval attachment of the polychaete *Hydroides elegans*. *Aquatic Biology*, **4**: 253–62.

Hurst, T. P., Kay, B. H., Brown, M. D., and Ryan, P. A. (2010) *Melanotaenia duboulayi* influence oviposition site selection by *Culex annulirostris* (Diptera: Culicidae) and *Aedes notoscriptus* (Diptera: Culicidae) but not *Culex quinquefasciatus* (Diptera: Culicidae). *Environmental Entomology*, **39**: 545–51.

Jang, M. H., Ha, K., Joo, G. J., and Takamura, N. (2003) Toxin production of cyanobacteria is increased by exposure to zooplankton. *Freshwat. Biol.*, **48**: 1540–50.

Jang, M. H., Jung, J. M., and Takamura, N. (2007) Changes in microcystin production in cyanobacteria exposed to zooplankton at different population densities and infochemical concentrations. *Limnology and Oceanography*, **52**: 1454–66.

Jensen, R. A. and Morse, D. E. (1990) Chemically-induced metamorphosis of polychaete larvae in both the laboratory and ocean environment. *Journal of Chemical Ecology*, **16**: 911–30.

Joint, I., Tait, K., and Wheeler, G. (2007) Cross-kingdom signalling: exploitation of bacterial quorum sensing molecules by the green seaweed *Ulva*. *Philosophical Transactions of the Royal Society B-Biological Sciences*, **362**: 1223–33.

Jüttner, F., Messina, P., Patalano, C., and Zupo, V. (2010) Odour compounds of the diatom *Cocconeis scutellum*: effects on benthic herbivores living on *Posidonia oceanica*. *Marine Ecology Progress Series*, **400**: 63–73.

Kicklighter, C. E., Germann, M., Kamio, M., and Derby, C. D. (2007) Molecular identification of alarm cues in the defensive secretions of the sea hare *Aplysia californica*. *Animal Behaviour*, **74**: 1481–92.

Krueger, D. A. and Dodson, S. I. (1981) Embryological induction and predation ecology in *Daphnia pulex*. *Limnology and Oceanography*, **26**: 219–23.

Kusch, J. (1993) Predator-induced morphological changes in *Euplotes* (Ciliata): Isolation of the inducing substance released from *Stenostomum sphagenetorum* (Turbellaria). *The Journal of Experimental Zoology*, **265**: 613–18.

Kusch, J. (1999) Self-recognition as the original function of an amoeban defense-inducing kairomone. *Ecology*, **80**: 715–20.

Kusch, J. and Heckmann, K. (1992) Isolation of the *Lembadion*-factor, a morphogenetically active signal, that induces *Euplotes* cells to change from their ovoid form into a larger lateral winged morph. *Developmental Genetics*, **13**: 241–6.

Laforsch, C.,Beccara, L.,and Tollrian, R. (2006) Inducible defenses: The relevance of chemical alarm cues in *Daphnia*. *Limnology and Oceanography*, **51**: 1466–72.

Lampert, W., Rothhaupt, K. O., and Von Elert, E. (1994) Chemical induction of colony formation in a green alga (*Scenedesmus acutus*) by grazers (*Daphnia*). *Limnology and Oceanography*, **39**: 1543–50.

Leblanc, G. A. (2007) Crustacean endocrine toxicology: a review. *Ecotoxicology*, **16**: 61–81.

Long, J. D., Smalley, G. W., Barsby, T., Anderson, J. T., and Hay, M. E. (2007) Chemical cues induce consumer-specific defenses in a bloom-forming marine phytoplankton. *Proceedings of the National Academy of Sciences of the United States of America*, **104**: 10512–17.

Loose, C. J., Von Elert, E., and Dawidowicz, P. (1993) Chemically induced diel vertical migration in *Daphnia*: a new bioassay for kairomones exuded by fish. *Archiv für Hydrobiologie*, **126**: 329–37.

Lürling, M. (2006) Effects of a surfactant (FFD-6) on *Scenedesmus* morphology and growth under different nutrient conditions. *Chemosphere*, **62**: 1351–8.

Machacek, J. (1991) Indirect effects of planktivorous fish on the growth and reproduction of *Daphnia galeata*. *Hydrobiologia*, **225**: 193–7.

Marko, M. D., Newman, R. M., and Gleason, F. K. (2005) Chemically mediated host-plant selection by the milfoil weevil: A freshwater insect-plant interaction. *Journal of Chemical Ecology*, **31**: 2857–76.

Matz, C., Webb, J. S., Schupp, P. J., Phang, S. Y., Penesyan, A., Egan, S., Steinberg, P., and Kjelleberg, S. (2008) Marine biofilm bacteria evade eukaryotic predation by targeted chemical defense. *Plos One*, 3(7): e2744.

Mirza, R. S. and Chivers, D. P. (2001) Are chemical alarm cues conserved within salmonid fishes? *Journal of Chemical Ecology*, **27**: 1641–55.

Miyakawa, H., Imai, M., Sugimoto, N., Ishikawa, Y., Ishikawa, A., Ishigaki, H., Okada, Y., Miyazaki, S., Koshikawa, S., Cornette, R., and Miura, T. (2010) Gene up-regulation in response to predator kairomones in the water flea, *Daphnia pulex*. *BMC Developmental Biology*, 10: 45–54.

Moomjian, L., Nystrom, S., and Rittschof. D. (2003) Behavioral responses of sexually active mud snails: Kairomones and pheromones. *Journal of Chemical Ecology*, **29**: 497–501.

Painter, S. D., Clough, B., Garden, R. W., Sweedler, J. V., and Nagle, G. T. (1998) Characterization of *Aplysia* attractin, the first water-borne peptide pheromone in invertebrates. *Biological Bulletin*, **194**: 120–31.

Pfeiffer, W. (1974) Pheromones in fish and amphibia. In M.C. Birch, (ed) *Pheromones*, pp. 269–96. North-Holland Publishing Company, Amsterdam, London.

Pfeiffer, W., Riegelbauer, G., Meier G., and Scheibler, B. (1985) Effect of hypoxanthine-3(N)-oxide and hypoxanthine-1(N)-oxide on central nervous excitation of the Black Tetra *Gymnocorymbus-ternetzi* (Characidae, Ostariophysi, Pisces) indicated by dorsal light response. *Journal of Chemical Ecology*, **11**: 507–23.

Pijanowska, J. (1997) Alarm signals in *Daphnia*? *Oecologia*, **112**: 12–16.

Pohnert, G. and Von Elert, E. (2000) No ecological relevance of trimethylamine in fish - *Daphnia* interactions. *Limnology and Oceanography*, **45**: 1153–6.

Ponnusamy, L., Xu, N., Nojima, S., Wesson, D. M., Schal, C., and Apperson, C. S. (2008) Identification of bacteria and bacteria-associated chemical cues that mediate oviposition site preferences by *Aedes aegypti*. *Proceedings of the National Academy of Sciences of the United States of America*, **105**: 9262–7.

Rahman, Y. J., Forward, R. B., and Rittschof, D. (2000) Responses of mud snails and periwinkles to environmental odors and disaccharide mimics of fish odor. *Journal of Chemical Ecology*, **26**: 679–96.

Rejmankova, E., Higashi, R. M., Roberts, D. R., Lege M., and Andre, R. G. (2000) The use of solid phase microextraction (SPME) devices in analysis for potential mosquito oviposition attractant chemicals from cyanobacterial mats. *Aquatic Ecology*, **34**: 413–20.

Rittschof, D. and Cohen, J. H. (2004) Crustacean peptide and peptide-like pheromones and kairomones. *Peptides*, **25**: 1503–16.

Rittschof, D., Tsai, D. W., Massey, P. G., Blanco, L. , Kueber G. L., and Haas, R. J. (1992) Chemical mediation of behavior in hermit crabs - alarm and aggregation cues. *Journal of Chemical Ecology*, **18**: 959–84.

Rohl, I., Schneider, B., Schmidt B., and Zeeck, E. (1999) L-Ovothiol A: The egg release pheromone of the marine polychaete *Platynereis dumerilii*: Annelida: Polychaeta. *Zeitschrift fur Naturforschung C-A Journal of Biosciences*, **54**: 1145–7.

Rundle, S. D. and Brönmark, C. (2001) Inter- and intraspecific trait compensation of defence mechanisms in freshwater snails. *Proceedings of the Royal Society of London Series B- Biological Sciences*, **268**: 1463–8.

Schaefer, A. L., Greenberg, E. P., Oliver, C. M., Oda, Y., Huang, J. J., Bittan-Banin, G., Peres, C. M., Schmidt, S., Juhaszova, K., Sufrin J. R., and Harwood, C. S. (2008) A new class of homoserine lactone quorum-sensing signals. *Nature,* **454**: 595–600.

Selander, E., Thor, P., Toth G., and Pavia, H. (2006) Copepods induce paralytic shellfish toxin production in marine dinoflagellates. *Proceedings of the Royal Society B-Biological Sciences,* **273**: 1673–80.

Shabani, S., Kamio, M., and Derby, C. D. (2008) Spiny lobsters detect conspecific blood-borne alarm cues exclusively through olfactory sensilla. *Journal of Experimental Biology,* **211**: 2600–8.

Silberbush, A., Markman, S., Lewinsohn, E., Bar, E., Cohen, J. E. and Blaustein, L. (2010). Predator-released hydrocarbons repel oviposition in a mosquito. *Ecology Letters* **13**: 1129–38.

Sleeper, H. L., Paul, V. J., and Fenical, W. (1980) Alarm pheromones from the marine opisthobranch *Navanax inermis. Journal of Chemical Ecology,* **6**: 57–70.

Snell, T. W., Kubanek, J., Carter, W., Payne, A. B., Kim, J., Hicks, M. K., and Stelzer, C. P. (2006) A protein signal triggers sexual reproduction in *Brachionus plicatilis* (Rotifera). *Marine Biology,* **149**: 763–73.

Steinke, M., Stefels, J., and Stamhuis, E. (2006) Dimethyl sulfide triggers search behavior in copepods. *Limnology and Oceanography* **51**: 1925–30.

Stibor, H. and Lüning, J. (1994) Predator-induced phenotypic variability in the pattern of growth and reproduction in *Daphnia hyalina* (Crustacea: Cladocera). *Functional Ecology,* **8**: 97–101.

Sumper, M., Berg, E., Wenzl, S., and Godl, K. (1993) How a sex-pheromone might act at a concentration below 10(-16) M. *EMBO Journal,* **12**: 831–6.

Takeichi, M., Hirai Y., and Yusa, Y. (2007) A water-borne sex pheromone and trail following in the apple snail, *Pomacea canaliculata. Journal of Molluscan Studies,* **73**: 275–8.

Tang, K. W., Smith, W. O., Elliott, D. T., and Shields, A. R. (2008) Colony size of *Phaeocystis antarctica* (Prymnesiophyceae) as influenced by zooplankton grazers. *Journal of Phycology,* **44**: 1372–8.

Ting, J. H. and Snell, T. W. (2003) Purification and sequencing of a mate-recognition protein from the copepod *Tigriopus japonicus. Marine Biology,* **143**: 1–8.

Tollrian, R. and von Elert, E. (1994) Enrichment and purification of *Chaoborus* kairomone from water - further steps toward its chemical characterization. *Limnology and Oceanography,* **39**: 788–96.

Uchida, H., Yasumoto, K., Nishigami, A., Zweigenbaum, J. A., Kusumi, T., and Ooi, T. (2008). Time-of-flight LC/ MS identification and confirmation of a kairomone in *Daphnia magna* cultured medium. *Bulletin of the Chemical Society of Japan,* **81**: 298–300.

van Donk, E., Lürling M., and Lampert, W. (1999) Consumer-induced changes in phytoplankton: inducibility, costs, benefits, and the impact on grazers. In R. Tollrian, R. and C. D. Harvell (eds) *The Ecology and Evolution of Inducible Defenses,* pp 89–103. Princeton University Press, Princeton, New Jersey.

van Gremberghe, I., Vanormelingen, P., van der Gucht, K., Mancheva, A., D'hondt, S., De Meester L., and Vyverman, W. (2009) Influence of *Daphnia* infochemicals on functional traits of *Microcystis* strains (Cyanobacteria). *Hydrobiologia,* **635**: 147–55.

van Holthoon, F. L., van Beek, T. A., Lürling, M., Van Donk, E., and De Groot, A. (2003) Colony formation in *Scenedesmus*: a literature overview and further steps towards the chemical characterisation of the *Daphnia* kairomone. *Hydrobiologia,* **491**: 241–54.

Von Elert, E. and Franck, A. (1999) Colony formation in *Scenedesmus*: grazer-mediated release and chemical features of the infochemical. *J. Plankton Res.,* **21**: 789–804.

Von Elert, E. and Loose, C. J. (1996) Predator-induced diel vertical migration in *Daphnia* - enrichment and preliminary chemical characterization of a kairomone exuded by fish. *Journal of Chemical Ecology,* **22**: 885–95.

Von Elert, E. and Pohnert, G. (2000) Predator specificity of kairomones in diel vertical migration of *Daphnia*: a chemical approach. *Oikos,* **88**: 119–28.

Von Elert, E. and Stibor, H. (2006) Predator-mediated life history shifts in *Daphnia*: enrichment and preliminary chemical characterisation of a kairomone exuded by fish. *Archiv für Hydrobiologie,* **167**: 21–35.

Von Frisch, K. (1941) Über einen Schreckstoff der Fischhaut und seine biologische Bedeutung. *Zeitschrift für Vergleichende Physiologie,* **29**: 46–145.

Yasumoto, K., Nishigami, A., Aoi, H., Tsuchihashi, C., Kasai, F., Kusumi, T., and Ooi, T. (2008a) Isolation and absolute configuration determination of aliphatic sulfates as the *Daphnia* kairomones inducing morphological defense of a phytoplankton - Part 2. *Chemical & Pharmaceutical Bulletin,* **56**: 129–32.

Yasumoto, K., Nishigami, A., Aoi, H., Tsuchihashi, C., Kasai, F., Kusumi T., and Ooi, T. (2008b) Isolation of new aliphatic sulfates and sulfamate as the *Daphnia* kairomones inducing morphological change of a phytoplankton *Scenedesmus gutwinskii. Chemical & Pharmaceutical Bulletin,* **56**: 133–6.

Yasumoto, K., Nishigami, A., Yasumoto, M., Kasai, F., Okada, Y., Kusumi, T., and Ooi, T. (2005) Aliphatic sulfates relased from *Daphnia* induce morphological defense of phytoplankton: isolation and synthesis of kairomones. *Tetrahedron Letters,* **46**: 4765–7.

Zeeck, E., Harder T., and Beckmann, M. (1998a) Uric acid: The sperm-release pheromone of the marine polychaete *Platynereis dumerilii*. *Journal of Chemical Ecology,* **24**: 13–22.

Zeeck, E., Müller, C. T., Beckmann, M., Hardege, J. D., Papke, U., Sinnwell, V., Schroeder, F. C., and Francke, W. (1998b) Cysteine-glutathione disulfide, the sperm-release pheromone of the marine polychaete *Nereis succinea* (Annelida: Polychaeta). *Chemoecology,* **8**: 33–8.

Zimmer, R. K., Schar, D. W., Ferrer, R. P., Krug, P. J., Kats, L. B., and Michel, W. C. (2006) The scent of danger: Tetrodotoxin (TTX) as an olfactory cue of predation risk. *Ecological Monographs,* **76**: 585–600.

CHAPTER 3

Pheromones mediating sex and dominance in aquatic animals

Thomas Breithaupt and Jörg D. Hardege

Social interactions between aquatic animals are mediated by visual, mechanical (acoustic, tactile, hydrodynamic), electrical, and chemical stimuli Bradbury and Vehrencamp 2011. Animals actively or inadvertently release body fluids and secret chemicals from specific glands. These can be carried over long distances, and provide a wealth of information for a receiver about attributes of the sender, such as sex, species, social status, individuality, sexual receptivity, immune status, and many others. Hence, it is not surprising that some of these compounds have evolved into chemical signals, 'pheromones'.

In spite of being very widespread, chemical communication is difficult to study. Chemical signals in water generally go unnoticed by researchers since the compounds are not visible and signal release does not usually involve eye-catching body movements, such as territorial marking in dogs. Most of what we know about aquatic chemical signals comes from studying the behavioural and physiological responses of the receiver. Only recently has progress in visualization techniques allowed us to observe chemical signals, and to study the behaviour of both the sender and receiver in natural interactions (Breithaupt and Eger 2002; Barata *et al.* 2007). Chemical communication occurs in various behavioural contexts of aquatic animals, including courtship, animal contests, mother–offspring interactions and aggregation behaviour (Breithaupt and Thiel 2011). The goal of this chapter is to review the current knowledge on pheromones of aquatic animals. We will limit ourselves to the signals used in courtship and animal fights as these have been investigated in the most depth in the context of sexual selection.

Species from many different animal taxa have been found to use pheromones underwater. These include reptiles (Mason and Parker 2010), amphibians (Woodley 2010), fish (Stacey and Sorensen 2009), crustaceans (Breithaupt and Thiel 2011), molluscs (Susswein and Nagle 2004), rotifers (Snell 1998), and polychaetes (Hardege 1999). The aim of this chapter is not to summarize the results from species of all the different taxa, but to focus on some recurrent themes and mechanisms found across taxa in aquatic pheromone research. We refer the reader to the above review articles for detailed overviews of chemical communication in the respective taxa. Our examples will focus on fish, crustaceans, and polychaetes, reflecting either the best-studied groups (fish) or our own research bias (crustaceans, polychaetes).

3.1 What is a pheromone?

Pheromones are defined as chemical signals used for communication within a species (Wyatt 2009). Karlson and Lüscher (1959) introduced this term to label a category of 'substances which are secreted to the outside by an individual and received by a second individual of the same species, in which they release a specific reaction, for example, a definite behaviour [releaser pheromone] or a developmental process [primer pheromone]'. New research has shown that pheromones are not always species specific, that the responses to them can be modified by experience, and that they can be mixtures of different chemical components (Wyatt 2009). For example, some fish species share the same pheromonal compounds (Stacey and Sorensen 2009). Aquatic pheromones are either dispersed by the surrounding fluid and detected by olfactory receptors (soluble or volatile pheromones, 'distance pheromones'), or they are expressed at a surface (usually the outer body surface) and detected by contact (gustatory, taste) chemoreceptors ('contact pheromones') (Bauer 2011).

Pheromones undergo the same evolutionary selection mechanisms as other communication signals such as acoustic and visual displays. A signal is any act or structure (or chemical) which alters the behaviour of other organisms, which evolved because of that effect, and which is effective because the receiver's response has also evolved (Maynard Smith and Harper 2003). This definition emphasizes that signalling is adaptive for the signaller, and responding to the signal is adaptive for the receiver. Signallers that are exploited by the receiver are selected against and so are receivers that easily fall victim to deceit. In many cases, signal production involves extra cost. These costs indicate to the receiver the reliability of the signal and enhance likelihood of a response (Maynard Smith and Harper 2003; Zahavi and Zahavi 1997). Stacey and Sorensen (2009) argue that pheromones may not always be signals. Intermediate stages of chemical signal evolution ('spying', Fig. 3.3) may include specific adaptations of the receiver but not of the signaller (Stacey and Sorensen 2009). However, we suggest that there is no fundamental difference between pheromones and other signals. Sexual stimuli being 'spied' by the opposite sex will facilitate reproduction and therefore entail fitness benefits to both the receiver and the releaser. Similarly, visual stimuli, such as the extended abdomen of gravid female fish, may elicit strong courtship behaviour in the receiver (e.g. the famous zigzag dance of the male stickleback, Tinbergen 1948) and hence increase the reproductive success of the female even if the visual stimulus is unchanged.

Other semiochemicals (i.e. chemicals conveying information used in interactions between organisms; also called 'infochemicals'; Dicke and Sabelis 1988) such as 'kairomones' and 'allomones', discussed in other chapters of this book, have not evolved for communication purposes. These are also referred to as 'cues', that is features of the world used by an animal as a guide to future action. For example, chemicals (kairomones) leaking from an injured organism may attract predators. They are cues (for predators) and not signals as they have not evolved to attract predators. Signals may have originally evolved from 'cues' such as the hormonal pheromones in fish that evolve from unspecialized metabolites into courtship signals (see below).

3.2 Production, transmission, and reception

Aquatic animals use both distance and contact pheromones for sexual and aggressive communication. In crustaceans and fish, distance pheromones are often released in urine through the excretory pores (located anteriorly in crustaceans, Atema and Steinbach 2007; or caudally in fish, Appelt and Sorensen 2007), but urine is not always the carrier of the chemical signal (in crustaceans see e.g. Kamio et al. 2002). In fish, pheromones can be released through the gill epithelia (Vermeirssen and Scott 1996), faeces (Polkinghorne et al. 2001), or with the gametes (Stacey and Hourston 1982). In marine polychaetes pheromones are released through the gonopores together with the gametes and coelomic fluid (Zeeck et al. 1998a). Contact pheromones, present in crustaceans and rotifers, are located at different areas of the body surface and are suspected to be produced in tegumental glands (Kamiguchi 1972; Snell 2011; Caskey et al. 2009; Zhang et al. 2010). If pheromones are urine-borne, signallers potentially have control over the timing of release (Appelt and Sorensen 2007; Barata et al. 2007; Berry and Breithaupt 2010; Breithaupt and Eger 2002). The sphincter muscles of the excretory pores are under neuronal control and can be used to modulate urine output.

Once pheromones are released they are dispersed by ambient water movements, developing an odour plume whose shape and structure is continuously modified by the local flow conditions as it is carried downstream (see Chapters 1 and 7). In stagnant environments (lakes or ponds) odour dispersal will be relatively slow. Signallers that are walking or swimming leave a scented trail behind that facilitates detection as it can be used by receivers to track and find the signaller (e.g. pheromone trails in copepod crustaceans, Bagøien and Kiørboe 2005; and in marine polychaetes, Ram *et al.* 2008). In crustaceans, stationary senders generate directed water currents by ventilating or by fanning maxillipeds, walking legs, or pleopods to disperse the chemical signals (see Fig. 3.1; Breithaupt 2001; Atema and Steinbach 2007; Kamio *et al.* 2008). Actively flushing signals towards conspecifics appears to be a general strategy in many crustaceans as they are equipped with specialized fanning structures to generate water currents (see e.g. Breithaupt 2001; Cheer and Koehl 1987).

Crustaceans can actively increase the perception range by creating water currents that draw the molecules towards them (in lobsters: Atema and Steinbach 2007; crayfish: Denissenko *et al.* 2007; copepods: Moore *et al.* 1999). Active odour sampling by antennal flicking is an additional mechanism to enhance olfaction in crustaceans (Koehl 2006). While active odour acquisition strategies such as fanning or flicking are known from crustaceans, other aquatic animals such as polychaetes and fish are more mobile underwater and may enhance

their perceptive range by increased swimming. Pheromone detection appears to be mainly limited to olfactory organs such as the aesthetasc bearing antennules in crustaceans or the olfactory epithelium in the nasal cavity of fish (Derby and Sorensen 2008). In crustaceans and in most fish the olfactory fibres project into a specialized olfactory lobe in the brain (Caprio and Derby 2008). In marine polychaetes the pheromone detecting chemoreceptors are located on the parapodial cirri (Hardege 1999).

3.3   Sex pheromones in fish—spying males and the evolution of chemical communication

In the majority of fish species, fertilization takes place outside the body. Males and females shed their gametes into the surrounding water. For this mode of reproduction to be successful it is crucial that the timing of gonadal maturation, courtship, and spawning is perfectly matched between the

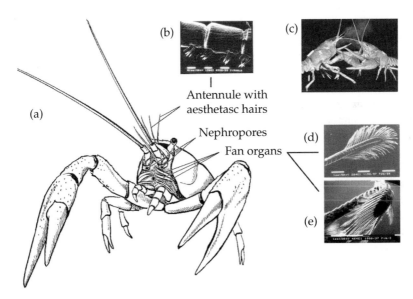

Figure 3.1   Crayfish structures involved in chemical communication (a). Urinary chemical signals are released from the nephropores and are detected by the chemoreceptors (aesthetasc hairs) on the antennules (inset b). Urine is carried by gill currents and forms urinary odour plumes which are directed towards the opponent's antennules (inset c). Fan organs located below the nephropores aid with odour acquisition of the receiver. During the power stroke the hairs on the distal flagella are extended (inset d). During the recovery stroke (inset e) the hairs are folded. This allows generation of uni-directional water currents (modified after Breithaupt 2001 with permission from *The Biological Bulletin*; inset image C from Breithaupt 2011 with permission from Springer).

two sexes. The synchronization of reproductive events is mediated by a sequence of internal (hormonal) and external (pheromonal) chemical signals. When exposed to female sexual stimuli the male's hormone level increases, leading to changes in gamete production and behaviour. The hormones involved in maturation of female gonads (gonadal steroids) and in ovulation (prostaglandins) are released into the surrounding water. They stimulate the olfactory lobe of the male and trigger hormonal (primer effects) and behavioural changes (releaser effects) in the male. These stimuli are called 'hormonal pheromones' as they contain compounds derived from hormones and act as olfactory signals (Stacey and Sorensen 2009). Olfactory reception of gonadal steroids and prostaglandins in males was demonstrated by electrophysiological studies of male olfactory epithelia. Such electro-olfactograms revealed that male receivers in many species from many taxonomic groups (e.g. Cypriniformes, Siluriformes, Characiformes, Salmoniformes, Perciformes) are highly sensitive to hormonal compounds of female signallers (Stacey and Cardwell 1995). In contrast, specialization of the signaller in the production and release of hormonal pheromones is not always evident. The role of sexes in hormonal pheromones differs between fish species depending on the mating system. For example, in gobies (order Perciformes) the males establish territories and attract females. It is the males that release steroidal pheromones and the females that develop high sensitivity detection and behavioural responses to the male compounds. See Stacey and Sorensen (2009) for an in depth review of hormonal pheromones including specializations of the different families of fish. Hormonal pheromones are released by four different routes: free (unmodified) steroids are released rapidly through the gills or with gonadal fluid; conjugated steroids are released more slowly through urine and bile (faeces), respectively; and prostaglandins are always released in the urine (Stacey and Sorensen 2009).

3.3.1 The time course of reproductive behaviour in cyprinid fish

The best studied model for the function of 'hormonal pheromones' is the goldfish *Carassius auratus*.

Many other cyprinid fish as well as herrings, catfishes, and characins use very similar mechanisms (Stacey and Sorensen 2009). In goldfish, the timing of spawning is controlled by the female and the male adjusts its hormonal maturation and behaviour to her (Fig. 3.2).

Recrudescent pheromone: ovarian maturation starts with vitellogenesis (yolk deposition) and gradually increasing plasma levels of the steroid 17β-Estradiol (E2). E2 stimulates urinary release of a still unidentified 'recrudescent pheromone' by the female, which attracts males and probably helps males to stay within aggregations containing females (Kobayashi *et al.* 2002; Fig. 3.2).

Preovulatory pheromones: after vitellogenesis is completed, external conditions (rising temperatures, detection of suitable spawning substrate) induce a sharp increase in plasma levels of luteinizing hormone (LH) in the female that lasts about 15 hours culminating in ovulation and spawning (Manukata and Kobayashi 2010; Fig. 3.2). The LH surge stimulates synthesis of the steroid hormone 17,20β-dihydroxy-4-pregnen-3-one (17,20β-P, the maturation inducing steroid) in the ovarian follicles, which induces oocyte maturation. 17,20β-P and its conjugate 17,20β-P sulfate are released into the water as 'preovulatory pheromones'. 17,20β-P is continuously released through the gills and mirrors internal concentration. Detection of picomolar concentrations induces LH surges in exposed males which then drives behavioural responses (males follow and chase females; Fig. 3.2) as well as an increase in male milt (sperm and seminal fluid) production (Dulka *et al.* 1987). Female release of the conjugate 17,20β-P sulfate occurs later in the LH surge. The sulfate conjugated steroid is released with the urine and elicits brief, acute bouts of male chasing and following (Poling *et al.* 2001). Whereas the increase in milt prepares the males for spawning, the male's behavioural responses of 'following' and 'chasing' may facilitate female ovulation.

Postovulatory pheromones: around the time of ovulation, females produce large amounts of the hormone prostaglandin $F_{2\alpha}$ ($PGF_{2\alpha}$) which initiates female spawning behaviour. $PGF_{2\alpha}$ and its metabolite 15-keto-prostaglandin $F_{2\alpha}$ ($15K\text{-}PGF_{2\alpha}$) are released in urinary pulses and act as 'postovulatory pheromones' with $15K\text{-}PGF_{2\alpha}$ being about 100 times more potent than the unmodified hormone.

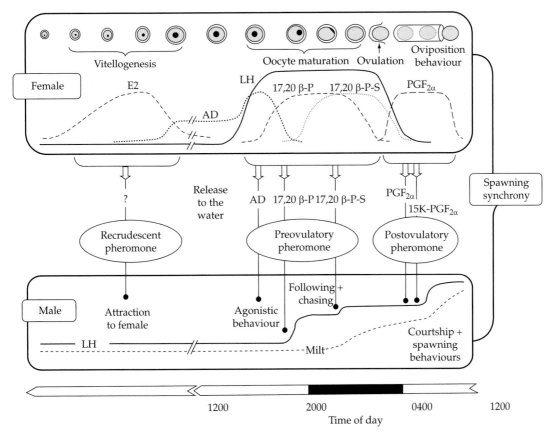

Figure 3.2   Time course of hormonal and pheromonal changes during reproductive behaviour in the goldfish *Carassius auratus*. Increase of female hormones E2 (17β-Estradiol), AD (Androstenedione), LH (luteinizing hormone), 17,20β-P and 17,20β-P-S (maturation inducing steroids), and PGF$_{2\alpha}$ (Prostaglandin F$_{2\alpha}$) precedes release of hormonal pheromones (recrudescent pheromone, preovulatory pheromone, postovulatory pheromone) which stimulate male maturation (LH rise, increase in milt) and elicits behavioural responses. See text for further explanations. Reprinted from Stacey and Sorensen (2009), with permission from Elsevier Press.

Olfactory detection of these pheromones by males is essential for reproduction. Males fail to mate if their olfactory system is blocked (Stacey and Kyle 1983). Female urinary release of postovulatory pheromone appears to be specialized for a signalling function. The frequency of receptive female's urine pulses increases in the presence of males but not with females (Appelt and Sorensen 2007). Females are more likely to urinate when they are rising into floating vegetation which is used as spawning substrate. Males are attracted to the spawning substrate by 15k-PGF$_{2\alpha}$ (Appelt and Sorensen 2007).

Male sex pheromones: male goldfish release large quantities of the steroid androstenedione (AD)

which stimulates aggressive behaviour in other males. The possibility that females use male odours for mate choice has not been examined in any detail, although common carp *Cyprinus carpio* have been shown to detect AD (Irvine and Sorensen 1993).

Evolution of chemical communication in fish: Stacey and Sorensen (2009) propose that the evolution of chemical communication in fish involves three distinct phases (Fig. 3.3). The ancestral phase involves release of a metabolic product which is undetectable to the receiver. In a second phase, *spying*, receivers will be selected to detect and adaptively respond to these metabolites. Male reception of unmodified female gonadal pheromones

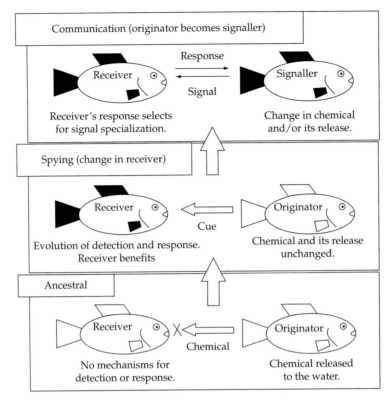

Figure 3.3 Proposed stages in the evolution from chemical cues to chemical communication signals in fish. Reprinted from Stacey and Sorensen, (2009), with permission from Elsevier Press.

(17,20β-P) may be a good example of spying. In a third phase, chemical communication is fully established with receivers detecting and responding to the stimulus and signallers having the specializations to produce it more effectively. The postovulatory pheromones (15K-PGF$_{2\alpha}$) released by ovulated female goldfish and the bile acids (3K-PS) released by spermiating male lampreys (see below) may be examples for such fully evolved communication signals. While in fish, discrimination of sex through pheromones is evident in many species, little is known about recognition of species through chemicals. Many hormonal pheromones are shared between closely related species and do not seem to provide species specific information. Differences in timing or space of reproduction as well as additional information from other chemical or non-chemical sources may ensure species specificity in such cases.

3.3.2 Exceptions from hormonal pheromone communication systems

While the majority of fish species investigated so far use hormonal pheromones there are exceptions. Some fish do not use hormonal pheromones to synchronize reproduction. Lampreys are jawless fish representing the most basal lineage of vertebrates. Adults are parasites. They use their jawless mouth to clamp onto a fish host and ingest their blood. They migrate into freshwater streams to spawn in the spring. Males build nests upon their arrival and release a sex pheromone that attracts ovulated females from downstream. This bile-acid derived pheromone 3-keto-petromyzonol sulfate (3K-PS) is produced in the liver of spermiating males and is released from glandular cells in the gills that appear to be specialized for pheromone release (Li *et al.* 2002; Siefkes *et al.* 2003).

The salmonids (order Salmoniformes) include species with males responding to female steroids and prostaglandins (Atlantic salmon and brown trout) as well as species that show no responses to prostaglandins (Pacific trout and Pacific salmon, *Oncorhynchus*). For example, members of the genus *Oncorhynchus* do not detect any of the known commercially available PGFs. However, they show a strong response to the amino-acid L-kynurenine, a non-hormonal tryptophane metabolite (Yambe *et al.* 2006). Female L-kynurenine appears to replicate in males of the more derived salmonid species of the genus *Oncorhynchus* the releaser effect that PGFs have in the more basal whitefish, Atlantic salmon, and brown trout.

3.4 Sex pheromones in crustaceans— indicators of female receptivity and triggers of mate guarding

Crustaceans undergo regular moults in order to grow. In many crustaceans, including amphipods, blue crabs, shore crabs, and lobsters, reproduction is closely linked to female moulting. Females are only receptive during a short time-interval following moult (Jormalainen 1998). In such cases pre-copulatory mate guarding is a common male mating strategy. By monopolizing a female that is about to moult the males ensures her presence during receptivity and maximizes the proportion of paternity in case of potential sperm competition (Jormalainen 1998). Sex pheromones emanate from pre-moult female crabs (*Pachygrapsus pachypus*) that attract males from a distance and elicit courtship behaviour in them (Ryan 1966). Initial studies of these pheromones suggested that the moult-inducing steroid hormone 20-hydroxyecdyson (20HE) was the active component (Kittredge and Takahashi 1972). However, these results were dismissed based on experiments on a wide range of species including shore crabs, lobsters, and blue crabs (Atema and Gagosian 1973; Gleeson *et al.* 1984; Hardege *et al.* 2002) showing that 20HE does not elicit pair formation. Kittredge and Takahashi's (1972) suggestion, despite its rejection in crustacean pheromone research, was pivotal in triggering studies into fish hormonal pheromones (see above). In shore crabs *Carcinus maenas*, 20HE, while not mediating reproduction, acts as a feeding deterrent in preventing cannibalism of the freshly moulted female by males (Hayden *et al.* 2007).

Sex pheromones in Brachyuran crabs: since Ryan's early studies there has been plenty of evidence for male responses to female sex pheromones in various orders of crustaceans (see Breithaupt and Thiel 2011). In blue crabs *Callinectes sapidus*, helmet crabs *Telmessus cheiragonus*, and shore crabs, female pheromones elicit male attraction and pre-copulatory mate guarding (male crabs 'cradle-carry' females; Gleeson *et al.* 1984; Kamio *et al.* 2002; Hardege *et al.* 2002). In shore crabs, uridine diphosphate (UDP) has been identified as one component of the urine-borne pheromone mixture and has been shown to elicit cradle-carrying (Hardege and Terschak 2011; Hardege *et al.* 2011). UDP is set free from UDP-N-actylglucosamine during the final step of chitin biosynthesis (Hardege and Terschak 2011). Female pheromone level is directly linked to moult stage. Female shore crabs moult in the summer whereas males moult in the spring. Therefore, perception of UDP in the summer is a good indicator for a male that a nearby female is about to moult, and hence is receptive. It is not clear whether female shore crabs have evolved any specialization for sending pheromone (urine). The signal also does not appear to be strictly species specific. Some other crab species such as *Chionoecetes opilio* and *Libinia emarginata* show a sexual response when stimulated by the shore crab pheromone (Bublitz *et al.* 2008). Blue crab males do not respond with sexual behaviour. In blue crabs, perception of receptive female's urine stimulates male attraction and cradle-carrying (Kamio and Derby 2011). If the female is not accessible to the male, the male conducts a courtship display called 'stationary paddling' where males stand high on legs with extended claws and paddle their last walking legs to generate a forward directed current. Initial experiments by Gleeson (1991) concluded that the female sex pheromone is a small urine-borne polar molecule which is detected by olfactory receptors (aesthetasc hairs) on the first antenna (antennules) of the males. Recent studies using male stationary paddling as a bioassay suggest that the pheromone is a multi-component mixture. Males responded to two different molecular size fractions (small <500 Da and medium 500–1000 Da; Kamio and Derby 2011).

Sex pheromones in lobsters: in lobsters (*Homarus americanus*) pre-copulatory mate guarding occurs when males and receptive females share a shelter during the time preceding and following mating (for up to two weeks). This extended cohabitation allows males to secure their paternity whereas females gain protection during the vulnerable post-moult time when their shell is still soft. In lobster, the female is attracted to the dominant males' shelter. Males only allow pre-moult females into their shelter and respond aggressively to inter-moult females and to males. Males recognize the odour of freshly moulted females to which they are attracted and display reduced aggression (Atema and Steinbach 2007). In interspecific groups of lobsters *H. americanus* and *H. gammarus*, males accept only the conspecific female and mating occurs exclusively between conspecifics (van der Meeren *et al.* 2008). Together, these studies suggest the existence of a female pheromone in lobsters, providing information about species, gender, and sexual receptivity. The molecular identity of this pheromone is unknown.

Crayfish sex pheromones: in crayfish, mating occurs when both sexes are in inter-moult (i.e. hard shelled). Mating is preceded by overt aggression between the sexes. Females release pulses of urine during aggressive acts. Males recognize female receptivity through these urinary components and attempt to turn and copulate (Berry and Breithaupt 2010). The high aggression in initial phases of crayfish mating may be a result of sexual conflict because, unlike soft shell maters, receptive female crayfish have many mating opportunities and do not benefit from mate guarding.

Sex pheromones in copepods: copepods are small planktonic crustaceans (1–10 mm) inhabiting almost all parts of marine and freshwater environments. They are the most abundant multicellular organism in the open ocean. Occurring at relatively low densities (a few individuals per cubic metre) and having poor visual senses, they face the challenge of detecting and finding a mate for reproduction (Bagøien and Kiørboe 2005). Detailed studies of mate finding behaviour in 3-D showed that females leave a chemical trail in their wake which is followed by males. Males, upon intercepting a female trail, change their direction and track even convoluted trails up to 31 s old and 17 cm long (Bagøien and Kiørboe 2005; see

also Yen and Lasley 2011). Male *Temora longicornis* follow artificial trails flavoured with conspecific odour more often than heterospecific trails (Yen and Lasley 2011). However, they do not discriminate between male and female trails, indicating that other cues such as contact pheromones (see below) are essential for the decision to mate.

Male sex pheromones: male pheromones are much less investigated. In crustaceans, males are generally the sex that initiates courtship. Males appear to be recognized and evaluated by females through multimodal stimuli. For example, female lobsters are attracted to the shelter of a dominant male. During shelter visits, both resident male and visiting female increase urine release. Blocking urine release of the resident male will decrease the attractiveness of the shelter to visiting females (Bushmann and Atema 2000). On the other hand, release of male urine alone will not attract a female to the shelter, suggesting that non-urinary chemical cues are also involved in mediating female attraction from a distance. Additional non-chemical cues, such as vision or touch, are necessary to stimulate female shelter visits (Bushmann and Atema 2000). Similarly, crayfish females respond to and recognize males only if they see and smell them. Chemical stimuli alone are not sufficient for female recognition of males (Aquiloni *et al.* 2009). There is some evidence from spiny lobsters and crayfish that chemical signals contribute to female assessment of male body size (Raethke *et al.* 2004; Aquiloni and Gherardi, 2008) but the chemicals need to be complemented by visual stimuli in order to allow female choice.

Contact pheromones: in many crustacean species, particularly those that occur in high population densities (aggregations or schools) such as certain decapod shrimps, males frequently encounter other members of the population and check them for sex and receptivity (Bauer 2011). In some caridean shrimp species (e.g. *Heptacarpus sitchensis*), only after touching a newly moulted female with his long antennae will the male seize her and immediately copulate (Bauer 2011). In the sub-millimetre sized copepods (e.g. the marine herpacticoid *Triticum vulgaris*) males recognize a female by probing her body with a series of antennal strokes. This probing behaviour determines his decision to mate guard or not (Snell 2011). In copepods, the molecules were identified as 70-kDa glycoproteins that are secreted and displayed on the body surface of females (Snell 2011). Glycoproteins were also found on the surface of caridean shrimp from the genus *Lysmata*, but behavioural experiments on the role of these molecules as a mate recognition pheromone in shrimps yielded contradictory results (Caskey *et al.* 2009; Zhang *et al.* 2010), calling for additional studies. Recently, the contact pheromones were identified in *Lysmata bogessii* as a blend of lipophilic cuticular hydrocarbons (CHCs) (Zhang *et al.* 2011). Of these, (Z)-9-octadecenamide is the key odour compound with hexadecanamide and methyl linoleate enhancing the bioactivity of the pheromone blend. Cuticular hydrocarbons are known from insects and can carry information about the sender's sex, receptivity, and species identity (Billeter *et al.* 2009).

3.5   Pheromones mediating dominance interactions

Pheromones play an important role in resolving aggressive interactions of aquatic animals. Physical combats can lead to injury inflicted by claws (as in decapod crustaceans) or jaws (as in fish). Conflicts are often resolved by ritualized behaviours including visual, tactile, and chemical signals. The role of pheromones is well understood in some decapod crustaceans and in cichlid fish.

Aggressive encounters are a regular element of social interactions in decapod crustaceans including lobsters, crayfish, crabs, hermit crabs, and stomatopods. Fighting serves to resolve conflicts over limiting resources such as food, shelter, and mates (Briffa and Sneddon 2006). Lobsters and stomatopods fight over shelter with residents aggressively defending their shelter against intruders. In hermit crabs, the gastropod shell protects the body and is subject to aggressive competition. Male crayfish and crabs fight over food items and access to females. Fights often go through a sequence of behavioural stages that increase in aggression as the fight proceeds (Atema and Steinbach 2007). Chemical signalling by urinary pulses is an integral part of fighting behaviour. In crayfish and lobsters, urine release coincides with aggressive acts, and the frequency of urine pulses increases with increasing level of aggression (Breithaupt and Atema 2000;

Breithaupt and Eger 2002). Visualization of urine signals in crayfish competitive interactions shows that only the combination of physical (e.g. claw strikes) and chemical (urine release) displays is effective in changing the behaviour of the receiver (Breithaupt and Eger 2002). A dominance hierarchy is established through repeated encounters and maintained through mutual chemical assessment. American lobsters and some hermit crabs recognize the individual scent of opponents they have fought before (Atema and Steinbach 2007). Wyatt (2011) termed those cues 'signature mixtures' as they vary between individuals and need to be learned (unlike pheromones). In American lobsters, both males and females regularly patrol nearby shelters occupied by conspecifics and exchange chemical signals with their residents (Atema and Steinbach 2007). In this scenario, with high site fidelity and repeated encounters between the same individuals, memory of individual odours provides an advantage in preventing costly fights. Hermit crabs aggregate at shell-supplying sites and interact in pairs over ownership of gastropod shells (Gherardi and Tricarico 2011). They quickly acquire the memory (within a few minutes) to classify a combatant as familiar and associate its odour with the quality of the owner's shell. In subsequent encounters they investigate conspecifics with high quality shells much more intensively than those with low quality shells (Gherardi and Tricarico 2011). Unlike American lobsters and hermit crabs, freshwater crayfish respond to urinary signals of dominant conspecifics even if they have not encountered those individuals before. In the latter, dominance appears to be mediated by a status indicating dominance pheromones (Breithaupt 2011). Here, the high population density and low site fidelity of crayfish populations make it less likely that individuals encounter the same opponent repeatedly (Breithaupt 2011). The chemical nature of the signature mixtures and dominance pheromones in crustaceans is still unknown.

Similar to decapod crustaceans, urinary signals appear to play an important role in competitive interactions of some fish species. In the Mozambique tilapia, a lek-breeding species where males aggregate in breeding areas and establish dominance hierarchies, urine signals mediate aggression between competing males (Barata *et al.* 2007). Males increase the frequency of urination when aggressive but not when submissive. Dominant males store more urine in their bladder than subordinates. The urine of dominant fish is also more potent in stimulating olfactory receptors of other males than that of subordinates. Males of this species appear to advertise their social rank by urine-borne dominance pheromones. The male pheromones may also be used by females to assess male quality. A urinary compound was purified from dominant males that elicited a strong response in female olfactory receptors (Barata *et al.* 2008). Based on female electro-olfactogram responses, this compound was characterized as a sulfated sterol-like molecule. The compound appears to be more abundant in the urine of dominant males than in that of subordinate males. Females responded to compounds that were more abundant in dominant males than in subordinate males. Other compounds, such as the steroid androstenedione which is released through the gills of sexually active carps and goldfish, have been suggested to regulate competitive interactions among males (Stacey and Sorensen 2009). However, similar to crustaceans, the chemical identity of pheromones regulating aggressive behaviour in fish is still elusive.

3.6 Pheromones mediating spawning without courtship — *Arenicola marina*

Arenicola marina, also known as the lugworm, is a polychaete species with a reproductive strategy where sexual partners have no direct contact during mating. In these worms, males shed gametes onto the surface of a beach at low tide and the incoming tide sweeps the gametes over the egg-laden burrows of females (Howie 1961). The release of gametes into the environment is clearly a risky strategy that requires the evolution of mechanisms to coordinate such behaviour. Lugworms utilize a complex cascade of endocrine and environmental timing mechanisms to ensure simultaneous maturity of the majority of individuals within a population (Bentley and Pacey 1992), and also use sex pheromones to fine-tune gamete release and fertilization (Hardege and Bentley 1997). Maturation of gametes in lugworms starts in the summer under the control of a neurohormone (prostomial maturation hormone)

that affects oocyte growth via a 'coelomic maturation factor' (CMF, Watson and Bentley 1997) that acts directly on the oocytes. Throughout the summer, male spermatogonia grow in the coelom in clusters of a few hundred sperm that are linked through cytoplasmic bridges. These sperm morulae stay intact within the males until the day of reproduction (Bentley *et al.* 1990) when the activity of a sperm maturation factor (SMF), identified by Bentley *et al.* (1990) to be a fatty acid, 8,11,14-eicosatrienoic acid, results in free spermatozoa.

Although hormone activity-driven maturation may well occur in a coordinated manner through endocrine or environmental cues, such as photoperiod (lunar periodicity) and a temperature decrease in autumn, none of these factors alone is sufficient to synchronize the broadcast spawning event (Bentley and Hardege 1996). Spawning behaviour in males is induced by the activity of SMF, whose injection into males results in them shedding sperm within an hour (Bentley and Pacey 1992). The signal that triggers the activation of this neurohormone is external — the detection of oocyte release-associated odour that stems from females having laid their oocytes in the burrows (Hardege *et al.* 1996). The chemical nature of this water-soluble, low molecular mass, female sex pheromone remains unknown but its activity as a primer pheromone results in males eventually shedding gametes. Interestingly, males also respond when detecting chemicals associated with sperm release of other mature males (Bentley and Hardege 1996), this resulting in a 'snowball effect' of synchronized mass spawning of a population (Hardege and Bentley 1997) due to male–male primer pheromone actions.

Although mass spawning may result in increased male–male competition (i.e. through sperm competition) it is thought to be beneficial to males as it increases the presence of male sex pheromones during a spawning event. Male sperm when shed onto the beach may be free spermatozoa, but these sperm are not motile. It is the increase of pH from the body fluid at pH 7.3 to seawater levels of around pH 8.2 that induces the sperm flagella apparatus to function and the sperm to become motile (Pacey *et al.* 1994). Dilution of sperm at the time of the incoming tide is rapid and females only have a window of opportunity of a few minutes to collect sperm to fertilize their oocytes, thus an active transport system for sperm is required (Williams *et al.* 1997). Female lugworms detect compounds associated with male spermatozoa and actively pump seawater (containing sperm) through their burrows (Hardege *et al.* 1996). The likelihood that females detect the spawning activity of single males or small numbers of males is low; as such epidemic spawning should increase an individual male's chances of reproductive success. Male lugworms may have adapted to this challenge by not shedding all spermatozoa in a single event, as they can spawn for up to four consecutive days (Hardege and Bentley 1997). Iteroparity, the annual repeat of the synchronized mass spawning, and the use of primer and releaser pheromones, may all be adaptations of lugworms to the challenge of broadcast spawning without any contact with the opposite sex (Hardege and Bentley 1997).

3.7 Pheromones mediating broadcast spawning

Exact timing and synchronization of reproduction should be a major requirement for broadcast spawning organisms that shed gametes into the free water column and as such have little or no influence as to the fate of their gametes and offspring. Broadcast spawning is nevertheless widespread in marine organisms, with the mass spawning of fish, corals, squid, or jellyfish. Timing is particularly important for species where reproduction is a single, once in a lifetime event, such as most nereidid polychaete worms. The environmental and endocrine factors that control such spectacular events as the mass spawning of Palolo worms, *Eunice viridis* (Caspers 1984) have therefore, and not surprisingly, been the focus of much research (see Bentley and Pacey 1992; Andries 2001) starting as early as 1913 when Lillie and Just described the spawning of *Nereis limbata*. Nereidids have three distinct phases of reproductive maturation and spawning: i) the timing of the maturation of a large proportion of a given population to reach maturity at particular times of the year; ii) the timing of the reproductive event for example through lunar periodicity; and iii) the fine timing of the reproductive event itself that ultimately culminates in the release of gametes. In Box 3.1 a detailed description of the phases is given.

Box 3.1 The phases of reproduction

(i) *Maturation to reproductive stage*: growth and maturation of *Nereis* is controlled via two hormones, a juvenile hormone (termed *Nereidine*) that declines and a maturation hormone, also known as feedback factor, that increases during maturation (Fischer 1984). For reproduction the immature specimen metamorphoses into the mature state, the heteronereid, which is characterized by enlarged eyes and specialized swimming setae as an adaptation for the pelagic phase of reproduction (Hauenschild and Fischer 1969). This heteronereid has only one objective, to shed gametes and ensure successful reproduction, and dies soon afterwards.

(ii) *Timing of the reproductive event:* Metamorphosis to the heteronereid appears mainly through the summer months with the spawning events being highly species-specific. This is based on day length, lunar periodicity, and other environmental cues including temperature, salinity, and food availability (see Hardege 1999 for review). For example, mass spawning events for the two species *N. succinea* and

P. dumerilii at the Isefjord, Denmark, occur at new moon and quarter moon respectively (Hardege *et al.* 1990).

(iii) *The spawning event:* The reproductive behaviour of the majority of nereidid species involves the sexual partners leaving their burrows, usually during the night, and swimming on or near the water surface in search of a sexual partner. This swarming behaviour may include females producing pheromone odour trails that males follow, such as in *Nereis succinea* (Ram *et al.* 2008), but ultimately leads to the sexual partners swimming around each other in narrow circles, also termed the 'nuptial dance' (Hauenschild and Fischer 1969) before releasing gametes. Following Lillie and Just's (1913) initial studies that suggested the involvement of sex pheromones in the stimulation of the spawning behaviour, attempts to identify these cues started as early as Townsend (1939), followed by studies by Boilly-Marer and Lassalle (1980) that also demonstrated the heterospecificity of the gamete release inducing cues.

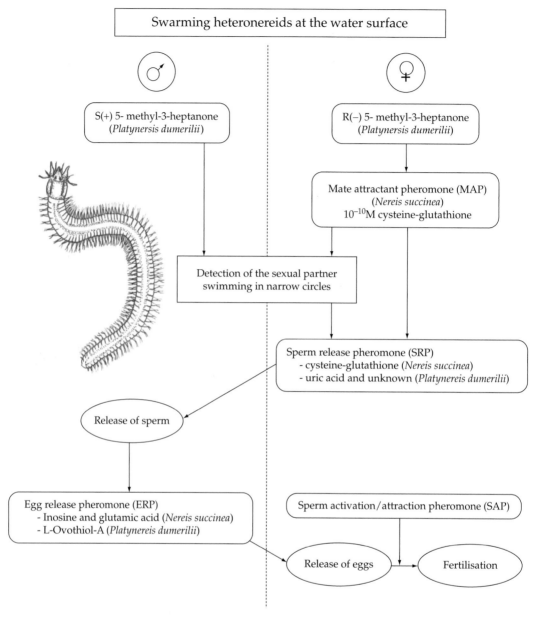

Figure 3.4 Model for the involvement of releaser pheromones in the control of the reproductive behaviour, the 'nuptial dance' and the release of gametes in *Nereis succinea* and *Platynereis dumerilii*. Figure courtesy of John A. Terschak.

Figure 3.4 shows the reproductive behaviour of the two nereidid species *Nereis succinea* and *Platynereis dumerilii* as well as the involvement of sex pheromones in the coordination of the various aspects of their reproductive behaviour. The mate-recognition pheromone in *Platynereis dumerilii* is a lipophilic compound, 5-methyl-3-heptanone (Zeeck *et al.* 1988), that, although found in coelomic fluid of other species such as *Nereis succinea*, seems not to induce the nuptial dance behaviour in any of these (Hardege 1999). Lipophilic ketones diffuse rapidly on the water surface, similar to oil films, but leave no directional information for the receiver. The ketone induces an increase in swim speed in the receiver at sub 10^{-11} M levels and typical swimming in narrow circles but no release of gametes at higher concentrations. Nereidid males generally show higher swim speeds than females and Ram *et al.* (2008) demonstrated that *N. succinea* males show trail-following behaviour using a tetrapeptide, the mixed disulfide of glutathione and cysteine, CySSG (termed Nereithione, Hardege *et al.* 1997) through which males significantly increase their chances to find a female.

Once mature heteronereids meet and perform the nuptial dance, further pheromones are released, usually at relatively high concentrations (10^{-4}–10^{-6} M), and induce shedding of gametes. Gamete release-inducing pheromones identified to date include uric acid in male (Zeeck *et al.* 1996) and 3,5-octadiene-2-one and L-Ovothiol-A in female *Platynereis dumerilii* (Röhl *et al.* 1999), and Nereithione in male (Zeeck *et al.* 1998a; Ram *et al.* 1999) and a mixture of inosine and glutamic acid in female *Nereis succinea* (Zeeck *et al.* 1998b).

The production of the female sex pheromone CySSG (Nereithione) from its likely precursors glutathione (GSH) and cysteine is a late event during the metamorphosis of *N. succinea* (Hardege *et al.* 2004), and is potentially biochemically costly since glutathione is a well known regulator of the oxidative status of cells, especially in oocytes where it protects eggs from the oxidative burst that occurs at fertilization (Shapiro 1991). To release a molecule that requires such an essential resource as glutathione may potentially make female pheromone release an indicator of female fitness (Hardege *et al.* 2004). High release rates (at 10^{-5} M level) and accurate timing are characteristics similar to the induced responses known from insect chemical defence systems (Eisner *et al.* 2005).

As shown in the recent review by Andries (2001) all identified sex pheromones in nereidids are relatively simple molecules that have higher response threshold concentrations (10^{-4}–10^{-6} M) than one would naively expect. This may be due to the need to be detected against significant background noise (Andries 2001), low receptor specificity, or amplification potential, or simply the semelparity (once in a lifetime reproduction) of the worms that would make mistakes extremely costly.

Alternatively, high release rates may be associated with the much lower biochemical costs of using a secondary metabolite (a compound that is available) compared to producing a chemically unique signalling compound. Few studies on marine organisms have looked at the potential costs/benefits of the concepts of using exclusive chemicals or using secondary metabolites. Simple physiological by-products might be less costly to use but can still be as specific as specialist compounds if combined with species- or even situation-specific timing of events. The complex endogenous rhythms and environmental cues utilized by nereidids may well function as pre-mating isolation preventing heterospecific activities of simple pheromones. The reproductive isolation of the two species *N. succinea* and *P. dumerilii* described at the Isefjord, Denmark (Zeeck *et al.* 1990) where there are subtle species specific differences such as the time of the spawning during the night, is a hint to how this mechanism works.

Interestingly, mass spawning in corals is an event where little to no species specificity of pheromones seems to occur. This has led to the theory of epidemic spawning resulting in predator swamping with the potential for cross-fertilization and hybridization as a cost that is less problematic. In fact hybridization is common amongst corals (Babcock *et al.* 1992).

3.8 Future perspectives and applications of pheromone research

Aquatic animals release pheromones through urine or faeces or from parts of the body surface such as gills or parapodia. The examples introduced here show some common characteristics of aquatic pheromones and highlight shortcomings in research on aquatic pheromones:

(i) Few aquatic pheromones have been identified and those compounds have yet to be studied for function–structure relations or common structural characteristics. The high chemical diversity of those compounds identified suggest that almost all compounds that can be detected by a receiver against the background noise are potential chemical signals if the receiver can associate the compound with a specific behavioural context. More chemical identifications are required with specific sets of related organisms to develop evolutionary concepts and phylogenetic trees of signalling cues.

(ii) Those components that are known are rarely produced *de novo* in specific glands but in most cases are metabolites of ongoing biochemical processes in the body. Examples include the identified sex pheromones in fish that derive from maturation hormones, in shore crabs where a moult cycle-related compound is used, and in polychaetes where compounds stem, for example, from the urea cycle and the nitrogen excretion pathways. Is this a common concept and does this restrict the diversity of chemicals used in aquatic systems, for example compared to insects? How much of this is due to structure–activity relationships?

(iii) In many aquatic organisms sex pheromones are released by the females, which is different from many other courtship signals, such as the acoustic and visual displays of birds and amphibians, which are mostly produced by the males and used by the female for mate selection (Andersson 1994). This may be due to the receiver driven evolution of chemical signals. In water, females cannot hide their reproductive state as the metabolites involved in gonadal maturation are released into the water and spread in the environment. Those males that evolve receptors and behaviours to recognize and locate receptive females will have a reproductive advantage over males that do not use this information. However, whether females can also use the provision of pheromones (from males) or the relative response level of males upon the female cues to test the quality of responding males still needs to be investigated. Only then we can start to understand the significance of female produced cues in mate choice.

(iv) Male pheromones, in turn, have been investigated in only a few fish and nereidid polychaete species. Male lampreys' pheromones attract females to the breeding site, but it is not known whether females use this information to assess male quality. In nereidids, male pheromones are only involved in the final step of spawning, the release of gametes after the partners have established direct contact. In the worms male cues presumably exist due to the semelparity that leads to females requiring

honest or convincing messages as to the presence of sperm/males ready to spawn before releasing eggs in a once in a lifetime spawn. Do these male signals provide honest information about male quality? Evidence from *Nereis* does not point towards this, since females release eggs towards any males as long as these produce male pheromones. How this works in other species remains unknown.

(v) The leakage of chemical compounds from aquatic organisms may often lead to eavesdropping by conspecifics and predators and lead to exploitation of the source organism. How signallers avoid this exploitation and use pheromones to their own benefit needs to be further explored. Can crustaceans, for example, control urine release or use timing mechanisms to optimize their use? Furthermore, it has not been investigated if and how aquatic organisms can cheat with their chemical signals, for example by withholding information that otherwise would be available to nearby organisms (*missed opportunities* Hasson 1994). Theoretically, pheromones are ideal to study signal production, costs, and honesty as they could be measured using analytical techniques, but to date little progress has been made in aquatic systems.

(vi) Finally, timing of the release of chemical signals is very important not only to the organisms, but also for the practical use of aquatic pheromones for conservation (e.g. monitoring and trapping of invasive species) and aquaculture. Mainly due to our limited knowledge of the chemistry of aquatic cues, this field is still in its infancy. The first successful applications in the control of invasive lampreys in the Great Lakes of Northern America are promising but success so far is limited to this one species. This ultimately also leads to the question of adaptation and evolution of chemical signals. If pheromones are markers for reproductive fitness of individuals, how will these be affected by pollution and stress? Will marine communication systems be affected by global change, such as increasing temperature and decreasing pH due to ocean acidification? Only when more chemical structures are identified can we begin to approach these questions and start to apply theories of animal communication that have been well established, for example, for visual cues.

References

Andries, J.C. (2001) Endocrine and environmental control of reproduction in Polychaeta. *Canadian Journal of Zoology*, **79**: 254–70.

Andersson, M. (1994) *Sexual Selection*. Princeton University Press, Princeton.

Appelt, C.W. and Sorensen P.W. (2007) Female goldfish signal spawning readiness by altering when and where they release a urinary pheromone. *Animal Behaviour*, **74**: 1329–38.

Atema, J. and Gagosian, R. (1973) Behavioral responses of male lobsters to ecdysones. *Marine Behaviour and Physiology*, **2**: 15–20.

Atema, J. and Steinbach, M.A. (2007) Chemical communication and social behavior of the lobster *Homarus americanus* and other decapod Crustacea. In J.E. Duffy and M. Thiel eds. *Evolutionary Ecology of Social and Sexual Systems: Crustaceans as Model Organisms*, pp. 115–44. Oxford University Press, Oxford.

Aquiloni, L. and Gherardi, F. (2008) Mutual mate choice in crayfish: large body size is selected by both sexes, virginity by males only. *Journal of Zoology*, **274**, 171–9.

Aquiloni, L., Massolo, A., and Gherardi, F. (2009) Sex identification in female crayfish is bimodal. *Naturwissenschaften*, **96**, 103–10.

Babcock, R.C., Mundy, C., Keesing, J., and Oliver, J. (1992) Predictable and unpredictable spawning events: in situ behavioural data from free-spawning coral reef invertebrates. *Invertebrate Reproduction and Development*, **22**: 213–28.

Bagøien, E. and Kiørboe, T. (2005) Blind dating – mate finding in planktonic copepods. I. Tracking the pheromone trail of Centropages typicus. *Marine Ecology Progress Series*, **300**: 105–15.

Barata, E., Hubbard, P., Almeida, O., Miranda, A., and Canario, A. (2007) Male urine signals social rank in the Mozambique tilapia (*Oreochromis mossambicus*). *BMC Biology*, **5**: 54.

Barata, E.N., Fine, J.M., Hubbard, P.C., Almeida, O.G., Frade, P., Sorensen, P.W., and Canario, A.V.M. (2008) A sterol-like odorant in the urine of Mozambique tilapia males likely signals social dominance to females. *Journal of Chemical Ecology*, **34**: 438–49.

Bauer, R.T. (2011) Chemical communication in decapods shrimps: the influence of mating and social systems on the relative importance of olfactory and contact pheromones. In T. Breithaupt and M. Thiel, eds. *Chemical Communication in Crustaceans*, pp. 277–96. Springer, New York.

Bentley, M.G. and Hardege, J.D. (1996) The role of a fatty acid hormone in the control of reproduction in the male lugworm. *Invertebrate Reproduction and Development*, **30**: 159–65.

Bentley, M.G. and Pacey, A.A. (1992) Physiological and environmental control of reproduction in polychaetes, *Oceanography and Marine Biology: An Annual Review*, **30**: 443–81.

Bentley, M.G., Clark, S., and Pacey, A.A. (1990) The role of arachidonic acid and eicosatrienoic acids in the activation of spermatozoa in *Arenicola marina* L. (Annelida: Polychaeta). *Biological Bulletin*, **178**: 1–9.

Berry, F. and Breithaupt, T. (2010) To signal or not to signal? Chemical communication by urine-borne signals mirrors sexual conflict in crayfish. *BMC Biology*, **8**: 25.

Billeter, J.C., Atallah, J., Krupp, J.J., Millar, J.G., and Levine, J.D. (2009) Specialized cells tag sexual and species identity in Drosophila melanogaster. *Nature*, **461**: 987–92.

Boilly-Marer, Y. and Lassalle, B. (1980) Electrophysiological responses of the central nervous system in the presence of homospecific and heterospecific sex pheromones in nereids (Annelida, Polychaeta). *Journal of Experimental Zoology*, **213**: 33–9.

Bradbury, J.W. and Vehrencamp, S.L. (2011) *Principles of Animal Communication*. Second Edition. Sinauer Associates, Inc., Sunderland, Massachusetts.

Breithaupt, T. (2001) Fan organs of crayfish enhance chemical information flow. *Biological Bulletin*, **200**: 150–4.

Breithaupt, T. (2011) Chemical communication in crayfish. In T. Breithaupt and M. Thiel, eds. *Chemical Communication in Crustaceans*, pp. 257–76. Springer, New York.

Breithaupt, T. and Atema, J. (2000) The timing of chemical signaling with urine in dominance fights of male lobsters (*Homarus americanus*). *Behavioural Ecology and Sociobiology*, **49**: 67–78.

Breithaupt, T. and Eger, P. (2002) Urine makes the difference: chemical communication in fighting crayfish made visible. *Journal of Experimental Biology*, **205**: 1221–31.

Breithaupt, T. and Thiel, M. (2011) *Chemical Communication in Crustaceans*. Springer, New York.

Briffa, M. and Sneddon, L.U. (2006) Physiological constraints on contest behaviour. *Functional Ecology*, **21**: 627–37.

Bublitz, R., Sainte-Marie, B., Newcombe-Hodgetts, C., Fletcher, N., Smith, M., and Hardege, J.D. (2008) Interspecific activity of sex pheromone of the European shore crab (*Carcinus maenas*). *Behaviour*, **145**: 1465–78.

Bushmann, P.J. and Atema, J. (2000) Chemically mediated mate location and evaluation in the lobster, *Homarus americanus*. *Journal of Chemical Ecology*, **26**: 883–99.

Caprio, J. and Derby, C. D. (2008). Aquatic animal models in the study of chemoreception. In A.I. Basbaum, A. Kaneko, G.M. Shepherd and G. Westheimer eds. *The*

Senses: A Comprehensive Reference, pp. 97–134. Academic Press, San Diego.

Caskey, J.L., Watson, G.M., and Bauer, R.T. (2009) Studies on contact pheromones of the caridean shrimp *Palaemonetes pugio*: II. The role of glucosamine in mate recognition. *Invertebrate Reproduction and Development*, **53**: 105–16.

Caspers, H. (1984) Spawning periodicity and habitat of the palolo worm *Eunice viridis* (Polychaeta: Eunicidae) in the Samoan Islands. *Marine Biology*, **79**: 229–36.

Cheer, A.Y.L. and Koehl, M.A.R. (1987) Paddles and rakes: fluid flow through bristled appendages of small organisms. *Journal of Theoretical Biology*, **129**: 17–39.

Denissenko, P., Lukaschuk, S., Breithaupt, T. (2007) Flow generated by an active olfactory system of the red swamp crayfish. *Journal of Experimental Biology*, **210**: 4083–91.

Derby, C.D. and Sorensen, P.W. (2008) Neural processing, perception and behavioral responses to natural chemical stimuli by fish and crustaceans. *Journal of Chemical Ecology*, **34**: 898–914

Dicke, M. and Sabelis, M.W. (1988) Infochemical terminology: based in cost-benefit analysis rather than origin of compounds? *Functional Ecology*, **2**: 131–9.

Dulka, J.G., Stacey, N.E., Sorensen, P.W., and Van der Kraak, G.J. (1987) A sex steroid pheromone synchronizes male-female spawning readiness in goldfish. *Nature*, **325**: 251–3.

Eisner, T., Eisner, M., and Siegler, M. (2005) *Secret Weapons: Defences of Insects, Spiders, Scorpions, and Other Many-Legged Creatures*. The Belknap Press of Harvard University Press, Cambridge.

Fischer, A. (1984) Control of oocyte differentiation in nereids (Annelida, Polychaeta) - facts and ideas. *Fortschritte der Zoologie*, **29**: 227–45.

Gherardi, F. and Tricarico, E. (2011) Chemical ecology and social behavior of Anomura. In T. Breithaupt and M. Thiel, eds. *Chemical Communication in Crustaceans*, pp. 297–312. Springer, New York.

Gleeson, R.A., Adams, M.A., and Smith, A.B. III. (1984) Characterization of a sex pheromone in the blue crab, *Callinectes sapidus*: crustecdysone studies. *Journal of Chemical Ecology*, **10**: 913–21.

Gleeson, R.A. (1991) Intrinsic factors mediating pheromone communication in the blue crab, *Callinectes sapidus*. In R.T. Bauer and J.W. Martin eds. *Crustacean Sexual Biology*. pp 17–32. Columbia University Press, New York.

Hardege, J.D. (1999) Nereidid polychaetes as model organisms for marine chemical ecology. *Hydrobiologia*, **402**: 145–61.

Hardege, J.D., Bartels-Hardege, H.D., Zeeck, E., and Grimm, F.T. (1990) Induction of swarming in *Nereis succinea*. *Marine Biology*, **104**: 291–5.

Hardege, J.D., Bentley, M.G., Beckmann, M., and Muller, C.T. (1996) Sex pheromones in marine polychaetes: volatile organic substances (VOS) isolated from *Arenicola marina*. *Marine Ecology Progress Series*, **139**: 157–66.

Hardege, J.D., Müller, C.T., and Beckmann, M.A. (1997) Waterborne female sex pheromone in the ragworm, *Nereis succinea* (Annelida, Polychaeta). *Polychaete Research*, **17**: 18–21.

Hardege, J.D. and Bentley, M.G., (1997) Spawning synchrony in *Arenicola marina*: evidence for sex pheromonal control. *Proceedings of the Royal Society of London B*, **264**: 1041–7.

Hardege, J.D., Jennings, A., Hayden, D., Mueller, C.T., Pascoe, D., Bentley, M.G., and Clare, A.S. (2002) A novel behavioural assay and partial purification of a female-derived sex pheromone in *Carcinus maenas*. *Marine Ecology Progress Series*, **244**: 179–89.

Hardege, J.D., Bartels-Hardege, H., Müller, C.T., and Beckmann, M. (2004). Peptide pheromones in female *Nereis succinea*. *Peptides*, **25**: 1517–22.

Hardege, J.D. and Terschak, J.A. (2011) Identification of crustacean sex pheromones. In T. Breithaupt and M. Thiel, eds. *Chemical Communication in Crustaceans*, pp. 373–92. Springer, New York.

Hardege, J.D., Bartels-Hardege, H.D., Fletcher, N., Terschak, J.A., Harley, M., Smith, M.A., Davidson, L., Hayden, D., Müller, C.T., Lorch, M., Welham, K., Walther, T., and Bublitz, R. (2011) Identification of a female sex pheromone in *Carcinus maenas*. *Marine Ecology Progress Series*, **436**: 177–89.

Hasson, O. (1994) Cheating signals. *Journal of Theoretical Biology*, **167**: 223–38

Hauenschild, C, and Fischer, A. (1969) *Platynereis dumerilii*. Gustav Fischer Verlag, Stuttgart.

Hayden, D., Jennings, A., Muller, C., Pascoe, D., Bublitz, R., Webb, H., Breithaupt, T., Watkins, L., and Hardege, J.D. (2007) Sex-specific mediation of foraging in the shore crab, *Carcinus maenas*. *Hormones and Behaviour*, **52**: 162–8.

Howie, D.I.D. (1961) The spawning mechanism in the male lugworm. *Nature*, **192**: 1100–1.

Irvine, I.A.S. and Sorensen, P.W. (1993) Acute olfactory sensitivity of wild common carp, *Cyprinus carpio*, to goldfish sex pheromones is influenced by gonadal maturity. *Canadian Journal of Zoology*, **71**: 2199–210.

Jormalainen, V. (1998) Precopulatory mate guarding in crustaceans: male competitive strategy and intersexual conflict. *The Quarterly Review of Biology*, **73**: 275–304.

Kamiguchi,Y. (1972) A histological study of the 'sternal gland' in the female freshwater prawn, *Palaemon paucidens*, a possible site of origin of the sex pheromone. *Journal of the faculty of science Hokkaido University, Series VI, Zoology*, **18**: 356–65.

Kamio, M., Matsunaga, S., and Fusetani, N. (2002) Copulation pheromone in the crab *Telmessus cheiragonus* (Brachyura: Decapoda). *Marine Ecology Progress Series*, **234**:183–90.

Kamio, M., Reidenbach, M.A., and Derby, C.D. (2008) To paddle or not: context dependent courtship display by male blue crabs, *Callinectes sapidus*. *Journal of Experimental Biology*, **211**: 1243–8.

Kamio, M. and Derby, C.D. (2011) Approaches to a molecular identification of sex pheromones in blue crabs. In T. Breithaupt and M. Thiel, eds. *Chemical Communication in Crustaceans*, pp. 393–412. Springer, New York.

Karlson, P. and Lüscher, M. (1959) 'Pheromones': a new term for a class of biologically active substances. *Nature*, **183**: 55–6.

Kittredge, J.S. and Takahashi, F.T. (1972) The evolution of sex pheromone communication in the Arthropoda. *Journal of Theoretical Biology*, **35**: 467–71.

Kobayashi, M., Sorensen, P.W., and Stacey, N.E. (2002) Hormonal and pheromonal control of spawning behavior in the goldfish. *Fish Physiology and Biochemistry*, **26**: 71–84.

Koehl, M.A.R. (2006) The fluid mechanics of arthropod sniffing in turbulent odor plumes. *Chemical Senses*, **31**: 93–105.

Li, W., Scott, A.P., Siefkes, M.J., Yan, H., Liu, Q., Yun, S.-S., and Gage, D.A. (2002) Bile acid secreted by male sea lamprey that acts as a sex pheromone. *Science*, **296**: 138–41.

Lillie, F.R. and Just, E.F. (1913) Breeding habits of the heteronereis form of *Nereis limbata* at Whitstable Mass. *Biological Bulletin*, **24**: 147–60.

Manukata, A. and Kobayashi, M. (2010) Endocrine control of sexual behaviour in teleost fish. *General and Comparative Endocrinology*, **165**: 456–68.

Mason, R.T. and Parker, M.R. (2010) Social behavior and pheromonal communication in reptiles. *Journal of Comparative Physiology A*, **196**: 729–49.

Maynard Smith, J. and Harper, D. (2003) *Animal Signals*. Oxford University Press, Oxford.

Moore, P.A., Fields, D.M., and Yen, J. (1999) Physical constraints of chemoreception in foraging copepods. *Limnology and Oceanography*, **44**: 166–77.

Pacey, A, Cosson, J., and Bentley, M.G. (1994) The acquisition of forward motility in the spermatozoa of the polychaete *Arenicola marina*. *Journal of Experimental Biology*, **195**: 259–80.

Poling, K.R., Fraser, E.J., and Sorensen, P.W. (2001) The three steroidal components of the goldfish preovulatory pheromone signal evoke different behaviors in males. *Comparative Biochemistry and Physiology B*, **129**: 645–51.

Polkinghorne, C.N., Olson, J.N., Gallaher, D.G., and Sorensen, P.W. (2001) Larval sea lamprey release two unique bile acids to the water at a rate sufficient to produce detectable riverine pheromone plumes. *Fish Physiology and Biochemistry*, **24**: 15–30.

Raethke, N., MacDiarmid, A.B., and Montgomery, J.C. (2004) The role of olfaction during mating in the southern temperate spiny lobster *Jasus edwardsii*. *Hormones and Behavior*, **46**: 311–18.

Ram, J.L., Müller, C.T., Beckmann, M., and Hardege, J.D. (1999) The spawning pheromone cysteine-glutathione disulfide ('nereithione') arouses a multicomponent nuptial behaviour and electrophysiological activity in *Nereis succinea* males. *FASEB Journal*, **13**: 945–52.

Ram, J.L., Fei, X., Danaher, S.M., Lu, S., Breithaupt, T., and Hardege, J.D. (2008) Finding females: pheromone-guided reproductive tracking behavior by male *Nereis succinea* in the marine environment. *Journal of Experimental Biology*, **211**: 757–65.

Röhl, I., Schneider, B., Schmidt, B., and Zeeck, E. (1999). L-Ovothiol A: the egg release pheromone of the marine polychaete *Platynereis dumerilii*: Annelida: Polychaeta. *Zeitschrift fuer Naturforschung*, **54**: 1145–7.

Ryan, E.P. (1966). Pheromone: Evidence in a decapod crustacean. *Science*, **151**: 340–1.

Shapiro, B.M. (1991) The control of oxidant stress at fertilization. *Science*, **252**: 533–6.

Siefkes, M.J., Scott, A.P., Zielinski, B., Yun, S.S., and Li, W. (2003) Male sea lampreys, *Petromyzon marinus* L., excrete a sex pheromone from gill epithelia. *Biology of Reproduction*, **69**: 125–32.

Snell, T.W. (1998) Chemical ecology of rotifers. *Hydrobiologia*, **387/388**: 267–76.

Snell, T.W. (2011) Contact chemoreception and its role in zooplankton mate recognition. In T. Breithaupt and M. Thiel, eds. *Chemical Communication in Crustaceans*, pp. 451–66. Springer, New York.

Stacey, N.E. and Cardwell, J.R. (1995) Hormones as sex pheromones in fish: widespread distribution among freshwater species. In F.W. Goetz and P. Thomas eds. *Proceedings of the Fifth International Symposium on the Reproductive Physiology of Fish*, pp. 244–8. Symposium 95, Austin, Texas.

Stacey, N.E. and Hourston, A.S. (1982) Spawning and feeding behaviour of captive Pacific herring, *Clupea harengus pallasi*. *Canadian Journal of Fisheries and Aquatic Sciences*, **39**: 489–98.

Stacey, N.E. and Kyle, A.L. (1983) Effects of olfactory tract lesions on sexual and feeding behavior of goldfish. *Physiology and Behavior*, **30**: 621–8.

Stacey, N. and Sorensen, P. (2009) Hormonal pheromones in fish. In D.W. Pfaff, A.P. Arnold, A.M. Etgen, S.E. Fahrbach, and R.T. Rubineds. *Hormones, Brain and Behavior, 2nd Edition*, pp. 639–81, Elsevier Press, San Diego.

Susswein, A.J. and Nagle G.T. (2004) Peptide and protein pheromones in molluscs. *Peptides*, **25**: 1523–30.

Tinbergen, N. (1948) Social releasers and the experimental method required for their study. *The Wilson Bulletin*, **60**: 6–51.

Townsend, G. (1939) On the nature of the material elaborated by fertilizable *Nereis* eggs inducing spawning of the male. *Biological Bulletin*, **77**: 306–7.

van der Meeren, G.I., Chandrapavan, A., and Breithaupt, T. (2008) Sexual and aggressive interactions in a mixed species group of lobsters *Homarus gammarus* and *H. americanus*. *Aquatic Biology*, **2**: 191–200.

Vermeirssen, E.L.M. and Scott, A.P. (1996) Excretion of free and conjugated steroids in rainbow trout (*Oncorhynchus mykiss*): evidence for branchial excretion of the maturation-inducing steroid, 17,20β-dihydroxy-4-pregnen-3-one. *General and Comparative Endocrinology*, **101**: 180–94.

Watson, G.J. and Bentley, M.G. (1997) Evidence for a coelomic maturation factor controlling oocyte maturation in the polychaete *Arenicola marina* (L.). *Invertebrate Reproduction and Development*, **31**: 297–306.

Williams, M.E., Bentley, M.G., and Hardege J.D. (1997) Assessment of field fertilization success in the infaunal polychaete *Arenicola marina* (L.). *Invertebrate Reproduction and Development*, 31, 189–98.

Woodley, S.K. (2010) Pheromonal communication in amphibians. *Journal of Comparative Physiology A*, **19**: 713–27.

Wyatt, T.D. (2009) Fifty years of pheromones. *Nature*, **457**: 262–3

Wyatt, T.D. (2011) Pheromones and behavior. In T. Breithaupt and M. Thiel, eds. *Chemical Communication in Crustaceans*, pp. 23–38. Springer, New York.

Yambe, H., Kitamura, S., Kamio, M., Yamada, M., Matsanuga, S., Fusetani, N., and Yamazaki, F. (2006) L-Kynurenine, an amino acid identified as a sex pheromone in the urine of ovulated female masu salmon.

Proceedings of the National Academy of Science, **103**: 15370–4.

Yen, J. and Lasley, R. (2011) Chemical communication between copepods: finding a mate in a fluid environment. In T. Breithaupt and M. Thiel, eds. *Chemical Communication in Crustaceans*, pp. 177–97. Springer, New York.

Zahavi, A. and Zahavi, A. (1997) *The Handicap Principle: A Missing Piece in Darwin's Puzzle*. Oxford University Press, Oxford.

Zeeck, E., Hardege, J.D., Bartels-Hardege, H., and Wesselmann, G. (1988) Sex pheromone in a marine polychaete: determination of the chemical structure. *Journal of Experimental Zoology*, **246**: 285–92.

Zeeck, E., Hardege, J.D., and Bartels-Hardege, H.D. (1990) Sex pheromones and reproductive isolation in two nereid species, *Nereis succinea* and *Platynereis dumerilii*. *Marine Ecology Progress Series*, **67**: 183–8.

Zeeck, E., Harder, T., Beckmann, M., and Müller, C.T. (1996) Marine gamete-release pheromones. *Nature*, **382**: 214.

Zeeck, E., Müller, C.T., Beckmann, M., Hardege, J.D., Papke, U., Sinnwell, V., Schroeder, F.C., and Francke, W. (1998a) Cysteine-glutathione disulfide, the sperm-release pheromone of the marine polychaete *Nereis succinea* (Annelida: Polychaeta). *Chemoecology*, **8**: 33–8.

Zeeck, E., Harder, T., and Beckmann, M., (1998b) Inosine, L-glutamic acid and L-glutamine as components of a sex pheromone complex of the marine polychaete *Nereis succinea* (Annelida: Polychaeta). *Chemoecology*, **8**: 77–84.

Zhang, D., Zhu, J., Lin, J., and Hardege, J.D. (2010) Surface glycoproteins are not the contact pheromones in the Lysmata shrimp. *Marine Biology*, **157**: 171–6.

Zhang, D., Terschak, J.A., Harley, M.A., Lin, J., and Hardege, J.D. (2011) Simultaneous hermaphroditic shrimp use lipophilic cuticular hydrocarbons as contact sex pheromones. *PLOS ONE*, 6(4): e17720. doi:10.1371/journal.pone.0017720.

CHAPTER 4

Chemical signals and kin biased behaviour

Gabriele Gerlach and Cornelia Hinz

4.1 Living with relatives

In aquatic organisms, the mobility of larvae and their often week-long dispersal phases underpinned the assumption that their group composition was a random mix. According to this theory the place of arrival for settling larvae and their encountered social environment would be determined by prevailing currents, with almost no possibility for larvae to influence their future destiny. This has now been contradicted by population genetic studies. The application of genetic markers, such as DNA microsatellites, has often revealed a small scale population structure with genetically more similar animals living more closely together than others. In the terrestrial environment kin-structured groups—often established by related females and their progeny—are very common. Our review should make it clear that such a kin structure is also prevalent in the aquatic environment, indicating that corresponding social behaviour should be considered to explain population structure and survival.

Kin-structured associations require behavioural and sensory mechanisms to explain how related organisms can stay together or find each other despite being exposed to currents. Behavioural tests have shown that developing larvae gain surprisingly good swimming capabilities so that they cannot be considered passive particles (Bellwood and Fisher 2001). This enables them to conduct at least vertical movements in the water column, which helps them avoid drifting away or losing contact with group members. In addition, waterborne substances released via urine, gills, or mucus carry information about sex, dominance status, reproduc-

tive stage, individuality, and kinship and can aid orientation and the finding of a desirable social environment. These chemical cues influence social interactions in groups, establishment of mating pairs, and the relationship between parents and offspring.

In this chapter we will explain the theoretical selective advantage and evolution of such behaviour and the different mechanisms of kin recognition; we further show how common kin-structured associations are in different aquatic taxa. We will then consider the model organisms, zebrafish and stickleback, to explain how important chemical cues are for recognition and chemical communication.

4.1.1 Kin selection

Besides understanding the mechanisms of kin recognition, the question arises: why is it important to be with kin? The answer is that cooperative behaviour can be observed in kin compared to non-kin groups, which favours reproduction and survival in kin groups. W.D. Hamilton (1964) proposed the concepts of *kin selection* and *inclusive fitness* which incorporate the potential costs and benefits of performing a given act to an individual. Kin selection is the evolutionary mechanism that selects for those behaviours that increase the inclusive fitness of the donor (increase number of own offspring and/or offspring of related individuals). Individuals may increase their inclusive fitness by behaving cooperatively towards related individuals. This theory can explain why behavioural differences exist in the treatment of kin versus non-kin. Even if some behaviour is costly for the donor it might be an evolutionary stable strategy if it is directed towards kin, but selection will act against it if the benefits are allocated to non-kin.

While kin selection has been shown in many terrestrial species such as ground squirrels (*Spermophilus richardsonii*) (Davis 1982), lions (Bertram 1976), pied kingfisher (*Ceryle rudis*) (Reyer 1984), and many social insects (Breed 2003), evidence is scarcer in the aquatic environment; probably due to difficulties of experimental manipulation than due to the lack of kin selection.

4.1.2 Kin recognition

Kin recognition is defined as the 'recognition of and discrimination toward various categories of kin' (Wilson 1987). Kin recognition refers to the underlying cognitive processes, while kin discrimination refers to the observable behavioural patterns made. Different kin recognition mechanisms are used depending on the ecology and sociality of the species (Wilson 1987; Sherman *et al.* 1997). Four general categories of kin recognition mechanisms have been suggested. The first category includes acts of cooperative behaviour directed towards individuals in a particular geographic location, such as the *same nest site or territory*, and applies to species with very predictable spatial distribution. The second category of kin recognition is based on *familiarity* and prior association, that is individuals demonstrate nepotistic behaviour towards any familiar conspecific. Learned familiarity with nest or shoal mates may serve as a good indication that individuals are related, as long as the social system is sufficiently stable to avoid intermingling of unrelated individuals with siblings. The ability to identify even unfamiliar kin is attributed to a third kin recognition mechanism based on *phenotype-matching*, in which an individual learns a common phenotype for familiar kin from parents, siblings, or from itself (self-matching) during early development (Sherman *et al.* 1997) and later uses this template for comparison to phenotypes of unfamiliar conspecifics (Tang-Martinez 2001). Phenotype matching depends on a consistent correlation between phenotypic and genotypic similarity, so that detectable traits are more similar among close relatives than among distantly related or unrelated individuals (Holmes and

Sherman 1983). Kin recognition based on *recognition alleles* has been suggested to be independent of learning, where the phenotypic marker and the recognition of that marker both have a genetic basis.

4.1.3 Evidence for kin recognition and kin associations in the aquatic environment

There are many examples that show that aquatic invertebrates as well as vertebrates can differentiate between and prefer kin to non-kin. In the following, we want to give an overview of how frequent kin recognition and kin associations are in different taxa of aquatic organisms and about the role of chemical signals in the recognition process.

4.1.3.1 Invertebrates

Probably due to difficulties in mating and raising crustaceans in the lab there are only few studies on kin recognition in crustaceans. But population recognition has been demonstrated in the American lobster (*Homarus americanus*). Based on olfactory cues females preferred males of their own population compared to males of a genetically different population (Atema *et al.* in prep). The American lobster is also capable of individual recognition using chemical information. They remember the urine odour of other individuals after a brief fight, thereby avoiding renewed and likely futile fighting (Atema and Voigt 1995; Karavanich and Atema 1998). Males provide information about their dominance via chemical signals in the urine and can alter the amount and time of urine release (Breithaupt and Atema 2000). Among 20 different species of sponge dwelling shrimps (*Synalpheus*) there are some with non-dispersing larvae which form close genetically-related family groups, subject to kin selection, establishing a eusocial system (Duffy and Macdonald 2010).

To our knowledge there are only very few studies which show olfactory based kin recognition and kin aggregations in molluscs. Jüterbock (2010) showed population specific aggregation behaviour in the periwinkle *Littorina littorea*. Snails from different Eastfriesian Islands, Germany, prefer to aggregate with individuals from their own population, probably using chemical information contained in their mucus.

4.1.3.2 Amphibians

Kin recognition seems to be a widespread phenomenon in the tadpole stage of various anuran species (Blaustein and O'Hara 1986; Blaustein and Waldman 1992; Roche 1993) and salamanders (Walls and Roudebush 1991; Masters and Forester 1995; Harris *et al.* 2003). Larval four-toed salamanders, *Hemidactylum scutatum* (Harris *et al.* 2003), wood frog, *Rana sylvatica*, tadpoles (Waldman 1984; Cornell *et al.* 1989), cascades frog, *Rana cascadae*, tadpoles (Blaustein and Ohara 1981; Ohara and Blaustein 1981), and others preferentially associate with their siblings over non-siblings under experimental conditions. These findings are consistent with field observations. Tadpoles of American toads, *Bufo americanus*, and cascades frog, *Rana cascadae*, for example, form social aggregations with their relatives in the wild (Blaustein *et al.* 1988). The aggregation with relatives may have various benefits, such as earlier metamorphosis in *Bufo melanosticus* (Saidapur and Girish 2001) or reduced aggressiveness in fire salamander, *Salamandra infraimmaculata* (Markman *et al.* 2009). A good example of which factors might influence kin preferences can be found in the spadefoot toad's, *Scaphiopus bombifrons*, tadpoles, which may develop into herbivorous omnivores or cannibalistic carnivores. While the omnivores preferentially associate with siblings, carnivores from the same clutch prefer non-siblings (Pfennig *et al.* 1993) and therefore avoid feeding on relatives.

The mechanisms for kin recognition apparently differ between species. Familiarity, phenotype-matching, as well as the involvement of recognition alleles or a combination of different factors have been described (Blaustein *et al.* 1988). Self-referent phenotype matching, for which own body signals are used as a template, seems to be a common mechanism, because even tadpoles that grow up in isolation show kin recognition, like in *Bufo scaber* tadpoles (Gramapurohit *et al.* 2006). Studies on different species indicate that tadpoles recognize their kin by chemical cues (Gramapurohit *et al.* 2006; Eluvathingal *et al.* 2009). Waldman (1985) could show that these cues are detected by olfaction because tadpoles of the American toad, *Bufo americanus*, did not discriminate between kin and non-kin when their nares were blocked; with open nares they did.

Development by metamorphosis is unique to amphibians, resulting in the question of whether adult amphibians still discriminate between kin and non-kin and, if they do, by which mechanism. Waterborne olfactory cues that can be detected by water breathing larvae may not be detectable for air breathing adults. Some studies fail to find kin discrimination in late larval development (Blaustein *et al.* 1993b; Gramapurohit *et al.* 2006), others describe kin discrimination even after metamorphosis (Blaustein *et al.* 1984). Most kin recognition studies on adult amphibians concern brooding behaviour. Gibbons *et al.* (2003) demonstrated that *Plethodon cinereus* females cannibalize unrelated neonates significantly more often than their own.

4.1.3.3 Fish

In fish, the advantage of kin aggregation has been demonstrated in laboratory experiments in salmonids (Brown and Brown 1996). Kin groups had fewer aggressive interactions, used a greater proportion of 'threat' behaviour as opposed to fighting, and had improved growth. Wild juvenile subordinate Atlantic salmon gained foraging advantages by close association with dominant kin, which included higher territory and food sharing (Griffiths and Armstrong 2002). Zebrafish gained 15% more body length when raised in kin groups compared to non-kin groups (Gerlach *et al.* 2007). Improved growth rates in larvae or juvenile fish can correlate with increased fitness because faster-growing larvae, especially female larvae, reach sexual maturity faster and show higher levels of reproductive output. Increased growth rate is also considered an indicator of direct fitness for both sexes in a variety of fishes since it is related to higher over-wintering survival in juvenile coho salmon, *Oncorhynchus kisutch* (Dill *et al.* 1981) and yellow perch, *Perca flavescens* (Dill *et al.* 1981; Post and Evans 1989). Weight gain and longer body length are also correlated with increased social status and ability to obtain and defend resources in juvenile steelhead trout (*Oncorhynchus mykiss*) (Abbott and Dill 1985), decreased time to smoltification in Atlantic salmon (*Salmo salar*) (Metcalfe *et al.* 1989; Metcalfe and Thorpe 1990), and reduced risk of predation in Atlantic salmon, *Salmo salar* (Feltham 1990). Large dominant Atlantic salmon (*Salmo salar*) had greater reproductive success than smaller individuals, and juvenile growth rate was the main determinant of status (Garant *et al.* 2003).

However, under specific circumstances aggregation may also be costly, increasing the risk of predation or resource competition. Under winter conditions, when juvenile Atlantic salmon (*Salmo salar*) shelter inactively in streambed refuges and the potential for kin-biased behaviour is negligible, individuals avoid kin, probably to avoid imposing the costs of aggregation upon close relatives (Griffiths *et al.* 2003).

Many fish live in shoals, especially juveniles but also many adults, and there is evidence from field observations that these groups often are kin-structured. Kin associations have been demonstrated for example in many teleost species including Atlantic cod (*Gadus morhua*) (Herbinger *et al.* 1997), coho salmon (*Oncorhynchus kisutch*) (Quinn and Busack 1985), common shiners (*Notropis cornutus*) (Ferguson and Noakes 1981), Eurasian perch (*Perca fluviatilis*) (Gerlach *et al.* 2001), and threespine sticklebacks (*Gasterosteus aculeatus*) (FitzGerald and Morrissette 1992) though Peuhkuri and Seppä (1998) did not find high relatedness among individuals within stickleback schools. Black chin tilapia (*Sarotherodon melanotheron*) originating from riverine environments appeared to be panmictic, but samples from open waters, such as lagoons, showed sibling aggregations (Pouyaud *et al.* 1999). Multilocus fingerprinting techniques have revealed that wild guppy (*Poecilia reticulata*) school members are more closely related than the average of the population (Magurran *et al.* 1995). Within associations of the coral reef humbag damselfish (*Dascyllus aruanus*), Buston *et al.* (2009) found closely related individuals indicating that they might have recruited together. Eurasian perch, *Perca fluviatilis*, in Lake Constance aggregate in genetically structured groups containing close kin (Gerlach *et al.* 2001). Based on olfaction they preferred their own genetically distinct population over another population living on the opposite side of Lake Constance (Behrmann-Godel *et al.* 2006). Juvenile coho salmon, *Oncorhynchus kisutch*, (Quinn and Busack 1985) and Arctic charr, *Salvelinus alpinus*, (Olsén 1989a) prefer to associate with kin over non-kin under experi-

mental conditions. Nevertheless grouping with relatives can also be disadvantageous, because it may lead to inbreeding or inflict costs due to increasing the risk of predation or resource competition. Hence, kin preferences should be context dependent, as has been shown for example in rainbowfish, *Melanotaenia eachamensis* (Arnold 2000). Females of this species prefer to associate with female relatives, but not male relatives, demonstrating that the females are able to balance the advantages of nepotism and the costs of inbreeding. Evidence for inbreeding avoidance by kin recognition also comes from other species, such as the sexually mature female zebrafish, *Danio rerio*, that prefers the odour of unrelated males over brothers (Gerlach and Lysiak 2006). That inbreeding is not necessarily disadvantageous in general is shown in a study on the cichlid *Pelvicachromis taeniatus*, which performs biparental care. This kind of brood care requires highly synchronous behaviour which seems to be higher between relatives. Both sexes recognized kin and preferred to mate with close relatives (Thünken *et al.* 2007).

Besides its influence on shoaling and mate choice decisions, kin recognition also plays an important role in breeding behaviour, especially in brood-caring species. For example, male bluegill sunfish, *Lepomis macrochirus*, prefer to associate with fry of their own nest over foreign fry (Neff and Sherman 2003) and male threespined sticklebacks, *Gasterosteus aculeatus*, recognize foreign eggs in manipulated nests and cannibalize these clutches more often than sham-manipulated control clutches (Frommen *et al.* 2007a). Further, young convict cichlids, *Amatitlania siquia* (former *Cichlasoma nigrofasciatum*), can orient to their family solely on the basis of chemical cues and discern between the smell of their own family's tank and the smell of another family's tank (Wisenden and Dye 2009).

Independently of whether kin are preferred or avoided, the underlying mechanism of kin recognition is the same. The most common mechanism by which fish learn to recognize their relatives seems to be phenotype matching because individuals of various taxa recognize even unfamiliar kin (Brown and Brown 1993; Gerlach and Lysiak 2006). Though familiarity with kin is not obligatory it might enhance kin recognition, as Gerlach and Lysiak (2006) demonstrated for zebrafish, *Danio rerio* (see Fig. 4.1). The predominant cues by which fish recognize their kin seem to be mostly chemically based and are probably received by olfaction (Quinn and Busack 1985; Hiscock and Brown 2000; Mann *et al.* 2003; Mehlis *et al.* 2008). Besides olfactory cues, visual cues might also play a role in kin recognition, for example female rainbowfish, *Melanotaenia eachamensis*, do not discriminate among groups of different levels of relat-

edness when confronted with chemical cues alone (Arnold 2000) and zebrafish, *Danio rerio*, are able to discriminate kin from non-kin by olfactory cues alone, but need visual contact with their relatives during early development to develop this kin recognition ability (Gerlach *et al.* 2010 in prep). Further information and examples on kin recognition, its underlying mechanisms, and the social context in which it has been observed are listed in Table 4.1.

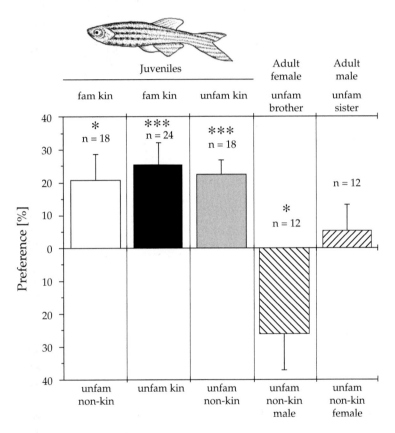

Figure 4.1 Olfactory preference of zebrafish tested in an Atema choice flume (see Chapter 5 for additional information). Testfish: juveniles, 21 days old, adult males and females. To obtain stimulus water for each of the experiments, we placed equal numbers of fish (n = 9, 21 days old) from a kin group overnight in separate 9-L aquaria. For testing olfactory preference of adult females and males stimulus water was used in which a single male and female had been held for 12 hrs. During odour choice tests, a single testfish was placed into the flume and had five minutes to acclimate before measurements began. The test consisted of two 3-minute periods where odour stimuli were presented on alternate sides of the flume to control for possible side bias of the fish. After the first 3-minute period, another 2-minute acclimation time was allowed when stimuli were alternated in the flume, and the test continued for another 3 minutes. From position records taken every 10 seconds, we calculated how often each fish selected a given stimulus water. We calculated the difference between the total number of observations in which a test fish was in the lane with stimulus A or B, and tested whether that difference was significant from zero using two tailed paired t tests (Program JMP, SAS Institute Inc. 1995). If the fish did not have a preference for one of the odour stimuli or was unable to detect a difference between them, we would expect a random distribution on either side of the flume, which would result in no (zero) difference in distribution frequency between treatments. Bars pointing into upper or lower direction show a preference for the stimuli at the top or at the bottom of the graph; * indicates statistical significance, p < 0.05, *** p < 0.001.

Table 4.1 Summary of taxa living in the aquatic environment in which olfactory-based kin recognition or kin associations have been observed.

Taxon	Mechanism	Signals	Social context	Literature
Crustaceans				
Tenagonysis lasmania		chemical?	recognition of offspring	(Johnston and Ritz 2005)
Hermit crab (*Pagurus longicarpus*)		chemical? and tactile	recognition after fighting	(Gherardi and Atema 2003; Gherardi and Atema 2005)
American lobster (*Homarus americanus*)	recognition of familiar individuals	chemical		(Breithaupt and Atema 2000)
Molluscs				
Littorina littorea	recognition of familiar individuals	chemical?	recognition of own population	(Jüterbock 2010)
freshwater snail (*Physa acuta*)		chemical?	female mate choice	(Facon et al. 2006)
Amphibians				
Tiger salamander (*Ambystoma tigrinum nebulosum*)	phenotype-matching	chemical	group association/cannibalism avoidance	(Pfennig et al. 1994; Pfennig 1999)
Four-toed salamander (*Hemidactylium scutatum*)		chemical and/or visual	group association	(Harris et al. 2003)
Red-legged frog (*Rana aurora*)	phenotype-matching	chemical	group association	(Blaustein et al. 1993a)
Cascades frog (*Rana cascadae*)	self-referent phenotype-matching or genetic	chemical	group association	(Blaustein and O'Hara 1982; Blaustein and Waldman 1992)
African clawed frog (*Xenopus laevis*)	self-referent MHC type matching	chemical and/or visual	group association	(Villinger 2007; Villinger and Waldman 2008)
Bufo melanostictus		chemical	group association	(Saidapur and Girish 2000; Eluvathingal et al. 2009)
Bufo scaber	phenotype-matching	chemical	group association	(Gramapurohit et al. 2006; Gramapurohit et al. 2008)
American toad (*Bufo americanus*)		chemical	group association	(Waldman 1985)
Western toad (*Bufo boreas boreas*)		chemical and/or visual	group association	(Blaustein and Ohara 1982)
Spadefoot Toad (*Scaphiopus bombifrons*)		chemical and/or visual	group association/cannibalism avoidance	(Pfennig et al. 1993)
Spadefoot Toad (*Scaphiopus multiplicatus*)	diet influence	chemical or visual (or limited tactile)	group association	(Pfennig 1990)
Fish				
Zebrafish (*Danio rerio*)	phenotype-matching	chemical and/or visual	group association	(Mann et al. 2003; Gerlach and Lysiak 2006; Gerlach et al. 2007; Gerlach et al. 2008)
Black perch (*Embiotoca jacksoni*)		chemical and/or visual	group association; reduced aggressiveness toward familiars	(Sikkel and Fuller 2010)
Threespine stickleback (*Gasterosteus aculeatus*)	yes	chemical	group association; inbreeding avoidance; egg recognition by males	(FitzGerald and Morrissette 1992; Utne-Palm and Hart 2000; Reusch et al. 2001; Frommen and Bakker 2004; Frommen et al. 2007b; Mehlis et al. 2008)

Species	Mechanism	Cue	Function	References
Bluegill sunfish (*Lepomis macrochirus*)	self-referent phenotype-matching	chemical	group association; offspring recognition	(Lacy and Sherman 1983; Neff and Gross 2001; Neff 2003; Neff and Sherman 2003; Neff and Sherman 2005; Hain and Neff 2006, 2007, 2009)
Rainbowfish (*Melanotaenia eachamensis*)		chemical and/or visual	group association	(Arnold 2000)
Cichlid (*Neolamprologus pulcher*)	phenotype-matching	chemical and/or visual	group association, cooperative breeding	(Dierkes et al. 2005; Stiver et al. 2008; Le Vin et al. 2010)
European Perch (*Perca fluviatilis L.*)		chemical	group association	(Behrmann-Godel et al. 2006)
Pelvicachromis taeniatus	self-recognition	chemical	active inbreeding	(Thünken et al. 2007; Thünken et al. 2009)
Guppy (*Poecilia reticulata*)	phenotype-matching & familiarity	chemical and/or visual	group association	(Griffiths and Magurran 1999; Hain and Neff 2007; Evans and Kelley 2008)
Coho salmon (*Oncorhynchus kisutch*)	phenotype-matching	chemical	group association	(Quinn and Busack 1985)
Rainbow trout (*Oncorhynchus mykiss*)	phenotype-matching	chemical	group association	(Brown and Brown 1992; Brown et al. 1993; Johnsson 1997)
Brown trout (*Salmo trutta*)		chemical	group association	(Olsén et al. 1996; Ojanguren and Brana 1999; Carlsson and Carlsson 2002)
Atlantic salmon (*Salmo salar*)		chemical	group association	(Brown and Brown 1992; Moore et al. 1994; O'Connor et al. 2000; Rajakaruna et al. 2001; Griffiths and Armstrong 2002; Griffiths et al. 2003; Rajakaruna and Brown 2006; Rajakaruna et al. 2006)
Arctic charr (*Salvelinus alpinus*)	learning	chemical	group association	(Olsén 1989b; Brown et al. 1996; Olsen and Winberg 1996; Olsén and Jarvi 1997; Olsén et al. 1998; Olsén et al. 2002)
Brook trout (*Salvelinus fontinalis*)		chemical	group association	(Hiscock and Brown 2000; Fraser et al. 2005; Rajakaruna and Brown 2006; Rajakaruna et al. 2006)

4.2 Chemical components involved in kin recognition

Little is known about the nature of the olfactory signals of kin recognition. Genes that code for major histocompatibility complex (MHC)-molecules, which play a fundamental role in the discrimination of self and non-self in the immune system, are thought to also play a role in social recognition (Boehm and Zufall 2006). The MHC-genotype influences kin and mating preferences in different fish taxa (Olsén *et al.* 1998; Olsén *et al.* 2002; Milinski 2003). Peptide-ligands, that bind to MHC-molecules and are known to occur in urine, influence olfactory mate-choice decisions in sticklebacks (Milinski *et al.* 2005), and therefore might represent components of the signals for kin recognition as well. Probably, the signals for this recognition process are very complex and are influenced by different factors. For example, Olsén *et al.* (2003) could show that diet influences individual odours in MHC-identical siblings of juvenile Arctic charr, *Salvelinus alpinus*, demonstrating that kin recognition in fish still raises many open questions and must be part of future research.

4.2.1 The role of MHC-molecules in the immune system

Genes of the major histocompatibility complex (MHC) are essential for discrimination of self and non-self in the immune system of vertebrates. Transplantation experiments on tissue-graft rejection in the 1940s suggested the existence of so-called transplantation antigens. Today we know that these antigens are proteins coded by the MHC. Besides their influence in the context of tissue-graft acceptance or rejection, these proteins play a key-role in the defense against pathogens of the immune system. Two classes of MHC-proteins can be distinguished in specific immune defense. Class I proteins can be detected on the cell surface of nearly all body cells. Class II proteins are typically on the cell surface of antigen-presenting cells.

The extraordinary genetic diversity of the MHC is shown at several levels. A number of functionally equivalent loci exist within the MHC and each of these loci shows a remarkable allelic polymorphism

of sequences coding for the peptide-binding cleft. Furthermore, these loci are co-dominantly expressed and there is a high level of heterozygosity at these loci in populations under natural conditions (Eggert *et al.* 1998).

4.2.2 Influence of MHC-genotype on social interactions

Based on the high MHC polymorphism the MHC-genotype is a good indicator of relatedness. MHC similarity is higher between relatives compared to non-relatives. If individuals 'knew' about their own MHC-genotype and could perceive signals about the genotypes of other individuals, they would be able to use this information in social interactions such as cooperation with MHC-similar close relatives or inbreeding avoidance when choosing MHC-dissimilar mates. Besides inbreeding avoidance, MHC-based mate choice offers the opportunity to maximize pathogen resistance within offspring by choosing the optimal mate. The variety of MHC-proteins that differ in the structure of their peptide-binding clefts results in resistance against many different pathogens. However, studies showed that an optimal instead of a maximal MHC diversity is advantageous (Bonneaud *et al.* 2004; Ilmonen *et al.* 2007). Therefore, individuals should always favour a mate that differs in its genotype from their own and complements their own set of alleles to reach an optimum in the offspring.

There is indeed evidence that individuals are able to compare their own MHC-genotype to the genotype of others. The MHC influences mate-choice decisions and kin recognition in different species, including fish and amphibians. The nature of the signals in this recognition process is highly debated. Several hypotheses have been proposed, for example MHC-proteins, their fragments, and products of the MHC-dependent microflora occur in body fluids and might act as possible signals in this context. To date, peptides that bind to the MHC-proteins are considered likely candidates for representing kin odour, or components of kin odour. MHC-proteins that are no longer bound to the cell membrane lose their 3-D structure, and as a consequence the associated peptide is released and becomes accessible for binding to other molecules, such as an olfactory receptor. A variety of different peptides, derived from different proteins, can bind to one MHC-protein, but the whole peptide-sequence is not associated with the binding pocket. Peptides differ strongly in their complete amino acid sequence, but share common anchor-residues. Only these anchor-residues bind to the MHC-protein and directly reflect the structure of the peptide-binding cleft (Boehm and Zufall 2006).

Neither the exact source and composition of olfactory signals nor the mechanisms behind how these signals are perceived are well understood. Furthermore, kin discrimination by cues correlated with the MHC-genotype seems to be influenced by additional factors, such as diet or an individual's bacterial flora, making MHC studies very complex. To provide a small insight into this complexity we will give an overview of some important studies that have investigated the role of MHC on social interactions in amphibians and fish.

4.2.3 MHC and social interactions in amphibians and fish

4.2.3.1 MHC and kin recognition in amphibians
For amphibians the influence of MHC-genotype on kin recognition was studied in tadpoles of the African clawed frog *Xenopus laevis*. Villinger and Waldman (2008) analysed polymorphisms in DNA sequences coding for the MHC class I peptide-binding region. They could demonstrate that tadpoles of this species discriminate between familiar full-siblings based on differences in the MHC-haplotype. Tadpoles preferred siblings that shared the same MHC-haplotypes to those with no similarity. Since familiarity was the same between the tested tadpoles, the authors assumed that tadpoles use self-referencing for the recognition mechanism and compare their own MHC-genotype to the genotypes of conspecifics.

4.2.3.2 MHC and kin recognition in salmonids and zebrafish
In fish, the relationship of MHC-genotype and kin recognition has been most intensively studied in salmonids. Juvenile arctic charr (*Salvelinus alpinus* (L.) (Olsén *et al.* 1998a, 1998), atlantic salmon (*Salmo salar*), and brook trout (*Salvelinus fontinalis*)

(Rajakaruna *et al.* 2006) prefer the odor of MHC class II similar siblings over MHC class II dissimilar siblings in olfactory choice tests. Furthermore, Olsén and Winberg (1996) could show that juvenile *Salvelinus alpinus* that had grown up isolated since fertilization did not show this discrimination. The results indicate that kin recognition is based on a learning process during development.

We investigated the underlying mechanism of this learning process in zebrafish (*Danio rerio*), finding that kin recognition in zebrafish is based on phenotype matching, allowing individuals to recognize even unfamiliar kin by olfactory cues (Gerlach and Lysiak 2006) (see Fig. 4.1). Juveniles imprint on an olfactory template of their relatives at day 6 post fertilization (Gerlach *et al.* 2008). Earlier or later exposure did not result in kin recognition. The exposure to non-kin odour during this sensitive phase did not result in imprinting on the odour cues of unrelated individuals, suggesting a genetic predisposition to kin odour.

In *Salmo salar*, *Salvelinus fontinalis*, and *Salvelinus alpinus* juveniles do not discriminate between kin and non-kin when the non-kin stimulus was more MHC class II similar to the test fish than the kin-stimulus (Olsén *et al.* 1998; Olsén *et al.* 2003; Rajakaruna *et al.* 2006), indicating a predominant role of MHC class II genes in kin recognition. But if the kin and non-kin stimulus showed no MHC class II similarity to the test fish, juveniles did discriminate between kin and non-kin. Therefore, additional factors for kin recognition must exist in salmonids, for instance MHC class I allele similarity or the involvement of other factors. Some species such as catfish *Pseudoplatystoma coruscans* (Giaquinto and Volpato 2001), Nile tilapia, *Oreochromis niloticus* (Giaquinto *et al.* 2010), and Whitecloud mountain minnows, *Tanichthys albonubes*, (Webster *et al.* 2008) prefer conspecifics fed the same or a better diet as the testfish.

4.2.3.3 MHC and mate choice in sticklebacks
The influence of MHC on reproduction was studied intensively in the threespined stickleback (*Gasterosteus aculeatus*). Diversities of class I and class II genes interfered with individual reproductive parameters, such as male colouration (Jager *et al.* 2007), and mate-choice decisions are strongly

influenced by olfactory signals correlated with the MHC. Reusch *et al.* (2001) found that gravid female sticklebacks chose males with a large number of MHC class II alleles in odour choice tests, as we should expect because a high level of heterozygosity results in high pathogen resistance. They did not detect a preference in females for MHC class II dissimilar males. Contrarily, Aeschlimann *et al.* (2003) observed female sticklebacks favouring the odour of MHC class II dissimilar males. Additionally, they detected that females did not prefer a large number of MHC class II alleles *per se*, but complemented their own set of alleles with that of the male to achieve an optimum for the offspring.

Milinski *et al.* (2005) investigated the nature of the olfactory signals for MHC-based mate choice. They tested for the influence of MHC peptide ligands and found that mate-choice decisions of female sticklebacks could be modified by these peptides. The attractiveness of males with suboptimal numbers of MHC class II alleles could be increased when peptides were added to the natural odour, while attractiveness of males with super-optimal numbers of alleles decreased. The results show that peptide ligands are important for the perception of MHC-signals in mate choice and might play a comparable role in kin recognition.

Similarly as for kin recognition, for mate choice additional factors next to the MHC might influence individual preferences. Sommerfeld *et al.* (2008) measured olfactory attractiveness of males independently of the olfactory MHC-signal through different seasons and reproductive conditions. They found that males are highly attractive during nest maintenance and that the MHC-signal is not sufficient for female attraction. Milinski *et al.* (2010) showed that an additional signal for 'maleness' is needed and that the MHC-signals are conditional on reproductive state. What these additional signals may look like and which mechanisms underlie the production, perception, and processing of MHC-signals is still unknown and will be part of future research.

4.3   Concluding remarks

In this chapter we have shown evidence that shoals, groups, and colonies of aquatic organisms often do not consist of a random mix of conspecifics but rather of more or less related individuals that stay together and colonize the same habitat. Many field studies of terrestrial animals have shown that group members benefit from kin associations. The fact that there is less evidence in aquatic systems is probably due to the difficulties of observing such behaviour rather than to the absence of kin selection. The ability to identify related individuals may have consequences reaching beyond the occurrence of kin-structured groups and kin-biased behaviour and may expand to population recognition. Optimal outbreeding by choosing genetically similar but not very closely related mating partners supports advantageous genetic adaptations to the local environment. If locally well-adapted populations become further isolated, for instance by geographic or climatic factors, genetic divergence can lead to speciation. The combination of behavioural experiments and new sequencing techniques opens the exciting possibility of detecting and studying such microevolutionary processes in the future.

References

Abbott, D.H. and Dill, L.M. (1985) Patterns of aggressive attack in juvenile steelhead trout (*Salmo gairdneri*). *Canadian Journal of Fishery and Aquatic Sciences*, **42**: 1702–6.

Aeschlimann, P.B., Haberli, M.A., Reusch, T.B.H., Boehm, T., and Milinski, M. (2003) Female sticklebacks *Gasterosteus aculeatus* use self-reference to optimize MHC allele number during mate selection. *Behavioral Ecology and Sociobiology*, **54**: 119–26.

Arnold, K.E. (2000). Kin recognition in rainbowfish (*Melanotaenia eachamensis*): sex, sibs and shoaling. *Behavioral Ecology and Sociobiology*, **48**: 385–91.

Atema, J. and Voigt, R. (1995) Behavior and Sensory Biology. In Factor, J.R. (ed.) *Biology of the Lobster*. Academic Press.

Behrmann-Godel, J., Gerlach, G., and Eckmann, R. (2006). Kin- and Population recognition in sympatric Lake Constance perch (*Perca fluviatilis*): can assortative shoaling drive population divergence? *Behavioral Ecology and Sociobiology*, **59**: 461–8.

Bellwood, D.R. and Fisher, R. (2001). Relative swimming speeds in reef fish larvae. *Marine Ecology Progress Series*, 211, 299–303.

Bertram, B.C.R. (1976) Kin selection in lions and in evolution. In Bateson, P.P.G. and Hinde, R.A. (eds.) *Growing*

Points in Ethology, pp. 281–301.Cambridge, Cambridge University Press.

Blaustein, A.R. and O'Hara, R.K. (1981). Genetic control for sibling recognition. *Nature*, **290**: 246–8.

Blaustein, A.R. and O'Hara, R.K. (1982). Kin recognition cues in *Rana cascadae* tadpoles. *Behavioral and Neural Biology*, **36**: 77–87.

Blaustein, A.R., O'Hara, R.K., and Olson, D.H. (1984) Kin preference behavior Is present after metamorphosis in *Rana cascadae* frogs. *Animal Behaviour*, **32**: 445–50.

Blaustein, A.R. and O'Hara, R.K. (1986) An Investigation of kin recognition in Red-legged frog (*Rana aurora*) tadpoles. *Journal of Zoology*, **209**: 347–53.

Blaustein, A.R., Porter, R.H., and Breed, M.D. (1988). Kin recognition in animals - empirical-evidence and conceptual issues. *Behavior Genetics*, **18**: 405–7.

Blaustein, A.R. and Waldman, B. (1992). Kin recognition in anuran amphibians. *Animal Behaviour*, **44**: 207–21.

Blaustein, A.R., Yoshikawa, T., Asoh, K., and Walls, S.C. (1993a) Ontogenetic shifts in tadpole kin recognition: loss of signal and perception. *Anim Behav*, **46**: 525–38.

Blaustein, A.R., Yoshikawa, T., Asoh, K., and Walls, S.C. (1993b) Ontogenetic shifts in tadpole kin recognition: loss of signal and perception. *Animal Behaviour*, **46**: 525–38.

Boehm, T. and Zufall, F. (2006) MHC peptides and the sensory evaluation of genotype. *Trends in Neurosciences*, **29**: 100–7.

Bonneaud, C., Sorci, G., Morin, V., Westerdahl, H., Zoorob, R., and Wittzell, H. (2004) Diversity of Mhc class I and IIB genes in house sparrows (*Passer domesticus*). *Immunogenetics*, **55**: 855–65.

Breed, M.D. (2003) Nestmate recognition assays as a tool for population and ecological studies in eusocial insects: A review. *Journal of the Kansas Entomological Society*, **76**: 539–50.

Breithaupt, T. and Atema, J. (2000) The timing of chemical signaling with urine in dominance fights of male lobsters (*Homarus americanus*). *Behavioral Ecology and Sociobiology*, **49**: 67–78.

Brown, G.E. and Brown, J.A. (1992) Do rainbow trout and Atlantic salmon discriminate kin? *Canadian Journal of Zoology*, **70**: 1636–40.

Brown, G.E. and Brown, J.A. (1993) Do kin always make better neighbors - the effects of territory quality. *Behavioral Ecology and Sociobiology*, **33**: 225–31.

Brown, G.E., Brown, J.A., and Crosbie, A.M. (1993). Phenotype matching in juvenile rainbow trout. *Animal Behaviour*, **46**: 1223–5.

Brown, G.E. and Brown, J.A. (1996) Kin discrimination in salmonids. *Reviews in Fish Biology and Fisheries*, **6**: 201–19.

Brown, G.E., Brown, J.A., and Wilson, W.R. (1996) The effects of kinship on the growth of juvenile Arctic charr. *Journal of Fish Biology*, **48**: 313–20.

Buston, P.M., Fauvelot, C., Wong, M.Y.L., and Planes, S. (2009) Genetic relatedness in groups of the humbug damselfish *Dascyllus aruanus*: small, similar-sized individuals may be close kin. *Molecular Ecology*, **18**: 4707–15.

Carlsson, J. and Carlsson, J.E.L. (2002) Micro-scale distribution of brown trout: an opportunity for kin selection? *Ecology of Freshwater Fish*, **11**: 234–9.

Cornell, T.J., Berven, K.A., and Gamboa, G.J. (1989) Kin recognition by tadpoles and froglets of the wood frog *Rana sylvatica*. *Oecologia*, **78**: 312–16.

Davis, L.S. (1982) Sibling recognition in Richardson's ground squirrels (*Spermophilus richardsonii*). *Behavioral Ecology and Sociobiology*, **11**: 65–70.

Dierkes, P., Heg, D., Taborsky, M., Skubic, E., and Achmann, R. (2005) Genetic relatedness in groups is sex-specific and declines with age of helpers in a cooperatively breeding cichlid. *Ecology Letters*, **8**: 968–75.

Dill, L.M., Ydenberg, R., and Fraser, A.H.G. (1981) Food abundance and territory size in juvenile coho salmon (*Onchorhynchus kisutch*). *Canadian Journal of Zoology*, **59**: 1801–9.

Duffy, J.E. and Macdonald, K.S. (2010) Kin structure, ecology and the evolution of social organization in shrimp: a comparative analysis. *Proceedings of the Royal Society B-Biological Sciences*, **277**: 575–84.

Eggert, F., Muller-Ruchholtz, W., and Ferstl, R. (1998) Olfactory cues associated with the major histocompatibility complex. *Genetica*, **104**: 191–7.

Eluvathingal, L.M., Shanbhag, B.A., and Saidapur, S.K. (2009) Association preference and mechanism of kin recognition in tadpoles of the toad *Bufo melanostictus*. *Journal of Biosciences*, **34**: 435–44.

Evans, J.P. and Kelley, J.L. (2008) Implications of multiple mating for offspring relatedness and shoaling behaviour in juvenile guppies. *Biology Letters*, **4**: 623–6.

Facon, B., Ravigne, V., and Goudet, J. (2006) Experimental evidence of inbreeding avoidance in the hermaphroditic snail *Physa acuta*. *Evolutionary Ecology*, **20**: 395–406.

Feltham, M.J. (1990) The diet of red-breasted mergansers (*Mergus serrator*) during the smolt run in NE Scotland: the importance of salmon (*Salmo salar*) smolts and parr. *Journal of Zoology*, **222**: 285–92.

Ferguson, M.M. and Noakes, D.L.G. (1981) Social grouping and genetic variation in common shiners *Notropis cornutus* (Pisces Cyprinidae). *Environmental Biology of Fishes*, **6**: 357–60.

Fitzgerald, G.J. and Morrissette, J. (1992) Kin recognition and choice of shoal mates by threespine sticklebacks. *Ethology, Ecology and Evolution*, **4**: 273–83.

Fraser, D.J., Duchesne, P., and Bernatchez, L. (2005) Migratory charr schools exhibit population and kin associations beyond juvenile stages. *Molecular Ecology,* **14:** 3133–46.

Frommen, J.G. and Bakker, T.C.M. (2004) Adult three-spined sticklebacks prefer to shoal with familiar kin. *Behaviour,* **141:** 1401–9.

Frommen, J.G., Brendler, C., and Bakker, T.C.M. (2007a) The tale of the bad stepfather: male three-spined sticklebacks *Gasterosteus aculeatus* L. recognize foreign eggs in their manipulated nest by egg cues alone. *Journal of Fish Biology,* **70:** 1295–301.

Frommen, J.G., Luz, C., and Bakker, T.C.M. (2007b). Kin discrimination in sticklebacks is mediated by social learning rather than innate recognition (vol 113, pg 276, 2007). *Ethology,* **113:** 414–414.

Garant, D., Dodson, J.J., and Bernatchez, L. (2003) Differential reproductive success and heritability of alternative reproductive tactics in wild Atlantic salmon (*Salmo salar* L.). *Evolution,* **57:** 1133–41.

Gerlach, G., Schardt, U., Eckmann, R., and Meyer, A. (2001) Kin-structured subpopulations in Eurasian perch (*Perca fluviatilis* L.). *Heredity,* **86:** 213–21.

Gerlach, G. and Lysiak, N. (2006) Kin recognition and inbreeding avoidance in zebrafish, *Danio rerio,* is based on phenotype matching. *Animal Behaviour,* **71:** 1371–7.

Gerlach, G., Hodgins-Davis, A., Macdonald, B., and Hannah, R.C. (2007) Benefits of kin association: related and familiar zebrafish larvae (*Danio rerio*) show improved growth. *Behavioral Ecology and Sociobiology,* **61:** 1765–70.

Gerlach, G., Hodgins-Davis, A., Avolio, C., and Schunter, C. (2008). Kin recognition in zebrafish: A 24-hour window for olfactory imprinting. *Proceedings of the Royal Society, London B,* **275:** 2165–70.

Gherardi, F. and Atema, J. (2003) Individual recognition in hermit crabs: Learning and forgetting the opponents identity. *Integrative and Comparative Biology,* **43:** 854–854.

Gherardi, F. and Atema, J. (2005) Memory of social partners in hermit crab dominance. *Ethology,* **111:** 271–85.

Giaquinto, P.C. and Volpato, G.L. (2001) Hunger suppresses the onset and the freezing component of the antipredator response to conspecific skin extract in pintado catfish. *Behaviour,* **138:** 1205–14.

Giaquinto, P.C., Berbert, C.M.D., and Delicio, H.C. (2010) Female preferences based on male nutritional chemical traits. *Behavioral Ecology and Sociobiology,* **64:** 1029–35.

Gibbons, M.E., Ferguson, A.M., Lee, D.R. ,and Jaeger, R.G. (2003). Mother-offspring discrimination in the red-backed salamander may be context dependent. *Herpetologica,* **59:** 322–33.

Gramapurohit, N.P., Veeranagoudar, D.K., Mulkeegoudra, S.V., Shanbhag, B.A., and Saidapur, S.K. (2006) Kin recognition in *Bufo scaber* tadpoles: ontogenetic changes and mechanism. *Journal of Ethology:* **24,** 267–74.

Gramapurohit, N.P., Shanbhag, B.A. and Saidapur, S.K. (2008) Kinship influences larval growth and metamorphic traits of *Bufo scaber* in a context-dependent manner. *Journal of Herpetology,* **42:** 39–45.

Griffiths, S.W. and Magurran, A.E. (1999) Schooling decisions in gunnies (*Poecilia reticulata*) are based on familiarity rather than kin recognition by phenotype matching. *Behavioral Ecology and Sociobiology,* **45:** 437–43.

Griffiths, S.W. and Armstrong, J.D. (2002) Kin-biased territory overlap and food sharing among Atlantic salmon juveniles. *Journal of Animal Ecology,* **71:** 480–6.

Griffiths, S.W., Armstrong, J.D., and Metcalfe, N.B. (2003). The cost of aggregation: juvenile salmon avoid sharing winter refuges with siblings. *Behavioral Ecology,* **14:** 602–6.

Hain, T.J.A. and Neff, B.D. (2006) Promiscuity drives self-referent kin recognition. *Current Biology,* **16:** 1807–11.

Hain, T.J.A. and Neff, B.D. (2007) Multiple paternity and kin recognition mechanisms in a guppy population. *Molecular Ecology,* **16:** 3938–46.

Hain, T.J.A. and Neff, B.D. (2009) Kinship affects innate responses to a predator in bluegill *Lepomis macrochirus* larvae. *Journal of Fish Biology,* **75:** 728–37.

Hamilton, W.D. (1964). The genetical evolution of social behaviour. I. *Journal of Theoretical Biology,* **7:** 1–16.

Harris, R.N., Vess, T.J., Hammond, J.I., and Lindermuth, C.J. (2003) Context-dependent kin discrimination in larval four-toed salamanders *Hemidactylium scutatum* (Caudata: Plethodontidae). *Herpetologica,* **59:** 164–77.

Herbinger, C.M., Doyle, R.W., Taggart, C.T., et al. (1997) Family relationships and effective population size in a natural cohort of Atlantic cod (*Gadus morhua*) larvae. *Canadian Journal of Fisheries and Aquatic Science,* **54** Suppl.1: 11–18.

Hiscock, M.J. and Brown, J.A. (2000) Kin discrimination in juvenile brook trout (*Salvelinus fontinalis*) and the effect of odour concentration on kin preferences. *Canadian Journal of Zoology,* **78:** 278–82.

Holmes, W. and Sherman, P. (1983) Kin recognition in animals. *The American Scientist,* **71:** 46–55.

Ilmonen, P., Penn, D.J., Damjanovich, K., Morrison, L., Ghotbi, L., and Potts, W.K. (2007) Major histocompatibility complex heterozygosity reduces fitness in experimentally infected mice. *Genetics,* **176:** 2501–8.

Jager, I., Eizaguirre, C., Griffiths, S.W., et al. (2007) Individual MHC class I and MHC class IIB diversities are associated with male and female reproductive traits in the three-spined stickleback. *Journal of Evolutionary Biology,* **20:** 2005–15.

Johnsson, J.I. (1997). Individual recognition affects aggression and dominance relations in rainbow trout, *Oncorhynchus mykiss*. *Ethology*, 103: 267–82.

Johnston, N.M. and Ritz, D.A. (2005) Kin recognition and adoption in mysids (Crustacea: Mysidacea). *Journal of the Marine Biological Association of the United Kingdom*, 85, 1441–7.

Jüterbock, A. (2010) Population dynamics of the periwinkle *Littorina littorea* in the East Frisian Wadden Sea. Diploma thesis University of Oldenburg, Germany.

Karavanich, C. and Atema, J. (1998) Olfactory recognition of urine signals in dominance fights between male lobster, *Homarus americanus*. *Behaviour*, 135: 719–30.

Lacy, R.C. and Sherman, P.W. (1983) Kin Recognition by Phenotype Matching. *American Naturalist*, 121: 489–512.

Le Vin, A.L., Mable, B.K. and Arnold, K.E. (2010). Kin recognition via phenotype matching in a cooperatively breeding cichlid, *Neolamprologus pulcher*. *Animal Behaviour*, 79, 1109–14.

Magurran, A.E., Seghers, B.H., Shaw, P.W., and Carvalho, G.R. (1995) The behavioral diversity and evolution of guppy *Poecilia reticulata*, populations in Trinidad. *Advances in the Study of Behavior* 24: 155–202.

Mann, K.D., Turnell, E.R., Atema, J. and Gerlach, G. (2003) Kin recognition in juvenile zebrafish (*Danio rerio*) based on olfactory cues. *The Biological Bulletin*, 205: 224–5.

Markman, S., Hill, N., Todrank, J., Heth, G., and Blaustein, L. (2009) Differential aggressiveness between fire salamander (*Salamandra infraimmaculata*) larvae covaries with their genetic similarity. *Behavioral Ecology and Sociobiology*, 63: 1149–55.

Masters, B.S. and Forester, D.C. (1995) Kin Recognition in a Brooding Salamander. *Proceedings of the Royal Society of London Series B-Biological Sciences*, 261, 43–8.

Mehlis, M., Bakker, T.C.M., and Frommen, J.G. (2008) Smells like sib spirit: kin recognition in three-spined sticklebacks (*Gasterosteus aculeatus*) is mediated by olfactory cues. *Animal Cognition*, 11: 643–50.

Metcalfe, N.B., Huntingford, F.A., Graham, W.D., and Thorpe, J.E. (1989) Early social status and the development of life-history strategies in Atlantic salmon. *Proceedings of the Royal Society of London B*, 256: 7–19.

Metcalfe, N.B. and Thorpe, J.E. (1990) Determinants of geographic variation in the age of seaward migrating salmon, *Salmo salar*. *Journal of Animal Ecology*, 59: 135–45.

Milinski, M. (2003) The function of mate choice in sticklebacks: optimizing MHC genetics. *Journal of Fish Biology*, 63: 1–16.

Milinski, M., Griffiths, S., Wegner, K.M., Reusch, T.B.H., Haas-Assenbaum, A., and Boehm, T. (2005) Mate choice decisions of stickleback females predictably modified by MHC peptide ligands. *Proceedings of the National Academy of Sciences of the United States of America*, 102: 4414–18.

Milinski, M., Griffiths, S.W., Reusch, T.B.H., and Boehm, T. (2010) Costly major histocompatibility complex signals produced only by reproductively active males, but not females, must be validated by a 'maleness signal' in three-spined sticklebacks. *Proceedings of the Royal Society B-Biological Sciences*, 277: 391–8.

Moore, A., Ives, M.J., and Kell, L.T. (1994) The role of urine in sibling recognition in Atlantic salmon *Salmo salar* (L) Parr. *Proceedings of the Royal Society of London Series B-Biological Sciences*, 255: 173–80.

Neff, B.D. and Gross, M.R. (2001) Dynamic adjustment of parental care in response to perceived paternity. *Proceedings of the Royal Society of London Series B-Biological Sciences*, 268: 1559–65.

Neff, B.D. (2003) Paternity and condition affect cannibalistic behavior in nest-tending bluegill sunfish. *Behavioral Ecology and Sociobiology*, 54: 377–84.

Neff, B.D. and Sherman, P.W. (2003) Nestling recognition via direct cues by parental male bluegill sunfish (*Lepomis macrochirus*). *Animal Cognition*, 6: 87–92.

Neff, B.D. and Sherman, P.W. (2005) In vitro fertilization reveals offspring recognition via self-referencing in a fish with paternal care and cuckoldry. *Ethology*, 111: 425–38.

O'Connor, K.I., Metcalfe, N.B., and Taylor, A.C. (2000) Familiarity influences body darkening in territorial disputes between juvenile salmon. *Animal Behaviour*, 59: 1095–101.

O'Hara, R.K. and Blaustein, A.R. (1981) An investigation of sibling recognition in *Rana cascadae* tadpoles. *Animal Behaviour*, 29: 1121–6.

Ojanguren, A.F. and Brana, F. (1999) Discrimination against water containing unrelated conspecifics and a marginal effect of relatedness on spacing behaviour and growth in juvenile brown trout, *Salmo trutta* L. *Ethology*, 105: 937–48.

Olsén, K.H. (1989a) Sibling recognition in juvenile Arctic charr, *Salvelinus alpinus* (L). *Journal of Fish Biology*, 34: 571–81.

Olsén, K.H. (1989b) Sibling recognition in juvenile Arctic charr, *Savelinus alpinus* (L.). *Journal o Fish Biology*, 34: 571–81.

Olsén, K.H., Jarvi, T., and Lof, A.C. (1996) Aggressiveness and kinship in brown trout (*Salmo trutta*) parr. *Behavioral Ecology*, 7: 445–50.

Olsén, K.H. and Winberg, S. (1996) Learning and sibling odor preference in juvenile arctic char, *Salvelinus alpinus* (L.). *Journal of Chemical Ecology*, 22: 773–86.

Olsén, K.H. and Jarvi, T. (1997) Effects of kinship on aggression and RNA content in juvenile Arctic charr. *Journal of Fish Biology*, **51**: 422–35.

Olsén, K.H., Grahn, M., Lohm, J., and Langefors, A. (1998). MHC and kin discrimination in juvenile Arctic charr, *Salvelinus alpinus* (L.). *Animal Behaviour*, **56**: 319–27.

Olsén, K.H., Grahn, M., and Lohm, J. (2002) Influence of MHC on sibling discrimination in Arctic charr, *Salvelinus alpinus* (L.). *Journal of Chemical Ecology*, **28**: 783–95.

Olsén, K.H., Grahn, M., and Lohm, J. (2003) The influence of dominance and diet on individual odours in MHC identical juvenile Arctic charr siblings. *Journal of Fish Biology*, **63**: 855–62.

Olsén, K.H., Grahn, M., Lohm, J., and Langefors, A. (1998a) MHC and kin discrimination in juvenile Arctic charr, *Salvelinus alpinus* (L.). *Animal Behaviour*, **56**: 319–27.

Peuhkuri, N. and Seppa, P. (1998) Do three-spined sticklebacks group with kin? *Annales Zoologici Fennici*, **35**: 21–7.

Pfennig, D.W. (1990) 'Kin recognition' among spadefoot toad tadpoles: a side-effect of habitat selection? *Evolution*, **44**: 785–98.

Pfennig, D.W., Reeve, H., and Sherman, P.W. (1993) Kin recognition and cannibalism in spadefoot toad tadpoles. *Animal Behaviour*, **46**: 87–94.

Pfennig, D.W., Sherman, P.W., and Collins, J.P. (1994) Kin recognition and cannibalism in polpheinc salamanders. *Behavioral Ecology*, **5**: 225–32.

Pfennig, D.W. (1999) Cannibalistic tadpoles that pose the greatest threat to kin are most likely to discriminate kin. *Proceedings of the Royal Society of London Series B-Biological Sciences*, **266**: 57–61.

Post, J.R. and Evans, D.O. (1989) Size-dependent over-winter mortality of young-of-the-year yellow perch (*Perca flavescens*): laboratory, in situ enclosure and field experiments. *Canadian Journal of Fishery and Aquatic Sciences*, **46**: 1958–68.

Pouyaud, L., Desmarais, E., Chenuil, A., Agnese, T.F., and Bonhomme, F. (1999) Kin cohesiveness and possible inbreeding in the mouthbrooding tilapia *Sarotherodon melanotheron* (Pisces Cichlidae). *Molecular Ecology*, **8**: 803–12.

Quinn, T.P. and Busack, C.A. (1985) Chemosensory recognition of siblings in juvenile coho salmon (*Oncorhynchus kisutch*). *Animal Behaviour*, **33**: 51–6.

Rajakaruna, R.S., Brown, J.A., Kaukinen, K., and Miller, K.M. (2001) MHC genes influence the kin recognition in juvenile Atlantic salmon. *American Zoologist*, **41**: 1562–1562.

Rajakaruna, R.S. and Brown, J.A. (2006) Effect of dietary cues on kin discrimination of juvenile Atlantic salmon (*Salmo salar*) and brook trout (*Salvelinus fontinalis*). *Canadian Journal of Zoology*, **84**: 839–45.

Rajakaruna, R.S., Brown, J.A., Kaukinen, K.H., and Miller, K.M. (2006). Major histocompatibility complex and kin discrimination in Atlantic salmon and brook trout. *Molecular Ecology*, **15**: 4569–75.

Reusch, T.B.H., Haberli, M.A., Aeschlimann, P.B., and Milinski, M. (2001) Female sticklebacks count alleles in a strategy of sexual selection explaining MHC polymorphism. *Nature*, **414**: 300–2.

Reyer, H.U. (1984) Investment and relatedness: a cost/benefit analysis of breeding and helping in the pied kingfisher. *Animal Behaviour*, **32**: 1163–78.

Roche, J.P. (1993) The benefits of kin recognition in tadpoles: a review of the literature. *Maine Naturalist*, **1**: 13–20.

Saidapur, S.K. and Girish, S. (2000) The ontogeny of kin recognition in tadpoles of the toad *Bufo melanostictus* (Anura; Bufonidae). *Journal of Biosciences*, **25**: 267–73.

Saidapur, S.K. and Girish, S. (2001) Growth and metamorphosis of *Bufo melanostictus* tadpoles: Effects of kinship and density. *Journal of Herpetology*, **35**: 249–54.

SAS Institute Inc. (1995) *JMP statistics and graphics guide (Version 3.1)*. SAS Institute inc., Cary, NC.

Sherman, P.W., Reeve, H.K., and Pfennig, D.W. (1997) Recognition systems. In Krebs, J.R. and Davies, N.B. (eds.) *Behavioral Ecology: An Evolutionary Approach*. pp. 69–96. Blackwell Scientific, Oxford UK.

Sikkel, P.C. and Fuller, C.A. (2010) Shoaling preference and evidence for maintenance of sibling groups by juvenile black perch *Embiotoca jacksoni*. *Journal of Fish Biology*, **76**: 1671–81.

Sommerfeld, R.D., Boehm, T., and Milinski, M. (2008) Desynchronising male and female reproductive seasonality: dynamics of male MHC-independent olfactory attractiveness in sticklebacks. *Ethology Ecology & Evolution*, **20**: 325–36.

Stiver, K.A., Fitzpatrick, J.L., Desjardins, J.K., Neff, B.D., Quinn, J.S., and Balshine, S. (2008) The role of genetic relatedness among social mates in a cooperative breeder. *Behavioral Ecology*, **19**: 816–23.

Tang-Martinez, Z. (2001) The mechanisms of kin discrimination and the evolution of kin recognition in vertebrates: A critical re-evaluation. *Behavioural Processes*, **53**: 21–40.

Thünken, T., Bakker, T.C.M., Baldauf, S.A., and Kullmann, H. (2007) Active inbreeding in a cichlid fish and its adaptive significance. *Current Biology*, **17**: 225–9.

Thünken, T., Waltschyk, N., Bakker, T.C.M., and Kullmann, H. (2009) Olfactory self-recognition in a cichlid fish. *Animal Cognition*, **12**: 717–24.

Utne-Palm, A.C. and Hart, P.J.B. (2000) The effects of familiarity on competitive interactions between three-spined sticklebacks. *Oikos*, **91**: 225–32.

Villinger, J. (2007) Kin recognition and MHC discrimination in African clawed frog (*Xenopus laevis*) tadpoles. Canterbury, University Of Canterbury.

Villinger, J. and Waldman, B. (2008) Self-referent MHC type matching in frog tadpoles. *Proceedings of the Royal Society B-Biological Sciences*, **275**: 1225–30.

Waldman, B. (1984) Kin recognition and sibling association among wood frog (*Rana sylvatica*) Tadpoles. *Behavioral Ecology and Sociobiology*, **14**: 171–80.

Waldman, B. (1985) Olfactory basis of kin recognition in toad tadpoles. *Journal of Comparative Physiology a-Sensory Neural and Behavioral Physiology*, 156, 565–77.

Walls, S.C. and Roudebush, R.E. (1991) Reduced aggression toward siblings as evidence of kin recognition in cannibalistic salamanders. *American Naturalist*, **138**: 1027–38.

Webster, M.M., Adams, E.L., and Laland, K.N. (2008) Diet-specific chemical cues influence association preferences and prey patch use in a shoaling fish. *Animal Behaviour*, **76**: 17–23.

Wilson, E.O. (1987) Kin recognition: an introductory synopsis. In Fletcher, D.E. and Michener, C.D. (eds.) *Kin Recognition in Animals*. John Wiley & Sons Ltd, Chichester.

Wisenden, B. and Dye, T. (2009) Young convict cichlids use visual information to update olfactory homing cues. *Behavioral Ecology and Sociobiology*, **63**: 443–9.

CHAPTER 5

The use of chemical cues in habitat recognition and settlement

Gabriele Gerlach and Jelle Atema

In this chapter we address how dispersed marine larvae search for their settlement habitat. While early studies were inspired by observed non-random associations between settled larvae and conspecifics, or associations between sedentary epibionts and their basibionts (e.g. Crisp 1974), later studies focused on the question of whether passive hydrodynamic dispersal could be tempered by active habitat selection (see for review Pawlik 1992; Rodriguez *et al*. 1993). Here, we focus on the active sensory and specifically olfactory processes involved in larval recruitment in coral reef fishes and marine invertebrates generated by the ongoing discovery of a non-random distribution of animals leading to population structure in situations where genetic mixing was expected.

In the marine environment, the pelagic phase of larvae can last from several days (clownfish) to several months (some crustaceans) (see Shanks 2009, for review). Initially, it was assumed that such dispersal would lead to a random settlement of larvae completely dependent on hydrography and storm events. There is now, however, mounting evidence that the case is more complex. In fish and crustacea, later stage larvae and juveniles become powerful swimmers and develop multiple senses with which they can actively influence dispersal and search for an appropriate habitat to settle. The emphasis on chemical information in this chapter is given by the intent of the book, although it should not mislead us to think that other sensory information is not important. It is clear that odour *per se* has no direction, but an odour concentration gradient has vector properties. However, these vectors are difficult to extract from typically chaotic dispersal fields in a time period useful to animals. For navigation, other senses than odour are typically involved, particularly vision and lateral line flow senses (see Chapter 1). Finally, we use the term 'odour' to refer to signal chemicals regardless of the chemoreceptor system used to detect them.

The well documented homing migrations of salmon, eels, and several other species show that mobile juveniles and adults recognize natal habitat cues and migrate to parental spawning grounds in rivers (see Chapter 6), but even here, the open ocean phase of this process and the underlying mechanisms of their habitat choice remain poorly known and difficult to study. Consequently, much of our knowledge about population structure, particularly in species with dispersing eggs, spores, or larvae is based on genetic markers. In addition, the origin of individual settling fish larvae can be determined directly from otolith tagging and less directly from otolith microchemistry, whereas modelling larval dispersal can include data from sea sampling.

Otolith tagging and microchemistry studies indicated that as many as 15–60% of a juvenile cohort may recruit to their natal habitat despite having pelagic larval durations of 3 to 7 weeks (Jones *et al*. 1999; Swearer *et al*. 1999). Genetic markers (micro-satellites) provide an alternative means of finding evidence for homing by assigning larvae to their natal population by comparing allele frequencies of progeny and adult populations. These methods confirmed high self-recruitment in the panda clownfish (*Amphiprion polymnus*) in Papua New Guinea (Saenz-Agudelo *et al*. 2009) and was used to demonstrate that 40 to 60% of larval cardinalfish *Ostorhinchus doederleini* could be assigned to the reef population where they were settling (Gerlach et al. 2007).

Although advection by ocean currents is a major force in larval dispersal and retention in lee-side eddies appears a common feature for retention of passive propagules (Lobel and Robinson 1986), the overall retention has been shown to be far greater than predicted by (purely passive) advection models (Armsworth 2000; Cowen et al. 2000; Wolanski et al. 2000). Dispersal models are better at predicting observed recruitment when they include a behavioural component (Armsworth 2001; Cowen et al. 2006; Paris et al. 2007; Burgess et al. 2009; and see Kingsford et al. 2002 for review). Paris and Cowen (2004) proposed a biophysical retention mechanism based on an ontogenetic shift in larval abundance toward deeper water. This vertical migration would allow them to take advantage of home-directed currents at different depths. This requires active swimming and, indeed, impressive larval swimming capabilities have been shown in laboratory 'tread mills' (Fisher et al. 2000; Stobutzki and Bellwood 1997). Choice of favourable currents based on their chemical composition was proposed as a mechanism to limit dispersal; this hypothesis was based on the demonstrated ability of settling larvae to recognize odour differences between reefs (Gerlach et al. 2007). Very little is known about the composition of such environmental chemical compounds. Reef odour differences can either originate from the environment and/or from conspecifics. The possibility that conspecific adults provide the odour signal that attracts the settlers has been tested and shown in some coral reef fishes (Døving et al. 2006; Lecchini et al. 2005). The reef's entire assemblage of macro- and microfauna, including corals, algae, and bacteria contribute to a unique cocktail of odours that is constantly released. This would allow for the possibility of not only species selection but also local population selection.

A tempting example of the latter is found in American lobster, Homarus americanus, a species with larval dispersal and swimming capabilities nearly identical to many reef fishes, but in addition also has great adult mobility (in contrast with settled adult reef fishes). Despite their great dispersal potential, lobster population structure exists on a scale of only tens of kilometres, seen with both population-genetic and morphometric markers (Atema et al. unpubl.). Perhaps not surprisingly, the animals themselves can detect this local difference by odour (Atema et al. unpubl.); preference for local conspecifics may be a mechanism for the maintenance of local population structure.

5.1 Olfactory driven choice of settlement habitat in invertebrates

Cyphonaute larvae of the bryozoan Membranipora membranacea are able to explore substrates in all directions in flow velocities that are faster than their locomotion speeds, with a preferred upstream motion (Abelson 1997). This behaviour has been largely ignored to date despite its potential ecological importance. For instance, upstream exploration may enable larvae to locate specific, obligatory settlement sites by tracking waterborne chemicals to their sources. In the sessile marine tunicates, Phallusia mammillata, larval selection of a suitable substrate is a critical factor determining the distribution of species (Groppelli et al. 2003). Under laboratory conditions, larvae discriminated between siliceous and carbonaceous stones on the basis of their silica content, and larvae attached to siliceous stones grew faster and had a wider area of contact with the substrate than those that grew on carbonaceous stones. Silica is a mineral that can be discriminated by the chemosensory palps of ascidian larvae during substrate choice. This suggests that among other environmental factors the mineral composition of the substrate can contribute to regulating the spatial distribution of tunicate communities, although the specific chemical signal involved is still unknown.

Juveniles of the benthic polychaete Arenicola marina move until they recognize a substrate suitable for settlement (Hardege et al. 1998). As in a number of marine invertebrates, inter- and intraspecific interactions are of importance in selecting a habitat, and specific chemical factors are involved. Juvenile conspecifics were attracted to each other during settlement, but avoided sediments that had been populated by adult conspecifics. Attraction and larval settlement were induced by a number of synthetic chemicals, including fatty acids, amino acids, and tripeptides, and by the presence of the alga Rhodomonas sp. Sediment enriched with tetramarin, a commercial trout diet, induced similar rates of settlement suggesting that it can be induced non-specifically by water soluble compounds that indicate organically enriched and therefore potentially favourable sites for settlement. In *A. marina*, brominated aromatics (e.g. 3,5-dibromo-4-hydroxybenzoic acid) inhibit settlement and function as negative cues.

Coral settlement can be induced by several species of crustose coralline algae (CCA); however, these algae also employ physical and biological anti-settlement defenses that vary in effectiveness (Harrington *et al.* 2004). Distinct settlement patterns were observed when coral larvae were provided with a choice of substrates. The authors used settling larvae of two common reef building coral species (*Acropora tenuis* and *A. millepora*) and for settlement habitat five species of CCA (*Neogoniolithon fosliei, Porolithon onkodes, Hydrolithon reinboldii, Titanoderma prototypum*, and *Lithoporello melobesioides*) that co-occur on reef crests and slopes of the Great Barrier Reef, Australia. Settlement on the most preferred substrate, the CCA species *T. proto-typum*, was 15 times higher than on *N. fosliei*, the least preferred substrate. The rates of post-settlement survival of the coral settlers varied between CCA species in response to their anti-settlement strategies (shedding of surface cell layers, overgrowth, and potential chemical deterrents). Rates of larval settlement, post-settlement survival, and the sensitivity of larvae to chemical extracts of CCA were all positively correlated across the five species of CCA.

Barnacles using internal fertilization are an interesting study object for settlement choice because once the cyprid larva is attached to the settlement substrate any chance to make further sexual contact to conspecifics is unequivocally limited to neighbours (Fig. 5.1). The cyprid's first antennae secrete glue that attaches the shell to the substrate. Choosing a site with insufficient access to food means starving, and even a penis that in some species exceeds seven body lengths cannot guarantee reproduction when the nearest neighbour is farther away. Unlike corals and other species with dispersing gametes, barnacles are constrained by their penis length for fertilization. In the barnacle *Balanus improvisus* conspecific extract increased intensity of surface exploration on all types of substrate (Berntsson *et al.* 2004). Larvae of the major fouling barnacle *Balanus*

amphitrite respond behaviourally to the physical and chemical characteristics of the substrate (Dreanno *et al.* 2006a); biogenic chemical cues have been implicated. The chemical basis of barnacle gregariousness was first described more than 50 years ago (Knight-Jones and Crisp 1953). The biological cue to gregarious barnacle settlement was later identified to be a settlement-inducing protein complex (SIPC) (Dreanno *et al.* 2006b). The SIPC shares a 30% sequence homology with the thioester-containing family of proteins that includes the alpha(2)-macroglobulins. The cDNA (5.2 kb) of the SIPC encodes a protein precursor comprising 1,547 aa with a 17-residue signal peptide region. Although the SIPC has been regarded as an adult cue that is recognized by the cyprid at settlement, it is also expressed in juveniles and larvae, where it may function in larva–larva settlement interactions.

In aquarium experiments post-larvae of nine crustacean species (*Calappa calappa, Pachygrapsus planifrons, Xanthidae* sp., *Lysiosquillina maculata, L. sulcata, Raoulserenea* sp. *Stenopus hispidus, Palaemonidae* sp., and *Panulirus penicillatus*), were tested for habitat choice (Lecchini *et al.* 2010). Six species made active habitat choices among the four habitats tested (live coral, dead coral, macroalgae, and sand), but the presence or absence of conspecifics on the habitat did not influence their selective choice. Sensory experiments found that four species differentiated between their preferred habitat versus another habitat and two species differentiated between conspecifics and heterospecifics using visual and/or olfactory cues.

Experiments on habitat selection by early benthic juvenile spiny lobster, *Panulirus interruptus*, were performed under laboratory conditions to test natural substrate selection and whether selection of substrata is affected by odour signals (de Lara *et al.* 2005). Despite the abundance of different macrophytes as habitat, 93% of the juvenile lobsters were found at the base of the blades of seagrass, *Phyllospadix spp.*, in the intertidal zone at 0 to 3 m, and such preferences were affected by odour signals.

American lobster (*Homarus americanus*) post-larvae (i.e. settlers) select particular substrates (Botero and Atema 1982) and may use odour as a mechanism of orientation (Boudreau *et al.* 1993). In a Y-maze apparatus, both swimming direction

Figure 5.1 Reproductive behaviour of barnacles. In some species the penis length of barnacles can reach several times the size of the body, an adaptation to internal fertilization in an immobile species. Barnacles release an agglomeration signal to avoid colonization beyond this reach.

(upstream or downstream) and arm selection (control or experimental) were strongly influenced by the chemical nature of the substrate stimulus tested. They concluded that distance chemoreception may play a role in habitat selection by lobster post-larvae at settlement. Social behaviour of adult lobsters is often based on olfactory signals (Atema and Steinbach 2007) including individual recognition (Atema and Voigt 1995; Karavanich and Atema 1998). Lobsters also recognize and prefer the odour of their own population to a genetically different population (Atema *et al.* in prep.) suggesting that settling larvae could use local population odour to settle and thus maintain population structure. Settling post-larvae of the European lobster (*Homarus gammarus*) preferred the chemical cues that they experienced during hatching (Hubatsch and Gerlach in prep). This ability might lead them to natal habitat, which might be especially important for the Helgoland

lobster population because other suitable habitat is hundreds of kilometres away.

5.2 Habitat recognition in coral reef fish

One of the great mysteries of coral reef fish ecology is how larvae that are dispersed in the ocean locate the reefs and the patches of coral reef habitat on which they settle. It is likely that this is a process of several stages on different spatial scales. Some sensory cues have been elucidated for scales of a few metres, but the larger scales remain subject to inference and speculation. It also remains possible that larval navigation is insignificant at the scale of kilometres. Tracking larval dispersal and recruitment behaviour in the open ocean is difficult. Leis and Carson-Ewart (2003) observed coral reef fish larvae they had captured near the reef and released—sometimes even at night—some kilometres away from reefs. Such brave experiments do allow a brief

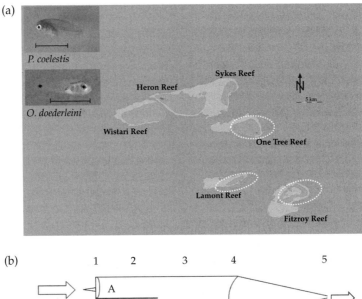

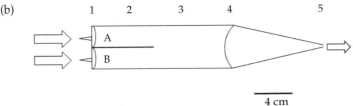

Figure 5.2 (a) Larval reef fish of two species, the cardinalfish *Ostorhinchus doederleini* and the neon damselfish *Pomacentrus coelestis*, were caught before and at settlement at different reefs of the Capricorn Bunker Group, Great Barrier Reef, Australia. We tested these settlers in an Atema choice flume for their olfactory preference of water from different reefs. In years when we found genetic differences between reef populations (2003 and 2005 for *O. doederleini* and 2006 for *P. coelestis*), we could assign the odor preferences shown by both species significantly to water from the reefs where they were caught. Dotted circles enclose genetically different populations that were tested for olfactory choice. Fish larvae photos G. Gerlach. (b) Atema choice flume (20 x 4 x 2.5 cm) 1 = collimator to homogenize inflow turbulence, 2 = barrier-separated channels to further laminate flow, 3 = test area, 4 = downstream fine mesh (0.5 mm) net to contain test fish; 5 = outflow. Timeline of each test: 5 min: acclimation in plain water; 2 min: odor stimulus x on side A, odor stimulus y on side B; 1 min: plain water; 2 min: odor stimulus x on side B, odor stimulus y on side A.

glimpse of the recruitment process. This approach has shown directional swimming in late-stage reef fish larvae (Leis and Carson-Ewart 2003).

Larval recognition of settlement habitat can be based on the detection of conspecifics and/or of characteristics of coral habitat. Larvae could use any combination of chemical, visual, acoustic, and hydrodynamic signals. In a study of 18 species, 13 chose their settlement habitat due to the presence of conspecifics and not based on the characteristics of coral habitat, and 5 species did not move toward their settlement habitat (e.g. *Scorpaenodes parvipinnis*, *Apogon novemfasciatus*) (Lecchini *et al.* 2005). Among the different sensory cues tested in the lab, two species (*Parupeneus barberinus* and *Ctenochaetus striatus*) used visual, chemical and—undefined—

'mechanical' cues, six used two types (e.g. *Myripristis pralinia*: visual and chemical cues; *Naso unicornis*: visual and 'mechanical' cues), and five used one type (e.g. *Chrysiptera leucopoma*: visual cues; *Pomacentrus pavo*: chemical cues). *Chromis. viridis* larvae responded positively to visual, acoustic/ vibratory, and olfactory cues expressed by conspecifics (Lecchini *et al.* 2005). Overall, larvae chose compartments of experimental arenas containing conspecifics in 75% of trials, and failed to show significant directional responses to heterospecifics or coral substrates.

These results demonstrate that coral reef fish larvae could use multiple sensory cues for effective habitat selection at settlement, including the ability to discriminate species-specific cues. It is likely that different senses provide superior information at different distances and signal-to-noise ratios.

Coral reef fishes *Ostorhinchus doederleini* (Apogonidae) and *Pomacentrus coelestis* (Pomacentridae) preferred the odour of reef lagoon water to ocean water (Atema *et al.* 2002) and the odour of their settlement reef to that of other nearby reefs (see Fig. 5.2 and Gerlach *et al.* 2007). These olfactory preferences were temporally stable but it is not known when the preference was established. When caught at or just before settlement *O. doederleini* and *P. coelestis* larvae, given a choice between water from their settlement reef and a neighbouring reef, preferred the settlement reef odour. This 'home' preference was maintained over time (5–9 days) regardless of whether the animals were kept in settlement reef water or the water of the neighbouring reef. The degree of preference declined gradually over days, equally in both odour exposure groups. The important result is that exposure to water from the neighbouring reef did *not* induce a switch in preference for this new reef odour suggesting a stable habitat odour preference at least in the late phase of settling (Miller-Sims *et al.* 2011).

Some species of coral reef fish use seagrass beds and mangroves as juvenile habitats. Once pelagic larvae of these fish have located a coral reef from the open ocean, they still have to find embayments or lagoons harbouring these juvenile habitats. Post-larvae of the French grunt (*Haemulon flavolineatum*), a reef fish that is highly associated with mangroves and seagrass beds during its juvenile life stage, chose significantly more often the water from mangroves and seagrass beds than water from the coral reef (Huijbers *et al.* 2008).

The settler transition from a pelagic to a benthic existence takes place over several days to weeks and constitutes significant metamorphosis from transparent larvae to pigmented juveniles with species-specific patterns. Several laboratory choice-chamber experiments were conducted to explore sensory capabilities and behavioural responses to ecological stimuli to better understand habitat selection by 'pre-metamorphic' (larval) and 'post-metamorphic' (juvenile) stages of a coral reef fish (Lecchini *et al.* 2007). *Thalassoma hardwicke* larvae were attracted to benthic macroalgae (*Turbinaria ornata* and *Sargassum mangarevasae*), while slightly

older post-metamorphic juveniles chose to occupy live coral colonies (*Pocillopora damicornis*). Habitat choices of larvae were primarily based upon visual cues and were not influenced by the presence of older conspecifics. In contrast, juveniles selected live coral colonies and preferred those occupied by older conspecifics; choices made by juveniles were based upon both visual and olfactory cues from conspecifics.

Larvae of the domino damselfish *Dascyllus albisella* preferred to settle with larger groups of conspecific juveniles rather than with single conspecifics, empty coral heads, or con-familial groups (Booth 1992). The laboratory experiment also indicated that preferences may have been established through visual cues, suggesting that vision may supplement chemical cues in facilitating larval settlement preferences. It appears that settlement patterns at least on a small (within-reef) scale are influenced by larval habitat preferences.

The natal origin of some clownfish (*Amphiprion polymnus*) juveniles was determined by mass-marking all larvae produced in a population via tetracycline immersion, and establishing parentage by DNA genotyping all potential adults and all new recruits arriving in the population (Jones *et al.* 2005). Many settled remarkably close to home, yet no individuals settled into the same anemone as their parents. After a 9–12 day larval pelagic duration, one-third of settled juveniles returned to a 2 hectare natal area, with many settling less than 100 m from their birth site. Different species of anemonefish inhabit specific anemones for which they show a lifelong olfactory preference (Arvedlund *et al.* 2000). The olfactory cues, to which some species of anemonefish larvae imprint, are secreted in the mucus on the tentacles and the oral disc of the host anemone. Close contact of the eggs of anemonefishes with the host's tentacles seems therefore to be important. Anemonefishes *Amphiprion akindynos*, *A. bicinctus*, *A. melanopus*, and *A. perideraion*, have distinctive spawning site preferences. Arvedlund *et al.* (1999) examined the host-selection made by *A. melanopus* that had been reared under constant conditions, but whose embryos had received 1 of 3 treatments: (1) contact with a known natural host sea anemone, *Entacmaea quadricolor*; (2) contact with the sea anemone *Heteractis malu*, which is not a host for *A.*

melanopus in nature, but is a host for anemonefish of other species; and (3) without a sea anemone (or chemical cues released from sea anemones) at any life stage. The study shows that olfaction, not vision, is used by juvenile *A. melanopus* to recognize host anemones. Anemonefish larvae can also use different chemical cues to recognize habitat for settlement. Recently settled anemonefish, *Amphiprion percula*, exhibited a strong preference for water treated with leaves from rainforest vegetation, which may explain the high levels of self-recruitment on island reefs (Dixson *et al.* 2008). It is also the best documented case of identified chemical cues strong enough that they might operate over large distances.

Habitat recognition does not end with settlement. If artificially displaced, adult cardinalfish will return to their home site (Marnane 2000), but the sensory background of such homing was not further analysed. A different study also on cardinalfish suggests that odour might be important (Døving *et al.* 2006). They showed that adult/juvenile five-lined cardinalfish (*Cheilodipterus quinquelineatus*) and splitbanded cardinalfish (*Apogon compressus*) preferred artificial reef sites that had previously been occupied by conspecifics to reef sites inhabited by another species (*Apogon leptacanthus*). Remarkably, they also preferred the scent released by artificial reefs previously occupied by conspecifics of their own reef site, to reef sites previously occupied by conspecifics of another site. Thus, olfactory preferences can even extend to subpopulations, as shown in the lobsters discussed above, and may hold settlement cues for returning larvae.

5.2.1 Imprinting

These examples indicate that larvae can use chemical cues to find suitable habitat for settlement and evidence is mounting that larvae can return to *natal* habitat and perhaps even their own subpopulations. If so, then how do larvae learn and remember the smell of the spawning grounds that they left shortly after hatching?

Such recognition processes might be based on olfactory imprinting as a form of robust long-term memory (e.g. of family members and/or natal habitats) that is acquired within a certain time window during early development and persists throughout

life. For instance, in salmon, the response properties of the peripheral olfactory system become physiologically tuned to home-stream odours so that after several years in the open ocean sexually mature fish can find their natal stream for spawning (Dittman and Quinn 1996; Scholz *et al.* 1976; Chapter 6). In salmon, imprinting involves changes in the peripheral nervous system; specific genes are expressed in the olfactory epithelium during the time of imprinting (Hino *et al.* 2007). Larval anemonefish (*Amphiprion melanopus*) have to be exposed to their specific anemone during the hour of hatching to later recognize and prefer this anemone (Miller-Sims PhD thesis). Similarly, although not a coral reef inhabitant, zebrafish (*Danio rerio*) imprint on olfactory cues of kin only during a 24-hour window at day 6 post fertilization (Gerlach *et al.* 2008) (see Chapter 4 for details). It may well be that imprinting on environmental chemical cues follows a similar process.

5.2.2 Climate change and pollution impair habitat recognition

Processes induced by anthropogenic activities and climate change might change capabilities to orientate by olfaction (Chapter 17). Munday *et al.* (2009) showed that ocean acidification from elevated levels of atmospheric carbon dioxide (CO_2) abolishes the ability of *Amphiprion* larvae to detect olfactory cues from adult habitats. Larval clownfish reared in control seawater (pH 8.15) discriminated between a range of cues that could help them locate reef habitat and suitable settlement sites. This discriminatory ability was disrupted when larvae were reared in conditions simulating CO_2-induced ocean acidification. Larvae reared at levels of ocean pH that could occur later this century (pH 7.8) became strongly attracted to olfactory stimuli they normally avoided, and they no longer responded to any olfactory cues when reared at the pH levels (pH 7.6) that might be attained later into the next century on a business-as-usual trajectory of carbon-dioxide emissions. If acidification continues unabated, the impairment of sensory ability may reduce the population sustainability of many marine species, with potentially profound consequences for marine diversity. However, the possibility that animals might adapt to changing pH levels has not been explored in these cases.

In a freshwater species, the swordtail fish, *Xiphophorus birchmanni*, chemically-mediated species recognition can be hindered by anthropogenic disturbance to the signalling environment (Fisher *et al.* 2006). Wild-caught females showed a strong preference for conspecifics when tested in clean water, but failed to show a preference when tested in stream water subject to sewage effluent and agricultural runoff. When exposed to elevated concentrations of humic acid (HA), female *X. birchmanni* lost their preference for conspecific male chemical cues, while visual mating preferences and motivation to mate were retained. Natural levels of HA reflect this process and have been observed to fluctuate as much as 5 mg l^{-1} over a matter of hours, and can range from trace amounts to over 200 mg l^{-1} (Steinberg 2003). Testing water was buffered and the effects of HA on pH were minimal: pH without supplemental HA was 7.8 ± 0.2, pH with 200 mg l^{-1} HA was 7.4 ± 0.1; both are well within the range that has been measured in the field.

Humic acid (HA) is a pervasive, naturally occurring plant derivative found in aquatic and terrestrial environments. Adult zebrafish (*Danio rerio*) are not accustomed to humic acid and their ability to discriminate between conspecific and heterospecific urinary chemical cues is impaired by high levels of HA (200 mg l^{-1}) (Fabian *et al.* 2007). These findings suggest that, in addition to human-produced synthetic compounds, changes in the abundance of naturally occurring substances may also negatively impact olfactory-based behaviours in aquatic species by altering the sensory environment.

5.3 Concluding remarks

In conclusion, extensive evidence shows the importance of chemical information in the settlement processes employed by pelagic larvae. This is not surprising since chemical signals are produced by nearly everything and every organism. Nearly unlimited amounts of different mixtures are generated constantly, thus providing an enormous variety of cues that organisms can use to identify important sources. In addition, water is an excellent solvent and a powerful transporter of chemical signals. Life as we imagine it has always been defined by water. This includes the many still unknown mechanisms by which organisms locate odour sources, from reefs to individual mates and prey. Future research will show how common olfactory guided orientation might be, not only in coral reefs but in all aquatic environments, from freshwater ponds and rivers to thermal vents in the deep sea, where chemical guidance of migrating and settling organisms can be expected.

References

Abelson, A. (1997) Settlement in flow: Upstream exploration of substrata by weakly swimming larvae. *Ecology,* **78**: 160–6.

Armsworth, P.R. (2000) Modelling the swimming response of late stage larval reef fish to different stimuli. *Marine Ecology Progress Series,* **195**: 231–47.

Armsworth, P.R. (2001) Directed motion in the sea: efficient swimming by reef fish larvae. *Journal of Theoretical Biology,* **210**: 81–91.

Arvedlund, M., Bundgaard, I., and Nielsen, L.E. (2000) Host imprinting in anemonefishes (Pisces: Pomacentridae): does it dictate spawning site preferences? *Environmental Biology of Fishes,* **58**: 203–13.

Arvedlund, M., Mccormick, M.I., Fautin, D.G., and Bildsoe, M. (1999) Host recognition and possible imprinting in the anemonefish *Amphiprion melanopus* (Pisces: Pomacentridae). *Marine Ecology Progress Series,* **188**: 207–18.

Atema, J., Kingsford, M., and Gerlach, G. (2002) Larval reef fish could use odour for detection, retention and orientation to reefs. *Marine Ecology Progress Series,* **241**: 151–60.

Atema, J. and Steinbach, M.A. (2007) Chemical communication and social behavior of the lobster, *Homarus americanus*, and other Decapod Crustacea. In Duffy, J.E. and Thiel, M. (Eds.) *Evolutionary Ecology of Social and Sexual Systems: Crustaceans as Model Organisms.* Oxford University Press, New York NY, USA.

Atema, J. and Voigt, R. (1995) Behavior and sensory biology. In Factor, J.R. (Ed.) *Biology of the Lobster.* Academic Press, New York.

Berntsson, K.M., Jonsson, P.R, Larsson, A.I., and Holdt, S. (2004) Rejection of unsuitable substrata as a potential driver of aggregated settlement in the barnacle *Balanus improvisus Marine Ecology Progress Series,* **275**: 199–210.

Booth, D.J. (1992) Larval settlement-patterns and preferences by Domino damselfish (*Dascyllus albisell* Gill). *Journal of Experimental Marine Biology and Ecology*, **155**: 85–104.

Botero, L. and Atema, J. (1982) Behavior and substrate selection during larval settling in the lobster *Homarus americanus*. *J. Crust. Biol.* **2**: 59–69.

Boudreau, B., Bourget, E., and Simard, Y. (1993) Behavioral-Responses of Competent Lobster Postlarvae to Odor Plumes. *Marine Biology*, **117**: 63–9.

Burgess, S.C., Hart, S.P., and Marshall, D.J. (2009) Pre-settlement behavior in larval bryozoans: The roles of larval age and size. *The Biological Bulletin*, **216**: 344–54.

Cowen, R.K., Lwiza, K.M.M., Sponaugle, S., Paris, C.B., and Olson, D.B. (2000) Connectivity of marine populations: Open or closed? *Science*, **287**: 857–9.

Cowen, R.K., Paris, C.B., and Srinivasan, A. (2006) Scaling of connectivity in marine populations. *Science*, **311**: 522–7.

Crisp, D.J. (1974) Factors influencing the settlement of marine invertebrate larvae. In Grant, P.T. and Mackie, A.M. (Eds.) *Chemoreception in Marine Organisms*. Academic Press, New York.

De Lara, V.C.F., Butler, M., Hernandez-Vazquez, S., Del Proo, S.G., and Serviere-Zaragoza, E. (2005) Determination of preferred habitats of early benthic juvenile California spiny lobster, *Panulirus interruptus*, on the Pacific coast of Baja California Sur, Mexico. *Marine and Freshwater Research*, **56**: 1037–45.

Dittman, A.H. and Quinn, T.P. (1996) Homing in Pacific salmon: Mechanisms and ecological basis. *Journal of Experimental Biology*, **199**: 83–91.

Dixson, D.L., Jones, G.P., Munday, P.L., *et al.* (2008) Coral reef fish smell leaves to find island homes. *Proceedings of the Royal Society B-Biological Sciences*, **275**: 2831–9.

Døving, K.B., Stabell, O.B., Östlund-Nilsson, S., and Fisher, R. (2006) Site fidelity and homing in tropical coral reef cardinalfish: are they using olfactory cues? *Chemical Senses* **31**: 265–72.

Dreanno, C., Kirby, R.R., and Clare, A.S. (2006a) Locating the barnacle settlement pheromone: spatial and ontogenetic expression of the settlement-inducing protein complex of *Balanus amphitrite*. *Proceedings of the Royal Society B-Biological Sciences*, **273**: 2721–8.

Dreanno, C., Matsumura, K., Dohmae, N., *et al.* (2006b) An alpha(2)-macroglobulin-like protein is the cue to gregarious settlement of the barnacle *Balanus amphitrite*. *Proceedings of the National Academy of Sciences of the United States of America*, **103**: 14396–401.

Fabian, N.J., Albright, L.B., Gerlach, G., Fisher, H.S., and Rosenthal, G.G. (2007) Humic acid interferes with species recognition in zebrafish (*Danio rerio*). *Journal of Chemical Ecology*, **33**: 2090–6.

Fisher, H.S., Wong, B.B.M., and Rosenthal, G.G. (2006) Alteration of the chemical environment disrupts communication in a freshwater fish. *Proceedings of the Royal Society B-Biological Sciences*, **273**: 1187–93.

Fisher, R., Bellwood, D.R., and Job, S.D. (2000) Development of swimming abilities in reef fish larvae. *Marine Ecology Progress Series*, **202**: 163–73.

Gerlach, G., Atema, J., Kingsford, M.J., Black, K.P., and Miller-Sims, V. (2007) Smelling home can prevent dispersal of reef fish larvae. *Proceedings of the National Academy of Sciences*, **104**: 858–63.

Gerlach, G., Hodgins-Davis, A., Avolio, C., and Schunter, C. (2008). Kin recognition in zebrafish: A 24-hour window for olfactory imprinting. *Proceedings of the Royal Society, London B*, **275**: 2165–70.

Groppelli, S., Pennati, R., and Al., E. (2003) Observations on the settlement of *Phallusia mammillata* larvae: effects of different lithological substrata. *Italian Journal of Zoology*, **70**: 321–6.

Hardege, J.D., Bentley, M.G., and Snape, L. (1998) Sediment selection by juvenile *Arenicola marina*. *Marine Ecology Progress Series*, **166**: 187–95.

Harrington, L., Fabricius, K., De'ath, G., and Negri, A. (2004). Recognition and selection of settlement substrata determine post-settlement survival in corals. *Ecology*, **85**: 3428–37.

Hino, H., Iwai, T., Yamashita, M. and Ueda, H. (2007) Identification of an olfactory imprinting-related gene in the lacustrine sockeye salmon, *Oncorhynchus nerka*. *Aquaculture*, **273**: 200–8.

Huijbers, C.M., Mollee, E.M., and Nagelkerken, I. (2008) Post-larval French grunts (*Haemulon flavolineatum*) distinguish between seagrass, mangrove and coral reef water: Implications for recognition of potential nursery habitats. *Journal of Experimental Marine Biology and Ecology*, **357**: 134–9.

Jones, G.P., Milicich, M.J., Emslie, M.J., and Lunow, C. (1999) Self-recruitment in a coral reef fish population. *Nature*, **402**: 802–4.

Jones, G.P., Planes, S., and Thorrold, S.R. (2005) Coral reef fish larvae settle close to home. *Current Biology*, **15**: 1314–18.

Karavanich, C. and Atema, J. (1998) Olfactory recognition of urine signals in dominance fights between male lobster, *Homarus americanus*. *Behaviour*, **135**: 719–30.

Kingsford, M.J., Leis, J.M., Shanks, A., Lindeman, K.C., Morgan, S.G., and Pineda, J. (2002) Sensory environments, larval abilities and local self-recruitment. *Bulletin of Marine Science*, **70**: 309–40.

Knight-Jones, E.W. and Crisp, D.J. (1953) Gregariousness in barnacles in relation to the fouling of ships and to anti-fouling research. *Nature*, **171:** 1109–10.

Lecchini, D., Mills, S.C., Brie, C., Maurin, R., and Banaigs, B. (2010) Ecological determinants and sensory mechanisms in habitat selection of crustacean postlarvae. *Behavioral Ecology*, **21:** 599–607.

Lecchini, D., Osenberg, C.W., Shima, J.S., Mary, C.M., and Galzin, R. (2007) Ontogenetic changes in habitat selection during settlement in a coral reef fish: ecological determinants and sensory mechanisms. *Coral Reefs*, **26:** 423–32.

Lecchini, D., Shima, J., Banaigs, B. and Galzin, R. (2005). Larval sensory abilities and mechanisms of habitat selection of a coral reef fish during settlement. *Oecologia*, **143:** 326–34.

Leis, J.M. and Carson-Ewart, B.M. (2003) Orientation of pelagic larvae of coral-reef fishes in the ocean. *Marine Ecology Progress Series*, **252:** 239–53.

Lobel, P.S. and Robinson, A.R. (1986) Transport and entrapment of fish larvae by mesoscale eddies and currents in Hawaiian waters. *Deep Sea Research*, **33:** 483–500.

Marnane, M.J. (2000) Site fidelity and homing behaviour in coral reef cardinalfishes. *Journal of Fish Biology*, **57:** 1590–600.

Miller-Sims, V. (2007). Olfactory homing and dispersal of coral reef fishes in the Great Barrier Reef, Australia. PhD thesis, Boston University.

Miller-Sims, V., Gerlach, G., Kingsford, M.J. and Atema, J. (2011) Reef odour imprinting: Coral reef fish demonstrate stable olfactory preference for their settlement reef. *Marine and Freshwater Behaviour and Physiology*, **44:** 133–41.

Munday, P.L., Dixson, D.L., Donelson, J.M., *et al.* (2009) Ocean acidification impairs olfactory discrimination and homing ability of a marine fish. *Proceedings of the National Academy of Sciences of the United States of America*, **106:** 1848–52.

Paris, C.B., Cherubin, L.M., and Cowen, R.K. (2007) Surfing, spinning, or diving from reef to reef: effects on population connectivity. *Marine Ecology Progress Series*, **347:** 285–300.

Paris, C.B. and Cowen, R.K. (2004) Direct evidence of a biophysical retention mechanism for coral reef fish larvae. *Limnology and Oceanography*, **49:** 1964–79.

Pawlik, J.R. (1992) Chemical ecology of the settlement of benthic marine invertebrates. *Oceanography and Marine Biology*, **40:** 273–335.

Rodriguez, S.R., Ojedal, F.P., and Inestrosa, N.C. (1993). Settlement of benthic marine invertebrates. *Marine Ecology Progress Series*, **97:** 193–207.

Saenz-Agudelo, P., Jones, G.P., Thorrold, S.R., and Planes, S. (2009). Estimating connectivity in marine populations: an empirical evaluation of assignment tests and parentage analysis under different gene flow scenarios. *Molecular Ecology*, **18:** 1765–76.

Scholz, A.T., Horrall, R.M., Cooper, J.C., and Hasler, A.D. (1976) Imprinting to chemical cues: the basis for home stream selection in salmon. *Science*, **192:** 1247–9.

Shanks, A.L. (2009) Pelagic larval duration and dispersal distance revisited. *The Biological Bulletin*, **216:** 373–85.

Steinberg, C.E.W. (2003) *Ecology of Humic Substances in Freshwaters*. Springer: New York, NY.

Stobutzki, I. and Bellwood, D.R. (1997) Sustained swimming abilities of the late pelagic stages of coral reef fishes. *Marine Ecology Progress Series*, 149: 35–41.

Swearer, S.E., Caselle, J.E., Lea, D.W., and Warner, R.R. (1999) Larval retention and recruitment in an island population of a coral-reef fish. *Nature*, **402:** 799–802.

Wolanski, E., Spagnol, S., Thomas, S., *et al.* (2000) Modelling and visualizing the fate of shrimp pond effluent in a mangrove-fringed tidal creek. *Estuarine Coastal and Shelf Science*, **50:** 85–97.

CHAPTER 6

Migration and navigation

Ole B. Stabell

6.1 Introduction

Migration is a widespread phenomenon among aquatic animals. In a broad context, migration is defined as 'the act of moving from one spatial unit to another' (Baker 1978). However, a restriction to the definition should be that the migration is adaptive, meaning that 'it is a response to natural selection and that accidental displacement is excluded' (Smith 1985). Depending on the animal size, therefore, migrations may range from less than a metre up to thousands of kilometres. A general feature of such migratory occurrences is that the animals regularly, or at some specific life stages, will return to distinct local sites—often for the purpose of seeking shelter or for reproduction. The sensory system used for finding the way may vary, but in many instances chemical sensing appears paramount in lower vertebrates as well as in invertebrates.

Navigation is defined as 'the orientation towards a destination, regardless of its direction, by means of other than the recognition of landmarks' (Allaby 2005); that is, 'it refers to a situation in which an animal determines the position of a given point in space' (Smith 1985). In a biological context, orientation is defined as 'a change in position by an animal (or plant) in response to an external stimulus' (Allaby 2005). It should be noted that for animals migrating over long distances, the use of orientation and navigation in the way carried out by humans (i.e. dependence on exact chronometers for determination of longitude, and thereby starting position) is doubtful (Døving and Stabell 2003). However, orientation skills used according to the biological definition, leading to navigation-like outcomes, are still at hand. In this chapter I will consider some fascinating ways that animals use chemical cues to find their way home.

6.2 Bottom-dwelling animals

A range of invertebrate species that are permanently tied to the bottom substrate are capable of displaying homing. For instance, such behaviour is present in crustacean decapods (e.g. some shrimps, lobsters, crayfish, hermit crabs, and crabs) where burrows and shelters are the primary homing goals following feeding excursions. It has been suggested that chemical cues may play a role in the homing of these benthic decapods, but direct evidence for this is still lacking (Vannini and Cannicci 1995). On the other hand, in intertidal molluscs like chitons (sea cradles) and gastropods, chemical senses are essential for finding the way home (Chelazzi 1990). It has long been recognized that snails are able to deposit pedal mucus trails, both on land and in water (see Kohn 1961). By making chemical trails on the substrate while moving, the animals may subsequently retrace that trail when returning to their preferred site of residence. In the following, we will have a closer look on how this is carried out.

6.2.1 Invertebrate homing to specific sites after foraging

Chitons and limpets are typical intertidal animals that are found mainly on rocky shores within the tidal zone (Fig. 6.1a,b). Many species within these groups display homing behaviour to a more or less permanent shelter after feeding excursions. Depending on the species, feeding excursions may take place both at high and low tide, meaning that grazing in some species may also occur when exposed to air. Among intertidal limpets, distinct homing behaviour has been long known, and alert observers find that limpets grow their scales in close association with a home spot so that the shell edges form negative images of the substrate, that is, they fit perfectly into the shape of the substrate. Such morphological adjustments may be important in escaping predator attacks, but it has also been postulated that this exact fit between the shell and the substratum leads to a seal that prevents desiccation during intertidal periods of exposure to air (Hartnoll and Wright 1977). In chitons, the species representing the most determined homers are able to relocate an almost permanent and quasi-personal scar or crevice for hiding (see Chelazzi 1990 and references therein).

The chiton *Acanthopleura gemmata* is one such a precise homer that migrates up to 3 m away from its scar to feed on algal grounds. Its behaviour has been followed in experiments showing that each individual follows its own outgoing trail when returning home (Chelazzi *et al.* 1987). After an outward directed movement away from its scar, an animal will normally make a loop and regain contact

with the outgoing trail and retrace it back home. If part of the outgoing path is interrupted, for instance by brushing, the animals will start searching movements, and if contact is re-established they will continue to retrace the outgoing trail back home (Fig. 6.1a). There seems to be a high coincidence between the path traces made by the same chiton on consecutive nights. In the few cases where the outgoing and the return paths of the previous night markedly diverged, the chiton followed the return trail from the previous night on its outward journey (Chelazzi *et al.* 1987). This implies that trail information is long-lasting (i.e. at least 24 hrs).

Individual specificity also appears to be present in the trail information. Among chitons that were experimentally displaced when foraging away from home, and placed in a different area on a trail belonging to a conspecific snail, very few were able to follow the conspecific trail to the 'new' home. However, control animals that were displaced to be replaced on their own trails showed a high homing performance (Chelazzi *et al.* 1987). In limpets, long-lasting trails have actually been demonstrated in *Siphonaria alternata* (Cook 1969), and individual

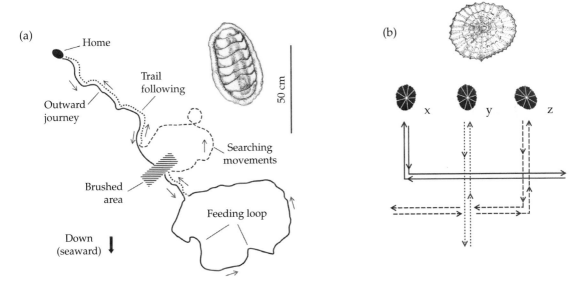

Figure 6.1 Homing experiments with intertidal molluscs. (a) Return to home scar by a chiton after feeding excursion. The animal will regain contact with the outward directed trail after a feeding loop (solid line, bottom of picture). If part of the trail is interrupted (shaded rectangle), re-contact with the trail may be obtained by searching movements. Stippled line shows trail-following to home scar after re-establishing contact with own trail (modified from Chelazzi *et al.* 1987, fig. 3). (b) Experiments with limpets demonstrating use of individual trails (x,y,z) during outward feeding excursions and homing (modified from Funke 1968, fig. 10a). Both figures reproduced with kind permission of Springer Science and Business Media.

trails have been long known in the two *Patella* species *P. caerulea* and *P. vulgata* (Funke 1968). Individual trails may be particularly important in limpets, taking into consideration the fit of their very personal scar (Fig. 6.1b). Trail-following mechanisms in limpets have been found that are very similar to those already described for chitons (Cook *et al.* 1969; Della Santina 1994). Cook *et al.* (1969), further, found that limpets other than the occupants could also recognize scars, suggesting that some specific chemical labelling is taking place at such spots. At the home spot *Patella* always manoeuvres exactly into the same position. Often it has to turn or glide around slightly, and will finally settle down with small movements of its shell (Funke 1968). The sensory mechanisms used for such fine-tuned adjustments are not known.

6.2.2 Trail making and trail detection in benthic invertebrates

Marine snails, like the intertidal common periwinkle (*Littorina littorea*), possess strong shells and do not display homing to hiding places. However, male periwinkles are able to follow pheromones embedded in the mucus trails of females (Erlandsson and Kostylev 1995). Periwinkles may also follow trails of conspecific snails for the purpose of finding food. They are able to discriminate trails by their age, and are even able to sort between fed and hungry conspecifics from information deposited in the trail (Edwards and Davies 2002).

Some freshwater pulmonate gastropods, like the Bladder snail (*Physa acuta*), carry out repeated migrations to replenish air, by the use of chemical trails (Wells and Buckley 1972). This was found in experiments using a Y-tube design with photocells to detect snail movements (Fig 6.2a). The experiment revealed that Bladder snails made an initial random choice of one of the arms, but subsequently continued to prefer that arm on their repeated surface tours. If the flasks and Y-tubes were rotated 180° while the snails were down in the flask, the snails continued to run the same arm as before (Fig. 6.2b). Thus, the stimuli determining arm preference were to be found within the maze. If the water in the tubes was siphoned out and replaced by fresh water between the runs, the previously preferred

side was still selected, but if the Y-tube and cork were turned 180° between runs with the flask in a fixed position, the performance was disrupted (Fig. 6.2c). Also, if the Y-tubes were scrubbed out between runs the next ascent was equally likely to be on either side of the Y-maze. Following the first choice after scrubbing, the snail would eventually settle down to run consistently on the last chosen side, suggesting that some chemical information had been added to the tube surfaces. When the preferred side of the maze was blocked below the surface, the snail made repeated attempts to pass before, seemingly only by chance, it ascended the open arm to replenish air. In later surfacing attempts, the snail would follow a conservative approach by first ascending the blocked arm, and then descending along its old route to the open arm.

The evidence from these experiments strongly suggested that snails were somehow marking the inside of the apparatus and that the sensory detection of the trail was most probably by contact chemoreception bound to a narrow area of the snail's

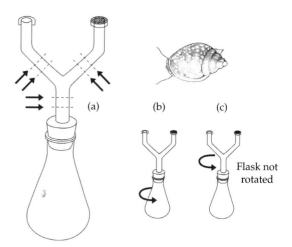

Figure 6.2 Trail experiment with the freshwater pulmonate Bladder snail, *Physa acuta*. (a) The apparatus used for trail-following experiments. Photocells in pairs (arrows and stippled lines in main stem and arms) were used to detect snail movements. (b) Experimental design where both Y-tube and flask were rotated 180° between runs. The snail will continue to follow the original arm choice, regardless of the pointing direction of the arm. (c) Experimental design where Y-tube and cork were turned 180° between runs with the flask in a fixed position. Trail is interrupted between Y-tube and the flask, and the snail becomes disoriented. Modified from Wells and Buckley (1972); figs 1, 3, and 7. Reproduced with kind permission of Elsevier.

foot. Further experiments demonstrated that the snails were willing to follow trails laid by conspecifics, and that they were able to determine the direction in which these individuals were moving (Wells and Buckley 1972). As opposed to limpets and chitons, the trails of Bladder snails were found to have a life span of only about 30 min.

6.2.3 Mechanisms of chemosensory trail detection in molluscs

In homing experiments with limpets, tests have been carried out to eliminate external cues, such as topographic memory, reverse-displacement, and paths in algal mats made by the radula, further suggesting that retracing previously laid mucus trails is the ruling mechanism (Cook 1971). Gastropods possess the ability of distant chemoreception, probably through the sensory cells at the cephalic tentacles or in the mantle cavity, while contact chemoreception may be handled by sensory cells in the mantle fringe and the foot (Kohn 1961; Croll 1983; Emery 1992). Recent studies on sensory organs of gastropod molluscs have primarily dealt with histology and ultra-structure of receptor cells, and the use of the various cell types in contact chemoreception has mainly been the subject of speculation (e.g. Göbbeler and Klaussmann-Kolb 2007). Accordingly, the mechanisms underlying chemical detection of trails in migrating molluscs, as well as the chemical properties of trails, are still open fields of research.

6.3   Free-swimming animals

Reproductive homing after feeding excursions is also a common occurrence in free-moving lower vertebrates, and it is often carried out over long distances. Green sea turtles return to nesting sites on natal beaches (Mortimer and Portier 1989; Allard *et al.* 1994), and in migratory marine fish like cod (*Gadus morhua*) and halibut (*Hippoglossus stenolepis*), homing has been suggested as the prime stock-separating mechanism (Svedäng *et al.* 2007; Loher 2008). Also in migratory freshwater fish, like muskellunge (*Esox masquinongy*) and catfish (*Pseudoplatystoma corruscans*), there is a strong tendency to utilize natal nursery areas for repro-

duction (Crossman 1990; Pereira *et al.* 2009). However, the sensory system(s) used for homing in these species are still a matter of debate. On the other hand, among stationary, non-migrating species, olfactory controlled site fidelity has been demonstrated. These species stay permanently in a specific area, and a large proportion will return to the previously occupied area if removed. Semi-aquatic animals, like freshwater turtles, also display site fidelity and home to the original site of residence if artificially displaced (e.g. Lebboroni and Chelazzi 2000; Smar and Chambers 2005). Among anuran amphibians, several species display a conservative choice of spawning area after feeding excursions on land (Sinsch 1990). In the common frog (*Rana temporaria*), for instance, the patch where egg clutches are deposited is predictable (within metres) in clusters of marsh-ponds over years (personal observation). Olfaction has been suggested as the orientation mechanism in semi-aquatic urodeles migrating to spawning sites (Madison 1969; McGregor and Teska 1989; Joly and Miaud 1993), and olfactory cues may also be important for detecting the breeding pond in anurans (Sinsch 1990).

Many migrating species of fish display homing behaviour to particular areas of lakes or streams for reproduction, and by far the most studied of these species are the salmonids (i.e. several species of salmon, trout, and char). After hatching, young salmonids will stay in the vicinity of the spawning ground, normally for at least a year. The young fish will sooner or later migrate to the sea to feed and stay there for at least a year before returning to spawn. On the other hand, populations of several salmonid species may take on feeding migrations only in freshwater, and remain in lakes or streams during their entire growth period. A common feature for all migrating salmonids is their dependence on the olfactory sense for finding the way (reviewed by Stabell 1984, 1992), even when displaced considerable distances away from the river mouth (i.e. up to 100 km away; Toft 1975). Another feature is a striking resemblance to bottom-dwelling animals, in that stream-dwelling fish also carry out substrate marking. How then is substrate marking used by fish with regard to stationary behaviour and finding the way in streams?

6.3.1 Stationary behaviour and site fidelity

Stream-dwelling fish display resident behaviour, and occupy territories that have generally been regarded as feeding territories. They also move between territories, but individual fish of most species will move over relatively short distances and in this way they may occupy a limited area of a stream, commonly denoted its 'home range' or 'home area'. If artificially displaced within the stream, that is away from the home area, a fish will normally return home irrespective of whether it has been displaced upstream or downstream (Gerking 1959; Gunning 1959). In brown trout (*Salmo trutta*) the 'home range' of individual fish may be in the region of 10 to 25 metres, and this species has been used to investigate the behavioural and sensory mechanisms underlying homing in stream-dwelling fish (Halvorsen and Stabell 1990).

In order to investigate homing in fish, brown trout in a Norwegian stream were individually tagged and transplanted 200 m, either upstream or downstream. Half of the fish displaced in both directions were rendered anosmic by heat treatment (i.e. could not use olfaction), while the other half was sham-burned on the snout (controls; Halvorsen and Strabell 1990). After nine weeks the stream was electro-fished, and the sections of fish recapture were recorded. The results showed that anosmic fish stayed in the area of release, whether displaced upstream or downstream (Fig. 6.3). On the other hand, a large proportion of the sham treated fish returned and was recaptured in their original 'home area' or in adjacent sections. The control fish were presumably homing by mechanisms of rheotropism (Arnold 1974), that is by an upstream movement (positive rheotactic response) towards attractive odours in their original home area, and a downstream movement (negative rheotactic response) due to lack of such chemical signals. Interestingly, the anosmic fish did not move downstream (as anticipated), indicating that presence or absence of home-area-specific odours are detected against a background bouquet of odours rather than within the bouquet as such. This distinction may be of importance for understanding the basic mechanisms underlying the complex events present in the homing behaviour of fish. A second experiment

revealed that trout were able to combine the two rheotactic alternatives (Halvorsen and Stabell 1990). If fish were transplanted from a tributary brook of the stream to an area of the main stem of the stream 125–150 m above the tributary outlet, some fish also returned to their home sections of the tributary (Fig. 6.4). To reach their home area the fish had to carry out an initial downstream movement followed by an upstream movement when reaching the tributary outlet, demonstrating that a combination of rheotactic responses had been used.

Behavioural patterns of stream-dwelling trout were further studied with regard to chemical properties and attractiveness of home-stream water (Arnesen and Stabell 1992). In preference tests in the laboratory, (Stabell 1987) it was shown that fish preferred water taken from the outlet of the home-stream over that taken from a neighbouring stream. In most cases, the fish also showed a preference for neutral water scented with fish from their home section over water scented by fish from distant sections of the home-stream. This implies that aspects of kin recognition (see Chapter 4) and substrate marking may be involved in the home site fidelity of trout. Strong indications for the presence of separate spawning demes of trout in the main stem of the study stream have subsequently been found by genetic studies (Carlsson *et al.* 1999).

6.3.2 Substrate marking by fish

Juvenile Atlantic salmon (*Salmo salar*) display territorial behaviour similar to that of resident brown trout mentioned previously. When three full sibling groups of salmon parr, each group originating from a separate river, were tested for odour preferences in the laboratory, the fish demonstrated preferences for species as well as strain odour over tap water (Stabell 1987). In competition tests, the odour of its own strain was preferred over that of alien conspecifics (i.e. another river strain). When extracts made from intestinal contents (faecal material) of fish from different strains were used in competition tests, the salmon parr demonstrated a preference for dissolved intestinal material taken from their own strain. These results suggested that faecal material, presumably deposited on the bottom substrate, contains attractive chemical components. In

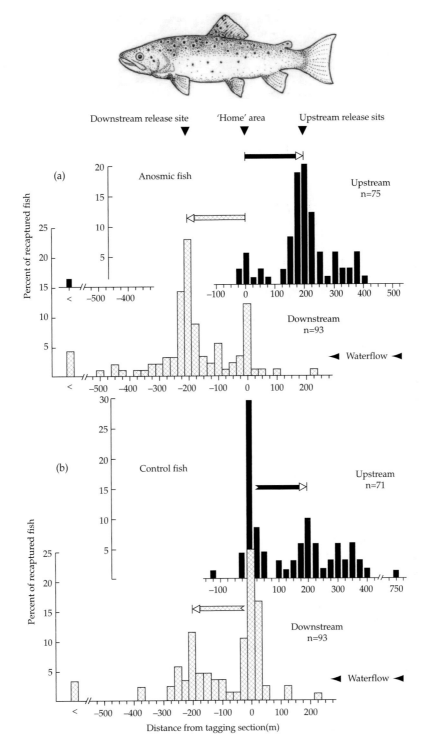

Figure 6.3 Transplantation experiments with resident brown trout (*Salmo trutta*) carried out within the main stem of the Nordre Finnvikelv stream, Norway. Fish were transplanted 200 metres away from their home area, either upstream (black bars) or downstream (hatched bars). Horizontal arrows give the transplantation directions, and bars shows the pooled results of trout distribution within each treatment group after nine weeks. (a) fish rendered anosmic by heat treatment, and (b) control fish sham-burned on snout. Reproduced from Halvorsen and Stabell (1990); fig. 2.

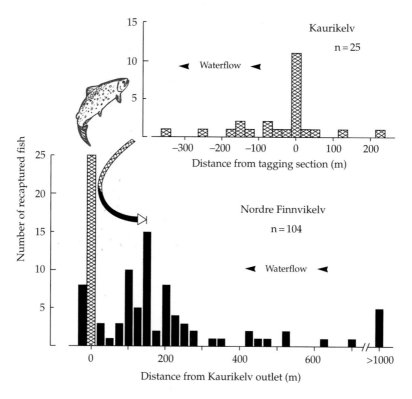

Figure 6.4 Transplantation experiments with resident brown trout (*Salmo trutta*) from the Kaurikelv tributary to the main stem of the Nordre Finnvikelv stream, Norway. Fish were released in Nordre Finnvikelv 125–150 metres above the tributary outlet. Black bars show the distribution of recaptures in Nordre Finnvikelv, and hatched bars show the distribution of fish recaptured in the Kaurikelv tributary, after nine weeks. Reproduced from: Halvorsen and Stabell (1990) fig. 3.

a previous behavioural study, chemical signals present in the intestinal content of juvenile Arctic char (*Salvelinus alpinus*) had been reported attractive to conspecific adults (Selset and Døving 1980), and such material was also attractive to juvenile char (Olsén 1987).

6.3.3 Feeding migration and homing for reproduction

Salmonid species of the genera *Salmo*, *Salvelinus* and *Oncorhynchus* are all denoted anadromous, that is they live in the ocean but spawn in freshwaters, but several of the species also have land-locked forms that display potamodromous migration patterns (i.e. they migrate between freshwaters). The complexity of the life-history patterns and migratory systems in the various species of anadromous salmonids has been described in detail by Nordeng

(1977, 1983, 1989a), Jonsson (1985), Stabell 1992, and Quinn (2005). Atlantic salmon smolt that migrate to the sea tend to migrate closer in time to siblings than unrelated fish, regardless of whether they have been raised together, suggesting a genetic base of discrimination (Olsén *et al.* 2004). In the sea, however, there are no indications of adult salmon shoaling with kin (Palm *et al.* 2008). The spawning migration from the sea back to their natal streams and nursery areas has been found to depend on an intact olfactory sense (Wisby and Hasler 1954; reviewed by: Stabell 1984, 1992).

In populations of some species, like rainbow trout (*Oncorhynchus mykiss*) and sockeye salmon (*Oncorhynchus nerka*), a more complex migration system exists, involving freshwater migration before the anadromous cycle (Northcote 1962, 1969; Brannon 1972; but see also: Nordeng 1989b; Stabell 1992). In those cases the fish will move from their incubation areas into a rearing lake as juveniles (fry), where they will stay until smolt transformation. Depending on the location of the nursery area, the lakeward migration may occur in both upstream and downstream direction, and fish from populations using downstream tributaries may even undertake a combined downstream/upstream movement to reach the rearing lake (Fig. 6.5). All three types of local migration may take place within a common local area (i.e. within a single lake system). After a growth period in the lake, the fish may (or may not) perform an anadromous run, and finally return to their natal rearing areas to spawn (Lindsey *et al.* 1959). Genetically controlled and rheotactic-influenced olfaction has been found as the

directing factor for salmon fry during lake-ward migration (Brannon 1972). Juvenile migrations to rearing lakes have also been described in potamodromous brown trout (Kaasa 1980), and may be present in anadromous Atlantic salmon as well (Halvorsen and Jørgensen 1996).

6.3.4 Homing theories

Two opposing theories have been introduced to explain the precise homing of salmonid fish to their native nursery areas; the *imprinting hypothesis* presented by Hasler and Wisby (1951), and the *pheromone hypothesis* presented by Nordeng (1971). Both hypotheses were subsequently extended and

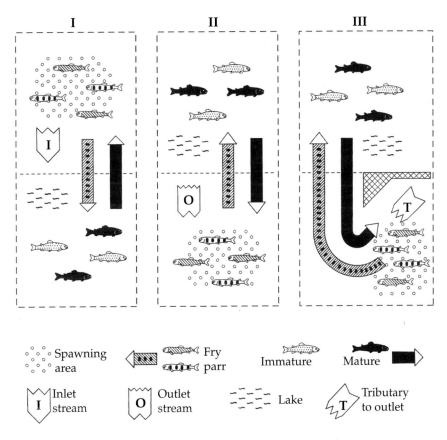

Figure 6.5 Schematic diagram of the lake-ward migration of potamodromous salmonids, migrating from upstream, downstream, or downstream tributary nursing areas, followed by spawning migrations in opposite directions as sexually mature fish. I: Population spawning in the lake inlet. II: Population spawning in the lake outlet. III: Population spawning in a tributary to the lake outlet. In anadromous systems, the immature fish in the lake may undergo a smolt transformation, resulting in a downstream, seaward migration, followed by upstream return from the sea as sexually mature fish. In such systems the return during spawning migrations will be directly to the nursery area of each population. Reproduced from Stabell (1992); fig. 2.

refined (Nordeng 1977; Hasler and Scholz 1983). The imprinting hypothesis postulates that waters from home-streams contain characteristic odours, originating in soil and vegetation, to which the young salmon become conditioned (or imprinted) while in the stream, and which they recognize and orient towards upon reaching their parent stream as mature migrants (see: Hasler and Scholz 1983, and references therein). The pheromone hypothesis proposed by Nordeng (1971, 1977) takes into consideration the presence of sympatric populations within a watershed, and postulates that mature fish are using attractants (i.e. population specific pheromones), released by young relatives in the home-stream for finding the way.

The common feature of the two hypotheses is that local orientation is dependent on the olfactory sense, but still they are essentially different (Nordeng 1989b), leading to conflicting opinions (Quinn 1990). It has for instance been argued that species of Pacific salmon have life cycles that are incompatible with the pheromone hypothesis (Smith 1985). Nordeng (1989a, b), however, presented arguments in favour of the pheromone hypothesis for the Pacific salmon species. On the other hand, it has been argued that the imprinting hypothesis, claiming imprinting at the time of smolting, 'may be a serious oversimplification of what actually takes place' (Quinn 1985). The ecological basis for such a critique is the fact that several Pacific salmonid species undergo smolt transformation in nursery lakes (see earlier part of this chapter), but still return to their natal areas as adults (Quinn 1985; Stabell 1992).

The return of adults to natal areas, following smolt transformation in lakes, has been explained by proposing that fry may learn the odours of their natal site (Quinn 1985), with the obvious result of actually undermining the basic idea of imprinting. It should here be noted that imprinting has been closely associated with a peak in thyroid hormones that occurs during smolt transformation (Hasler and Scholz 1983).

6.3.5 Chemical cues involved in site marking and site detection

In their work with homing mechanisms and pheromones in Arctic charr, Selset and Døving (1980)

discovered that intestinal contents as well as bile from young fish was attractive to mature fish of the same strain; that is population-specific odorants seemed to be found. Bile salts are secreted into the intestine of vertebrates for the purpose of emulsifying fat, and a major part of those compounds are reabsorbed in the intestine. However, a small fraction is also lost in the faeces. Døving *et al.* (1980) also found bile salts to be highly potent odorants, and salmonid pheromones were therefore suspected to be of bile-salt-like nature. The chemical structure of common bile salts has been found to vary among animal groups, and they have accordingly been suggested as potential evolutionary markers (Haslewood 1978). However, there is a potential for even larger structural variation among these steroid compounds (G.A.D. Haslewood, 1980: personal communication), and bile salt derivatives thereby represent a possible source of variation needed for population specific pheromones. It is here interesting to note that evidence has been provided by Foster (1985) that spawning site selection in lake trout (*Salvelinus namaycush*) can be induced by faecal residues of juvenile conspecifics.

Stabell *et al.* (1982) found variation in chemical compounds present in intestinal contents of different genetic strains of Atlantic salmon parr. Moreover, differences among strains were also found with respect to bile salts (Stabell *et al.* 1982). Subsequent electrophysiological tests with char (Fisknes and Døving 1982) revealed that the fractions displaying chemical variations in the study by Stabell *et al.* (1982), also contained the most potent olfactory material. The electrophysiological responses were found in the medial part of the olfactory bulb, in the same spatial area that previously had been found to evoke olfactory responses with bile salts (Døving *et al.* 1980). Furthermore, the responses were also evoked with chemical fractions equivalent to those attracting mature sea char (Selset and Døving 1980). In total, these results imply that the chemical compounds used by salmonids for substrate marking were most likely bile salt derivatives.

Supporting evidence that bile acids (e.g. bile salts) are functional chemical signals in salmonids, and that olfactory neurons are highly sensitive to such compounds, were also demonstrated in lake char (*Salvelinus mamaycush*) (Zhang *et al.* 2001; Zhang and Hara 2009). Some other convincing studies in support for migratory pheromones in fish have been carried out with sea lampreys (*Petromyzon marinus*) in the Great Lakes of North America. Sea lamreys begin their life as stream-dwelling, filter-feeding larvae that, after several years being about 120 mm in length, metamorphose into a parasitic phase that migrates out into the lake to find prey (Sorensen *et al.* 2003). The species invaded the Great Lakes about a century ago, and developed into a pest. In the lakes they are dispersed great distances as parasites, and do not displaying homing to natal streams (Bergstedt and Seelye 1995). Instead, they seem to locate spawning streams by innately recognized odorants common to Petromyzontid lampreys; that is to a common evolutionarily conserved pheromone (Fine *et al.* 2004). The pheromone is a mixture of compounds comprised of at least three bile acids, and these have been structurally identified (Sorensen *et al.* 2005; Hoye *et al.* 2007). In lamprey, the major pheromone blend consists of Petromyzonol sulphate (PS), Petromyzosterol disulfate (PSDS), and Petromyzonamine disulfate (PADS) (Fig. 6.6). Taken together, the studies on homing in salmonids and lampreys suggests that bile salt derivatives may constitute a common class of migratory pheromones used by fish, and that the responses to such odorants may be innately (i.e. genetically) controlled.

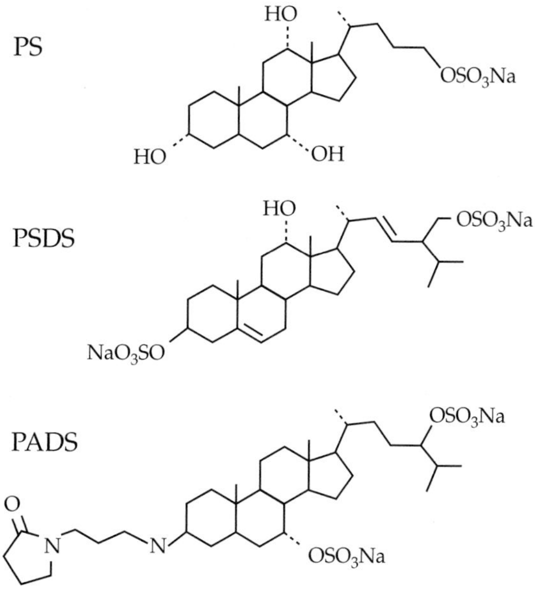

6.3.6 Long distance homing and navigation

Nordeng (1977) suggested a link between the outward migrating smolt and the homing migrants in salmonid fish. Downstream migration of smolts takes place in schools and occurs during spring and summer. In the Salangen river system, he found that downstream migration of Atlantic salmon smolt may commence on different dates in May, depending on when the water temperature reaches approximately 6 °C. Despite the variation on a year-to-year basis, the time lag between the initial descent of smolts and the first entry of ascending salmon into freshwater was found to be constant over several years, around 15 days (see Stabell 1992 for details). Accordingly, Nordeng (1977) proposed a pheromone hypothesis for the entire homeward migration, suggesting homeward navigation to be

Figure 6.6 The three major bile salt components released by larvae of the sea lamprey (*Petromyzon marinus*), and constituting the pheromone blend that attracts sexually mature sea lampreys. PS: Petromyzonol sulphate; PSDS: Petromyzosterol disulfate; PADS: Petromyzonamine disulfate. (After Haslewood (1978) and Hoye *et al.* (2007).)

an inherited response to population-specific pheromone trails in sea and coastal waters, released by descending smolt.

The mechanisms underlying long distance migration and 'navigation' in the sea have also been an area of controversy. Speculative proposals for guidance mechanisms in salmon have been repeatedly presented, without a sufficient basis in the ecology or sensory apparatus of these species. In particular, this has been the case for a variety of proposals related to 'compass orientation' and 'sequential imprinting' (e.g. Lohmann *et al.* 2008). In their ocean migration, salmon are not making 'shortcuts' along direct compass courses, but are following the ocean currents, although actively and at higher speed than the surrounding water (Royce *et al.* 1968). Neither are they carrying out searches along coastlines, but are following coastal currents more or less directly towards their goal from open waters. A nice example of this is the homeward migration from the Bering Sea of sockeye salmon in Bristol Bay, Alaska. Bristol Bay sockeye do not follow the shoreline but rather are concentrated near the surface several hundred kilometres offshore, coming closer to shore (about 40–80 km) as the bay narrows. The sockeye heading for five major fishing districts are initially fully mixed but gradually become more segregated as they leave the main migration corridor for their respective rivers (e.g. Quinn 2005).

Since fish lack a biological clock with the accuracy of a chronometer of the kind used by humans, compass navigation towards the home-stream seems impossible (Døving and Stabell 2003). However, fish may simply find their way home by rheotaxis in the sea. Following the discovery by Westerberg (1982) that salmon are using stratified temperature layers in the sea for navigation, Døving *et al.* (1985) performed ultrasonic tracking to study the behavioural and neural responses of Atlantic salmon during migration in the sea. They concluded that an olfactory discrimination of fine-scale hydrographic features may provide a necessary reference system for successful orientation in open waters by salmon. The fact that fish are able to detect infrasound (see Døving and Stabell 2003, and references therein) means that the salmon may detect the linear acceleration to which it is exposed when moving from one layer to an adjacent one. In open waters salmon are swimming with small-scale oscillations between layers. They may detect the relative movements of a layer by their auditory system, and use their olfactory sense to sense the chemical difference between layers. This further implies that fish may detect the movement of a particular layer and may orient in a counter-current direction of that layer. If this layer contains the characteristic odours of the home-stream, such rheotactic behaviour will eventually lead the fish to the sources of the attractive chemical signal (Døving and Stabell 2003).

6.4 Concluding remarks

For all species of aquatic invertebrates studied so far, chemical senses appear paramount for carrying out directed migrations. In animals like gastropods, that are permanently in contact with the substrate, tactile detection of chemical trails seems to be the ruling mechanism for finding the way. In the faster moving crustaceans, such as lobsters, crayfish, and crabs, however, olfactory-like sensing of dissolved

material may take place. In general, lower vertebrates seem to depend on the olfactory sense for performing migrations, but for semi-aquatic animals like amphibians 'the picture' is more complex. Olfaction seems paramount for migration in fish, even though they are in some cases moving over very long distances. Chemicals deposited on the substrate by fish for marking home area or spawning sites are also detected by olfaction. Apparently, therefore, animals have not 'lost the grip' of the substrate through evolution, but the way of detecting relevant substrate-bound chemicals has changed. The gustatory sense has been vaguely proposed as a participating sense in the sensory detection of home-stream water in fish, but so far no convincing evidence has been presented.

The compounds constituting the chemical signals used in migration and orientation, as well as their solubility and life spans (i.e. stability), are important aspects of chemical ecology. It would be useless for bottom-dwelling animals to deposit individual substrate trails if the longevity of the embedded compounds are not matched by the time it takes to return home. On the other hand, if the compounds used are too stable, there would eventually be a mess of trails in the animal's environment, presumably making the site non-navigable. So far, little is known about the chemical structures of gastropod trail signals, but with respect to fish, bile salts have been suggested as potential migration pheromones in salmonids, and in petromyzonids such compounds have actually been proven. Bile salts are extremely stable compounds that also display large structural variation. Because of their hydrophobic properties, bile salts display low solubility in water, and accordingly they are suitable compounds when it comes to longevity of chemical site marks. Their low solubility in water appears to be compensated for in salmonids as well as petromyzonids through extreme olfactory sensitivity to bile salts. Among the species studied in those families, detection thresholds lower than 10^{-13} molar have been found. For human detection of chemical signals used by aquatic animals when migrating, modern LC-MS systems (i.e. high performance liquid chromatography coupled with mass spectrometry) may bring this field into a new era.

Acknowledgements

During the writing of this chapter, my mentor, colleague, and friend Hans Nordeng, passed away on the 9th of May 2010. Hans was born on 19th November 1918, and devoted a major part of his life to work on the life history and homing in anadromous salmonids. In deep appreciation for his scientific contribution, and the ecological insight his work has provided, I dedicate this chapter to his memory.

References

Allaby, M., ed. (2005) *A Dictionary of Ecology*, 3rd ed., Oxford University Press Oxford.

Allard, M. W., Miyamoto, M. M., Bjorndal, K. A., Bolten, A. B. and Bowen, B. W. (1994) Support for natal homing in green turtles from mitochondrial DNA sequences. *Copeia*, 1994: 34–41.

Arnesen, A. M. and Stabell, O. B. (1992) Behaviour of stream-dwelling brown trout towards odours present in home stream water. *Chemoecology*, 3: 94–100.

Arnold, G. P. (1974) Rheotropism in fishes. *Biological Reviews of the Cambridge Philosophical Society*, 49: 515–76.

Baker, R. R. (1978) *The Evolutionary Ecology of Animal Migration*. Hodder and Stoughton Educational, London.

Bergstedt, R. A. and Seelye, J. G. (1995) Evidence for lack of homing by sea lampreys. *Transactions of the American Fisheries Society*, 124: 235–9.

Brannon, E. L. (1972) Mechanisms controlling migration of sockeye salmon fry. *International Pacific Salmon Fisheries Commission Bulletin*, 21: 1–86.

Carlsson, J., Olsen, K. H., Nilsson, J., Overli, O. and Stabell, O. B. (1999) Microsatellites reveal fine-scale genetic structure in stream-living brown trout. *Journal of Fish Biology*, 55: 1290–303.

Chelazzi, G. (1990) Eco-ethological aspects of homing behavior in molluscs. *Ethology Ecology & Evolution*, 2: 11–26.

Chelazzi, G., Dellasantina, P., and Parpagnoli, D. (1987) Trail following in the chiton *Acanthopleura gemmata*: operational and ecological problems. *Marine Biology*, 95: 539–45.

Cook, S. B. (1969) Experiments on homing in limpet *Siphonaria normalis*. *Animal Behaviour*, 17, 679–82.

Cook, S. B. (1971) A study of homing behavior in the limpet *Siphonaria alternata*. *Biological Bulletin*, 141: 449–57.

Cook, A., Bamford, O. S., Freeman, J. D. B., and Teideman, D. J. (1969) A study of homing habit of limpet. *Animal Behaviour*, 17: 330–9.

Croll, R. P. (1983). Gastropod chemoreception. *Biological Reviews of the Cambridge Philosophical Society*, 58: 293–319.

Crossman, E. J. (1990) Reproductive homing in muskellunge, *Esox masquinongy*. *Canadian Journal of Fisheries and Aquatic Sciences*, 47: 1803–12.

Della Santina, P. (1994) Homing pattern, activity area and trail following of the high shore Mediterranean limpet *Patella rustica* L (Mollusca Gastropoda). *Ethology Ecology & Evolution*, 6: 65–73.

Døving, K. B., Selset, R., and Thommesen, G. (1980) Olfactory sensitivity to bile acids in salmonid fishes. *Acta Physiologica Scandinavica*, 108: 123–31.

Døving, K. B., Westerberg, H., and Johnsen, P. B. (1985) Role of olfaction in the behavioral and neural responses of Atlantic salmon, *Salmo salar*, to hydrographic stratification. *Canadian Journal of Fisheries and Aquatic Sciences*, 42: 1658–67.

Døving, K. B. and Stabell, O. B. (2003) Trails in open waters: Sensory cues in salmon migration. In S. P. Collin and N. J. Marshall (eds.) *Sensory Processing in Aquatic Environments*, pp. 39–52. Springer-Verlag, New York.

Edwards, M. and Davies, M. S. (2002) Functional and ecological aspects of the mucus trails of the intertidal prosobranch gastropod *Littorina littorea*. *Marine Ecology-Progress Series*, 239: 129–37.

Emery, D. G. (1992) Fine structure of olfactory epithelia of gastropod mollusks. *Microscopy Research and Technique*, 22: 307–24.

Erlandsson, J. and Kostylev, V. (1995) Trail following, speed and fractal dimension of movement in a marine prosobranch, *Littorina littorea*, during a mating and a non-mating season. *Marine Biology*, 122: 87–94.

Fine, J. M., Vrieze, L. A., and Sorensen, P. W. (2004) Evidence that petromyzontid lampreys employ a common migratory pheromone that is partially comprised of bile acids. *Journal of Chemical Ecology*, 30: 2091–110.

Fisknes, B. and Døving, K. B. (1982) The olfactory sensitivity to group specific substances in Atlantic salmon (*Salmo salar* L.). *Journal of Chemical Ecology*, 8: 1083–92.

Foster, N. R. (1985) Lake trout reproductive behavior: Influence of chemosensory cues from young-of-the-year by-products. *Transactions of the American Fisheries Society*, 114: 794–803.

Funke, W. (1968). Heimfindevermögen und Ortstreue bei *Patella* L. (Gastropoda, Prosobranchia). *Oecologia* (*Berl.*), 2: 19–142.

Gerking, S. D. (1959) The restricted movement of fish populations. *Biological Reviews of the Cambridge Philosophical Society*, 34: 221–42.

Gunning, G.E. (1959) The sensory basis for homing in the longear sunfish, Lepomis magalotis megalotis

(Rafinesque). *Investigations of Indiana Lakes and Streams* **5**: 103–30.

Göbbeler, K. and Klussmann-Kolb, A. (2007) A comparative ultrastructural investigation of the cephalic sensory organs in Opisthobranchia (Mollusca, Gastropoda). *Tissue & Cell*, **39**: 399–414.

Halvorsen, M. and Jørgensen, L. (1996) Lake use by Atlantic salmon (*Salmo salar* L.) parr and other salmonids in nothern Norway. *Ecology of Freshwater Fish*, **5**: 28–36.

Halvorsen, M. and Stabell, O. B. (1990) Homing behaviour of displaced stream-dwelling brown trout. *Animal Behaviour*, **39**: 1089–97.

Hartnoll, R. G. and Wright, J. R. (1977) Foraging movements and homing in limpet, *Patella vulgata* L. *Animal Behaviour*, **25**: 806–10.

Hasler, A. D. and Wisby, W. J. (1951) Discrimination of stream odors by fishes and relation to parent stream behavior. *The American Naturalist*, **85**: 223–38.

Hasler, A. D. and Scholz, A. T. (1983) *Olfactory Imprinting and Homing in Salmon*, Springer-Verlag, Berlin.

Haslewood, G.A.D. (1978) The biological importance of bile salts. *Frontiers in Biology, vol.47*. North-Holland Publ. Comp., Amsterdam.

Hoye, T.R., Dvornikovs, V., Fine, J.M., Anderson, K.R., Jeffrey, C.S., Muddiman, D.C., Shao, F., Sorensen, P.W., and Wang, J. (2007) Details of the structure determination of the sulfated steroids PSDS and PADS: New components of the sea lamprey (*Petromyzon marinus*) migratory pheromone. *Journal of Organic Chemistry*, **72**: 7544–50.

Joly, P. and Miaud, C. (1993) How does a newt find its pond? The role of chemical cues in migrating newts (*Triturus Alpestris*). *Ethology Ecology & Evolution* **5**: 447–55.

Jonsson, B. (1985) Life history patterns of freshwater resident and sea-run migrant brown trout in Norway. *Transactions of the American Fisheries Society*, **114**: 182–94.

Kaasa, H. (1980). Stock characteristics and population specific migration in resident brown trout, *Salmo trutta* L., in the Eika watershed, Bø, Telemark County (In Norwegian). Cand. real. thesis. Institute of Zoology, University of Oslo, Norway.

Kohn, A. J. (1961) Chemoreception in gastropod molluscs. *American Zoologist*, **1**: 291–308.

Lebboroni, M. and Chelazzi, G. (2000) Waterward orientation and homing after experimental displacement in the European Pond Turtle, *Emys orbicularis*. *Ethology Ecology & Evolution*, **12**: 83–8.

Lindsey, C. C., Northcote, T. G. and Hartman, G. F. (1959) Homing of rainbow trout to inlet and outlet spawning streams at Loon Lake, British Columbia. *Journal of the Fisheries Research Board of Canada*, **16**: 695–719.

Loher, T. (2008) Homing and summer feeding site fidelity of Pacific halibut (*Hippoglossus stenolepis*) in the Gulf of Alaska, established using satellite-transmitting archival tags. *Fisheries Research*, **92**: 63–9.

Lohmann, K.J., Putman, N.F., and Lohmann, C.M.F. (2008) Geomagnetic imprinting: A unifying hypothesis of long-distance natal homing in salmon and sea turtles. *Proceedings of the National Academy of Science (PNAS)*, **105**: 19096–101.

Madison, D. M. (1969) Homing behavior of the red-cheeked salamander, *Plethodon jordani*. *Animal Behaviour*, **17**: 25–39.

McGregor, J. H. and Teska, W. R. (1989) Olfaction as an orientation mechanism in migrating *Ambystoma maculatum*. *Copeia*, **1989**: 779–81.

Mortimer, J. A. and Portier, K. M. (1989) Reproductive homing and internesting behavior of the Green Turtle (*Chelonia mydas*) at Ascension Island, South Atlantic Ocean. *Copeia*, **4**: 962–77.

Nordeng, H. (1971) Is the local orientation of anadromous fishes determined by pheromones? *Nature*, **233**: 411–13.

Nordeng, H. (1977) A pheromone hypothesis for homeward migration in anadromous salmonids. *Oikos*, **28**: 155–9.

Nordeng, H. (1983) Solution to the 'Char Problem' based on Arctic char (*Salvelinus alpinus*) in Norway. *Canadian Journal of Fisheries and Aquatic Sciences*, **40**: 1372–87.

Nordeng, H. (1989a) Migratory systems in anadromous salmonids. *Physiology and Ecology Japan, Special Volume*, **1**: 167–8.

Nordeng, H. (1989b) Salmonid migration: Hypotheses and principles. In E. L. Brannon and B. Jonsson, eds. *Proceedings of the Salmonid Migration and Distribution Symposium, June 1987*, pp. 1–8. School of Fisheries, Contrib. No. 793, University of Washington.

Northcote, T. G. (1962) Migratory behaviour of juvenile rainbow trout, *Salmo gairdneri*, in outlet and inlet spawning streams of Loon Lake, British Columbia. *Journal of the Fisheries Research Board of Canada*, **18**: 201–70.

Northcote, T. G. (1969) Patterns and mechanisms in the lakeward migratory behaviour of juvenile trout. In T. G. Northcote, ed. *Symposium on Salmon and Trout in Streams*, pp. 183–203. H.R. MacMillan Lectures in Fisheries, Univ. of British Columbia, Vancouver.

Olsén, K.H. (1987) Chemoattraction of juvenile Arctic charr (*Salvelinus alpinus* (L.)) to water scented by intestinal content and urine. *Comparative Biochemistry and Physiology A*. **87**: 641–3.

Olsén, K.H., Peterson, E., Ragnarsson, B., Lundquist, H., and Järvi, T. (2004) Downstream migration in Atlantic salmon (*Salmo salar*) smolt sibling groups. *Canadian Journal of Fisheries and Aquatic Sciences*, **61**: 328–31.

Palm, S., Dannewitz, J., Järvi, T., Koljonen, M.-L., Prestegaard, T., and Olsén, K.H. (2008) No indications of Atlantic salmon (*Salmo salar*) shoaling with kin in the Baltic Sea. *Canadian Journal of Fisheries and Aquatic Sciences*, **65**: 1738–48.

Pereira, L. H. G., Foresti, F., and Oliveira, C. (2009) Genetic structure of the migratory catfish *Pseudoplatystoma corruscans* (Siluriformes: Pimelodidae) suggests homing behaviour. *Ecology of Freshwater Fish*, **18**: 215–25.

Quinn, T. P. (1985) Salmon homing: is the puzzle complete? *Environmental Biology of Fishes*, **12**: 315–17.

Quinn, T. P. (1990) Current controversies in the study of salmon homing. *Ethology Ecology & Evolution*, **2**: 49–64.

Quinn, T. P. (2005). *The Behavior and Ecology of Pacific Salmon and Trout*. Bethesda, Md.: American Fisheries Society/University of Washington Press.

Royce, W.F., Smith, L.S., and Hartt, A.C. (1968) Models on oceanic migrations of Pacific salmon and comments on guidance mechanisms. *Fishery Bulletin*, **66**: 441–62.

Selset, R. and Døving, K. B. (1980) Behaviour of mature anadromous char (*Salmo alpinus* L.) towards odorants produced by smolts of their own population. *Acta Physiologica Scandinavica*, **108**: 113–22.

Sinsch, U. (1990) Migration and orientation in anuran amphibians. *Ethology Ecology & Evolution*, **2**: 65–80.

Smar, C. M. and Chambers, R. M. (2005) Homing behavior of Musk Turtles in a Virginia lake. *Southeastern Naturalist*, **4**: 527–32.

Smith, R. J. F. (1985). *The Control of Fish Migration*. Springer Verlag, Berlin.

Sorensen, P. W., Vrieze, L. A., and Fine, J. M. (2003). A multi-component migratory pheromone in the sea lamprey. *Fish Physiology and Biochemistry*, **28**: 253–7.

Sorensen, P. W., Fine, J. M., Dvornikovs, V., Jeffrey, C. S., Shao, F., Wang, J. Z., Vrieze, L. A., Anderson, K. R., and Hoye, T. R. (2005) Mixture of new sulfated steroids functions as a migratory pheromone in the sea lamprey. *Nature Chemical Biology*, **1**: 324–8.

Stabell, O. B. (1984) Homing and olfaction in salmonids: A critical review with special reference to the Atlantic salmon. *Biological Reviews of the Cambridge Philosophical Society*, **59**: 333–88.

Stabell, O. B. (1987) Intraspecific pheromone discrimination and substrate marking by Atlantic salmon parr. *Journal of Chemical Ecology*, **13**: 1625–43.

Stabell, O. B. (1992) Olfactory control of homing behaviour in salmonids. In T. J. Hara, ed. *Fish Chemoreception*, pp. 249–70. Chapman & Hall, London.

Stabell, O. B., Selset, R., and Sletten, K. (1982) A comparative chemical study on population-specific odorants from Atlantic salmon. *Journal of Chemical Ecology*, **8**: 201–17.

Svedang, H., Righton, D., and Jonsson, P. (2007) Migratory behaviour of Atlantic cod *Gadus morhua*: natal homing is the prime stock-separating mechanism. *Marine Ecology-Progress Series*, **345**: 1–12.

Toft, R. (1975) The significance of the olfactory and visual senses in the behaviour of spawning migration in Baltic salmon, (In Swedish), *Swedish Salmon Research Institute Report (LFI Meddelande)*.

Vannini, M. and Cannicci, S. (1995) Homing behaviour and possible cognitive maps in crustacean decapods. *Journal of Experimental Marine Biology and Ecology* **193**: 67–91.

Wells, M. J. and Buckley, S. K. L. (1972) Snails and trails. *Animal Behaviour*, **20**: 345–55.

Westerberg, H. (1982) Ultrasonic tracking of Atlantic salmon (*Salmo salar* L.) - II. Swimming depth and temperature stratification. *Report from the Institute of Freshwater Research Drottningholm*, **60**: 102–20.

Wisby, W. J. and Hasler, A. D. (1954) Effect of olfactory occlusion on migrating silver salmon (*Oncorhynchus kisutch*). *Journal of the Fisheries Research Board of Canada*, **11**: 472–8.

Zhang, C., Brown, S. B., and Hara, T. J. (2001) Biochemical and physiological evidence that bile acids produced and released by lake char (*Salvelinus namaycush*) function as chemical signals. *Journal of Comparative Physiology B*, **171**: 161–71.

Zhang, C. and Hara, T. J. (2009) Lake char (*Salvelinus namaycush*) olfactory neurons are highly sensitive and specific to bile acids. *Journal of Comparative Physiology A*, **195**: 203–15.

CHAPTER 7

Death from downstream: chemosensory navigation and predator–prey processes

Marc Weissburg

Aquatic predators rely greatly on chemosensory information to locate their prey. Because of the properties of water, chemical signals generally transmit information over longer distances than light, sound, or other mechanical stimuli, and chemical signals contain rich information about the identity and other characteristics of potential prey. It is therefore not surprising that chemosensory hunters are common, come from all important taxa in both marine and freshwater, and often exert large impact on their prey species. In fact, the effects of such predators often propagate to other community members beyond their target prey via trophic cascades or other indirect interactions.

Understanding chemosensory navigation is motivated by two essential issues. Understanding the ecological importance of chemosensory mediated predation requires a concomitant understanding of how animals find their prey. This knowledge forms a predictive framework for anticipating where, when, and why predation may be a potent force. Second, the transduction and integration of chemical signals sheds light on neural signal processing and neural systems. The fundamental similarity between chemosensory systems in humans and animals (including invertebrates) makes this an issue of practical importance, as does the possibility of using basic principles gleaned from animals in the construction of artificial chemosensory systems.

The goals of this chapter are to discuss how animals navigate through chemical plumes to find prey, and the ecological consequences of this process. The focus is on larger animals (10–1000 cm body size) that operate in flow velocities ranging from cm^{-s} to m^{-s}. Since the fluid environment exerts a major effect on the chemical signal structure and navigational strategies, I will emphasize tools to characterize the hydrodynamic/chemical signal conditions and how these conditions affect navigation. An important theme is examining which signals provide information from these plumes and what these information gathering mechanisms imply about the sensory system.

Addressing ecological questions requires data from natural environments. The second goal of this chapter is to relate our understanding of navigational strategies, and their constraints, to predator–prey dynamics in the field. A newly emerging body of literature couples data on fluid physical conditions to patterns of predation and provides linkages between information gathering and ecological processes. In addition, there is a substantial body of work on the effect of chemosensory predators that can be used to provide additional insights.

7.1 Plumes—a very brief review

The flow environment has a big effect on plume chemical signal structure, so that fluid dynamical parameters are used to provide insights into general plume properties, or as a way to roughly characterize the chemical signal environment when signals cannot be directly measured. The most general fluid dynamical parameter is the Reynolds number (Re), which describes the ratio of inertial to viscous forces in a fluid (Denny 1988; Webster and Weissburg 2009). Length and fluid velocity characterize the inertial forces, which are related to the magnitude of viscous forces characterized by fluid viscosity. Low Re represents viscous regimes where bulk fluid motion is dampened by the molecular 'stickiness' of the fluid. Re <<< 1 corresponds to a near absence of fluid motion, with diffusion as the only mechanism available to transfer a scalar quantity such as heat or molecules. Re ≈ 1–100 indicates the importance of both inertia and viscosity; the fluid moves, but viscosity quickly dampens the fluid energy so the flow is considered laminar. That is, flow stream-lines do not cross, the fluid moves nearly exclusively in one direction (i.e. downstream) and there is little mixing (which corresponds to nearly zero fluid motion in across-stream or vertical directions). The flow becomes more chaotic and random at increasingly higher Re, and fluid velocity is substantial in the vertical and across-stream directions.

Most macroscopic aquatic animals live in a turbulent world. Re ranges from 10^{-1}–10^8 depending on their movement speeds and sizes. However, many animals that navigate through odour plumes reside on the ocean (or river) bottom. In such cases, it may be more informative to use the roughness Reynolds number Re*, where the length scale is now the diameter of the substrate. Larger diameter substrates produce more turbulent flows, which is consistent with our intuition gained from watching water move over a surface; rougher surfaces produce more chaotic fluid motion.

Quantities like Re, and to a much larger extent Re*, are convenient ways to characterize the fluid environment in the lab, and a large body of laboratory experiments in flumes serves to link the fluid dynamical environment with odour plume structure. This can be especially helpful since direct measurement of odour plume signal properties is technically challenging, so that easily obtained proxy measurements have great utility. Generally speaking, odour plumes in turbulent conditions are composed of odour filaments intermingled with unscented fluid (Fig. 7.1; Moore and Atema 1991; Murlis 1986; Weissburg 2000). The filamentous nature of these plumes results in substantial spatial and temporal unpredictability. Gradients of odour concentration tend to be sharper in the cross-stream and vertical dimensions relative to the along-stream axis (Webster *et al.* 2001). Here, intense odour filaments surrounded by clean fluid may persist as they are transported downstream. From a temporal perspective, the filamentous nature creates substantial fluctuations. Although odour concentration decreases with downstream and cross-stream distance from the source, long time periods frequently are required to resolve these gradients, particularly in the along-stream dimension (Webster and Weissburg 2001). Greater turbulence produces more mixing of chemical plumes with the surrounding fluid, causing the plume to become more widely dispersed and diluted. This enhanced mixing also increases the homogenization of the plume so that average gradients are more easily discerned.

Unfortunately, quantities like Re and Re* are of limited utility, particularly outside of the laboratory, where the objective is to create well behaved and reproducible flows. The relationships between these descriptors, the degree of fluid mixing, and the corresponding odour dynamics are extremely general. The relationship between the degree of turbulence and Re (or Re*) is heavily dependent on the geometry of the flow (Vogel 1994), which can be a problem even in simplified settings. Waves, complex surface features or topography, tides and so on all influence turbulent mixing in near shore coastal environments in ways that are not accurately captured using measures such as Re and Re*. Similar problems hold for most stream systems, where large flow obstructions and local variation of sediment size (fine sediments to large cobbles) produces non-equilibrium boundary layers (Hart and Finelli 1999) not desirable in lab settings. For that reason, investigators hoping to link environmental properties with chemosensory navigation have begun to utilize fluid dynamical measures more directly associated with turbulence (see Box 7.1). Turbulence is often defined in relation to velocity fluctuations, particularly since correlated fluctuations of different velocity components (e.g. along-stream and cross-stream) are indicative of coherent fluid structures (i.e. eddies) that create mixing. Thus, the root mean square (RMS, i.e. standard deviation) of the velocity component is possibly a more useful measure than Re or Re*.

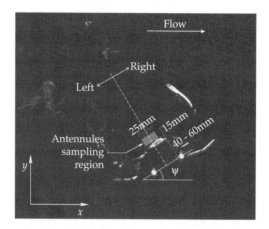

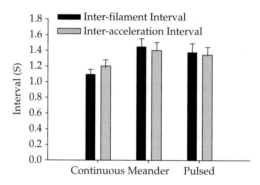

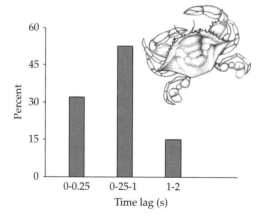

Figure 7.1 The effect of instantaneous odour filaments on upstream movements in foraging blue crabs revealed by simultaneous odour and animal visualization. Top panel shows odour filaments impinging on the region sampled by the crab's antennules. The two lights on the animals are markers for visualization, and yield the coordinates of the animal and body angle, ψ. Middle panel shows the relationship between odour pulse arrival and inter-acceleration intervals (mean ± std err) for three different plume conditions. The data indicate that the interval between odour filament arrivals corresponds to the interval between episodes of upstream acceleration, showing the perception of individual odour filaments produces upstream movement. Continuous refers to a plume released isokinetically into the flow, Meander is the same condition where an upstream cylinder was used to cause large scale cross-stream movement of the plume, Pulsed represents a puff of odour released at approximately 1 Hz. The bottom panel shows that crabs generally accelerate upstream within 1 s of when odour filaments reach the antennules (animals are not accelerating at the time of filament contact). Top panel from Dickman *et al.* (2009). Middle and bottom panels taken from data in Page (2011a), with permission.

Box 7.1 Calculating flow properties

Turbulence is often expressed as the RMS of the largest velocity component, that is, the component aligned to the bulk flow direction. More comprehensive calculations use all velocity components, such as in the calculation of turbulence intensity, TI:

$$TI = \sqrt{(RMS_u^2 + RMS_v^2 + RMS_w^2)/u^2 + v^2 + w^2}$$

Where u, v, and w represent along-stream, cross-stream, and vertical velocity components, respectively. These correspond to a standard x,y,z coordinate system, where the x axis aligns in the downstream direction, y is the cross-stream, and z the vertical dimension.

A related term is the Turbulent Kinetic Energy (TKE), which is given as:

$$\tfrac{1}{2}\left((RMS_u^2) + RMS_v^2 + RMS_w^2 \right)$$

Whereas TI is a measure of the magnitude of velocity fluctuations relative to the mean RMS, TKE is the mean kinetic energy per unit mass associated with eddies in turbulent flow. TI and TKE are closely associated, especially since u is often the dominant component of the flow, with v and w commonly being lower by at least 10-fold. Although these terms are all related, they vary independently across environments and there currently is no consensus as to which measure of fluid turbulence may be the most explanatory or predicative of chemical signal structure.

Emergent structures or complicated bottom topography can change the turbulence patterns so that comparisons across differing sites may be more accurate when all velocity components are represented either as TKE or TI. Note that these quantities will change with the vertical measurement location as a result of boundary layer flows (Schlichting, 1987). This is of particular concern for characterizing the signal environment of animals that move along the bed, since they are in the boundary layer region where flow velocity and turbulence change rapidly with vertical distance. Thus, it is important to perform measurements at heights that correspond to the chemosensory sampling organs of the tracking animal.

7.2 Navigational strategies

Considerable effort has been devoted to determining how animals extract distance and directional information from turbulent odour plumes. Most of these studies have involved larger organisms, particularly brachyurans, palinurids, and fish (e.g. Baker *et al.* 2002; Finelli *et al.* 2000; Horner *et al.* 2004; Jackson *et al.* 2007; Moore and Grills 1999). These animals move at velocities of the order of 10 cm s^{-1}, and are generally large relative to the cross-stream expanse

of the plume. Thus, they are constrained to sample the plume relatively coarsely in time, but may effectively sample the spatial distribution of odours across the main plume axis. Analysis of the temporal and spatial information in turbulent plumes, such as those emanating from discrete prey items, suggests that the time required to form accurate estimates of the local average concentration (at a given location) considerably exceeds that required for search (Webster and Weissburg 2001). The situation is similar for other time-varying parameters (e.g. variance, odour burst length, or concentration, etc.). In contrast to the difficulty of using temporal sampling to estimate average properties, spatial sampling may provide information on position relative to the plume relatively quickly (Atema 1996; Webster *et al.* 2001). Large animals may be able to position their sensors such that they experience considerable differences in odour stimulation as a result of being near or far away from the centre of the plume. Such differences are resolvable within a second or less under conditions typical of odour release from a discrete source (Atema 1996; Webster *et al.* 2001).

Numerous studies suggest that successful navigational strategies utilize the coarse temporal sampling to cue upstream movement in response to odour perception, with flow direction providing the information on the general direction to the source (i.e. upstream) (Baker *et al.* 2002; Weissburg and Zimmer-Faust 1993, 1994; Zimmer-Faust *et al.* 1995). The use of odour to elicit upstream movement occurs in conjunction with spatial sensing to detect signal contrast. Animals use differences in odour signal intensity perceived by different sensor populations (e.g. across legs on different body sides) to steer the animal towards the plume, by moving in the direction of the more strongly stimulated sensors (e.g. Page *et al.* 2011b).

Studies of aquatic chemosensory navigation consistently indicate that flow is necessary for successful target location. Behavioural studies indicate crustaceans, elasmobranchs, and gastropods require flow to locate odour sources (Brown and Rittschof 1984; Hodgson and Mathewson 1971; McLeese 1973; Weissburg and Zimmer-Faust 1994). In other cases, abolishing sensory input from mechanosensors disrupts olfactory orientation (Baker *et al.* 2002; Wyeth and Willows 2006). Simulations of navigation in realistic conditions generally require the use of both flow and odour information to produce *in silico* behaviour consistent with what we see in animals (Montgomery *et al.* 1999; Weissburg and Dusenbery 2002). These observations indicate that flow is a collimating stimulus that acts in conjunction with odour detection to produce movement upstream and towards the source. The benefit of utilizing both sensory mechanisms is intuitively obvious; odour is transported downstream so that going against the flow moves the searcher generally towards the source. Additionally, strategies overly reliant on odour may cause animals in turbulent conditions to follow spurious signals, such as chasing odour containing eddies that spin away from the main axis of the plume (Weissburg and Dusenbery 2002). Note that the coupling of odour information with flow permits a variety of responses that enable animals to narrow down their search space in order to acquire or reacquire contact with a plume. As described by Stabell (Chapter 6), salmon displaced upstream from their original location turn downstream in search of home. Foraging crustaceans often move cross-stream during a search in order to increase their chance of intercepting an odour plume (Weissburg and Zimmer-Faust 1993), which is analogous to the cross-wind casting seen

when moths lose contact with airborne plumes (Vickers and Baker 1996).

New evidence suggests that (similar to moths), individual odour filaments elicit upstream surges, and that sustained and rapid upstream progress is a function of high filament arrival rates (Fig. 7.1; Page *et al.* 2011a; Mead *et al.* 2003). In blue crabs, *Callinectes sapidus*, odour filament detection is binary with upstream movement produced in response to odour filaments with a supra-threshold concentration. Interestingly, thresholds are not fixed, but set relative to the prevailing average conditions. In addition, animals also encode something about the modal frequency of odour encounters in a plume specific way; whereas (relatively) long periods between odour arrival causes intermittent movement, (relatively) short intervals causes arrestment, presumably because animals interpret high stimulation frequency as indicative of being close to the source (Page *et al.* 2011a). Relative thresholds and frequency sensitivity and binary encoding are likely responses to the substantial variation in plume signal intensity. The average concentration experienced by animals varies by 10-fold even under the same physical conditions. Odour filament concentrations experienced by individuals may vary over 5 orders of magnitude (Page *et al.* 2011a).

With respect to steering, most observations indicate that navigating animals perform numerous small angle (0–45°) turns distributed symmetrically around 0° (e.g. Baker *et al.* 2002; Moore and Atema 1991; Weissburg and Zimmer-Faust 1994; Wyeth *et al.* 2006; Chapter 1). This pattern remains largely consistent across a range of physical conditions; even in situations that include large eddies where the local flow vectors are not aligned with the general current direction (Page 2009). The conclusion is that flow sets the general movement towards the source, but that steering is with respect to odour and not flow. Large animals frequently move to the time-averaged borders of the plume (Jackson *et al.* 2007; Zimmer-Faust *et al.* 1995), crossing or straddling this 'edge' where signal contrast between sensors closer to and farther from the centreline is greatest. The preferential location near regions of maximal contrast during navigation is indicative of spatial sensing. Spatial sensing using signal contrast also is consistent with observations that lobsters subjected to unilateral removal of chemosensory-bearing antennules turn towards the side of the remaining antennules when stimulated with odour (Devine and Atema 1982; Reeder and Ache 1980).

Recent evidence from experiments with blue crabs indicates that the location of the 'center-of-mass' (COM) of oncoming odour filaments with respect to the animal's body axis is a directional cue underlying this behaviour (Figure 7.2; Page *et al.* 2011b). Blue crabs foraging in turbulent plumes move in the direction of the COM, that is, animals move left when the COM of the incoming odour signals is to the left of the animal's midline. Animals also respond to changes in the COM by turning. A transverse shift in the COM of ca. 20 mm or more elicits a turn towards the appropriate direction (i.e. in the direction of the change of the COM shift) within around 2 s (Page 2009). Similar turning behaviour and response time courses have been seen in foraging lobsters, *Homarus americanus*, where odour spatial distributions were monitored with point sensors at the approximate location of the cephalic sampling organs (Atema 1996). In blue crabs, the probability of a correct turn increases with the degree of shift. Unlike upstream movement, steering is governed at least partially by information encoded in odour concentration; the probability of correct turn in response to a given COM increases with signal concentration (Jackson *et al.* 2007; Page 2011b). Animals in more turbulent conditions range farther from the plume centreline and display a greater frequency of large angle turns (Jackson *et al.* 2007; Page, 2009). This is to be expected if animals use spatial sensing to steer relative to the plume, as greater turbulence causes greater dispersion, and results in odour filaments that spin off from the plume to provide false information. Dramatic course corrections then ensue when animals recontact the main axis of the plume.

A priori considerations suggest that navigational strategies for large and rapidly mobile predators may not be the same as those used by smaller and/or slow moving creatures (Weissburg 2000). Slower moving animals (i.e. average velocities on the order of 1 mm s^{-1}) may use mechanisms that depend on temporal sampling in order to discern whether they are moving towards the source, since the ability to resolve local average conditions is related to sampling frequency. A slow moving animal has more ability to sample local conditions relative to animals that move quickly through the plume, and might thus be able to get accurate estimates of odour intensity that can be used to move to the source. As noted, this is unlikely to be achieved by coarse temporal sampling of rapidly mobile animals, given the unpredictability of turbulent signals.

Tests are few, but kinematic analysis of odour tracking suggests that slow moving gastropods of the genus *Busycon* may utilize temporal comparisons of odour intensity (Ferner and Weissburg 2005). This strategy also would be useful for the freshwater snails discussed by von Elert (Chapter 2), particularly because lysed algae would release attractants very close to the bed in the so-called viscous region of the boundary layer (Weissburg 2000). This is a very low flow region where the lack of turbulence creates conditions with an expected smooth gradient and where temporal sampling would provide accurate information. The echinoderm *Asterias forbesi* tracks chemical cues easily (Moore and Lepper 1997), and their slow speed suggests they also may respond to average gradients as opposed to individual filaments, but the lack of experiments in controlled and quantified stimulus environments makes this a tentative conclusion. Note that because time averagers are not dependent on individual odour filaments, they are not strongly affected by turbulence should it occur. In fact, time averaging strategies can operate better under moderate levels of fluid mixing (Ferner and Weissburg 2005), possibly because mixing causes faster convergence of average conditions; less sampling is therefore required to discern local odour plume properties.

The use of spatial sensing requires that the animal's sensory span (e.g. the distance between the chemosensors on the walking legs of a crab) be large relative to the scale of variation of the plume (Jackson *et al.* 2007; Webster *et al.* 2001). Thus, sampling widely enough to discern concentration differences is more difficult when animals are small. Insects such as moths face a similar problem, which they solve by using an endogenous internal mechanism to turn alternately left vs. right upon successive contact with odour filaments (Vickers 2000). This may be part of the repertoire of small aquatic animals as well, although direct tests are lacking.

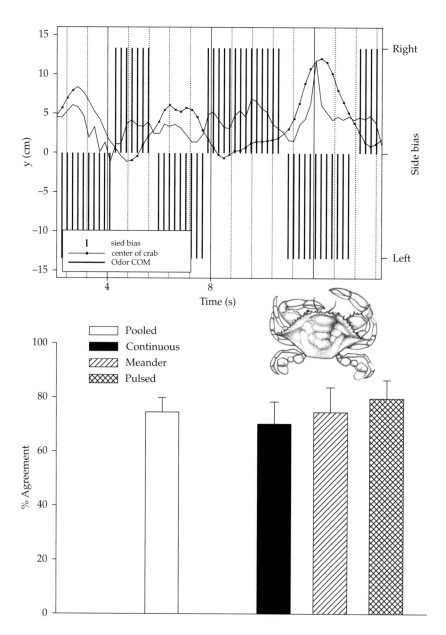

Figure 7.2 Relationship between odour filament spatial properties and turning behaviour. The top panel shows the X,Y coordinates of blue crabs navigating through plumes and the odour signal centre of mass (COM), obtained using simultaneous odour and animal visualization. Note that animals move in the direction of the side bias, which represents whether the signal is to the left or right of the animal's midline. Bottom panel shows the agreement between a turn and the location of the COM within one second prior to the turn. Each datum point represents the mean correct turn percentage for an individual animal path, yielding a mean correct turn percentage (± std err) over all animals in a given plume type, and pooled across all types. Continuous refers to a plume released isokinetically into the flow, Meander is the same condition where an upstream cylinder (10 cm dia) caused large scale cross-stream movement of the plume, Pulsed represents a puff of odour released at approximately 1 Hz.Top panel from Dickman et al. (2009). Bottom panel taken from data in Page (2011b), with permission.

The extent to which animal size and scale determines navigational strategies remains unclear, and resolution of this question awaits a broader and more systematic survey of different types of organisms. Weissburg (2000) suggests non-dimensional parameters, based on animal size, speed, sampling frequency, and scales of plume spatial variation and size, that represent a rough guide to sensory mechanisms. One conclusion of this analysis is that nearly all studies of navigational behaviour in multi-cellular organisms occur with animals that move rapidly and hence are expected to sample poorly in time (e.g. blue crabs, lobsters, moths). That is, the chemosensory sampling frequency of crustaceans seems to be of the order of 0.5–4 Hz and average movement speeds are of the order of 10 cm s^{-1}. A crab moving through a plume 1 metre long therefore obtains only limited information in time, and this level of sampling is insufficient to resolve the average concentrations in highly variable plumes (Webster and Weissburg 2001).

Studies on rapidly moving chemosensory foragers have extensively examined animals that are either large or small relative to the plume width, which are represented by the well-studied aquatic and terrestrial arthropods, respectively, as well as more limited work on fish. As noted, studies on slow moving creatures are advancing at a glacial pace, with only a handful of kinematic studies on gastropods and echinoderms.

Using non-dimensional parameters to relate biological to physical time and space scales provides a framework for analysis so that very different animals may be compared. In this respect, the goal is similar to that sought by the purely fluid dynamical non-dimensionalizations that inspired this approach; it is the relationship among a series of variables that is important for describing the system and which is encapsulated in the appropriate non-dimensional parameter. Although this approach has proved useful, it is not clear whether the length and time scales that have been suggested are universally appropriate. For instance, new observations suggest that olfactory navigation is common in procellariiform sea birds (Nevitt 2000). One suspects, given the vast travel distances and large scales, that these animals use time averaging to estimate odour concentrations and move to the source .

The plume parameters suggested by Weissburg (2000), which relate to very fine scale spatial variation, may not be appropriate here. A similar phenomenon occurs with respect to characterizing fluid flow, where different flow scenarios require different length scales for calculating Re (Vogel 1994). A variety of non-dimensional parameters have been used to examine general aspects of biological–physical interactions (Webster and Weissburg 2009), and this approach should be encouraged as a way to develop universal principles that are independent of biological particulars.

7.3 Ecological consequences

Odour guided navigation is a fundamental tactic for aquatic animals to find mates, prey, and habitats. Given that the prevailing physical conditions set limits on these abilities, the success of these endeavours will differ for animals in differing environments, with attendant ecological ramifications. Establishing connections between ecological processes, navigational abilities, and the physical environment is currently at the forefront of our investigations, with only a handful of studies on predator–prey interactions. For mate or habitat location, we have mainly been concerned with documenting the relevance of chemosensory information to finding mates or dwelling sites (Gleeson 1982; Horner *et al*. 2006; Ratchford and Eggleston 2000) as opposed to examining mechanisms of navigation and how they are influenced by the physical environment. Systems where we have good basic knowledge of the guidance mechanism for social signals (i.e. pheromones) are confined to moths (e.g. Vickers 2000) and copepods (e.g. Doall *et al*. 1998; Tsuda and Miller 1998). In contrast there are remarkably few studies on navigation to social odours and there exists a considerable opportunity to examine the ecological significance of the navigation process for these activities.

Early attempts to gain insights into predator–prey interactions from studies of navigation focused mainly on the fact that enhanced mixing diminishes the success of prey location, suggesting that turbulence constitutes a refuge from predation (e.g. Weissburg and Zimmer-Faust 1993). Although this general conclusion is still valid, its apparent simplicity is belied by at least two factors. The first is that, as noted above, the strategies of different predators may not be altered by mixing to the same extent. Second, an important factor not recognized in these early discussions is that predation may reflect bidirectional information gathering between predators and prey, which are potentially subject to the same environmental effects as their consumers. The predation imposed by a consumer in any given habitat will reflect the efficiency of predator navigation relative to the ability of prey to detect the predator and act to minimize predation risk; a maximally efficient predator will still go hungry if the prey detects the predator first and responds evasively.

Field studies examining the influence of the hydrodynamic environment on predation confirm that slow moving predators such as gastropods are not affected in the same way by turbulence as are rapidly moving predators such as crustaceans. Whereas high levels of fluid mixing reduces the success of blue crab foragers, whelks such as *Busycon*, that operate in similar environments, do not suffer the same deleterious effects (Ferner *et al.* 2009; Powers and Kittinger 2002). Whether or not high levels of fluid mixing constitute an effective refuge from predation seems dependent on the extent to which consumers employ a time-averaging strategy. This parallels the observations that, in contrast to the effects on crustaceans, turbulence does not diminish (and may actually enhance) navigational success of slow moving gastropods. Rapidly vs. slowly moving foragers may constitute pairs of predators for which prey defenses (i.e. refuge habitats) are antagonistic.

Many prey species assess predator presence using chemical cues from predators or injured conspecifics (Kats and Dill 1998). Prey that sense predators often change their behaviour to become more cryptic, which in the case of the bivalve *Mercenaria mercenaria* is accomplished by shutting down the siphonal pumping that releases the metabolites used by predators to navigate (Smee and Weissburg 2006). Bidirectional information gathering between predators and prey introduces substantial complexities into the relationship between mixing and predation success, and extrapolating field effects from the results of single-species investigations of chemosensory ability is not necessarily accurate when there is mutual sensing between predators and prey. For instance, the ability of *M. mercenaria* to sense blue crab or gastropod predators declines with increased mixing (Ferner *et al.* 2009; Smee and Weissburg 2006). The intensity and type of predatory effect is therefore a function of which species is better at detecting the other, as opposed to the acuity or detection distance of an individual species.

Field studies in which flow properties were manipulated and measured show the interactions between *M. mercenaria* and blue crab predators across a gradient in turbulent mixing. Patches of bivalves were placed in a variety of physical conditions in the estuary surrounding Skidaway Island, Georgia, and the fluid environment characterized by long-term measurements of turbulence intensity, TI. As noted above, bivalve prey, like their crustacean consumers, operate most effectively under conditions of low mixing. Increased mixing sets clear limits on the ability of prey to sense predators and respond appropriately in both the lab and the field (Ferner *et al.* 2009; Smee *et al.* 2008, 2010). These interactions produce the somewhat paradoxical result that predators locate and consume prey less frequently in the least turbulent environments, apparently because prey have a sensory advantage under these conditions (Fig. 7.3; Smee *et al.* 2010). Intermediate turbulence levels diminish the capabilities of both predators and prey, but shift the advantage to crab predators, which enjoy greatest predatory success under these conditions. Greater mixing again reduces predatory success, which reflects the further erosion of predator sensory abilities. The resulting hump-shaped function relating predatory success to turbulence indicates the changing interplay of sensory acuity between predators and prey across the environmental gradient. Note that the most effective environments for predators (low mixing) are not those that maximize predatory success, and that environments in which we expect predators to perform less well (intermediate mixing) produce the highest predation rates.

The environmental modulation of reciprocal predator–prey information gathering also has consequences for whether predators exert their effects via consumptive or non-consumptive mechanisms (Fig. 7.4; Smee *et al.* 2010). Prey that detect predators can change behaviour, life history, morphology, and

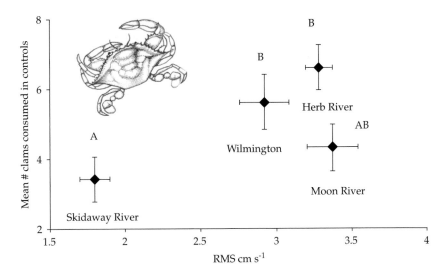

Figure 7.3   Relationship of flow and turbulence to predatory success of naturally foraging blue crabs. The number of clams consumed in plots of bivalves placed in the field was maximal at intermediate turbulence intensities as measured over several tidal cycles with an acoustic Doppler velocimeter. Numbers indicate mean flow velocity (± one std dev, which is equivalent to RMS turbulence), and letters indicate treatments that are significantly different based upon a 1-way ANOVA with Tukey-Kramer post hoc tests. From Smee *et al.* (2010).

so on, which often affects prey populations and other associated organisms as strongly as if the prey were consumed (Preisser *et al.* 2005). When *M. mercenaria* stop pumping upon detecting a predator, they also cease to feed, and predator presence (as opposed to consumption) reduces clam growth (Nakaoka 2000). Thus, although predators may not directly consume prey when turbulence is low, their presence may be detected by clam prey and result in substantial effects nonetheless. The effects of predators then shift to direct consumption as turbulence increases. Finally the total effect of predators declines when turbulence is sufficiently high to greatly affect the ability of consumers to find their prey. Note, though, that the previous scenario is contingent on the relative advantages that prey enjoy over their consumers at low mixing levels. If this advantage is not present, or if it is the consumer that has the advantage at low mixing levels, then there will be a different sequence of shifts between non-consumptive and consumptive effects across the environmental gradient.

The conceptual framework outlined by Smee *et al.* (2010) is related to the early seminal ideas on

how local environmental forces modulate activity levels and vulnerability of prey relative to their consumers (i.e. consumer and prey stress models; Menge and Sutherland 1987). Whereas those early models considered the capacity of the environment to inflict lethal or sub-lethal injury, the current framework is focused on 'sensory stress', or how the environment diminishes the capacity for information gathering. Turner has proposed a related idea, where the capacity of predators to affect prey behaviour via chemical signal transmission depends on the predator 'sphere of influence' (Turner and Montgomery 2003), which is a function of cue persistence and prey sensory acuity. The role of the physical environment in modulating chemical sensation to alter the overall effect of predators, and in shifting consumer pressure from consumptive to non-consumptive effects, will likely be an important area of research in the future. Like Turner and Peacor (Chapter 10), we submit that integrating studies of perceptual processes with approaches from population and community ecology is a powerful and rich synthesis that is both necessary and desirable.

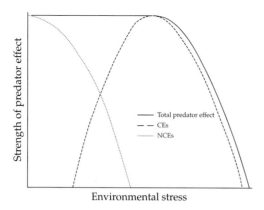

Figure 7.4 The interaction between the types of predator effects and level of environmental stressors. This model suggests that environmental stress (i.e. any physical parameter or process, such as turbulence, which can affect sensory perception) may alter predation intensity at stress levels that are not injurious, but that affect the ability of predator and prey to obtain information. In this model, consumptive effects (CEs) of predators on their prey are low in low-stress environments as these conditions maximize the ability of prey to avoid consumers. However, prey become more vulnerable to predators as stress increases and the CEs of predators increase, until stress exceeds a threshold that diminishes predator ability to forage. Non-consumptive effects (NCEs) occur when prey react to predators by altering their behaviour, morphology, and/or habitat selection. These effects are greatest when conditions maximize the ability of prey to detect and respond to predatory threats. From Smee *et al*. (2010).

7.4 Chemosensory guidance at different scales

Predators in the real world operate in a landscape in which there are chemical plumes at wide varieties of scales emanating simultaneously from different sources, and that convey information about both potential risks and rewards. This is a far cry from most prevailing laboratory simplifications of this process, and we may be ignoring some potentially significant ecological effects from these interactions.

First, large aggregations of animals may constitute a broad scent plume that attracts consumers and directs their attention to other odour sources. Foraging sea birds must certainly follow such plumes to locate plankton swarms from a great distance away (Nevitt 2000). Similarly, oyster reefs represent an enormous cloud of metabolites that may indicate a general direction towards resources, and organisms homing to those signals may cue in on individual prey items as they approach. Patches of

M. mercenaria are more vulnerable close to oyster reefs, and the chemicals emitted by live oyster reefs attract blue crab and whelk (*Busycon*) predators to the vicinity of nearby infaunal bivalves (Wilson, unpub. obs). Although the enhanced mixing produced by oyster reefs negatively affects chemosensory guidance, the chemicals emitted by the reef are strongly attractive (Wilson, unpub. obs) and the overall effect of the reef is to increase the vulnerability of nearby clams. Animals responding to a vague cloud emanating from one source (i.e. the reef) may therefore move into an area where they are able to discern individual prey items of another species (i.e. clams). We have little knowledge of how plumes of differing scale affect olfactory navigation, despite the fact that they may produce considerable indirect interactions among prey species.

Second, the existence of odour plumes at scales ranging from kilometres to microns is particularly challenging, since many organisms must surely track through plumes at a number of these scales to complete their navigation. Consider, for example, the homing and migration of salmon (Chapter 6) as they move through broad freshwater plumes, eventually to locate one particular stream within a drainage system. Larval coral reef fish sense the odour of their reef system during their planktonic phase to return to their home reef after dispersing many kilometres away (Chapter 5) and also may use odour signals to choose a specific settlement site. Thus, the chemical information used to accomplish a given task (e.g. locating a settlement site) can easily incorporate sensing plumes across scales of 3 orders of magnitude (100 m to 1 m) or more. In addition to integrated data on plume signal structure across these scales, we require knowledge of how the particular organism responds to features of different scales. Both because of our biases and our practical limitations, we tend to concentrate on animal responses to cm–m scale odour signals (but c.f. Hadfield and Koehl 2004; Nevitt 2000; Woodson *et al*. 2005, for studies at larger and smaller scales).

Third, animals navigating to food sources do so within a background of odours, including social signals (e.g. Horner *et al*. 2006; Kamio *et al*. 2008) as well as those from predators or injured conspecifics (Kats and Dill 1998), which indicate predation risk. The simultaneous perception of such cues may

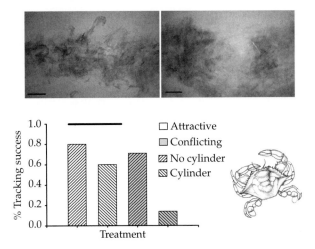

Figure 7.5 Turbulent mixing supresses the ability of crab foragers to distinguish attractive vs. aversive plumes. Top panel: Photographs illustrating mixing of dual plumes in 5 cm s⁻¹ flow (left) and in the same flow condition with a 10 cm dia cylinder upstream to create mixing (right). Odour sources were 8 cm apart with the origin approximately 25 cm upstream of the edge of the photo. Flow is from left to right. Note that the less mixed (left) plume shows a strong filamentous structure even far downstream. Where the two plumes intersect individual odour filaments are clear. Scale is 10 cm. Lower panel shows that enhanced mixing created by the cylinder suppresses the ability of crabs to discern an attractive plume (shrimp metabolites) when an aversive plume (injured blue crab metabolites) also is present. Attractive and Conflicting represent dual plume treatments, consisting of shrimp metabolites/asw and and shrimp metabolites/injured blue crab metabolites respectively. Crabs always track to the shrimp metabolites even in the presence of a cylinder (crabs were initally released 1.25 m from sources). However, crabs in the aversive condition do not track to the shrimp metabolite plume when mixing was increased by the cylinder. Numbers below the figure show the the results of a G-test. N = 12–15 trials per condition. See also colour Plate 3.

influence the ability and propensity to navigate, particularly if attractive cues (i.e. food or mating) are intermingled with aversive cues (i.e. a predator or injured conspecific). Aversive odours, when mixed with attractive food scents, often suppress feeding or food-searching (Ferner *et al*. 2009; Tomba *et al*. 2001; Vadas *et al*. 1994). However, in natural situations, odour from different sources can intermingle while still preserving distinct odour filaments from each source (Crimaldi and Browning 2004). With apologies to James Bond, useful signals are stirred rather than mixed. Experiments with insects indicate that these animals can, under the appropriate conditions, discern and react to attractive odour sources (conspecific pheromones) in the presence of inhibitory pheromone components (Fadamiro *et al*. 1999; Lui and Haynes 1992), if the two odours are released from different locations. Navigation to the attractive blend is suppressed when the release conditions (e.g. spacing of the two sources or their release frequencies) diminish the temporal and/or spatial discreteness of the filaments reaching the animal. Erosion of the structure

and blending of the different odour sources will occur naturally with sufficient levels of turbulent mixing. In fact, blue crabs will navigate in attractive odour plumes in the presence of aversive cues released from a different source, but this behaviour is suppressed when turbulence is enhanced (Fig. 7.5; M. Weissburg, unpub. obs). Thus, in nature, we expect that the fluid environment will interact with levels of aversive cues to govern navigation to food and possibly other attractive sources. This constitutes another opportunity to investigate the coupling between biological and physical processes in determining critical ecological activities.

7.5 Concluding remarks

Chemosensory guidance plays a significant role in determining the ability of aquatic creatures to find food, mates, and dwelling sites. Our understanding of the basic navigational mechanisms is sufficient to give us insights into where and when animals can use chemical signals to locate prey. The ability to generate such insights depends on information regarding the intensity of turbulent flow, as well as the relationship of animal temporal and spatial sampling abilities to the temporal and spatial scales of odour cue variation. Roughly speaking, animals that move quickly are probably restricted to sampling intermittently in time. Consequently, they require discrete, highly concentrated odour filaments in order to navigate efficiently, and generally are negatively affected by turbulent flow. Animals that sample slowly in time are better equipped to forage in turbulent conditions because this strategy does not force them to rely on discrete, concentrated odour filaments that are disrupted by mixing.

Understanding the interactions between animal sensory abilities, navigational performance, and environmental properties potentially can provide information on the impact of both consumptive and non-consumptive predator effects. It is quite likely that the intensity of these effects is altered across environmental gradients that influence the ability of predators to navigate effectively in turbulent chemical plumes. Studies that examine how predators with given sensory strategies affect prey in quantified field conditions provide a framework to anticipate how patterns of community regulation are shaped by the ability of predators (and prey) to gather information. Such studies require interdisciplinary techniques, but the success of other interdisciplinary programmes within chemical ecology argues for the value of these efforts. I remain confident that mechanistic studies of animal navigational mechanisms, combined with appropriate field studies, will link the inner and outer ecologies of animals that interact using information from fluid-borne cues.

References

Atema, J. (1996) Eddy chemotaxis and odor landscapes: exploration of nature with animal sensors. *Biological Bulletin,* **191**: 129–38.

Baker, C. F., Montgomery, J. C., and Dennis, T. E. (2002) The sensory basis of olfactory search beahvior in the banded kokopu (*Galaxias fasciatus*). *J. Comp. Physiol.A.* **188**: 533–60.

Brown, B. and Rittschof, D. (1984) Effects of flow and concentration of attractant on newly hatched oyster drills *Urosalpinx cinerea* (Say). *Marine and Behavioral Physiology,* **11**: 75–93.

Crimaldi, J. P. and Browning, H. S. (2004) A proposed mechanism for turbulent enhancement of broadcast spawning efficiency. *Journal of Marine Systems,* **49**: 3–18.

Denny, M. W. (1988) Biology and mechanics of the wave-swept environment. Princeton University Press, Princeton, NJ.

Devine, D. V. and Atema, J. (1982) Function of chemoreceptor organs in spatial orientation of the lobster, *Homarus americanus*: differences and overlap. *Biological Bulletin,* **163**, 144–53.

Dickman, B. D., Webster, D. R., Page, J. L., and Weissburg, M. J. (2009) Three-dimensional odorant concentration measurements around actively tracking blue crabs. *Limnology and Oceanography: Methods,* **7**: 96–108.

Doall, M. H., Colin, S. P., Strickler, J. R., and Yen, J. (1998) Locating a mate in 3D: the case of *Temora longicornis*. *Philosophical Transactions of the Royal Society of London B,* **353**: 681–89.

Fadamiro, H. Y., Cosse, A. A., and Baker, T. C. (1999) Fine-scale resolution of closely spaced pheromone and antagonist filaments by flying male *Helicoverpa zea*. *Journal of Comparative Physiology a-Sensory Neural and Behavioral Physiology,* **185**: 131–41.

Ferner, M. C., Smee, D. L., and Weissburg, M. J. (2009) Habitat complexity alters lethal and non-lethal olfactory interactions between predators and prey. *Marine Ecology Progress Series,* **374**: 13–22.

Ferner, M. C. and Weissburg, M. J. (2005) Slow-moving predatory gastropods track prey odors in fast and turbulent flow. *Journal of Experimental Biology,* **208**: 809–19.

Finelli, C. M., Pentcheff, N. D., Zimmer, R. K., and Wethey, D. S. (2000) Physical constraints on ecological processes: A field test of odor-mediated foraging. *Ecology,* **81**: 784–97.

Gleeson, R. A. (1982) Morphological and behavioral identification of the sensory structures mediating pheromone reception in the blue crab, *Callinectes sapidus*. *Biological Bulletin* **163**: 162–171.

Hadfield, M. G. and Koehl, M. A. R. (2004) Rapid behavioral responses of an invertebrate larva to dissolved settlement cue. *Biological Bulletin,* **207**: 28–43.

Hart, D. D. and Finelli, C. M. (1999) Physical-biological coupling in streams: The pervasive effects of flow on benthic organisms. *Annual Review of Ecology and Systematics,* **30**: 363–95.

Hodgson, E. S. and Mathewson, R. F. (1971) Chemosensory orientation of sharks. *Annals of the New York Academy of Science,* **188**: 174–82.

Horner, A. J., Weissburg, M. J., and Derby, C. D. (2004). Dual antennular chemosensory pathways can mediate orientation by Caribbean spiny lobsters in naturalistic flow conditions. *Journal of Experimental Biology,* **207**: 3785–96.

Horner, A. J., Nickles, S. P., Weissburg, M. J., and Derby, C. D. (2006) Source and specificity of chemical cues mediating shelter preference of Caribbean spiny lobsters (*Panulirus argus*). *Biological Bulletin*, 211:128–39.

Jackson, J. L., Webster, D. R., Rahman, S., and Weissburg, M. J. (2007) Bed roughness effects on boundary-layer turbulence and consequences for odor tracking behavior of blue crabs (*Callinectes sapidus*). *Limnology and Oceanography*, 52: 1883–97.

Kamio, M., Reidenbach, M. A., and Derby, C. D. (2008) To paddle or not: context dependent courtship display by male blue crabs, *Callinectes sapidus*. *Journal of Experimental Biology*, 211: 1243–48.

Kats, L. B. and Dill, L. M. (1998) The scent of death: Chemosensory assessment of predation risk by prey animals. *Ecoscience*, 5: 361–94.

Lui, Y.-B. and Haynes, K. F. (1992) Filamentous nature of pheromone plumes protects integrety of signal from background chemical noise in cabbage looper moth *Trichoplusia ni*. *Journal of Chemical Ecology*, 18: 299–307.

McLeese, D. W. (1973) Orientation of lobsters (*Homarus americanus*) to odor. *Journal of the Fisheries Research Board of Canada*, 30: 838–40.

Mead, K. S., Wiley, M. B., Koehl, M. A. R., and Koseff, J. R. (2003) Fine-scale patterns of odor encounter by the antenule of the mantis shrimp tracking turbulent plumes in wave-affected and unidirectional flow. *Journal of Experimental Biology*, 206: 181–93.

Menge, B. A. and Sutherland, J. P. (1987) Community regulation: variation, disturbance, competition, and predation in relation to environmental stress and recruitment. *American Naturalist*, 130: 730–57.

Montgomery, J. C., Diebel, C., Halstead, M. B. D., and Downer, J. (1999) Olfactory search tracks in the Antarctic fish *Trematomus bernachii*. *Journal of Comparative Physiology A*, 21: 151–4.

Moore, P. A. and Atema, J. (1991) Spatial Information in the 3-dimensional fine-structure of an aquatic odor plume. *Biological Bulletin*, 181: 408–18.

Moore, P. A. and Grills, J. L. (1999) Chemical orientation to food by the crayfish, *Orconectes rusticus*, influence by hydrodynamics. *Animal Behavior*, 58: 953–63.

Moore, P. A. and Lepper, D. M. E. (1997) Role of chemical signals in the orientation of the sea star *Asterias forbesi*. *Biological Bulletin*, 192: 410–17.

Murlis, J. (1986) The structure of odour plumes. In T. L. Payne, M. C. Birch, and C. E. J. Kennedy, (eds). *Mechanisms in Insect Olfaction*. Clarendon Press, Oxford.

Nakaoka, M. (2000) Nonlethal effects of predators on prey populations: Predator-mediated change in bivalve growth. *Ecology*, 81: 1031–45.

Nevitt, G. A. (2000) Olfactory foraging by Antarctic procellariiform seabirds: life at high Reynolds number. *Biological Bulletin*, 198: 245–53.

Page, J. L. (2009) The effects of plume property variation on odor plume navigation in turbulent boundary layer flows. PhD Dissertation, Biology, Georgia Institute of Technology, Atlanta.

Page, J.L., Dickman, B. D., Webster, D. R., and Weissburg, M. J. (2011a). Getting ahead: Context-dependent responses to odorant filaments drives along-stream progress during odor tracking in blue crabs. *Journal of Experimental Biology*, 214: 1498–512.

Page, J.L., Dickman, B. D., Webster, D. R., and Weissburg, M. J. (2011b). Staying the course: Chemical signal spatial properties and concentration mediate cross-stream motion in turbulent plumes. *Journal of Experimental Biology*, 214: 1513–22.

Powers, S. P. and Kittinger, J. N. (2002) Hydrodynamic mediation of predator-prey interactions: differential patterns of prey susceptibility and predator success explained by variation in water flow. *Journal of Experimental Marine Biology and Ecology*, 273: 171–87.

Preisser, E. L., Bolnick, D. I., and Benard, M. F. (2005) Scared to death? The effects of intimidation and consumption in predator-prey interactions. *Ecology*, 86: 501–9.

Ratchford, S. G. and Eggleston, D. B. (2000) Temporal shift in the presence of a chemical cue contributes to a diel shift in sociality. *Animal Behaviour*, 59: 793–99.

Reeder, P. B. and Ache, B. W. (1980) Chemotaxis in the Florida spiny lobster, *Panulirus argus*. *Animal Behavior*, 28: 831–9.

Schlichting, H. (1987) Boundary layer theory. McGraw-Hill, New York, NY.

Smee, D. L., Ferner, M. C., and Weissburg, M. J. (2008) Alteration of sensory abilities regulates the spatial scale of nonlethal predator effects. *Oecologia*, 156: 399–409.

Smee, D. L., Ferner, M. C., and Weissburg, M. J. (2010) Hydrodynamic sensory stressors produce nonlinear predation patterns. *Ecology*, 91: 1391–400.

Smee, D. L. and Weissburg, M. J. (2006) Claming up: environmental forces diminish the perceptive ability of bivalve prey. *Ecology*, 87: 1587–98.

Tomba, A. M., Keller, T. A., and Moore, P. A. (2001) Foraging in complex odor landscapes: chemical orientation strategies during stimulation by conflicting chemical cues. *Journal of the North American Benthological Society*, 20: 211–22.

Tsuda, A., and C. B. Miller. 1998. Mate finding in *Calanus marshallae* (Frost). *Philosophical Transactions of the Royal Society of London B*. 353: 713–720.

Turner, A. M. and Montgomery, S. L. (2003) Spatial and temporal scales of predator avoidance: Experiments with fish and snails. *Ecology,* **84**: 616–22.

Vadas, R. L., Burrows, M. T., and Hughes, R. N. (1994) Foraging strategies of dogwelks, *Nucella lapillus* (L.): interacting effects of age, diet and chemical cues to the threat of predation. *Oecologia,* **100**: 439–50.

Vickers, N. J. (2000) Mechanisms of animal navigation in odor plumes. *Biological Bulletin,* **198**: 203–12.

Vickers, N. J. and Baker, T. C. (1996) Latencies of behavioral response to interception of filaments of sex pheromone and clean air influence flight track shape in *Heliothis virescens* (F) males. *Journal of Comparative Physiology a-Sensory Neural and Behavioral Physiology,* **178**: 831–47.

Vogel, S. (1994) *Life in Moving Fluids.* Princeton University Press, Princeton.

Webster, D. R., Rahman, S., and Dasi, L. P. (2001) On the usefulness of bilateral comparison to tracking turbulent chemical odor plumes. *Limnology and Oceanography,* **46**: 1048–53.

Webster, D. R. and Weissburg, M. J. (2001) Chemosensory guidance cues in a turbulent odor plume. *Limnology and Oceanography,* **46**: 1048–53.

Webster, D. R. and Weissburg, M. J. (2009) The hydrodynamics of chemical cues among aquatic organisms. *Annual Review of Fluid Mechanics,* **41**: 73–90.

Weissburg, M. J. (2000) The fluid dynamical context of chemosensory behavior. *Biological Bulletin,* **198**: 188–202.

Weissburg, M. J. and Dusenbery, D. (2002) Behavioral observations and computer simulations of blue crab movement to a chemical source in controlled turbulent flow. *Journal of Experimental Biology,* **205**: 3397–8.

Weissburg, M. J. and Zimmer-Faust, R. K. (1993) Life and death in moving fluids: Hydrodynamic effects on chemosensory-mediated predation. *Ecology,* **74**: 1428–43.

Weissburg, M. J. and Zimmer-Faust, R. K. (1994) Odor plumes and how blue crabs use them to find prey. *Journal of Experimental Biology,* **197**: 349–75.

Woodson, C. B., Webster, D. R., Weissburg, M. J., and Yen, J. (2005) Response of copepods to physical gradients associated with structure in the ocean. *Limnology and Oceanography,* **50**: 1552–64.

Wyeth, R. C. and Willows, A. O. D. (2006) Odours detected by rhinophores mediate orientation to flow in the nudibranch mollusc, *Tritonia diomedea. Journal of Experimental Biology,* **209**: 1441–53.

Wyeth, R. C., Woodward, O. M., and Willows, A. O. D. (2006) Orientation and navigation relative to water flow, prey, conspecifics, and predators by the nudibranch mollusc *Tritonia diomedea. Biological Bulletin,* **210**: 97–108.

Zimmer-Faust, R. K., Finelli, C. M., Pentcheff, N. D., and Wethey, D. S. (1995) Odor plumes and animal navigation in turbulent water flow. A field study. *Biological Bulletin,* 188: 111–16.

CHAPTER 8

The taste of predation and the defences of prey

Linda Weiss, Christian Laforsch, and Ralph Tollrian

During the past century enormous progress has been made in the understanding of how predator–prey interactions are shaping ecosystems. We have obtained a broader knowledge of the effect of predation on cascading trophic interactions in food webs and recently it has become increasingly clear that in aquatic ecosystems such processes are immensely affected by chemical signals (Chapter 10). This is because prey organisms can receive and assess information on predation risk via chemical cues released by predators or con-specifics and change their behaviour, morphology, and/or life history accordingly, which is crucial for the survival of the individual organism and the population's fitness.

We here give examples of different modes of predator detection and describe a variety of defence mechanisms in marine and freshwater species that are induced by chemical cues. Subsequently, we highlight the progress that has been made in understanding the cellular, neuronal, and physiological mechanisms that underlie predator detection and defence structure development in a model organism, the crustacean zooplankton *Daphnia*.

8.1 Predation drives evolution of prey

Predation as a selection factor is a primary force driving the evolution of prey phenotypes and is thus acknowledged to be essential in shaping ecosystem structures. Predation events escalate along a series of steps starting with a search phase and initial prey encounter, detection, attack, capture, and finally prey ingestion (Fig. 8.1). Defence mechanisms have evolved for all these steps in prey

species. Many of these species-specific anti-predator responses start with the predator detection and identification, upon which specialized anti-predator responses are developed. The earlier the predator is perceived, the more likely is the individual's chance of survival. This predation cycle accounts for terrestrial and aquatic species, but in the following we will focus on predator detection in aquatic ecosystems with particular emphasis on freshwater sites (see Dicke and Grostal 2001 for terrestrial examples).

8.1.1 Modes of predator detection

Aquatic prey species have multiple senses that allow a reliable detection of predatory threats at different stages of the predatory cycle. Commonly one or a combination of the three major senses—mechanosensation, vision, and chemoreception—is required. A predator searching for food poses a *general* predation pressure on its prey. This *general* threat becomes an *acute* hazard during predator encounter, detection, and attack. Whereas a *general* predation risk is indicated by predator chemical cues, an *acute* predation event has to be visually or mechanosensory perceived. The combination of senses to assess predation and its current risk increases the accuracy of the elicited response. Mechanosensation performed by the sensory sensilla in arthropods or the lateral line organ of fish is used to detect an *acute* predator attack. Organisms can detect movements and vibration of organisms near them. Upon stimulation specialized escape reflexes may be elicited (Domenici and Blake 1997). Mechanoreceptors are commonly very sensitive

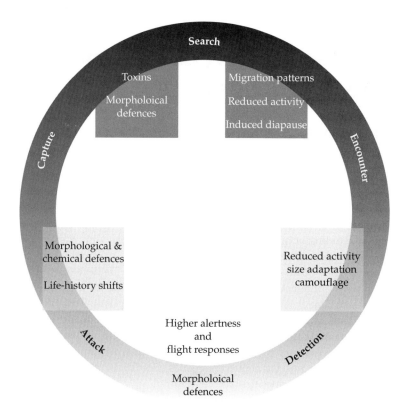

Figure 8.1 Predation cycle. The individual steps of a predation event start with the search for prey and escalate along a series of steps including: encounter, detection, attack, and capture. After successful prey ingestion this cycle repeats itself. A variety of defence mechanisms have been shown to operate at different stages of the predation cycle. Toxins may cause sickness and thereby decrease the motivation to search for prey. Morphological defences could lead to digestive handicaps. These defences cause a prolonged search phase, which is beneficial for the prey population rather than the ingested organism. Prey can decrease the chance of predator encounter by behavioural adaptations such as seeking shelter, adaptive migration patterns, or induced diapause. Camouflage is a prominent example of defence as it lowers the chance of detection by visual predators. Attacks can be defended through flight responses that are especially effective if the prey shows an increased state of alertness. Morphological defences can pose handling disabilities on the predator and thereby prevent a successful capture, e.g. gape-limitation of fish that are less successful in predating spined-morphs of *Daphnia lumholtzi*. Life history shifts towards an increase in somatic growth reduces predation risk through gape-limitation in *Chaoborus* predated *Daphnia pulex*. Illustration by L. Weiss (modified from Jeschke *et al.* 2009).

towards the slightest stimulation. Nevertheless, information on who or what is causing the water to move is limited. Distinction of predators from random water movement was shown in copepods (Hwang and Strickler 2001). These are able to estimate hydrodynamic water flows and respond to an approaching predator with a specific escape response. This pattern recognition is considered 'primitive' (Hwang and Strickler 2001) when compared to visual predator detection. Moreover mechanoreception is not reliable in the avoidance

of predation at an early stage of the predation cycle.

Vision is a light dependent sense and is thus used more frequently in terrestrial organisms for fast and reliable predator detection. Limnetic ecosystems are often marked by turbidity or unlit environmental conditions. Moreover, in some species vision is not used in environmental image formation (e.g. some insect larvae, zooplankton, molluscs; Dodson *et al.* 1994) suggesting that it is less suitable as a mode of *general* predator detection in these species as it impedes predator identification. During a predator attack this sense becomes more important and together with mechanoreception it may allow a successful escape.

Chemoreception is performed by the olfactory and gustatory systems. These evolutionary conserved senses comprise the major chemosensory pathways (Wisenden 2000; Firestein 2001) and are most important in the detection of predators through chemical signals. These chemical cues are water-soluble substances that allow aquatic organisms to communicate, interact, and orientate within their ecosystem. Such signals offer certain advantages. Assessment of chemical cues is common as they represent an authentic source of information, which can be transmitted over temporal and spatial distances (Tollrian and Harvell 1999). Furthermore, in contrast to visual or mechanical stimuli, chemical cues allow measurement of gradients of, for example, chemical cues that may be more highly concentrated in areas where predators are more abundant or reside longer. These areas can then be avoided by prey species. Chemosensory predator recognition is even more essential when predators are cryptic or display an ambush foraging strategy (Chivers et al. 1996b). The perception of chemical cues has the potential to perceive information about a *general* predation threat. This may put the prey organism into a certain state of alertness, a specific behavioural response, as well as elicit determined morphological and life history changes.

8.1.2   Chemical cues as indicators of *general* and *acute* predation threat

Chemical cues in the broadest sense can be understood as odours with different blends that are more or less specific for each type of predator and probably even predator species. They may for example originate from the predator itself or from predator-associated bacteria that release specific metabolites (i.e. kairomones of fish; Ringelberg and Van Gool 1998). At each step along the predation cycle (Fig. 8.1) chemical cues may be released and prey species have adapted to this most diverse set of chemical cues by forming explicit anti-predator defences. Frequently, the strength of the anti-

predator response increases with higher concentrations of the predator odour. Often trade-offs between predator avoidance and fitness values such as body size, lipid content, and fecundity are optimized.

The various chemical cues known up to now may contain specific or unspecific information about the potential predation risk. Conspecific alarm cues are released from damaged cells (e.g. club cells in fish) or tissues upon injury (Mathis et al. 1995; Chivers et al. 1996a). These alarm cues are commonly emitted during the last step of the predation cycle, that is successful capture and prey ingestion. However, it was speculated that without a distinctive predator cue being released, there is no information on the predator's activity level. Thus responses to conspecific alarm cues were shown to be weaker in comparison to responses of so-called predator-labelled cues (Schoeppner and Relyea 2009; Ferrari et al. 2010). Alarm cue labelling by conspecific-consuming predators may form a blend of chemicals that unequivocally indicates active predation.

Kairomones are defined as predator cues that are beneficial to the receiver but not to the sender in the context of information transfer. These chemicals might be advantageous to the emitter in a different context, for example as a pheromone in intraspecific communication. Thus, kairomones are not necessarily bound to any foraging behaviour. It is assumed that evolution of sensitivity to these cues is favoured if their production cannot be eluded, for example if they origin as metabolites from physiological processes. Otherwise coevolution acting on the predator would potentially have stopped their production. Likewise, an induced defensive response may be effective along several steps of the predation cycle and increases the chance of survival (Fig. 8.1; reviewed in Dicke and Grostal 2001; see Chapter 9).

In freshwater ecosystems a few reports exist that describe some of the chemical characteristics of these predator specific cues, whereas in marine systems this information is still scarce (Chapter 2), but urgently needed for a detailed understanding of the evolution of chemical communication as well as the mechanisms of predator detection and development of inducible defences.

8.1.3 Predator-induced defences

Predator-induced defence mechanisms are regarded as a key component shaping the complex interactions of food web dynamics by dampening predator–prey oscillations (Verschoor *et al.* 2004). Inducible defences are known from many different taxa and thus appear to be among the ecological factors that promote persistence and stability in multi-trophic communities (Vos *et al.* 2004). We will here define the different types of inducible defences and explain how they provide a reliable protection against predation risks (see Chapter 12).

Inducible defences are elicited at an early stage of the predation cycle, as soon as information about a *general* predation risk is available, and will reduce the risk of predator encounter. The defence may also be operational at later stages of the predation cycle, that is during an *acute* predation event. For example, defensive traits such as thorns have the potential to decrease the chance of successful capture. Inducible defences could be categorized as life history shifts, morphological adaptations, and behavioural responses. Life history refers to a set of evolved strategies that influence reproductive success. Changes of life history parameters often occur as resource allocations where somatic growth is traded with reproduction. This results in earlier maturation at a smaller body size and, as a consequence, resources that are saved on somatic growth can be invested in higher fecundity. Such trade-offs are common as no organism can be superior in all traits at all times. Inducible morphological defences can be seen in the development of spines, thorns, plasticity of tail shape, and body forms that are best adapted to thwart predators, whereas behavioural adaptations comprise predator avoidance strategies such as seeking shelter and a general decrease in activity and increased alertness.

The described categories of inducible defences can be deployed at several points in the predation cycle. A predator's search for prey poses a general predation pressure on all prey species in the community. Predator encounter can be decreased by behavioural anti-predator response and can be seen in *diel migration* patterns (e.g. *vertical*, *horizontal*, or *reverse* migration; Fig. 8.1; reviewed by Burks *et al.* 2002). *Diel vertical migration* is a behavioural pattern commonly found in lakes with a hypolimnetic refuge zone that provides shelter for zooplankton. Prey organisms exposed to fish kairomone reside in cooler and darker strata of the lake (hypolimnon) in the daytime and therefore reduce predation pressure of visually foraging predators. Active migration to the surface waters (epilimnon) is only performed during the night (Lampert 1993). This defence comes at the cost of a delayed maturation that depends on the decreased temperature at the bottom of the lake (Loose *et al.* 1994). Shallow lakes do not provide a depth refuge and therefore predator shelter is found here in the littoral zones of the lake by day (Burks *et al.* 2002). This diel horizontal migration has so far not been well documented in deeper, stratified lakes (Vetti Kvam and Kleiven 1995; Burks *et al.* 2002).

Other behavioural changes include periodicity adaptations where prey arrest their nocturnal behaviour to become erratically active in their search for food. This reduces the likelihood of predator encounter (Fig. 8.1; Pettersson *et al.* 2001), but comes at the cost of reduced feeding. Further, reduced swimming speed in the presence of ambush predators reduces the chance of being detected (Pijanowska and Kowalczewski 1997). The risk of predator detection can also be reduced by shifts in life history parameters. If somatic growth is traded for reproduction, a reduced body size with an increased number of offspring is acquired, which decreases the chance of being detected by visually hunting predators (Roff 1992; Fig. 8.1). On the other hand, changes in life history that result in the production of a decreased number of larger offspring may impose foraging difficulties in gape- limited predators (see Chapter 12).

Morphological structures in the form of extensions such as spines, enlarged body appendages, or strengthened shells and carapaces, become effective during prey attack and prey capture. Often these structures pose handling difficulties for the predator (Fig. 8.1). Although the development of morphological defences is costly they serve as advantageous defences against multiple predators (Laforsch and Tollrian 2004b) and at many different stages of the predation cycle.

8.1.3.1 Diversity of inducible morphological defenses

A great variety of inducible defenses have been reported from almost all taxa living in the marine and limnetic ecosystem. They are reported from unicellular organisms up to vertebrates. Grazing experiments with the bacteriovorus flagellate *Ochromonas* sp. showed that freely suspended cells of the freshwater bacterium *Flectobacillus* sp. were highly vulnerable to grazers, whereas induced filamentous cells were resistant to grazing (Corno and Jürgens 2006).

Protists such as *Euplotes* sp. develop winged morphotypes which effectively protect them from being eaten by the gape-limited predator *Lembadion* (Kusch and Heckmann 1992) or *Paramecium* (Altwegg *et al.* 2004). Rotifers are capable of defending themselves against *Asplanchna* by the formation of elongated spines (Gilbert 2009). Sponges, such as *Anthosigmella varians*, show morphological plasticity against angelfish by increasing the spiculae density when exposed to predation (Hill and Hill 2002) and corals, such as the Caribbean sea fan *Gorgonia ventalina* L., have specific defenses against fungal infections, macroalgae or *Millepora alciconis* overgrowth, resulting in an increased proportion of sclerites that reduce subsequent tissue damage (Alker *et al.* 2004). In molluscs inducible defenses against shore crab (*Carcinus maenas*) predation have been described, for example in the blue mussel *Mytilus edulis* and in the snail *Littorina* spec. These organisms respond to predation by producing a thicker shell and, in addition, *Mytilus* produces more byssal threads and thereby attaches more firmly to the substrate (Leonard *et al.* 1999). Nudibranch predation triggers rapid induction of defensive spines in bryozoans (Harvell 1984). Inducible defenses have also been described in crustaceans, such as the barnacles of the northern Gulf of California, which show a dimorphic shell shape against predatory snails (Lively 1986). Predation by the snail *M. lugubris lugubris* induces the development of a narrow operculum and/or a bent morph and this results in reduced predation rate by *M. lugubris lugubris*, which forages by ramming its shell spine into the barnacle operculum (Jarrett 2009).

In vertebrates, a prominent example of an inducible defense is found in the crucian carp, *Carassius carassius*, that attains a deeper body shape in the presence of the gape limited, piscivorous pike *Esox lucius* (Brönmark and Miner 1992). Lately, predator induced plasticity was also reported in the stickleback, *Gasterosteus aculeatus*. Sticklebacks exposed to predatory perch grew faster and developed an asymmetric morphology (Frommen *et al.* 2011). In a number of anuran species, tadpoles respond to predation of dragonfly larvae by increasing their tail depth and its pigmentation, possibly deflecting the attention of the predator attack towards the tail, providing a greater chance of escape after attack (Van Buskirk, and Relyea 1998; Relyea 2001).

8.1.4 Multi-predator environments

Most examples of inducible defences in prey species have been investigated in two-species interactions. Thereby, the costs and benefits of inducible defences against a specific predator with a specific foraging strategy were examined. However, in nature most environments bear threats from a diversity of predators with various foraging strategies. Multiple predators have significant impacts on prey, which cannot be predicted simply by summing the effects of single predator types (Sih *et al.* 1998). An inducible defence avoiding one predator could expose the organism to another predator and thereby increase predation pressure. For example, freshwater snails (*Helisoma trivolvis*) can either produce invasion-resistant shells for defence against water bugs or crush-resistant shells as defence against crayfish, but not both (Hovermann *et al.* 2005). Accordingly, individuals with an undefended morphology may be more successful in a multi-predator environment than individuals with the defended morphology. Consequently, to elucidate the accurate advantage of inducible defences in nature it is necessary to study defences in the context of all relevant predators within a habitat.

Over the past decade only a few studies have addressed inducible defence strategies in multi-predator environments (Sih *et al.* 1998; Relyea *et al.* 2003; Laforsch and Tollrian 2004b; Kishida and Nishimura *et al.* 2005). Defences against various predators include adaptations of life history parameters, behaviour, and morphology. These traits are deliberately adapted in order to provide defence towards several predation threats at the same time.

So far, four different strategies have been identified in multi-predator environments: the inducible defence is *directed against the predator posing the highest risk of mortality* (Sih *et al.* 1998). The inducible defence can serve as *a multi-tool for protection against several predators with different foraging strategies*. The additive benefits may increase the adaptive value and thus facilitate the evolution and persistence of this generalized defence (Laforsch and Tollrian 2004b). The inducible defence can be *fine-tuned and have traits that serve as protection against several predators*. Here, an additional trait may be present that acts as protection against several predators, which would not be formed in a single-predator environment (Lakowitz *et al.* 2008). *Conflicting defences* occur in situations where prey defences against one predator put the prey at greater risk of being killed by another predator and vice versa. Further, compensatory prey defences include an overall activity reduction, which does not occur if only one type of predator is present. This is accompanied with a decreased search for food and especially with a drastic reduction of mating activity. Therefore this type of defence replaces defensive behaviours that are observed against single predators (Krupa and Sih 1998; Sih *et al.* 1998).

Investigation of inducible defences in the light of multi-predation scenarios is a rather young subject of investigation when compared to the large number of reports that exist on inducible defences described in single-predator situations. Clearly, this scientific field deserves further investigation. Furthermore, it has to be considered that one premise for acting in such an environment is the capability to accurately perceive and distinguish between different predator types in a complex medium containing thousands of chemical cues.

8.2 *Daphnia* as a model organism for studies of the ecology and evolution of phenotypic plasticity

Determination of predator density and impact as well as adequate predator identification is crucial for the survival of prey organisms. Only an accurate assessment of the predation risk allows a forceful and cost-benefit optimized development of defensive traits. Therefore predator detection can be

regarded as a key process responsible for an increased chance of prey survival. Over the past few years attempts on the description of chemical cues have been made and research on the costs and benefits of inducible defences has further advanced. The waterflea *Daphnia* spp. has a particularly long history of ecological research. It is ecologically relevant as it forms the trophic link between primary producers and higher trophic levels in limnetic ecosystems. In this freshwater crustacean we find the ability to specifically react to environmental changes including variable predation risks. This allows the investigation of predator-induced life-history shifts, adaptive morphologies and behaviour within one taxon. Thus *Daphnia* has been acknowledged as a useful tool to investigate the ecology and evolution of inducible defences. Additionally *Daphnia* has a small size, short generation times, and a parthenogenic reproductive mode which allows breeding of multiple clonal strains in the laboratory. Subsequent to the publication of the *Daphnia* 'ecoresponsive' genome (Colbourne *et al.* 2011) this freshwater crustacean was announced as a model species in life science research (http://www.nih.gov/science/models). The genetic analysis that includes descriptions and comparison of genome structure and gene flow/genetic drift resulting from natural selection will provide a detailed understanding of adaptive evolutionary responses at the population level (Tollrian and Leese 2010).

However, for a detailed understanding of ecosystems and communities it is crucial to understand information exchange between individuals and elucidate the responses and pathways within individuals. Modern molecular tools that require genome data (e.g. estimation of differential gene expression) are timely and allow a first insight into cell proliferation processes and signalling cascades as well as the determination of target sites controlling the development of plasticity (Tollrian and Leese 2010).

8.2.1 Inducible defences in *Daphnia*

Distinct adaptive phenotypes have been described in *Daphnia* in response to the presence of various predator types (Figs. 8.2, 8.3) and *Daphnia* has therefore been acknowledged as a useful tool to investigate the ecology and evolution of inducible defences. Factors favouring the evolution of inducible defences include fluctuating predator populations (e.g. depending on seasonal changes) and costs that are associated with defences (Chapter 12). Defences are only formed when they are needed and costs for defences are saved when they are not essential (Tollrian and Harvell 1999; Auld *et al.* 2010). This can be seen in the well-described predator–prey relationship between *Daphnia pulex* and the phantom midge larvae *Chaoborus*, where seasonal changes in the abundance of *Chaoborus* explain the occurrence of alternative morph types in *Daphnia pulex* (Cyclomorphosis). Furthermore, a reliable cue, which indicates the level of predation risk, needs to be available. During active feeding *Chaoborus* larvae release kairomones, which elicit the formation of neckteeth in *Daphnia pulex*, in a life stage where they are most vulnerable to *Chaoborus* predation (Fig. 8.4). The chemical nature of this predator cue has been characterized as a small (<500 Da), heat-stable molecule carrying a carboxyl-group (see Chapter 2).

The response strength in *Daphnia* is positively correlated to the concentration of the kairomone and resembles the optimization of morphological costs with predation risk. Inducible defences need to be effective and increase the individual's chance of survival, and the neckteeth formed by *D. pulex* indeed appear to interfere with the feeding apparatus of *Chaoborous*. Handling disabilities in the heteropteran *Notonecta* sp. caused by morphological defences were also shown as a result of crest development in its prey, *Daphnia longicephala* (Grant and Bayly 1981). Other examples describe spine formation against planktivorous fish in *Daphnia lumholtzi* (Tollrian 1994) or crown of thorns development against *Triops* predation (Fig. 8.3; Petrusek *et al.* 2009). *D. cucullata* develop helmets that serve as a multi-purpose tool and protect from the predation of various predators, including *Chaobrous flavicans*, *Leptodora kindtii*, and *Cyclops* sp. (Laforsch and Tollrian 2004b, Fig. 8.2). In addition to these obvious defence structures, hidden morphological plasticity was also revealed in *D. pulex* and *D. cucullata*, which show a strengthening of their armour, providing a physical protection against mechanical challenges, when exposed to predator kairomones (Laforsch *et al.* 2004).

These factors explaining the evolution of morphological defences also account for changes in life

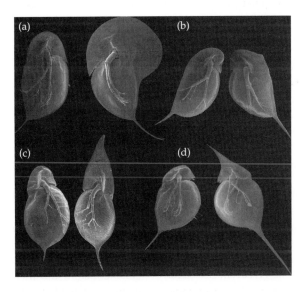

Figure 8.2   Inducible morphological defences in various *Daphnia* species. Scanning electron micrographs from undefended (left) and defended morphs (right) (a) *Daphnia longicephala*—crests act against *Notonecta* predation; (b) *Daphnia pulex*—neckteeth act against *Chaoborus* predation; (c) *Daphnia cucullata*—helmets and an increased tail-spine length act against *Chaoborus*, *Cyclops*, and *Leptodora* predation; (d) *Daphnia lumholtzi*—helmets and tail spines act against fish predation. Photograph by C. Laforsch.

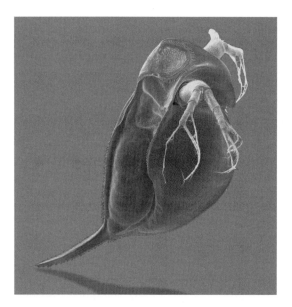

Figure 8.3 Scanning electron micrograph of predator induced *D. atkinsoni*. 'Crown of thorns' act against predation of *Triops cancriformis*. Photograph by C. Laforsch.

history and behaviour. Different strains of *Daphnia* show a large array of life history adaptations to vertebrate and invertebrate predators. These life history shifts are mostly considered to be adaptive and can be explained by prey size selectivity of predators (Marcotte and Browman 1986). For example, *Daphnia pulex* is capable of adapting its body size to *Chaoborus* and fish predation. *Chaoborus* larvae only prey on juvenile daphnids due to the gape limitation of their mouthparts and, consequently, an increase in body size leads to a 'size refuge' in *Daphnia*. Therefore, resources are shifted from reproduction to somatic growth (Tollrian and Dodson 1999). However, against fish predation, resources are shifted from somatic growth to reproduction (Stibor and Lüning 1994). This results in a smaller size at sexual maturity but an increased fecundity with more but smaller eggs. *Daphnia* clones that reproduce early at a smaller body size are less conspicuous and therefore will be favoured under fish predation (Fig. 8.4; Weber and Declerck 1997). Another adaptive life history defence under fish predation is the production of resting eggs (Ślusarczyk 1995). As resting eggs can pass the digestive tract, they allow a temporal escape.

In addition, predator-induced alterations in the behaviour of *Daphnia* have been reported. A pre-exposure to kairomones significantly increased survival rates of *Daphnia* when confronted with actively foraging invertebrate and vertebrate predators (Pijanowska *et al.* 2006) suggesting an increased internal state of alertness. Moreover, slower swimming speeds were reported for daphnids exposed to *Chaoborus*; a behaviour suggested to lower the chance of prey detection in the mechanoreceptive predator (Pijanowska and Kowalczewski 1997; Dodson *et al.* 1997).

The induction of *diel vertical migration* DVM and *diel horizontal migration* DHM by predator cues has been reported as another adaptive behaviour in *Daphnia* (Fig. 8.5; Tollrian and Harvell 1999; Loose *et al.* 1993; Dawidowicz and Loose 1992; Vetti Kvam *et al.* 1995). Whereas diel vertical migration describes the active migration of daphnids to the deeper and cooler strata of the lake to avoid fish predation, diel horizontal migration describes the migration to the littoral zones to avoid *Chaoborus* predation (Vetti Kvam *et al.* 1995; Wojtal-Frankiewicz *et al.* 2010). All the above-described types of inducible defences can only be formed upon predator perception and the ability to sense a predator is realized via chemical perception of kairomones.

8.2.2 Mechanisms of kairomone perception and information processing in *Daphnia*

Daphnia are living in an olfactory sea and they show tremendous ability to detect con-and heterospecifics by chemical communication in order to interact with their environment (Tollrian and Dodson 1999). Precise predator detection and identification is crucial for the development of a protective phenotype. Several studies (Miyakawa *et al.* 2010; Oda 2011) suggest that downstream of kairomone reception a series of biological reactions is initiated involving neuronal components, which are subsequently converted into endocrine signals. These induce changes in the expression of morphogenetic factors and thereby promote the development of defended morphs. However, the exact mechanisms are still elusive to a large extent. In the following paragraphs we will give a brief description of the sensory ecology of the *Daphnia* nervous

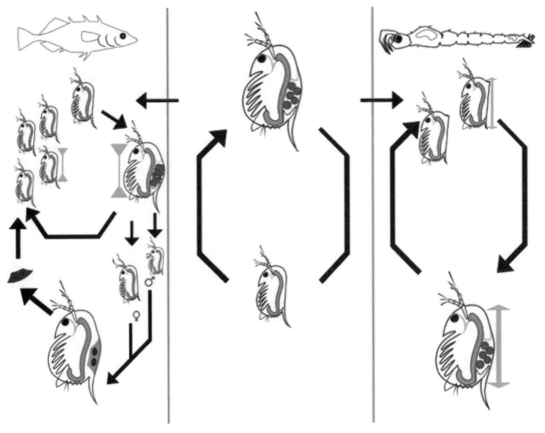

Figure 8.4   Reproduction cycle and inducible defence strategies in *Daphnia pulex*. In the absence of a predator stress (middle panel) *Daphnia* reproduce through parthenogenesis resulting in the production of female clones. In the presence of the predatory *Chaoborus* larvae (right panel) *Daphnia pulex* forms specialized morphological spines in the neck region, so called neckteeth. In addition reproduction is traded with somatic growth resulting in an increased body size but a decreased number of offspring. Under fish predation (left panel) resource allocation is shifted from somatic growth to reproduction, resulting in earlier sexual maturity at a smaller size and an increased number of smaller offspring. In addition resting egg production has been reported. Resting eggs are capable of passing through the fish's digestive tract and hence allow the survival of the clone. Illustration by L. Weiss.

system and discuss its involvement in predator–prey interactions.

8.2.3   Receptors and integration of chemical cues

Most crustaceans live in aquatic systems and therefore they rely on chemical cues in order to communicate with their environment. Sensory sensilla that contain sensory cells with chemoreceptors are distributed on all body parts, but are concentrated on the aesthetascs (Fig. 8.6). The neurons stimulated by chemoreceptors of the asthetascs project to the deutocerebrum of the central nervous system (Hallberg

et al. 1992). In cladocerans the modes of chemical reception and the locus of kairomone detection remain largely unknown.

The availability of the *Daphnia* genome allows insight into the mechanisms of kairomone detection; it allows a search for the presence of olfactory and gustatory receptor genes that are known from hexapod species. Blast searches in combination with EST tiling arrays and PCR amplification revealed the presence of 58 Grs that are encoded on the draft *Daphnia* genome sequence (Peñalva-Arana *et al.* 2009). *Daphnia*, like terrestrial hexapods, possess seven transmembrane domain (7 TMs) spanning gustatory receptors. This chemoreceptor

Figure 8.5 Diel vertical migration (DVM) in *Daphnia*. Vertical migration is induced by the presence of visual foragers such as fish in order to avoid predator encounter. *Daphnia* remain in cooler and darker water strata during the day and migrate to warm and nutritional water strata during the night. Illustration by L. Weiss and R. Tollrian. See also colour Plate 4.

superfamily consists of the gustatory receptor Gr family, which contains most of the protein diversity of the odorant receptor family and was found to be a novel development of terrestrial hexapod species. It is suggested that *Daphnia* chemically perceive the presence of a predator using the gustatory receptor type described, that is they 'taste' their predators (Peñalva-Arana *et al.* 2009).

From an evolutionary perspective olfaction and taste are old senses and common in other arthropod taxa and may also be present in *Daphnia* to some extent, that is it may have receptors responsible for the detection of odour/taste molecules. The different receptors show an affinity for a range of molecules and, conversely, a single molecule may bind to a number of olfactory receptors that show different affinities to the different cues. Upon receptor activation a G-protein coupled signal transduction is activated, which ultimately produces a nerve impulse. The integration of a signal is an interpretation of complex patterns of activated neurons in

higher brain areas. Signal integration allows the regulation and fine-tuning that underlies vital responses. With regards to predator detection this allows differentiation of various tastes or odours that may even have different blends, which are characteristic for kairomones. Moreover, stimuli resulting from multiple predators may require cross-talk between different signal transduction pathways in order to differentiate between several kairomones and evoke the correct defensive trait. However, the exact mechanisms of neuronal signal integration remain elusive.

8.2.4 Neuronal signalling

Several reports have documented that kairomone perception requires neuronal integration (Hanazato and Dodson 1992). Subsequently the cholinergic and gabaergic nervous system has been shown to be involved in the formation of inducible morphological defence structures (Barry 2002). Moreover, it

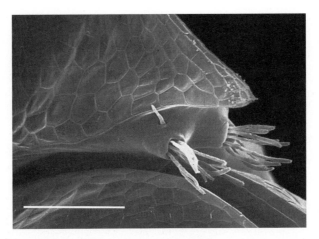

Figure 8.6   First antenna of *Daphnia magna*. The aesthetascs are carrying chemoreceptors that are suggested to be used for predator detection. Scale bar = 50μm. Photograph by C. Laforsch.

was shown that physostigmine modulates the response of kairomone stimulation in *Daphnia pulex* (Barry 2002). Physostigmine is an inhibitor of the enzyme acetylcholine esterase. This enzyme degrades and hence terminates stimulation of acetylcholine that is bound to the membrane at the postsynaptic side. An inhibition of the acetylcholine esterase thus results in a prolonged coupling of acetylcholine to its receptor and extends the stimulation of the postsynaptic membrane, resulting in increased postsynaptic spike rates. In *Daphnia pulex* this modulates the response and neckteeth expression is significantly increased (Barry 2002). Nevertheless, the kairomone was always a mandatory component and neckteeth expression could only be induced by physostigmine in the presence of the kairomone (Barry 2002). The application of GABA agonists that imitate the inhibitory gabaergic actions decreased the expression of neckteeth and at low kairomone concentrations may even have the ability to inhibit the expression of neckteeth completely (Barry 2002). Further, neuro-stimulatory substances were not able to elicit neckteeth expression in the absence of the kairomone. It was thus speculated that cholinergic stimulation is not the only component in this transduction cascade, but instead involves further stimulatory agents. In addition, the stimulation of the endocrine system was suggested to result in an enabling of the chitinase pathway to induce carapace appendages and/

or control ecdysone and juvenile hormones synthesis to promote moulting and maturity (Miyakawa *et al*. 2010; Oda *et al*. 2011).

8.2.5   Cellular components

Somatic polyploidy describes a mechanism including endopolyploidy and polyteny that involves increases in the quantity of nuclear DNA in somatic cells without subsequent cytokinetic divisions (Otto 2007). Polyploid cells are frequently described for invertebrates in playing important roles in physiological, developmental, and structural tasks (Gregory 2005). Due to increased DNA content they indicate regions of extremely elevated protein expression levels, and *Daphnia* from all genera that show the ability to perform inducible morphological defences appear to have polyploid cells (Fig. 8.7). Species that show the ability to form larger defence structures, such as crests or helmets, have an increased number of polyploid cells in comparison to species that produce smaller structures, such as neckteeth (Beaton and Hebert 1997). Polyploid cells were reported to control the mitotic activity of the surrounding cells (Beaton and Hebert 1997), which suggests that these cells serve as central control sites. The exact involvement of polyploid cells in this morphologically plastic tissue has yet to be established, but their spatial arrangement coincides with regions of active cell divisions. In *Daphnia*

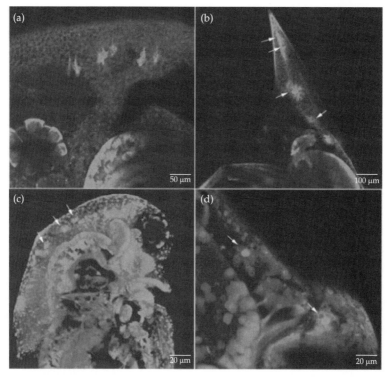

Figure 8.7 Confocal laser scanning microscope images of polyploid cells stained with nissl fluorescence. Polyploid cells (white arrows) are associated with morphological defence structures in predator exposed *Daphnia* species. (a) *Daphnia longicephala*, (b) *Daphnia lumholtzi*, (c) *Daphnia pulex*, (d) magnification of polyploid cells (white arrows) in *Daphnia pulex* associated with the neckteeth region. Photograph by L. Weiss. See also colour Plate 5.

longicephala (Fig. 8.7 A), for example, polyploid cells line the margin of the crest, whereas in *Daphnia pulex* (Fig. 8.7 C, D) polyploid cells are located below the region in which neckteeth are developed (Beaton and Hebert 1997). *Daphnia lumholtzi* has been shown to carry polyploid cells in its spine base and in the head spine (Fig. 8.7 B). The number of polyploid cells is rather species-specific and is not altered during the process of helmet or neckteeth development. This indicates that this 'physiological machinery' is running on standby and can be quickly turned on when needed. However, genetic products and the neuronal patterns organizing the formation of all forms of inducible defences still remain unknown.

8.2.6 Developmental time frames

Daphnia have developmental time frames in which defence formation occurs (Tollrian and Dodson 1999). In *Daphnia pulex* defences are formed in life stages where predation risk is high (first to fourth juvenile instar). During embryonic development there is apparently a sensitivity for kairomone stimulation (Imai *et al.* 2009) and significant morphological changes were observed in the last embryonic stage in helmet formation in *Daphnia cucullata* and in neck pedestal development in *Daphnia pulex* (Laforsch and Tollrian 2004a). In contrast, no morphological changes could be found during embryogenesis between induced and non-induced *Daphnia lumholtzi* and *Daphnia longicephala* as well as *Daphnia ambigua*. The strategies for initiating the defensive traits differ among *Daphnia* species depending on their natural predatory environment (Laforsch and Tollrian 2004a). Species that form large helmets require several moults before the helmet is fully developed, whereas the small neckteeth are formed more rapidly. However, there is always a delay between the occurrence of the cue and the functional formation of the defence. These lag times represent a disadvantage of inducible defences. Prey animals may compensate for this disadvantage by combining different defences with different lag times (Tollrian and Dodson 1999). Furthermore, it has been shown that reaction delays are overcome in situations where the maternal environment is a good predictor for the offspring's environment by the formation of transgenerational-induced defences (Agrawal *et al.* 1999).

8.3   Synopsis and future directions

A fundamental new topic of evolutionary ecology now focuses on the understanding of the genetic background of phenotypic plasticity. Sequencing of ecological model species genomes (such as *Daphnia*) will help to elucidate the interplay of changing environments and the effects on genome structures and gene regulation. Applying novel molecular tools will allow us to examine the nature of these adaptations and the genetic consequences at the individual and population level. This will yield information on the evolution of phenotypic traits. As we show here, predator detection and identification, which precedes the development of an adequate inducible defence, is crucial for an organism to increase its survival chances. Therefore, future challenges will lie in the identification of the physiological and metabolic mechanisms underlying phenotypic plasticity. Understanding the mechanisms of predator detection will provide a more detailed insight into the molecular processes of chemical communication. Elucidation of the adaptive physiology and metabolisms resulting from predator detection will provide knowledge of how cost-benefit-optimized responses are fine-tuned and how they are constrained. Nevertheless, in order to draw a complete picture of how chemical communication affects ecosystems, we still have a substantial need for the identification of the chemical nature of the information cues. Most kairomones are still only partly characterized. For the investigation of the pathways underlying the development of inducible defences this knowledge is crucial. As long as we do not know the chemicals responsible for these drastic physiological changes, we lack important information in all respects, ranging from the ecology of food web interactions and the shaping of ecosystem structures to the molecular level of predator detection and differential gene expression.

References

Agrawal, A.A., Laforsch, C., and Tollrian, R. (1999) Transgenerational induction of defences in animals and plants, *Nature*, **401**(6748): 60–3.

Alker, A.P., Kim, K., Dube, D.H., and Harvell, C.D. (2004) Localized induction of a generalized response against multiple biotic agents in Caribbean sea fans, *Coral Reefs*, **23**(3): 397–405.

Altwegg, R., Marchinko, K.B., Duquette, S.L., and Anholt, B.R. (2004) Dynamics of an inducible defence in the protist Euplotes, *Archiv für Hydrobiologie*, **160**(4): 431–46.

Auld, J.R., Agrawal, A.A., and Relyea, R.A. (2010) Re-evaluating the costs and limits of adaptive phenotypic plasticity, *Proceedings. Biological sciences/The Royal Society*, **277**(1681): 503–11.

Barry, M.J. (2002) Progress toward understanding the neurophysiological basis of predator-induced morphology in *Daphnia pulex*, *Physiological and biochemical Zoology: PBZ*, **75**(2): 179–86.

Beaton, M.J. and Hebert, P.D.N. (1997) The cellular basis of divergent head morphologies in *Daphnia*, *Limnology and Oceanography*, **42**(2): 346–56.

Burks, R.L., Lodge, D.M., Jeppesen, E., and Lauridsen, T.L. (2002) Diel horizontal migration of zooplankton: costs and benefits of inhabiting the littoral. *Freshwater Biology*, **47**(3): 343–65.

Brönmark, C. and Miner, J.G. (1992) Predator-induced phenotypical change in body morphology in crucian carp. *Science*, **258**(5086): 1348–50.

Chivers, D.P., Brown, G.E., and Smith, R.J.F. (1996a) The evolution of chemical alarm signals: attracting predators benefits alarm signal senders, *American Naturalist*, 148 (4): 649–59.

Chivers, D.P., Wisenden, B.D., and Smith, R.J.F. (1996b) Damselfly larvae learn to recognize predators from chemical cues in the predator's diet, *Animal Behaviour*, **52**(2): 315–20.

Colbourne, J.K., Prender, M.E., Gilbert, D.; Kelly, T.W., Tucker, A., Oakley, T.H., Tokishita, S. Aerts, A., Arnold, G.J., Basu, K.M., Bauer, D.J., Caceres, C.E., Carmel, L., Casola, C., Choi, J.-H., Detter, J.C., Dong, Q., Dusheyko, S., Eads, B.D., Fröhlich, T., Geiler-Samerotte, K.A., Gerlach, D., Hatcher, P., Jogdeo, S., Krijgsveld, J., Kriventseva, E.V., Kültz, D., Laforsch, C., Lindquist, E., Lopez, J., Manak, J. R., Muller J., Pangilinan, J., Patwardhan R.P., Pitluck, S., Pritham, E.J.,; Rechtsteiner, A., Rho, M., Rogozin, I.B., Sakarya, O., Salamov, A.,

Schaack, S., Shapiro, H., Shiga, Y., Skalitzky, C., Smith, Z., Souvorov, A., Sung, W., Tang, Z., Tshuchiya D., Tu, H., Vos, Harmjan, Wang, M., Wolf, Y.I., Yamagata, H., Yamada, T., Ye, Y., Shaw J.R., Andrews, J., Crease, T.J., Tang H., Lucas, S.M., Robertson, H.M., Bork, P. Koonin E.V., Zdobnov, E.M., Grigoriev, I.V. Lynch M. (2011) The ecoresponsive genome. *Science,* 331(6017): 555–61.

Corno, G. and Jürgens, K. (2006) Direct and indirect effects of protist predation on population size structure of a bacterial strain with high phenotypic plasticity. *Applied and environmental microbiology,* 72(1), pp. 78–86.

Dawidowicz, P. and Loose, C.J. (1992) Metabolic costs during predator-induced diel vertical migration of *Daphnia. Limnology and Oceanography,* 37(8): 1589–95.

Dicke, M. and Grostal, P., (2001) Chemical detection of natural enemies by arthropods: an ecological perspective. *Annual Review of Ecology and Systematics,* 32: 1–23.

Dodson, S.I., Tollrian, R., and Lampert, W. (1997) *Daphnia* swimming behaviour during vertical migration. *Journal of Plankton Research,* 19(8): 969.

Domenici, P. and Blake, R.W. (1997) The kinematics and performance of fish fast-start swimming. *Journal of Experimental Biology,* 200: 1165–78.

Dodson, S.I., Crowl, T.A., Peckarsky, B.L., Kats, L.B., Covich, A.P., and Culp, J.M. (1994) Non-visual communication in freshwater benthos: an overview. *Journal of the North American Benthological Society,* 13(2): 268–82.

Ferrari, M.C.O., Wisenden, B.D., and Chivers, D.P. (2010) Chemical ecology of predator-prey interactions in aquatic ecosystems: a review and prospectus. *Canadian Journal of Zoology,* 88(7): 698–724.

Firestein, S. (2001) How the olfactory system makes sense of scents. *Nature,* 413(6852): 211–18.

Frommen, J.G., Herder, F., Engqvist, L., Mehlis, M., Bakker, T.C.M., Schwarzer, J., and Thünken, T. (2011) Costly plastic morphological responses to predator specific odour cues in three-spined sticklebacks (*Gasterosteus aculeatus*). *Evolutionary Ecology,* 25: 1–16.

Gilbert, J.J. (2009) Predator-specific inducible defences in the rotifer *Keratella tropica. Freshwater Biology,* 54(9):1933–46.

Grant, J.W.G. and Bayly, I.A.E. (1981) Predator induction of crests in morphs of the *Daphnia carinata* king complex. *Limnology and Oceanography,* 26(2): 201–18.

Gregory, T.R. (2005) *The Evolution of the Genome,* Elsevier Academic, Burlington, MA.

Hallberg, E., Johansson, K.U., and Elofsson, R. (1992) The aesthetasc concept: structural variations of putative olfactory receptor cell complexes in Crustacea. *Microscopy Research and Technique,* 22(4): 325–35.

Hanazato, T. and Dodson, S.I. (1992) Complex effects of a kairomone of *Chaoborus* and an insecticide on *Daphnia pulex. Journal of Plankton Research,* 14(12): 1743.

Harvell, C.D. (1984) Predator-induced defense in a marine bryozoan. *Science,* 224(4655): 1357–9.

Hill, M.S. and Hill, A.L. (2002) Morphological plasticity in the tropical sponge *Anthosigmella varians*: responses to predators and wave energy. *The Biological Bulletin,* 202(1): 86–95.

Hoverman, J.T., Auld, J.R., and Relyea, R.A. (2005) Putting prey back together again: integrating predator-induced behavior, morphology, and life history. *Oecologia,* 144(3): 481–91.

Hwang, J.S. and Strickler, J.R. (2001) Can copepods differentiate prey from predator hydromechanically? *Zool. Stud,* 40(1): 1–6.

Imai, M., Naraki, Y., Tochinai, S., and Miura, T. (2009) Elaborate regulations of the predator-induced polyphenism in the water flea *Daphnia pulex*: kairomone-sensitive periods and life-history trade-offs. *Journal of experimental zoology. Part A, Ecological Genetics and Physiology,* 311(10): 788–95.

Jarrett, J.N. (2009) Predator-induced defense in the barnacle *Chthamalus fissus. Journal of Crustacean Biology,* 29(3): 329–33.

Jeschke, J.M. and Tollrian, R. (2000) Density-dependent effects of prey defences. *Oecologia,* 123(3): 391–6.

Jeschke, J.M., Laforsch, C., and Tollrian, R. (2008) Animal prey defenses. In Jorgensen S.E. and Fath B.D. (eds.) *General Ecology,* Elsevier, Oxford, pp. 189–94.

Kishida, O. and Nishimura, K. (2005) Multiple inducible defences against multiple predators in the anuran tadpole, *Rena pirica. Evolutionary Ecology Research,* 7(4): 619–31.

Krupa, J.J. and Sih, A. (1998) Fishing spiders, green sunfish, and a stream-dwelling water strider: male-female conflict and prey responses to single versus multiple predator environments. *Oecologia,* 117(1): 258–65.

Kusch, J. and Heckmann, K. (1992) Isolation of the Lembadion-factor, A morphogenetically active signal, that induces Euplotes cells to change from their ovoid form into a larger lateral winged morph. *Developmental Genetics,* 13(3): 241–6.

Laforsch, C., Ngwa, W., Grill, W., and Tollrian, R. (2004) An acoustic microscopy technique reveals hidden morphological defences in *Daphnia. Proceedings of the National Academy of Sciences of the United States of America,* 101(45): 15911–14.

Laforsch, C. and Tollrian, R. (2004a) Embryological aspects of inducible morphological defences in *Daphnia*. *Journal of Morphology*, **262**(3): 701–7.

Laforsch, C. and Tollrian, R. (2004b), Inducible defences in multi-predator environments: cyclomorphosis in *Daphnia cucullata*. *Ecology*, **85**(8): 2302–11.

Lakowitz, T., Brönmark, C., and Nyström, P. (2008) Tuning in to multiple predators: conflicting demands for shell morphology in a freshwater snail. *Freshwater Biology*, **53**(11): 2184–91.

Lampert, W. (1993) Ultimate causes of diel vertical migration of zooplankton: new evidence for the predator-avoidance hypothesis. *Ergebnisse der Limnologie ERLIA 6*, **39**: 79–88.

Leonard, G.H., Bertness, M.D., and Yund, P.O. (1999) Crab predation, waterborne cues, and inducible defenses in the blue mussel, *Mytilus edulis*. *Ecology*, **80**(1): 1–14.

Lively, C.M. (1986) Predator-induced shell dimorphism in the acorn barnacle *Chthamalus anisopoma*. *Evolution; International Journal of Organic Evolution*, **40**(2): 232–42.

Loose, C.J., Von Elert, E., and Dawidowicz, P. (1993) Chemically induced diel vertical migration in *Daphnia*: a new bioassay for kairomones exuded by fish. *Archiv für Hydrobiologie. Stuttgart*, **126**(3): 329–37.

Loose, C.J. a Dawidowicz, P. (1994) Trade-offs in diel vertical migration by zooplankton: the costs of predator avoidance. *Ecology*, **75**: 2255–63.

Marcotte, B.M. and Browman, H.I. (1986) Foraging behaviour in fishes: perspectives on variance. *Environmental Biology of Fishes*, **16**(1): 25–33.

Mathis, A., Chivers, D.P., and Smith, R.J.F. (1995) Chemical alarm signals: predator deterrents or predator attractants? *American Naturalist*, **145**(6): 994–1005.

Miyakawa, H., Imai, M., Sugimoto, N., Ishikawa, Y., Ishikawa, A., Ishigaki, H., Okada, Y., Miyazaki, S., Koshikawa, S., Cornette, R., and Miura, T. (2010) Gene up-regulation in response to predator kairomones in the water flea, *Daphnia pulex*. *BMC Developmental Biology*, **10**: 45.

Oda, S., Kato, Y., Watanabe, H., Tatarazako, N., and Iguchi, T. (2011) Morphological changes in *Daphnia galeata* induced by a crustacean terpenoid hormone and its analog. *Environmental Toxicology and Chemistry/SETAC*, **30**(1): 232–8.

Otto, S.P. (2007) The evolutionary consequences of polyploidy. *Cell*, **131**(3): 452–62.

Peñalva-Arana, D.C., Lynch, M., and Robertson, H.M. (2009) The chemoreceptor genes of the waterflea *Daphnia pulex*: many Grs but no Ors. *BMC Evolutionary Biology*, **9**: 79.

Petrusek, A., Tollrian, R., Schwenk, K., Haas, A., and Laforsch, C. (2009) A 'crown of thorns' is an inducible defence that protects *Daphnia* against an ancient predator. *Proceedings of the National Academy of Sciences of the United States of America*, **106**(7): 2248–52.

Pettersson, L.B., Andersson, K., and Nilsson, K. (2001) The diel activity of crucian carp, *Carassius carassius*, in relation to chemical cues from predators, *Environmental Biology of Fishes*, **61**(3): 341–5.

Pijanowska, J. and Kowalczewski, A. (1997) Predators caninduce swarming behaviour and locomotory responses in *Daphnia*. *Freshwater Biology*, **37**(3): 649–56.

Pijanowska, J., Dawidowicz, P., and Weider, L.J. (2006) Predator-induced escape response in *Daphnia*. *Archiv für Hydrobiologie*, *167*, **1**(4): 77–87.

Relyea, R.A. (2001) The lasting effects of adaptive plasticity: predator-induced tadpoles become long-legged frogs. *Ecology*, **82**(7): 1947–55.

Relyea, R.A. (2003) How prey respond to combined predators: a review and an empirical test. *Ecology*, **84**(7): 1827–39.

Ringelberg, J. and Van Gool, E. (1998) Do bacteria, not fish, produce 'fish kairomone'?. *Journal of Plankton Research*, **20**(9): 1847–52.

Roff, D.A. (1992) The evolution of life histories: theory and analysis. *Reviews in Fish Biology and Fisheries*, **3**(4):384–5

Schoeppner, N.M. and Relyea, R.A. (2009) Interpreting the smells of predation: how alarm cues and kairomones induce different prey defences. *Functional Ecology*, **23**(6): 1114–21.

Sih, S., Englund, E., and Wooster, W. (1998) Emergent impacts of multiple predators on prey. *Trends in Ecology and Evolution*, **13**: 350–5.

Ślusarczyk, M. (1995) Predator-induced diapause in *Daphnia*. *Ecology*, **76**(3): 1008–13.

Stibor, H. and Lüning, J. (1994) Predator-induced phenotypic variation in the pattern of growth and reproduction in *Daphnia hyalina* (Crustacea: Cladocera). *Functional Ecology*, **8**(1): 97–101

Tollrian, R. (1994) Fish-kairomone induced morphological changes in *Daphnia lumholtzi* (Sars). *Archiv fur Hydrobiologie*, 130, pp. 69–75.

Tollrian, R.T. and Dodson, S.D. (1999) Inducible defenses in Cladocera: constraints, costs and multipredator environments. In R.T. Tollrian and D.H. Harvell (eds), *The Ecology and Evolution of Inducible Defenses*, Princeton University Press, New Jersey, pp. 177–202.

Tollrian, R. and Harvell, C.D. (1999) *The Ecology and Evolution of Inducible Defences*, Princeton University Press, N.J.

Tollrian, R. and Leese, F. (2010) Ecological genomics: steps towards unravelling the genetic basis of inducible defences in *Daphnia*. *BMC Biology*, **8**: 51.

Van Buskirk, J. and Relyea, R.A. (1998) Selection for phenotypic plasticity in *Rana sylvatica* tadpoles. *Biological Journal of the Linnean Society*, **65**(3): 301–28.

Verschoor, A.M., Vos, M., and Van Der Stap, I. (2004) Inducible defences prevent strong population fluctuations in bi-and tritrophic food chains. *Ecology letters*, **7**(12): 1143–8.

Vetti Kvam, O. and Kleiven, O.T. (1995) Diel horizontal migration and swarm formation in *Daphnia* in response to *Chaoborus*. *Hydrobiologia*, **307**: 177–84.

Vos, M., Kooi, B.W., DeAngelis, D.L., and Mooij, W.M. (2004) Inducible defences and the paradox of enrichment. *Oikos*, **105**(3): 471–80.

Weber, A. and Declerck, S. (1997) Phenotypic plasticity of *Daphnia* life history traits in response to predator kairomones: genetic variability and evolutionary potential, *Hydrobiologia*, **360**(1): 89–99.

Wisenden, B.D. (2000) Olfactory assessment of predation risk in the aquatic environment. *Philosophical Transactions of the Royal Society of London. Series B, Biological Sciences*, **355**(1401): 1205–8.

Wojtal-Frankiewicz, A., Frankiewicz, P., Jurczak, T., Grennan, J., and McCarthy, T. (2010) Comparison of fish and phantom midge influence on cladocerans diel vertical migration in a dual basin lake. *Aquatic Ecology*, **44**: 243–54.

CHAPTER 9

The evolution of alarm substances and disturbance cues in aquatic animals

Douglas P. Chivers, Grant E. Brown, and Maud C.O. Ferrari

A major theme of this book is to understand the importance of chemical information in mediating interactions among aquatic organisms. Predation is one such interaction, in which chemical information is paramount. In the previous chapters, we saw evidence that many of the behavioural, morphological, and life history defences against predators were mediated by chemical information, including predator odours and chemicals released from injured or disturbed prey. In this chapter, we consider the evolution of these chemicals. The early work of Von Frisch (1941) elegantly demonstrated that European minnows (*Phoxinus phoxinus*) exhibited a marked fright reaction when exposed to chemical substances released from injured conspecifics. He coined the term '*Schreckstoff*' (German for 'fear substance') to describe the chemical cues responsible for inducing the fright response. Over the past 70 years, a great number of ecologists have conducted studies to identify such alarm substances among different groups of aquatic animals. These alarm substances have received a variety of names, including damage-released alarm cues, alarm cues, fear substances, injured conspecific cues, alarm pheromones, and chemical alarm signals. We will refer to these chemicals as 'alarm substances' or 'alarm cues' hereafter. Hazlett (1985) was the first to identify another class of chemical cues released by prey animals during a predator–prey encounter that likewise warned prey of danger. This group of chemicals, known as disturbance cues, are very different in that they do not depend on damage to be released from the sender.

The primary focus of much of the research on both alarm substances and disturbance cues has been to understand how these chemicals mediate predator–prey interactions, but the selection pressures leading to the evolution of such cues has received surprisingly little attention. Several reviews have summarized the role of alarm substances and disturbance cues in mediating anti-predator defences (e.g., Brown and Chivers 2005; Chivers and Smith 1998; Ferrari *et al.* 2010, Chapters 7, 8, 10 and 12 this volume). Our goal here is not to repeat this information, but rather to consider the selection pressures that led to the evolution of both classes of chemicals. In this chapter, we will first briefly discuss the taxonomy, ecology, and chemistry of alarm substances, followed by what is known to date on their evolution. We will highlight several hypotheses put forward to explain the evolution of alarm cues in a predation context (e.g. kin selection and secondary predator attraction), followed by recent findings suggesting that such cues may in fact have evolved as part of the innate immune system. We will try to include examples from a diversity of taxa, but of necessity will concentrate primarily on fish, given that almost all of the studies examining evolution are restricted to this group. The second part of the review will also briefly discuss the taxonomy and ecology of disturbance cues, and will encompass a brief section on their evolution. The differences in the mechanisms of release of alarm substances and disturbance cues may have considerable implications for how the cues evolved.

9.1 Alarm substances

9.1.1 The taxonomic distribution of alarm substances

Spurred on by the work of Von Frisch (1941), dozens of studies have demonstrated the existence of alarm substances in a wide diversity of taxa, including sponges, flatworms, annelids, molluscs, insects, crustaceans, echinoderms, fish, and amphibians (Chivers and Smith 1998; Ferrari *et al.* 2010 for reviews). Pfeiffer (1977) was among the first to systematically test different species of fish for the presence of alarm substances. The major conclusion of his work was that fear substances were restricted to members of the superorder Ostariophysi, a large group of fish (approximately 64% of all freshwater species), which includes the minnows, characins, catfish, loaches, and suckers. All of these species are known to possess large epidermal club cells, which contain the alarm cues (Fig. 9.1). There are no ducts leading from the cells to the exterior of the fish, hence the only way for the cues to be released is through mechanical damage to the skin (Smith 1992).

The last couple of decades have seen researchers identify similar epidermal cells in other groups of fish, including the poeciliids, percids, esocids, gobids, and cottids (see Ferrari *et al.* 2010 for a review). Yet, other groups of fish, such as the salmonids, respond to conspecific skin extracts with a dramatic anti-predator response, but lack the specialized epidermal cells that are found in the skin of Ostariophysan fish (Smith 1992).

9.1.2 The conservation of alarm substances

A common characteristic of alarm substances is that they appear to be evolutionarily conserved within phylogenetically related groups. Schutz (1956) first observed that cyprinid fish exhibit increased antipredator responses to the alarm substances of closely-related donors. In addition, he noted that the strength of the response was inversely related to taxonomic distance between the alarm substance donor and receiver. Similarly conserved responses have been found for the response to purine-N-oxides in cyprinids, characins, and siluriforms. Conserved responses have also been shown in non-Ostariophysan fish as well: centrarchids, poecillids, sticklebacks, and salmonids (Ferrari *et al.* 2010). In all cases, the general trend is an inverse relationship between the strength of the behavioural response and the degree of relatedness between alarm substance donor and receiver. More recently, Brown *et al.* (2010) have even shown population specificity in Trinidadian guppy (*Poecilia reticulata*) alarm substances. In all cases, the conserved responses to heterospecific alarm substances are independent of previous experience (i.e. not learned, Pollock *et al.* 2003). Such a degree of conservation may be attributed to the fact that alarm substance production may be under genetic constraints. Species which are

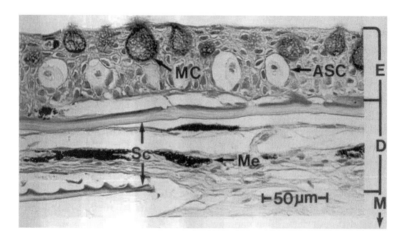

Figure 9.1 A cross-section of fathead minnow skin showing the epidermal (E), dermal (D), and muscle (M) layers. Club cells or alarm substance cells (ASC), mucous cells (MC), melanophores (Me), and dermal scales (Sc) are indicated with arrows. This section was stained with periodic acid-Schiff's reagent (PAS), and then counterstained with haematoxylin (H). Club cells are PAS-H negative and appear white with dark central nuclei, while mucous cells are generally PAS-H positive and appear dark (from Hugie 1990).

phylogenetically closely related are more likely to share common metabolic mechanisms and produce alarm substances that are chemically similar. As argued above for the Ostariophysan system (Kelly *et al.* 2006), alarm substances that are structurally similar should be readily detected by heterospecific receivers (Brown *et al.* 2010). Moreover, closely-related species should be expected to share similar life histories, habitat preferences, and foraging preferences. Recent studies have suggested that diet may have a significant effect on either the quality or quantity of alarm substance production (Brown *et al.* 2004b; Wisenden and Smith 1997) and this may contribute to the apparent conservation of alarm substances across closely related groups. Cross-species responses to alarm cues among closely related species have also been documented in other taxa (i.e. insects, larval amphibians, Ferrari *et al.* 2010).

9.2 The chemistry of alarm substances

The pioneering work of Pfeiffer *et al.* (1985) was the first to attempt to characterize the active ingredient of alarm substances of Ostariophysan fish. They concluded that the active ingredient might be hypoxanthine-3-N-oxide (Fig. 9.2), a small molecule characterized by a purine skeleton with a nitrogen-oxide (N-O) functional group at the '3' position. The N-O functional group, bound to the purine skeleton, was found to be the chief molecular trigger eliciting the behavioural response (Brown *et al.* 2000; Brown *et al.* 2003). Moreover, any purine skeleton with an N-O functional group appeared to elicit a behavioural response; purine molecules lacking an N-O functional group did not (Brown *et al.* 2000; Brown *et al.* 2003). Kelly *et al.* (2006) suggested that the alarm cues within Ostariophysian fish were actually comprised of species-specific blends of purine molecules and associated carrier proteins, which share a common N-O functionality. The so-called purine ratio hypothesis (Kelly *et al.* 2006) suggests that the well-documented conservation of alarm cues might be the result of species-specific differences in the relative proportions of carrier compounds associated with a common molecular trigger.

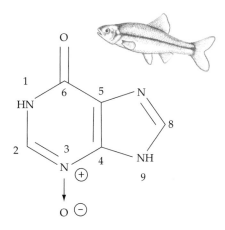

Figure 9.2 Hypoxanthine-3-*N*-oxide, alarm cue of Ostariophysan fish, with standard numbering scheme shown. Structurally similar compounds lacking the N-O functional group in the 3' position are ineffective at eliciting a behavioural response (from Brown *et al.* 2000).

Wisenden *et al.* (2009) have shown that one or more proteins may be required to elicit an alarm reaction in cyprinid fishes. Others have likewise shown that the active chemicals are proteinaceous molecules greater than 1,500 Da (Kasumyan and Ponomarev 1987; Lebedeva and Burlakov 1975). Clearly, we need more work on identifying the chemical nature of alarm substances. It is also important to note that despite the early stages of work examining the chemistry of the Ostariophysan alarm substance system, no work has been conducted to attempt to understand the chemistry of alarm substances of non-Ostariophysan fishes.

Dozens of studies have documented that tadpoles show behavioural responses to alarm substances from injured conspecifics (Ferrari *et al.* 2010). Fraker *et al.* (2009) attempted to identify the active component of alarm substances from the skin of ranid tadpoles. They argued that these chemicals may not actually require damage to be released, but rather may be actively secreted. In their experiment, tadpole extracts obtained from tadpoles previously euthanized with a deep-anaesthetic, Benzocaine, did not elicit anti-predator behaviour. They argue that the release of the chemicals is dependent on the animal being conscious and that deep anaesthesia results in the loss of consciousness. Moreover, they argue that the act of euthanizing tadpoles without anaesthesia necessarily involves capturing the animal and that this act can be enough to release the alarm cues. Fraker *et al.* (2009) went further to demonstrate that the alarm cues may comprise two polypeptides, each shorter than 10 kDa. While their results are consistent with the hypothesis that tadpoles need to be conscious for alarm substances to be released, another possible explanation is that Benzocaine interacts with the alarm substances, rendering them inactive. If consciousness is the key, then different methods of euthanasia (using chemical anaesthetics with different properties and action modes, or via cold shock or flash freezing) need to be tested. Further evidence is needed before we consider that the cues are actively secreted. If it turns out that this is the case, then the potential selection pressures leading to their evolution may be quite different (see below) than for alarm substances that require mechanical damage to be released.

9.3 The ecology of alarm substances

In recent years, there has been a wealth of research demonstrating the form and function of the behavioural response of prey individuals who detect an alarm substance (signal receivers). As this has been extensively reviewed elsewhere (Chivers and Smith 1998; Ferrari *et al.* 2010; Wisenden 2000), we will only briefly highlight some of this research. The detection of conspecific and heterospecific cues has been shown to provide reliable information regarding both the presence and intensity of local predation threats. Both laboratory and field studies showed dramatic short-term increases in species-typical anti-predator behaviour upon detection of conspecific and/or heterospecific alarm cues. In addition to simply eliciting a behavioural response, studies in several taxa have shown that the intensity of anti-predator response behaviour exhibited is proportional to the concentration of alarm substance detected (e.g., insects, fish, and amphibians, Ferrari *et al.* 2010). Conspecific alarm substances detected below the concentration needed to elicit an overt behavioural response, are known to provide information resulting in subtle changes in vigilance and foraging behaviour. For example, Brown *et al.* (2004a) exposed glowlight tetras (*Hemigrammus erythrozonus*) to 0.1 nM or 0.4 nM concentrations of H_3NO or a distilled water control and then exposed

them to the visual cue of an alarmed conspecific. While those exposed to the high concentration of H_3NO exhibited a strong increase in anti-predator behaviour, those exposed to the 0.1 nM concentration did not differ from a distilled water control. Upon detecting the visual cue, tetras exposed to distilled water exhibited a weak increase in predator avoidance, whereas those exposed to the sub-threshold concentration of H_3NO exhibited a significantly stronger increase in response to the visual cue. These results suggest that concentrations below the level needed to elicit a full-blown behavioural response may still provide valuable information and increase vigilance towards secondary sources of information and other potential threats.

Alarm substances have also been shown to play a critical role in the process of learned predator recognition (see Brown and Chivers 2005 for reviews; Ferrari *et al.* 2010). Many aquatic prey species do not show any evidence of innate recognition of novel predators, but rather must learn the chemical and/or visual cues associated with potential threats. A wealth of studies have shown that prey exposed to alarm substances paired with visual and/or chemical cues of a novel predator can subsequently recognize the predator as a threat. More interestingly, the intensity of risk associated with the learned predator is mediated by the concentration of alarm substances at the time of learning: if predator cues are paired with a high concentration of alarm substances, then the predator is subsequently recognized as a high threat. On the other hand, if the predator cues are paired with a low concentration of alarm cues, then the predator is recognized as a low threat. In other words, the intensity of response to the alarm substances dictates the risk intensity associated with the learned predator. In addition to recognizing novel predators, alarm substances can mediate the labelling of habitats as risky. For instance, a number of prey species avoid areas where alarm substances have previously been detected (Chivers and Smith 1998). As such, it is abundantly clear that the receivers of conspecific and heterospecific alarm substances are able to assess local predation risk and adjust their behavioural activities accordingly.

There is a growing body of evidence that predator avoidance elicited by the detection of an alarm substance increases the probability of prey surviving an encounter with a potential predator. For example, Mathis and Smith (1993) have shown that fathead minnows are significantly more likely to survive a staged encounter with a predatory northern pike when previously exposed to conspecific alarm substances. Similar enhanced survival benefits have been shown in other species (e.g., Mirza and Chivers 2001).

A common critique of studies examining the function of chemical alarm cues has been the fact that the bulk of earlier studies have been laboratory based. Magurran *et al.* (1996) argued that the observed response to alarm cues under laboratory conditions may reflect an artificially high level of perceived risk on the part of focal animals. Under laboratory conditions, prey individuals are typically not confronted with the ecological challenges of foraging, defending territories, or courting under the risk of predation, and such ecological challenges have been suggested to reduce the benefits associated with responding to alarm cues under natural conditions (Magurran *et al.* 1996). Hartman and Abrahams (2000) have argued that visual cues should be a more reliable cue for assessing local predation threats and that aquatic prey should only respond to chemical cues when visual cues are unavailable. However,

despite these critiques, a growing body of evidence shows that chemical alarm cues do indeed provide valuable threat-assessment information under fully natural conditions. For example, Brown and Godin (1999) demonstrated that wild Trinidadian guppies exhibited risk-aversive behaviour when approaching realistic predator models paired with guppy alarm cues versus the model paired with stream water. Using underwater video-cameras, both Wisenden *et al.* (2004) and Friesen and Chivers (2006) demonstrated that natural populations of several cyprinids and stickleback actively avoided areas labelled with conspecific alarm cues. Kim *et al.* (2009) and Leduc *et al.* (2010) have shown that juvenile Atlantic salmon (*Salmo salar*) exhibit strong antipredator responses to conspecific alarm cues in natural streams, under conditions of both high (daytime) and low (night time) visibility (Fig. 9.3). Finally, wild juvenile Atlantic salmon have been shown to exhibit a two-fold reduction in the size of their defended foraging territories in response to repeated exposure to conspecific alarm cues, even in streams with relatively high current velocities (~40 to 50 cm s^{-1}) (Kim *et al.* 2011a; Kim *et al.* 2011b). Together, these studies demonstrate that alarm substances function under a wide range of natural conditions.

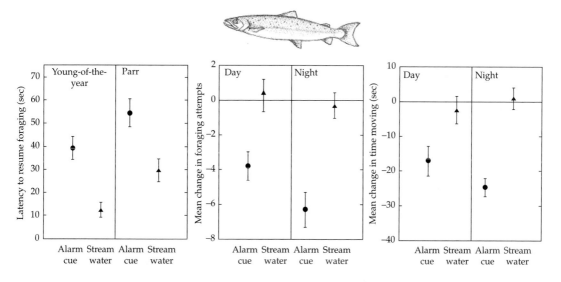

Figure 9.3 Anti-predator response of wild juvenile Atlantic salmon tested in natural streams. The left panel shows the mean (± SE) latency to resume foraging following exposure to conspecific alarm cues (circles) or stream water (diamonds) by two age classes of salmon. The centre and right panels show the mean (± SE) change in foraging attempts and time moving for young-of-the-year salmon tested under high visibility (day) and low visibility (night) conditions. Modified from Kim *et al.* (2009) and Leduc *et al.* (2010).

Prey animals can reduce their risk of predation not only by changing behaviour, but also by altering their morphology. Indeed, damage-release alarm cues have been shown to mediate changes in body morphology. For example, crucian carp *Carassius carassius* (Stabell and Lwin 1997) and goldfish, *Carassius auratus* (Chivers *et al.* 2008), alter their body length-to-depth ratio in response to alarm substances. Growing a deeper-body reduces the likelihood of capture by gape-limited predators. Alarm cues also induce adaptive changes in the shell morphology of molluscs (Hoverman and Relyea 2009).

Changes in life history characteristics are also mediated by exposure to alarm cues. For example, Chivers *et al.* (2001) demonstrated that frog embryos alter their hatching time when exposed to chemical alarm cues and snails exposed to crayfish feeding on snails are significantly larger and older at the age of first reproduction and at death (Crowl and Covich 1990). For a comprehensive review of behavioural, morphological, and life history defences of prey animals to alarm cues, see the recent review by Ferrari *et al.* (2010).

9.4 The evolution of alarm substances

While there are a great number of studies examining the ecological role of alarm substances, we know very little about the selection pressures leading to their evolution. Smith (1992) identified 16 potential hypotheses to explain the evolution of alarm and distress signals in animals (see Table 9.1). To date, only a handful of these hypotheses have received any empirical testing with respect to alarm substances. Many of the hypotheses rely on the fact that the prey can voluntarily release the signal. While this may be the case for disturbance signals which do not require mechanical damage for their release (see below), many of the hypotheses cannot be applied to damage-release cues because of the very nature of their release mechanism.

Table 9.1 Proposed benefits to the performer of conspicuous responses to predators, including releasing chemical alarm substances and disturbance cues. Taken from Smith (1992).

Selection	Receivers	Benefits to sender
Individual	None	Sensory assessment of predator
		e.g. ambush avoidance, visual ranging
Individual	Other prey	Reduce future attacks through reduced predator success in area and/or on prey type
		Initiate group crypsis
		e.g. silence or hiding, reducing chances of predator detecting group
		Initiate formation of tighter group (flock or shoal), improving survival chance of signaller
		Manipulate other prey into fleeing and distracting the predator
		Save group members or 'dear enemy' neighbours that benefit senders
		Save mates, avoid costs of mate replacement or brood loss
		Save offspring
		Save commensals, retain benefits of commensal relationship
Individual	Predator	Pursuit invitation: lure predator into unsuccessful attacks
		Pursuit inhibition: signal predator that attacks will be unsuccessful
		Confuse or surprise predator
		Aposematism: signal predator that prey is dangerous or unpalatable
		Injure predator
		e.g. toxic chemicals
Individual	2nd degree predator	Attract predators that may disrupt that may disrupt attack by first predator
Kin	Relatives	Increase survival of related individuals

9.4.1 Predation-centred hypotheses

9.4.1.1 Kin selection

Much of the early work surrounding alarm substances was completed during a time when group selectionist arguments about the evolution of behaviour were commonplace. The notion that minnows in schools or groups of tadpoles could release a warning chemical that would benefit other members of the group fit well within this context. However, with ethologists and ecologists moving away from group selectionist arguments, there was an immediate need to explain individual benefits provided by the alarm cues. Why would any given individual invest in producing specialized club cells that appeared only to function as a warning to others? Not surprisingly, the first argument for the evolution of alarm substances was related to kin selection. An individual would be selected if it released chemicals that would warn nearby relatives of danger. While this is at first an attractive hypothesis, our confidence in this hypothesis quickly wanes when we consider the many thousands of species of animals that possess alarm cues. It is unlikely that kin associations could be commonplace among such a large number of diverse species. There is mixed support for the idea that many species of fish school with close relatives. For example, Naish *et al.* (1993) demonstrated that the genetic variability within schools of European minnows was not lower than that among schools, as would be predicted if individuals were preferentially associating with kin. Conversely, Gerlach *et al.* (2001) have shown kin-structured subpopulations in Eurasian perch (*Perca fluviatilis*). Moreover, Olsen *et al.* (2004) show kin structured smolt emigration in Atlantic salmon (*Salmo salar*). There is a wealth of studies demonstrating the adaptive value of kin biased associations (Brown and Brown 1996). The conservation of guppy alarm cues within the populations described above (Brown *et al.* 2010) may in fact be due to higher than average relatedness among shoals. While it is unlikely that kin selection can explain the evolution of club cells in fish, the role of kin selection in the evolution of alarm cues (i.e. the functional response to conspecific alarm cues) deserves further study in other taxa. For example, larval amphibians are well-known to form close association with kin (Blaustein and O'Hara 1986). Even if kin selection is not the prime selective pressure leading to the evolution of alarm cues, it may play some role in the secondary maintenance of alarm cue function.

9.4.1.2 Secondary predator attraction

Mathis *et al.* (1995) were the first to provide an alternative to the kin selection arguments. However, their arguments moved away from the notion that alarm substances evolved as alarm signals to warn shoal-mates of danger. Indeed, they suggested that the intended recipients of the alarm signals were not nearby shoal-mates, but rather nearby predators. If predators were attracted to alarm cues released by captured prey, then they may attempt to pirate the meal of the first predator leading to an escape opportunity for the prey. If big enough, they may even attempt to eat the first predator. This secondary predator attraction hypothesis, analogous to predator distress calls in birds and mammals, relies on the fact that 1) alarm cues are attractive to predators and 2) that predators would fight over the prey leading to an increased probability that the prey would escape. Mathis *et al.* (1995) demonstrated that extracts of minnow skin containing alarm substances were more attractive to predatory pike than skin extracts of minnows lacking alarm substances (i.e. from breeding male minnows). Moreover, pike were attracted to hypoxanthine-3-N-oxide in the lab and predatory beetles were attracted to minnow skin in field studies. Wisenden and Thiel (2002) provided additional evidence for fish predators being attracted to alarm substances in the wild using bait with or without alarm substance-enhancement. The question of whether secondary predators would attempt to pirate prey from other predators was tested by Chivers *et al.* (1996). They showed that the arrival of a second predator after the first predator has already grasped its prey allows the prey to escape around 40% of the time (compared to 0% in trials without the presence of a secondary predator). This was the first study to provide evidence of a direct benefit to the sender of the signal. While this explanation provides some satisfaction, the frequency with which secondary predators exploit alarm substances is unknown and possibly uncommon.

Chivers *et al.* (2007) provided a critical test of the importance of predation pressure in determining alarm cell investment in fish. They speculated that if predation was the main selective force behind the evolution of fish alarm substances, then manipulating predation pressure should lead to changes in the investment of alarm cells. Fish raised in high predation environments should invest heavily in alarm cells while those raised in low predation environments should invest less. Of course, this assumes that individual fish can facultatively adjust their alarm cell number. The work of Hugie (1990) had established that there were population differences in club cell number, and the work of Wisenden and Smith (1997) had established that fish alter club cell investment in response to changes in food availability and social conditions. Chivers *et al.* (2007) conducted a series of three large-scale experiments where they manipulated predation pressure. However, in none of the studies were they able to show that changing predation pressure led to a change in club cell number. These critical experiments left considerable doubt about the role of predation in the evolution and maintenance of club cells. They did, however, spur on research looking at other explanations for the evolution of the club cells, including the hypothesis that the cells evolved to reduce exposure to pathogens and parasites.

9.4.2 Immune system-centred hypotheses for the evolution of alarm substances

Given that club cells are located in the epidermis of fish, they are in a great position to act as a first line of defence against parasites and pathogens. Consequently, Chivers *et al.* (2007) tested whether exposure to pathogenic water moulds (*Saprolegnia parasitica* and *S. ferax*) would lead to an increase in club cell investment. The effects they observed were quite striking. Over a relatively short time period (11 days), fish exposed to water moulds began investing heavily in club cells (Fig. 9.4). Moreover, the growth of *Saprolegnia* on petri plates was reduced when skin extracts of minnows containing alarm cues was added to the plates. Subsequent experiments also showed that minnows exposed to larval trematodes every 4 days for 16 days also increased their alarm cell investment. James *et al.* (2009) conducted a similar experiment and found that exposure of trematodes to skin extracts of minnows reduced the ability of the trematodes to infect minnows. However, they did not find that minnows increased alarm cell investment upon exposure to the trematodes. They speculated that this difference might result from the evolutionary history of the host/pathogen relationship. Specifically, specialist parasites may migrate through the epidermis without damaging the club cells, hence the fish does not respond by increasing club cell investment.

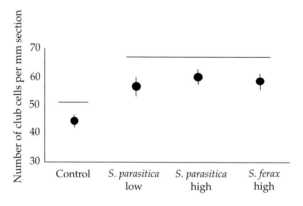

Figure 9.4 Mean (± SE) number of club cells per millimetre section for minnows raised in the presence of water moulds. Control = dilute salts (DSA and DSB) solution; Zoospore (*S. parasitica* and *S. ferax*) concentrations were 2,000 spores l⁻¹ (low) or 20,000 spores l⁻¹ (high) in dilute salts (DSA and DSB) media. Treatments under the same bar are not significantly different, those under different bars are significantly different at P < 0.05. Modified from Chivers *et al.* 2007.

If club cells evolved as part of the immune system, then stressing the immune system should lead to changes in club cell investment. Heavy metals are known to act as immunosuppressants in vertebrates (Sanchez-Dardon *et al.* 1999). With this in mind, Chivers *et al.* (2007) exposed minnows to low levels of cadmium while giving them a pathogen challenge. They found that exposure to Cd limited the ability of fish to increase club cell investment in response to a challenge with *Saprolegnia*. The most striking evidence for the immune function of club cells comes from Halbgewachs *et al.* (2009). They demonstrated that intraperitoneal injections of cortisol led to suppression of the immune system of minnows, as measured by a respiratory burst assay. Simultaneously, with the suppression of the immune system, there was a reduction in club cell investment. Iger and Abrahams (1990) showed that club cells would migrate towards a site of damage and display phagocytotic activity of cellular debris during the migration.

Finally, one study has extended the role of stress on club cell investment in fish. Manek *et al.* (unpub data) demonstrated that exposure of fish to UV radiation results in an increase in cortisol and a simultaneous reduction in club cells number. Taken together, these results suggested that the so-called 'alarm cells' of Ostariophysan fish likely evolved in an immune context rather than a predation context and that the alarm role likely evolved secondarily.

9.4.3 Multiple selection pressures?

At least within the Ostariophysan fish, it is becoming increasingly clear that 'alarm cues' may have initially evolved to serve other functions and their role as local risk assessment cues may have been secondarily selected. The notion of public information suggests that any publically available cue that is spatially and temporally associated with predation threats over evolutionary time scales may take on the function of an alarm cue (Wisenden and Chivers 2006). Given that the alarm cues are released following mechanical damage to the skin (Chivers and Smith 1998), any species-specific cue(s) subsequently released would provide reliable information regarding the presence of local predation threats. Nearby prey, which detect and respond to

these cues, should gain significant survival benefits (Mirza and Chivers 2001). As such, there would be strong selection pressure on the cue receivers to respond, independent of any benefits to the cue donors (senders). While not sufficient to account for the evolution of alarm cues *per se*, such secondary benefits may have played an important role in the fixation of the risk assessment and predator avoidance function.

While these arguments are clearly consistent with our understanding of the evolution of alarm substances from the point of view of the signal receiver, can the immune function hypothesis explain the observed degree of conservation? Given that the immune response to parasite and/or pathogen infection is genetically regulated, at least in part through the MHC complex (Eggert *et al.* 1999), there should be a strong correlation between phylogenetic relatedness and the chemical products of the immune system. As such, we should expect to find alarm substances that are chemically more similar among closely related species. Strong benefits to those detecting the alarm substance following a predation event would lead to the apparent evolutionary conservation of alarm substances.

9.5 Disturbance cues

9.5.1 The taxonomic distribution of disturbance cues

Hazlett (1985) was the first to document that prey animals that had been disturbed but not injured could release chemical cues that warned conspecifics of danger. In his experiment, virile crayfish (*Orconectes virilis*) exhibited an increase in vigilance behaviour upon exposure to water from a tank containing conspecific crayfish that were disturbed by a predator, but not conspecific crayfish that were not disturbed. Similar disturbance cues have been shown in hermit crabs (*Calcinus laevimanus*), sea urchins (*Strongylocentrotus franciscanus*), red-legged frog tadpoles (*Rana aurora*), as well as several species of fish, such as Iowa darters (*Etheostoma exile*), slimy sculpins (*Cottus cognatus*), brook trout (*Salvelinus fontinalis*), and convict cichlids (*Amatilania nigrofasciatus*) (see Ferrari *et al.* 2010 for a review). All of these studies indicate that prey exhibit anti-predator behaviour upon detecting the cues. However, the intensity of the behavioural response is reduced compared to detection of alarm substances.

9.5.2 The chemistry of disturbance cues

Several authors have argued that disturbance cues are nitrogenous wastes released as metabolic by-products through urine or across gill epithelia. Consequently, the cues should elicit a generalized non-species specific response. Indeed, consistent with this hypothesis, Vavrek *et al.* (2008) showed that two distantly related fish (convict cichlids and rainbow trout) respond to each other's disturbance cues. Moreover, crayfish exhibit an increase in vigilance in response to cues from leeches, newts, and fish (Hazlett 1985). Both crayfish (Hazlett 1990) and larval red-legged frogs (Kiesecker *et al.* 1999) release ammonia upon disturbance, and exposure to pulses of ammonia evoke anti-predator responses. Neither cichlids nor rainbow trout respond to pulses of ammonia (Vavrek and Brown 2009) but they do respond to urea (Brown *et al.* unpublished).

9.5.3 Evolution of disturbance signals

To date there are no experiments aimed at understanding the selective pressures leading to the evolution of disturbance cues. Given that the chemical cues are released by prey animals that detect a threat but have not been captured, there are many more possible selection pressures than for damaged-release alarm cues. Are the cues preferentially released when there are kin in the vicinity? Are they preferentially released depending on the number of conspecifics in the vicinity, leading to group crypsis or group evasion behaviours? Do disturbance cues function to deter predators from attacking? A quick examination of Smith (1992)'s 16 hypotheses indicates that this topic is ripe for experimentation.

9.6 Next steps

It appears unlikely that alarm substances of fish evolved as alarm signals in a traditional sense. Clearly, there is strong selection on the part of the receiver to detect cues that are inadvertently released when a nearby conspecific is captured, but there may be little selection for the sender to send the signal. What about other taxa? Tadpoles release alarm substances upon capture. Can we explain these responses based on selection on receivers to detect publicly available information, or has kin selection provided direct benefits to the sender? When cues do not require damage for their release, as is the case for disturbance cues, there are a great number of potential selection pressures that can be identified. Currently there is a scarcity of information from which to draw any conclusions about the evolution of disturbance cues.

For researchers that devoted much of their career to studying the ecological role of alarm substances, the question of the evolution of the substances and their chemistry appeared to be irrelevant. This is most unfortunate. In one sense, we do not need to know how alarm substances evolved to understand that they are responsible for mediating behavioural (Ferrari *et al.* 2005), morphological (Brönmark and Miner 1992; Chivers *et al.* 2008), and life history (Kusch and Chivers 2004) defences in prey. On the other hand, if we realize that immune function may be the driving force behind club cell evolution in many fish, then we may approach their study in a different way. In fact, there is likely a strong relationship between immune system-stress dynamics and predation dynamics. For example, we know that fish use the concentration of alarm cues during a learning paradigm as a proxy of the risk level posed by the predator (Ferrari *et al.* 2005). When they are exposed to a high concentration of alarm cues, they learn the predator is highly dangerous, but when they are exposed to a low concentration of alarm cues they learn the predator is less threatening. If the number of club cells in the fish skin is an indicator of a functioning immune system, then learning to recognize predators is actually a function of health.

Despite the fact that alarm cues appear in thousands of species of aquatic animals, and there are many hundreds of studies showing their role in risk assessment, we have done a poor job explaining their evolution. With a little creativity, many of the yet untested hypotheses could be examined relatively easily. For those interested in the evolution of alarm substances and disturbance cues, we caution against focusing on a single taxon. Researchers should embrace a comparative approach, allowing a broad perspective of possible selective pressures leading to the evolution of alarm chemicals.

References

Blaustein, A. R. and O'Hara, R. K. (1986) Kin recognition in tadpoles. *Scientific American*, **254**: 108–16.

Brönmark, C. and Miner, J. G. (1992) Predator-induced phenotypical change in body morphology in crucian carp. *Science*, **258**: 1348–50.

Brown, G. E. and Brown, J. A. (1996) Kin discrimination in salmonids. *Reviews in Fish Biology and Fisheries*, **6**: 201–19.

Brown, G. E. and Godin, J. G. J. (1999) Chemical alarm signals in wild Trinidadian gunnies (*Poecilia reticulata*). *Canadian Journal of Zoology*, **77**: 562–70.

Brown, G. E. and Chivers, D. P. (2005) Learning as an adaptive response to predation, in P. Barbosa and I. Castellanos (eds.), *Ecology of Predator-Prey Interactions*. Oxford University Press, Oxford, pp. 34–54.

Brown, G. E., Poirier, J. F., and Adrian, J. C. (2004a) Assessment of local predation risk: the role of sub-threshold concentrations of chemical alarm cues. *Behavioral Ecology*, **15**: 810–15.

Brown, G. E., Adrian, J. C., Smyth, E., Leet, H., and Brennan, S. (2000) Ostariophysan alarm pheromones: Laboratory and field tests of the functional significance of nitrogen oxides. *Journal of Chemical Ecology*, **26**: 139–54.

Brown, G. E., Adrian, J. C., Naderi, N. T., Harvey, M. C., and Kelly, J. M. (2003) Nitrogen oxides elicit antipredator responses in juvenile channel catfish, but not in convict cichlids or rainbow trout: Conservation of the ostariophysan alarm pheromone. *Journal of Chemical Ecology*, **29**: 1781–96.

Brown, G. E., Foam, P. E., Cowell, H. E., Fiore, P. G., and Chivers, D. P. (2004b) Production of chemical alarm cues in convict cichlids: the effects of diet, body condition and ontogeny. *Annales Zoologici Fennici*, **41**: 487–99.

Brown, G. E., Elvidge, C. K., Macnaughton, C. J., Ramnarine, I., and Godin, J. G. J. (2010) Cross-population responses to conspecific chemical alarm cues in wild Trinidadian guppies, *Poecilia reticulata*: evidence for local conservation of cue production. *Canadian Journal of Zoology*, **88**: 139–47.

Chivers, D. P. and Smith, R. J. F. (1998) Chemical alarm signalling in aquatic predator-prey systems: A review and prospectus. *Ecoscience*, **5**: 338–52.

Chivers, D. P., Brown, G. E., and Smith, R. J. F. (1996) The evolution of chemical alarm signals: Attracting predators benefits alarm signal senders. *American Naturalist*, **148**: 649–59.

Chivers, D. P., Zhao, X. X., Brown, G. E., Marchant, T. A., and Ferrari, M. C. O. (2008) Predator-induced changes in morphology of a prey fish: the effects of food level and temporal frequency of predation risk. *Evolutionary Ecology*, **22**: 561–74.

Chivers, D. P., Kiesecker, J. M., Marco, A., DeVito, J., Anderson, M. T., and Blaustein, A. R. (2001) Predator-induced life history changes in amphibians: egg predation induces hatching. *Oikos*, **92**: 135–42.

Chivers, D. P., Wisenden, B. D., Hindman, C. J., Michalak, T., Kusch, R. C., Kaminskyj, S. W., Jack, K. L., Ferrari, M. C. O., Pollock, R. J., Halbgewachs, C. F., Pollock, M. S., Alemadi, S., James, C. T., Savaloja, R. K., Goater, C. P., Corwin, A., Mirza, R. S., Kiesecker, J. M., Brown, G. E., Adrian, J. C., Krone, P. H., Blaustein, A. R., and Mathis, A. (2007) Epidermal 'alarm substance' cells of fishes maintained by non-alarm functions: possible defence against pathogens, parasites and UVB radiation. *Proceedings of the Royal Society B-Biological Sciences*, **274**: 2611–19.

Crowl, T. A. and Covich, A. P. (1990) Predator-induced life-history shifts in a freshwater snail. *Science*, **247**: 949–51.

Eggert, F., Muller-Ruchholtz, W., and Ferstl, R. (1999) Olfactory cues associated with the major histocompatibility complex. *Genetica*, **104**: 191–7.

Ferrari, M. C. O., Wisenden, B. D., and Chivers, D. P. (2010) Chemical ecology of predator-prey interactions in aquatic ecosystems: a review and prospectus. *Canadian Journal of Zoology*, **88**: 698–724.

Ferrari, M. C. O., Trowell, J. J., Brown, G. E., and Chivers, D. P. (2005) The role of learning in the development of threat-sensitive predator avoidance by fathead minnows. *Animal Behaviour*, **70**: 777–84.

Fraker, M. E., Hu, F., Cuddapah, V., McCollum, S. A., Relyea, R. A., Hempel, J., and Denver, R. J. (2009) Characterization of an alarm pheromone secreted by amphibian tadpoles that induces behavioral inhibition and suppression of the neuroendocrine stress axis. *Hormones and Behavior*, **55**: 520–29.

Friesen, R. G. and Chivers, D. P. (2006) Underwater video reveals strong avoidance of chemical alarm cues by prey fishes. *Ethology*, **112**: 339–45.

Gerlach, G., Schardt, U., Eckmann, R., and Meyer, A. (2001) Kin-structured subpopulations in Eurasian perch (*Perca fluviatilis* L.). *Heredity*, **86**: 213–21.

Halbgewachs, C. F., Marchant, T. A., Kusch, R. C., and Chivers, D. P. (2009) Epidermal club cells and the innate immune system of minnows. *Biological Journal of the Linnean Society*, **98**: 891–7.

Hartman, E. J. and Abrahams, M. V. (2000) Sensory compensation and the detection of predators: the interaction between chemical and visual information. *Proceedings of the Royal Society B-Biological Sciences*, **267**: 571–5.

Hazlett, B. A. (1985) Disturbance pheromones in the crayfish *Orconectes virilis*. *Journal of Chemical Ecology*, **11**: 1695–711.

Hazlett, B. A. (1990) Source and nature of disturbance-chemical system in crayfish. *Journal of Chemical Ecology*, **16**: 2263–75.

Hoverman, J. T. and Relyea, R. A. (2009) Survival trade-offs associated with inducible defences in snails: the roles of multiple predators and developmental plasticity. *Functional Ecology*, **23**: 1179–88.

Hugie, D. M. (1990) An intraspecific approach to the evolution of chemical alarm signalling in the Ostariophysii. M.Sc. Thesis (University of Saskatchewan).

Iger, Y. and Abrahams, M. (1990) The process of skin healing in experimentally wounded carp. *Journal of Fish Biology*, **36**: 421–37.

James, C. T., Wisenden, B. D., and Goater, C. P. (2009) Epidermal club cells do not protect fathead minnows against trematode cercariae: a test of the anti-parasite hypothesis. *Biological Journal of the Linnean Society*, **98**: 884–90.

Kasumyan, A. O. and Ponomarev, V. Y. (1987) Biochemical features of alarm pheromones in fish of the order Cypriniformes. *Journal of Evolutionary Biochemistry and Physiology*, **23**: 20–4.

Kelly, J. M., Adrian, J. C., and Brown, G. E. (2006) Can the ratio of aromatic skeletons explain cross-species responses within evolutionarily conserved Ostariophysan alarm cues?: testing the purine-ratio hypothesis. *Chemoecology*, **16**: 93–6.

Kiesecker, J. M., Chivers, D. P., Marco, A., Quilchano, C., Anderson, M. T., and Blaustein, A. R. (1999) Identification of a disturbance signal in larval red-legged frogs, *Rana aurora*. *Animal Behaviour*, **57**: 1295–300.

Kim, J. W., Grant, J. W. A., and Brown, G. E. (2011a) Do juvenile Atlantic salmon (*Salmo salar*) use chemosensory cues to detect and avoid risky habitats in the wild. *Canadian Journal of Fisheries and Aquatic Sciences*, **68**: 655–62.

Kim, J. W., Wood, J. L. A., Grant, J. W. A., and Brown, G. E. (2011b) The effects of acute and chronic increases in perceived predation risk on the territorial behaviour of juvenile Atlantic salmon (*Salmo salar*) in the wild *Animal Behaviour*, **81**: 93–9.

Kim, J. W., Brown, G. E., Dolinsek, I. J., Brodeur, N. N., Leduc, A. O. H. C., and Grant, J. W. A. (2009) Combined effects of chemical and visual information in eliciting antipredator behaviour in juvenile Atlantic salmon, *Salmo salar*. *Journal of Fish Biology*, **74**: 1280–90.

Kusch, R. C. and Chivers, D. P. (2004) The effects of crayfish predation on phenotypic and life-history variation in fathead minnows. *Canadian Journal of Zoology*, **82**: 917–21.

Lebedeva, N. Y. and Burlakov, A. B. (1975) Variation of the composition of water soluble proteins in the skin and mucous of the minnow, *Phoxinus phoxinus*, under the influence of a long period in the laboratory. *Journal of Ichthyology*, **15**: 163–4.

Leduc, A. O. H. C., Kim, J. W., Macnaughton, C. J., and Brown, G. E. (2010) Sensory complement model helps to predict diel alarm response patterns in juvenile Atlantic salmon (*Salmo salar*) under natural conditions. *Canadian Journal of Zoology*, 88: 398–403.

Magurran, A. E., Irving, P. W., and Henderson, P. A. (1996) Is there a fish alarm pheromone? A wild study and critique. *Proceedings of the Royal Society B-Biological Sciences*, **263**: 1551–6.

Mathis, A. and Smith, R. J. F. (1993) Chemical alarm signals increase the survival time of fathead minnows (*Pimephales promelas*) during encounters with northern pike (*Esox lucius*). *Behavioral Ecology*, **4**: 260–5.

Mathis, A., Chivers, D. P., and Smith, R. J. F. (1995) Chemical alarm signals - Predator deterrents or predator attractants. *American Naturalist*, **145**: 994–1005.

Mirza, R. S. and Chivers, D. P. (2001) Chemical alarm signals enhance survival of brook charr (*Salvelinus fontinalis*) during encounters with predatory chain pickerel (*Esox niger*). *Ethology*, **107**: 989–1005.

Naish, K. A., Carvalho, G. R., and Pitcher, T. J. (1993) The genetic structure and microdistribution of shoals of *Phoxinus phoxinus*, the European minnow. *Journal of Fish Biology*, **43**: 75–89.

Olsen, K. H., Petersson, E., Ragnarsson, B., Lundquist, H., and Jarvi, T. (2004) Downstream migration in Atlantic salmon (*Salmo salar*) smolt sibling groups. *Canadian Journal of Fisheries and Aquatic Sciences*, **61**: 328–31.

Pfeiffer, W. (1977) Distribution of fright reaction and alarm substance cells in fishes. *Copeia*, **1977**: 653–65.

Pfeiffer, W., Riegelbauer, G., Meier, G., and Scheibler, B. (1985) Effect of hypoxanthine-3(N)-oxide and hypoxanthine-1(N)-oxide on central nervous excitation of the black tetra *Gymnocorymbus ternetzi* (Characidae, Ostariophysi, Pisces) indicated by dorsal light response. *Journal of Chemical Ecology*, **11**: 507–23.

Pollock, M. S., Chivers, D. P., Mirza, R. S., and Wisenden, B. D. (2003) Fathead minnows, *Pimephales promelas*, learn to recognize chemical alarm cues of introduced brook stickleback, *Culaea inconstans*. *Environmental Biology of Fishes*, **66**: 313–19.

Sanchez-Dardon, J., Voccia, A., Hontel, A., Chilmonszyk, S., Dunier, M., Boemans, H., Blakley, B., and Fournier, M. (1999) Immunomodulation by heavy metals tested individually or in mixtures in rainbow trout (*Oncorhynchus mykiss*) exposed in vivo. *Environmental Toxicology and Chemistry*, **18**: 1492–7.

Schutz, F. (1956) Vergleichende Untersuchungen uber die Schrekreacktion bei Fischen und deren Verbreitung. *Zeitschrift Fur Vergleichende Physiologie*, **38**: 84–135.

Smith, R. J. F. (1992) Alarm signals in fishes. *Reviews in Fish Biology and Fisheries*, **2**: 33–63.

Stabell, O. B. and Lwin, M. S. (1997) Predator-induced phenotypic changes in crucian carp are caused by chemical signals from conspecifics. *Environmental Biology of Fishes*, **49**: 139–49.

Vavrek, M. A. and Brown, G. E. (2009) Threat-sensitive responses to disturbance cues in juvenile convict cichlids and rainbow trout. *Annales Zoologici Fennici*, **46**: 171–80.

Vavrek, M. A., Elvidge, C. K., DeCaire, R., Belland, B., Jackson, C. D., and Brown, G. E. (2008) Disturbance cues in freshwater prey fishes: do juvenile convict cichlids and rainbow trout respond to ammonium as an 'early warning' signal? *Chemoecology*, **18**: 255–61.

Von Frisch, K. (1941) Uber einen Schreckstoff der Fischhaut und seine Biologische Bedeutung. *Z. Vergl. Physiol.*, **29**: 46–149.

Wisenden, B. D. (2000) Olfactory assessment of predation risk in the aquatic environment. *Philosophical Transactions of the Royal Society of London Series B-Biological Sciences*, **355**: 1205–8.

Wisenden, B. D. and Smith, R. J. F. (1997) The effect of physical condition and shoalmate familiarity on proliferation of alarm substance cells in the epidermis of fathead minnows. *Journal of Fish Biology*, **50**: 799–808.

Wisenden, B. D. and Thiel, T. A. (2002) Field verification of predator attraction to minnow alarm substance. *Journal of Chemical Ecology*, **28**: 433–8.

Wisenden, B. D. and Chivers, D. P. (2006) The role of public chemical information in antipredator behaviour. In F. Ladich, *et al.* (eds.), *Communication in Fishes* Science Publishers, New Jersey, pp. 259–78.

Wisenden, B. D., Vollbrecht, K. A., and Brown, J. L. (2004) Is there a fish alarm cue? Affirming evidence from a wild study. *Animal Behaviour*, **67**: 59–67.

Wisenden, B. D., Rugg, M. L., Korpi, N. L., and Fuselier, L. C. (2009) Lab and field estimates of active time of chemical alarm cues of a cyprinid fish and an amphipod crustacean. *Behaviour*, **146**: 1423–42.

CHAPTER 10

Scaling up infochemicals: ecological consequences of chemosensory assessment of predation risk

Andrew M. Turner and Scott D. Peacor

Recent years have seen a shift in the way ecologists conceptualize species interactions in food webs, with a deeper appreciation for the potential influence of interactions other than the conspicuous act of predators eating prey. In particular, community ecologists are recognizing that the adaptive phenotypic responses of one species to another are important in shaping the magnitude and nature of species interactions. This development sets the stage for the integration of sensory biology and community ecology, as an understanding of sensory ecology is required to describe the ability of a species to perceive and respond to other species.

The inclusion of phenotypic responses in the conceptualization of ecosystems is a departure from the majority of food web theory that is widely used to manage natural resources and conserve biological diversity. The traditional approach to modelling ecological systems is to abstract a food web in which interaction coefficients between two species are fixed and form building blocks for the larger system. A change to the abundance of one species affects the abundance of a second species via a feeding relationship. The second species is in turn linked to other species via pair-wise predator–prey interactions, thus allowing numerical effects to cascade through the food web. Because the strength of an interaction is often dependent on the phenotypic traits of interacting species, shifts in trait values can also affect the abundance of species within food webs. For example, a change in the abundance of one species (i.e. a predator) may affect the traits of another species (i.e. the

behaviour, morphology, or life history of a prey), which can in turn be propagated to other species the prey interacts with, affecting their population size independent of any numerical effects of the predator on the prey. Changes in the fitness correlates or population density of interacting species, mediated by changes in their traits, are called non-consumptive effects (NCEs), and the indirect effects that may result are called trait-mediated indirect effects (TMIEs, Abrams 1995). NCEs and TMIEs have received significant study in recent years, and the emerging consensus is that in some systems they may be as important as the consumptive effects of predators in structuring populations and communities.

The central role of sensory information in NCEs and TMIEs has been under-appreciated and little studied. Trait shifts are only possible if organisms can assess critical features of the environment by gathering reliable information. Thus, the nature of species interactions depends in a fundamental way on the perception of sensory cues, yet the study of sensory biology is rarely linked with population and community ecology. Indeed, we are unaware of a single study that has examined how different sensory modes (e.g. visual, tactile, or chemical) could affect the strength of NCEs or TMIEs.

The contributions to this volume show that for aquatic systems, chemoreception is the most prevalent mechanism by which plants and animals gather information from their surroundings, and thus adjust their phenotypes in adaptive ways (see also

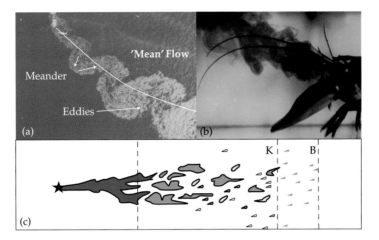

Plate 1 Odour dispersal. (a) Surface oil slick at 1 km scale showing three turbulence scales: curvature shows that this 'mean' flow is actually part of a large scale eddy; the flow meanders as a result of smaller eddies, which in turn show eddy fine structure (From Album of fluid motion). (b) Information current. Dye visualization of the lobster gill current, a turbulent jet carrying urine and pheromones up to 7 body lengths away from the animal. (c) Sketch of eddy cascade from coherent plume near source (star) to large eddies breaking down to ever smaller eddies up to the Kolmogorov scale (K) followed by odour patches up to the Batchelor scale (B). See Figure 1.2, page 5.

Plate 2 Aquatic noses. (a) Lobster olfactory (slanted, 1 mm long) and bimodal (vertical, 2 mm long) sensilla of the antennules. Micro-electrode (horizontal glass probe) matches the 20 μm diameter of olfactory sensilla to mimic their olfactory boundary (diffusion) layer, but its 100 x faster sampling rate allows odour measurement exceeding the scale of a lobster sensillum. (b) Sea star 'nose' of specialized tube feet, arm raised up out of the bottom boundary layer. (c) Catfish (*Ictalurus nebulosus*) with small forward-facing siphons as inflow nares (white arrows) and flat upward-facing outflow nares (black arrows) with valves barely visible behind nasal barbel (arrow head). (d) Olfactory rosette of catfish showing densely packed lamellae; odor receptor cilia located at inner lamellar surfaces: A = anterior, P = posterior, m = midline raphe (from Caprio and Raderman-Little (1978), with permission from Elsevier). See Figure 1.5, page 10.

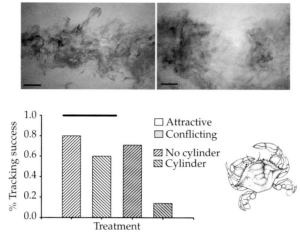

Plate 3 Turbulent mixing supresses the ability of crab foragers to distinguish attractive vs. aversive plumes. Top panel: Photographs illustrating mixing of dual plumes in 5 cm s[-1] flow (left) and in the same flow condition with a 10 cm dia cylinder upstream to create mixing (right). Odour sources were 8 cm apart and with the origin approximately 25 cm upstream of the edge of the photo. Flow is from left to right. Note that the less mixed (left) plume shows a strong filamentous structure even far downstream. Where the two plumes intersect individual odour filaments are clear. Scale is 10 cm. Lower panel shows that enhanced mixing created by the cylinder suppresses the ability of crabs to discern an attractive plume (shrimp metabolites) when an aversive plume (injured blue crab metabolites) also is present. Attractive and Conflicting represent dual plume treatments, consisting of shrimp metabolites/asw and and shrimp metabolites/injured blue crab metabolites respectively. Crabs always track to the shrimp metabolites even in the presence of a cylinder (crabs were initally released 1.25 m from sources). However, crabs in the aversive condition do not track to the shrimp metabolite plume when mixing was increased by the cylinder. Numbers below the figure show the the results of a G-test. N = 12-15 trials per condition. See Figure 7.5, page 107.

Plate 4 Diel vertical migration (DVM) in *Daphnia*. Vertical migration is induced by the presence of visual foragers, such as fish in order to avoid predator encounter. *Daphnia* remain in cooler and darker water strata during the day and migrate to warm and nutritional water strata during the night. Illustration by L. Weiss and R. Tollrian. See Figure 8.5, page 120.

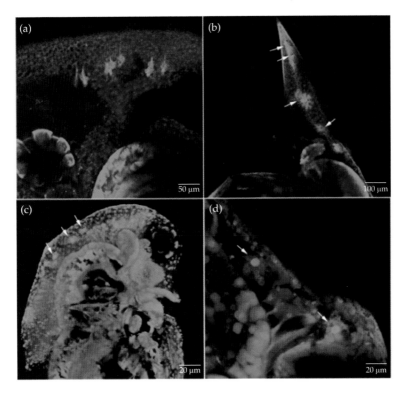

Plate 5 Confocal laser scanning microscope images of polyploid cells stained with nissl fluorescence. Polyploid cells (white arrows) are associated with morphological defence structures in predator exposed *Daphnia* species. (a) *Daphnia longicephala*, (b) *Daphnia lumholtzi* (c) *Daphnia pulex* (d) magnification of polyploid cells (white arrows) in *Daphnia pulex* associated with the neckteeth region. Photograph by L. Weiss. See Figure 8.7, page 122.

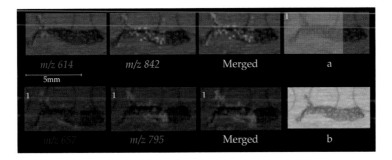

Plate 6 Differential localizations of ion masses associated with the sponge *Dysidea herbacea*; (a) shows the raster on the image; (b) shows the photomicrograph of the sponge-section itself (Esquenazi *et al.* 2008). See Figure 13.5, page 191.

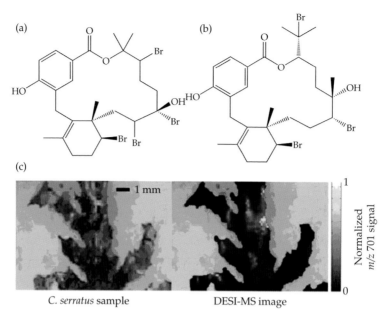

(a) (b)

(c)

C. serratus sample   DESI-MS image

Normalized m/z 701 signal

Plate 7 The relative concentrations of the bromophycolides (a) (above left) and (b) (above right) can be monitored on the surface of the red alga (c) *serratus*. The left picture shows a photography of the algal surface, the right one a distribution map of the metabolites that can be monitored by their [M+Cl-H]⁻ ions (Lane *et al.* 2009). See Figure 13.6, page 192.

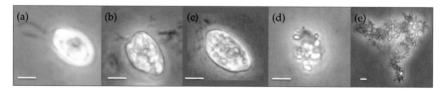

Plate 8 Lysis of the cryptophyte *Rhodomonas salina* after addition of algal allelochemicals. Scale bar: 5 μm. (a) control cell, (b) and (c) cells in osmotic shock, (d) cell membrane ruptured and cell content is released in the surrounding medium, (e) bacteria colonize algal cell remains. Photographs: P. Uronen. See Figure 14.1, page 197.

Brönmark and Hansson 2000; Hay and Kubanek 2002; Derby and Sorensen 2008). Chemoreception has several unique qualities relative to other sensory modalities, and we argue that these differences will likely affect species' interactions. Thus, the reception of chemical information may have large consequences for individual fitness, community structure, and ecosystem function.

Our goal here is to highlight examples from the freshwater and marine literature that illustrate how consideration of information flow can assist in constructing a useful theory of food web dynamics. The translation from infochemicals to community level processes spans several scales of time and ecological organization (Fig. 10.1). We provide a review of the stages in this translation, and we discuss the unique characteristics of infochemicals that must be considered in studying their influence on population and community level effects. Finally, in the Discussion, we ask whether the consequences of trait shifts induced by chemical cues are any different from the consequences of trait shifts induced by other sensory modes. Although we focus on the aquatic literature, similar interactions are also important in terrestrial ecosystems (e.g. Schmitz *et al.* 2004).

10.1   From trait response to community level effects: an overview of the process

We partition the effect of infochemicals on community and ecosystem level properties into multiple steps (Fig. 10.1). These steps have been studied to varying degrees, which we represent by arrow thickness. The first step is the creation of the infochemical, which may come from the prey as byproducts of predation or alarm signals, or may come directly from the predator (step a). This step is examined in Chapter 2. The second step is the direct effect of the chemical cues, which we divide into three broad categories. The presence of infochemicals may affect the stress level of individuals (step b, Slos and Stoks 2008; Fraker *et al.* 2009; Edeline *et al.* 2010), cause the prey to modify phenotypic traits—including life history, behaviour, morphology, and physiology—(step c, Lima 1998; Chapter 6), or induce changes in traits of other species in the system (step d).

The core concept underlying the direct influence of infochemicals on prey traits is that infochemicals provide information that indicates a change in the environment which in turn affects prey fitness. Changes in traits to address environmental changes

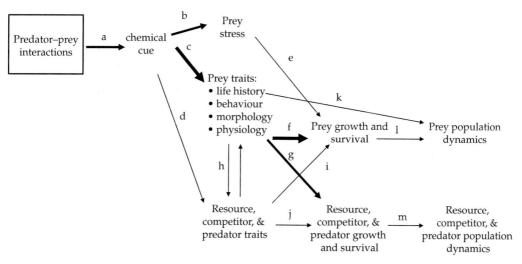

Figure 10.1   A heuristic model illustrating how trait shifts induced by infochemicals translate up to population-, community-, and ecosystem-level patterns. Thickness of arrows represents relative level of attention in the literature. Letters near arrows denote interactions discussed in text.

may in turn come at a cost due to intrinsic trade-offs associated with the level of trait expression (Chapter 12). The most studied trade-off is the conflict between energy gain and predation risk. Traits such as defensive morphology, habitat choice, activity level, or movement speed affect both the amount of resources acquired and the level of predation risk. If infochemicals indicate higher levels of predation risk, it may be adaptive to reduce time spent foraging even at the potential cost of reduced resource acquisition. These direct effects of the infochemicals on prey traits have been widely studied in diverse systems and taxa (e.g. Tollrian and Harvell 1999), and are the subject of other chapters in this book (e.g. Chapter 8).

The focus of our review is the consequences of changes in prey traits on individual fitness, populations, communities, and ecosystems. A change in prey traits will likely affect prey growth rate and survival (step f). For example, an increase in refuge use could affect prey feeding rates or a change in physiology could affect conversion efficiency of resources. The induced changes in prey traits may also affect other species in the system (step g). In this case, the predator is said to have a trait-mediated indirect effect (TMIE) on a species, as they occur through the predator causing a change in prey traits, rather than prey density (Abrams 1995; Werner and Peacor 2003). Less studied, though potentially very important, are trait changes in one prey species that trigger trait changes in other species (step h). Linked interactions of trait shifts occur when a predator induces a change in prey traits such as habitat preference, and this change in habitat preference in turn affects the traits (e.g. habitat preference) of the predator, resources, competitors, or other predators. These changes in species traits could reflect back and affect the prey's growth rate and survival (step i), or affect the growth and survival of the species the prey interacts with (step j).

Finally, predator-induced changes in prey traits could affect population and community level processes. Clearly, if prey life history is altered (e.g. reduced clutch size) this will affect population growth rate (steps k and l, respectively). Further, the indirect effects on other species in the system will also affect their demographics (step m). However, this is largely uncharted territory, as the vast majority of research has examined individual level responses. We also review recent work of the effect of predator-induced changes in prey traits on ecosystem level processes.

10.2 Population-level processes

In this section we discuss in more detail the manner in which infochemicals alter individual growth, reproduction, and survival, and the ensuing effects on population growth rates.

10.2.1 Predator effect on prey growth rate through induced changes in foraging

The deployment of anti-predator defenses usually entails significant costs, which often take the form of reduced foraging rates. When confronted with increased predation risk, prey are known to increase refuge use, spend a smaller portion of their time foraging, and to forage at slower rates (Lima 1998). The clear consequence is a reduction in prey growth rates. Such reductions have been demonstrated in many systems, and in lab, mesocosm, and field studies. To provide one example, Turner (1997) used chemical cues associated with predation to manipulate the level of risk perceived by physid snails reared in outdoor mesocosms. With increasing risk, refuge use increased and somatic growth declined in a monotonic manner, suggesting that snails are sensitive not just to the presence or absence of risk, but can perceive degrees of variation in risk and adjust their behaviour accordingly (see also Peacor and Werner 2001; Van Buskirk and Arioli 2002). In a field experiment (Turner and Montgomery 2003), snails were held in cages at different distances from a molluscivorous sunfish. Snails closest to the caged pumpkinseed sunfish used refuge habitats more and grew slower than did snails further from the caged predator (Fig. 10.2).

A reduction in foraging rate may not always be associated with reduction in growth rate or size. Simple models and experiments have demonstrated that the same proportional reduction in foraging rates can have very different effects on growth rate. This is because three distinct effects combine to affect the prey's growth rate. First, as focused on above, the reduction in foraging rate will directly

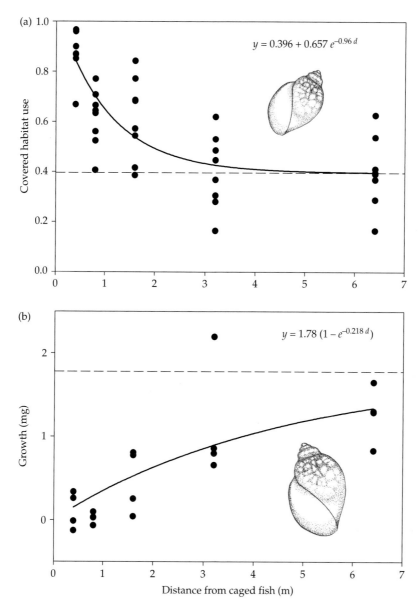

(a) The figure shows a plot with equation $y = 0.396 + 0.657\,e^{-0.96\,d}$, with y-axis "Covered habitat use" and x-axis "Distance from caged fish (m)".

(b) The figure shows a plot with equation $y = 1.78\,(1 - e^{-0.218\,d})$, with y-axis "Growth (mg)" and x-axis "Distance from caged fish (m)".

Figure 10.2 Results of a field experiment evaluating the influence of infochemicals on prey habitat use and growth (data from Turner and Montgomery 2003). (a) Dependence of predator avoidance behaviour on distance from a caged predator. Data are cage means for covered habitat use by the pulmonate snail *Physa acuta*. Predator is the pumpkinseed sunfish *Lepomis gibbosus*. Best fit parameters are for exponential decay model. (b) Dependence of *Physa* growth on distance from caged pumpkinseed sunfish. Data are cage means for growth over nine days. Best fit parameters are for an exponential rise to maximum model.

affect resource acquisition and negatively affect growth rate. Second, the infochemicals may reduce the foraging of both intra- and interspecific competitors, which will have an indirect positive effect on the focal individual's growth rate. Third, the composite reduction in foraging by all individuals will have a positive effect on resources levels. The combined effect is that predator presence can have negative, negligible, or even positive effects on prey growth rate. Peacor and Werner (2000, 2004) and

Turner (2004) present a model that examines how environmental context influences the net predator effect on growth. For example, a stronger negative effect on prey growth rate is expected for plentiful static resources (i.e. detritus) than a limited growing (i.e. plant) resource (Peacor and Werner 2004). Peacor (2002) showed that a predator-induced reduction in tadpole foraging can have a positive effect on prey growth rate, providing a clear example of context dependence. A net positive effect of the predator can arise when the reduction in prey foraging rate is accompanied by a proportionally larger increase in resource levels due to the non-linear nature of resource growth rates. Werner and Peacor (2006) varied resource growth rates by manipulating nutrient levels, and showed that infochemicals from caged dragonfly larvae had a strong negative effect on tadpole foraging, but had either no effect or a positive effect on tadpole growth rate, depending on nutrient levels.

10.2.2 Predator effect on within-population variation in growth and size

The overwhelming majority of studies of infochemicals on prey growth have focused on average effects. However, predation risk can also affect among-individual variation in body size, and variation in body size can have strong effects on population- and community-level properties (Wissinger et al. 2010). Peacor et al. (2007) performed experiments in which predation risk had both a negative and positive effect on within-population size variation. They present a conceptual model that predicted the results based on two mechanisms. First, if individuals respond differentially to the predator by reducing foraging rate to varying degrees, predation risk is predicted to increase individual size variation (see Turner and Mittelbach 1990 for an example). Second, predation risk will affect the size dependence of assimilation and respiration (and thereby the exponents in bioenergetics growth models), which can cause an increase or decrease in individual size variation. We are not aware of other studies that have reported these effects. However, this is likely due to the focus on mean effects, as we are aware that other researchers have observed such effects without publishing them. We see this as an important implication of infochemicals, as the resultant effect on variation in body size could both affect population level dynamics, but also offer a lens into the mechanisms by which predation risk affects foraging and growth rate (Peacor et al. 2007).

10.2.3 Predator effect on prey life history

Predator induced changes in prey life history or morphology can affect prey growth rate (Chapter 8). When pulmonate snails are exposed to chemical cues of feeding crayfish, they delay timing of reproduction, which affords higher growth rates and larger body size at maturity (Crowl and Covich 1990; Hoverman et al. 2005). Because crayfish are ineffectual predators of large bodied snails, this shift in life history is an adaptive response to risk of predation. In contrast, infochemicals from predators cause some species of larval anurans to metamorphose at a smaller size (Skelly and Werner 1990). In streams with fish predators the mayfly *Baetis bicaudatus* undergoes metamorphosis at a younger age and smaller size, thereby minimizing exposure to trout (Peckarsky et al. 2001).

10.2.4 Demographic consequences of trait shifts

In contrast to the fairly extensive literature concerning the consequences of predation risk on the individual components of fitness, there are few studies on the demographic consequences that may emerge from anti-predator defenses. In perhaps the only purely experimental study to directly measure demographic responses to a risk manipulation in a field setting, Boeing et al. (2005) measured population growth consequences of anti-predator defenses in the *Chaoborus–Daphnia* predator–prey system. *Daphnia* responds to *Chaoborus* kairomones with habitat shifts, growth of spines, and shifts in life history. Boeing et al. measured population growth rates of *Daphnia* stocked at low density in lake enclosures. Enclosures were treated with no predators, freely swimming *Chaoborus*, or a kairomone treatment in which *Chaoborus* were confined to a mesh tube apart from the *Daphnia*. They found that the anti-predator responses of *Daphnia* to *Chaoborus* kairomone led to a 32% reduction in population growth. Interestingly, free-swimming *Chaoborus* reduced population growth to a similar degree, suggesting that most of the net effect of *Chaoborus* on *Daphnia* population growth was levied via non-consumptive mechanisms.

For most study systems, direct observations of the demographic consequences of predation risk can only be studied at large spatial and temporal scales. This constraint limits the investigator's ability to manipulate predation risk, especially given the necessity of presenting prey with the perception of risk without imposing any actual mortality. An alternative approach is to measure individual demographic parameters under contrasting predation regimes, either in the field or in laboratory experiments, and then use demographic models to integrate these parameters and estimate the non-consumptive effects of predators on population growth. McPeek and Peckarsky (1998) used this approach to examine effects of stoneflies on mayfly prey and showed that non-consumptive effects on prey population growth were much larger than the direct consumptive effects. The non-consumptive effect was caused by a strong effect of stoneflies on mayfly larval growth and body size which in turn reduced adult fecundity. In contrast, for dragonfly predators feeding on damselfly larvae, the non-consumptive effects of predators were weaker than the direct consumptive effects, despite the strong effect of dragonflies on somatic growth of damselflies (McPeek and Peckarsky 1998). Non-consumptive

effects were weak because fecundity of adult damselflies was size-independent. Pangle *et al.* (2007) employed a similar demographic approach to evaluate the effect of the invasive invertebrate predator *Bythotrephes* on cladoceran zooplankton populations in the Laurentian Great Lakes. Zooplankton responded to *Bythotrephes* by occupying deeper depths where growth is slowed by colder water temperatures. For this system, the non-consumptive effects of *Bythotrephes* were larger than the consumptive effects. Given the challenges associated with manipulating risk on an appropriate scale, demographic approaches like these offer an effective alternative approach to partitioning the consumptive and non-consumptive effects of predators in a field setting.

10.3 Community- and ecosystem-level effects of infochemicals in aquatic systems

In this section we highlight empirical evidence that infochemicals can strongly influence the indirect effect of predators in food webs. Our synthesis is organized by the most studied forms of indirect interactions: trophic cascades, exploitative competition, and effects of multiple predators (Fig. 10.3). We provide some select examples, but note that additional examples are highlighted in recent reviews (Werner and Peacor 2003). The interactions in these simple three- species systems form the basis of trait-mediated indirect interactions in more

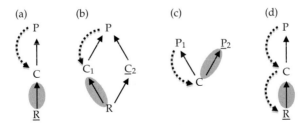

Figure 10.3 Diagrams of principal trait-mediated indirect interactions examined in the literature. P, C, and R represent predators, consumers (as prey), and resources. Solid arrows represent standard predator–prey consumptive interactions, and point in the direction of energy flow. The dashed arrow indicated a predator induced trait change in prey. The oval indicates the predator–prey interaction affected by the induced trait change, and the underlined species indicates the species affected by the indirect interaction. (a) A trophic cascade in which a predator affects a resource via changes in prey behaviour. (b) Effect of a predator on interspecific competition. In this example the predator induces a trait change in C_1, which has an indirect positive effect on C_2. The predator could also simultaneously have a TMII on C_1 through induced trait changes in C_1. (c) Facilitation or apparent competition through changes in prey traits. (d) Linked trait modifications: infochemicals from P affect traits of C, which in turn induces a trait modification by R.

complex webs (Byrnes *et al.* 2006; Werner and Peacor 2006), which have received little attention and are not reviewed here.

10.3.1 Trophic cascades

Studies of trophic cascades in aquatic ecosystems have revealed that a relatively small number of predators can have very large effects on the abundance and productivity of lower trophic levels. Trophic cascades occur when predators suppress foraging by prey, either by eating them or by impairing foraging via non-consumptive mechanisms, which in turn alters the dynamics of the prey's resources (Fig. 10.3a).

The Ostariophysan alarm system of cyprinid minnows provides one of the best studied examples of chemosensory evaluation of predation risk (e.g. Smith 1992) and, interestingly, minnows are also key players in several prominent examples of trophic cascades in nature. The classic study of trophic cascades in temperate lakes (Carpenter *et al.* 1987) was likely largely due to a TMIE (He and Kitchell 1990; Peckarsky *et al.* 2008). Despite the removal of 90% of the bass population from Peter Lake, the few remaining bass (~40 adults) induced a rapid habitat shift by the 45,000 minnows stocked into the lake. Cladoceran zooplankton were initially suppressed by minnows, but following the predator-induced habitat shift of minnows zooplankton reached a very high biomass that persisted through the summer. Thus, the cascading effects were primarily caused by a reduction in minnow foraging and the chemically-mediated fright response allowed a small number of predators to have large food web effects. In another study involving bass and cyprinids (Power *et al.* 1985), largemouth bass in a warm water stream induced stoneroller minnows to move to the shallow margins of pools or to pools lacking bass, thereby reducing herbivory. Other large scale manipulations of predation that provide evidence for chemically-mediated trophic cascades include Byrne *et al.* (2006) for kelp forests.

The whole-ecosystem studies described above provide strong evidence that behavioural responses to predators, induced by reception of infochemicals, play an important role in mediating trophic cas-

cades in natural systems. Experimental approaches using chemosensory manipulations of perceived risk help to partition consumptive effects (CEs) and non-consumptive effects (NCEs) and provide an unambiguous test of the role of infochemicals. In one approach, two treatments are applied that mimic the CE and NCE of a predator. For the CE treatment, prey are usually manually removed at a rate that simulates the consumptive effects. For the NCE treatment, the predator is confined in a way that allows chemical cues to be released into the environment, so prey perceive predation risk but there is no mortality. This approach has been used to show that the predator's NCE can dominate the net effect of the predator (VanBuskirk and Yurewicz 1998; Peacor and Werner 2001; Werner and Peacor 2006; Trussell *et al.* 2006b). For example, Trussell *et al.* (2006b) experimentally disentangled the CE and NCE in the green crab–carnivorous snail–barnacle system of the rocky intertidal zone. The CE was simulated by manually removing snails and the NCE was simulated by placing a crab in a container in an upstream section of the enclosure. Results showed that the NCE of the green crab was as strong as the CE on barnacle abundance.

More typically, researchers have explored whether NCEs can cause trophic cascades without examining the influence relative to consumptive effects. A number of such studies have been performed in mesocosm settings (e.g. Turner 1997; McCollum *et al.* 1998). Field manipulations of predator cues that provide evidence for trophic cascades mediated by chemically-induced trait shifts include experiments conducted in streams (McIntosh *et al.* 2004), lakes (Leibold 1991), ponds (Bernot and Turner 2001), the rocky intertidal (Trussell *et al.* 2004), and shallow marine habitats (Grabowski and Kimbro 2005).

10.3.2 Interspecific competition

Ecologists have long studied how predation and competition interact in shaping prey communities, and it is becoming clear that adaptive trait shifts can play an important role in shaping the nature of this interaction. The induction of adaptive trait shifts, mediated by chemical cues, can potentially intensify or weaken competition, and even alter the outcome of competitive interactions (Fig. 10.3b).

If two competing prey vary in the magnitude of their plastic response to predator cues, predator presence can alter the competitive interaction (Werner 1991; Werner and Anholt 1996). Peacor and Werner (2000, 2004) have developed a mathematical model to elucidate this process. Several factors must be considered, but the core of the problem is understanding how the predator presence will affect the relative rate resources are acquired by the different competitors. The effect of a predator on the growth rate of two competitors, one that responded strongly to the predator and one that did not, was examined as a function of the density of a reactive prey species. The model successfully predicted the NCE of a predator (dragonfly larvae) on reactive and unreactive competitors (small green frog tadpoles and bullfrog tadpoles, respectively). For the reactive competitor, a strong negative effect of predators diminished to no effect with increasing green frog tadpole density. However, for a nonreacting competitor, a negligible effect of predators increased to a strong positive effect with increasing green frog tadpole density.

In a study with several competitors, Werner and Peacor (2006) found that the effect of predator-induced reduction in tadpole foraging had markedly contrasting effects on different species of snails due to different diet overlaps. They further showed how the magnitude of the NCEs depended strongly on system productivity. Relyea (2000) showed that leopard frogs grew faster than wood frogs when reared together in the absence of predation risk, but when reared together in the presence of caged dragonflies or mudminnows, wood frogs grew faster than leopard frogs. Leibold (1991) used deep enclosures in lakes to test how fish cues affected the coexistence of *Daphnia pulicaria* and *Daphnia galeata*. In the absence of fish cues both species occupy the epilmnion, where *D. pulicaria* is the superior competitor and eventually excludes *D. galeata* from the system. In the presence of fish cues, *Daphnia pulicaria* is hypolimnetic, *D. galeata* epilimnetic, and the species coexist. Thus, fish cues induced habitat segregation and fostered coexistence of the two species. Finally, predators can intensify competition among prey by confining them to a common refuge, where resources will be rapidly depleted and diet overlap may be exacerbated (Mittelbach and Chesson 1987; Mittelbach 1988; Walters and Juanes 1993).

Studies like these highlight the sensitivity of interspecific competition to the presence of predation risk. Studies of species-specific variation in phenotypic response to infochemicals, conducted within taxonomically related guilds, generally show considerable interspecific variation in plasticity (e.g. Relyea and Werner 2000), so these sorts of trait-mediated effects on competitors may be quite common.

10.3.3 Effects of multiple predators: facilitation and inhibition

Two predator species occupying the same trophic level often compete for resources by consuming the same prey, but they may also interact by inducing trait shifts in prey that make food more or less available to each other (Sih *et al.* 1998; Fig. 10.3c). If prey employ similar anti-predator defenses against the two predators, then the induction of anti-predator defenses by one predator species will make prey less available to other predators (Soluk 1993). On the other hand, prey often exhibit highly specific responses to different species of predators. Predator-specific responses are expected when predators forage in a contrasting manner, or forage in different habitats (e.g. Turner *et al.* 2000). In this case, the induction of anti-predator defenses by one predator species could make the prey more vulnerable to other predators. Eklöv and VanKooten (2001) present the example of pike and perch feeding on roach. Pike forage in the littoral zone, where they ambush prey, whereas perch forage in open water. Roach respond to each predator by moving to the safer habitat. When faced with both predator species, roach mortality is higher than would be expected based on the additive effects of each predator alone, and pike feeding rates are higher with perch than when alone. Thus, conflicting avoidance strategies can turn potential competitors into facilitators.

10.3.4 Linked trait modifications

The systems reviewed above have one trait-mediated link between predator and prey, which in turn affects the abundance or growth rate of a third species in the system. However, changes in species traits can also elicit a change in the traits of the third species, which can in turn affect the abundance of other species in the food web (Fig. 10.3d). In this case there would be two trait-mediated links. In one empirical example of linked behavioural shifts, Romare and Hansson (2003) manipulated perceived risk within large enclosures placed into a lake in order to test whether predation risk alters the relationship between roach and their zooplankton resources. Confined pike induced roach to become less active and seek refuge in the vegetated littoral zone. This change in roach behaviour caused *Daphnia* to migrate from the vegetation and into the open water. Such tightly coupled behavioural interactions illustrates how chemically conveyed information can be transmitted through a food web (see also Huang and Sih 1991; Bollens *et al.* 2011).

10.3.5   Nutrient translocation

Ecologists have come to appreciate that animals shape the flow of nutrients and energy through ecosystems, and thus individual behaviour is inexorably linked to ecosystem dynamics (Vanni 2002; Schindler *et al.* 2003; Schmitz 2008). For example, in the course of consumption, egestion, excretion, and mortality, animals translocate nutrients among habitats or ecosystems (Kitchell *et al.* 1979). In many cases, patterns of animal movement are shaped by trade-offs associated with finding food and avoiding predators. The contributions to this volume establish the importance of chemosensory cues in the assessment of predation risk and food availability. Thus, it is likely that nutrient fluxes both within and among ecosystems are ultimately shaped by the flow of chemical information.

One important example involves the daily vertical migration (DVM) of zooplankton in lakes and oceans, induced by infochemicals associated with planktivorous fish or invertebrate predators (Dodson 1988). Herbivorous zooplankton play an important role in the remineralization of nitrogen and phosphorus. Even though both diel vertical migration and the role of zooplankton in nutrient remineralization have been extensively studied, the consequences of vertical migration for biogeochemical cycles have not yet received extensive empirical attention. Dini *et al.* (1987) studied a northern lake

and report that because *Daphnia* graze predominately in surface waters but take refuge from fish in deeper water, DVM accelerates the downward flux of phosphorus across the thermocline in summer. In contrast, Haupt *et al.* (2010) report that vertical migration by *Daphnia* resulted in the upward transport of phosphorus into the epilimnion. Because zooplankton often respond to variation in predation risk by altering the extent of DVM, we expect that a change in planktivory will alter the vertical flux of phosphorus, nitrogen, and carbon. Given the potentially important role of zooplankton in oceanic carbon fluxes, predator induced shifts in vertical migration could have significant consequences for oceanic carbon storage and global climate change (Bollens *et al.* 2011).

A number of other studies point toward the possible importance of predator-induced habitat shifts in shaping biogeochemical cycles. Stief and Holker (2006) found that midge larvae in the benthos of lakes play an important role in remineralization of organic material. When midges perceive predatory fish via chemical cues, they reduce foraging and retreat deeper into their burrows. This, in turn, accelerates organic processing rates and nutrient remineralization, in part because the midges oxygenate their burrows and alter oxidation-reduction potential of the sediments. On a larger spatial scale, many planktivores forage on zooplankton in the open water under crepuscular conditions, but take refuge in the littoral vegetation during the day (e.g Hall *et al.* 1979), and in doing so transfer nutrients from the pelagia to the littoral zone (Schindler *et al.* 1996). With a change in predation risk planktivores may alter their migratory patterns, which in turn can alter the pelagic to littoral nutrient flux. In stream systems, grazer behaviour, shaped by predation risk, is known to play an important role in regulating algal growth, which in turn may affect nutrient spiralling (McIntosh and Townsend 1996; Simon *et al.* 2004). Given the ubiquity of habitat shifts induced by predator cues and the demonstrated importance of animals in recycling nutrients, the emerging picture is that infochemicals could strongly influence nutrient cycling and translocation of nutrients in aquatic systems, but this hypothesis has received too little empirical attention.

10.3.6   Ecological efficiency across productivity gradients

Food web structure shapes the efficiency with which elevated primary productivity at the bottom of the food chain filters up to enhance secondary productivity at higher trophic levels (Leibold and Wilbur 1992). Because infochemicals associated with predation risk can cause large reductions in prey foraging effort or induce defenses that make prey less vulnerable, it is conceivable that the efficiency by which production at one trophic level is converted into growth and reproduction at the next trophic level will also be dependent on traits associated with predation risk. Several experiments, conducted in simple food chains, support this idea. Turner (2004) examined how predator-induced changes in prey foraging can affect trophic efficiency. He measured the growth rate of the pulmonate snail *Helisoma trivolvis* feeding on periphyton grown in mesocosms under two levels of nutrient enrichment. In the absence of predation risk, nutrient enrichment had a significant positive effect on growth of *H. trivolvis*, but when exposed to feeding crayfish, nutrient enrichment had no effect on snail growth rates (Fig. 10.4). In a similar experiment, Peacor (2002) grew tadpoles in mesocosms at three levels of nutrient enrichment, cross-factored with a dragonfly predation risk treatment. In contrast to the results of Turner (2004), nutrient enrichment had a stronger effect on tadpole growth in the presence of predators than in the absence of risk, illustrating the complicated context dependence of such interactions. Trussell *et al.* (2006a) examined the mechanisms of depressed trophic transfer efficiency for a carnivorous snail feeding on barnacles. Risk of crab predation reduced ecological efficiency, measured as tissue production, by 60–83%. This large reduction occurred both because snails had a lower food intake rate when under risk and because the growth efficiency (ratio of tissue production to caloric intake) was lower when under risk.

10.3.7   Theoretical implications

Whereas empirical work has focused primarily on short-term consequences of NCEs and TMIEs to

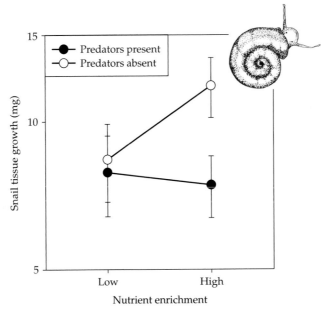

Figure 10.4   Effects of nutrient enrichment on grazer secondary production under two levels of predation risk (data from Turner 2004). Grazers are the pulmonate snail *Helisoma trivolvis* reared in mesocosms. Predation risk was manipulated by adding caged crayfish to mesocosms, and nutrient loading was manipulated by fertilizing high nutrient tanks at the rate of 50 μg/l P and low nutrient tanks at the rate of 5 μg/l P. Note that growth is shown on a logarithmic scale. Points represent mean growth ± 1 SE.

individual growth rate and survival, there has been considerable theoretical attention to long term population dynamics (for reviews see Bolker *et al*. 2003; Abrams 2010; Peacor and Cressler, in press). Trait responses to chemical cues (or other signals) introduce higher order interactions (HOIs) in ecological theory; the interaction between two species becomes a function of the density of a third species. For example, when predator presence causes a change in prey foraging behaviour, the function describing the interaction between the prey and its resource becomes a function of predator density. Such HOIs, and associated nonlinearities, are predicted to have large effects on model predictions, and this has been found when including NCEs and TMIEs in theory.

Any review of the theory in this discipline is in large part a review of the spearheading concepts developed by Peter Abrams. Abrams examined how the inclusion of prey balancing predation risk and resource acquisition can affect both short term indirect effects and long term population consequences under a number of assumptions and circumstances. Some counterintuitive results are predicted, including cases in which predators have a positive effect on prey population growth rate, and increases in resources can have a negative effect on prey population growth rate. The most general results concerning population dynamics are that adaptive prey trait-responses can have strong effects on stability and persistence, and that whether the effect will be stabilizing or destabilizing depends on the food web configuration. A universal result is that the curvature of the relationship between the trait change, and the cost and benefits of the trait change, strongly influence model predictions (review in Peacor and Cressler, in press). There has been limited theory examining how imperfect knowledge of predation risk will affect trait changes, and in turn population and community dynamics. It is likely that the sensory mode will have a large influence on perceptual ability, and hence will shape species interactions.

There is a large gap between theory and empirical work on NCEs and TMIEs. For example, there are few efforts to measure the parameters and relationships needed to develop ecological theory that includes NCEs and TMIEs (Bolker *et al*. 2003; Abrams 2010; Peacor and Cressler, in press). Further, we are unaware of any theoretical analysis that

examines how the sensory mode used by prey to detect the signal of predation risk (e.g. infochemicals) might affect population dynamics and community structures.

10.4 Scaling up the effects of infochemicals

The food web is used as a tool for understanding how perturbations to the abundance of a species can cascade through the community. We have reviewed evidence that species interactions do not necessarily involve feeding. Information in the form of chemicals moves from one individual to another, and the phenotypic responses have large ecological consequences on population-, community-, and ecosystem-level processes. Thus, indirect effects are transmitted through aquatic communities both by consumptive interactions and by the flow of chemical information. More generally, we suggest that the 'interaction web' of community ecology consists of demographic interactions and informatic interactions (i.e. an information web). Below we discuss how a consideration of the mechanism of information transfer will help ecologists to achieve a better mechanistic understanding of species interactions in food webs.

10.4.1 Does sensory mode affect species' interactions?

The unique properties of chemical information transfer in water make trait plasticity and thus trait-mediated indirect interactions widespread in aquatic ecosystems. Inducible defenses require a predator detection system that is reliable, informative, and importantly, a system that operates over a spatial or temporal range that is greater than the predator's food detection system, and thus provides separation from the predator (Chapter 6). In the absence of an effective predator detection system, selection would more often favour the evolution of fixed or constitutive defenses. In this case, species traits would be less variable, the outcome of species interactions more easily predictable, and the job of a community ecologist would be much easier.

Chemical cues are information rich, as they can carry much detail regarding predator identity or even predator diet. Chemical cues are also persistent, offering the important advantage of temporal separation of prey from predators. Auditory, visual, or mechanical cues are not persistent in time, so they offer only an instantaneous glimpse of information, whereas the chemical residue emanating from mobile predators persists for hours or even days (Turner and Montgomery 2003; Peacor 2006).

The available evidence suggests that in aquatic systems, chemical information transfer is indeed the primary mechanism by which prey assess the risk of predation. Our review of a database of non-lethal effects in freshwater and marine systems (Preisser 2007) shows that 72% of the studies of NCEs and TMIEs involve systems in which trait shifts are induced at least in part by infochemicals. Reviews by Brönmark and Hansson (2000), Hay and Kubanek (2002), and Dicke (2006) concur that for animals in aquatic ecosystems, chemoreception predominates over other sensory modes.

If we accept that TMIEs are common in aquatic systems because of the unique advantages of chemical information transfer, then an important question is whether TMIEs based on chemoreception are different in nature than those caused by other sensory modes. If chemical information transfer produces emergent community-level patterns that are different than those produced by other sensory modes, then community ecologists will benefit by paying closer attention to the mechanisms of predator detection. We next review two types of indirect species interactions in which a full understanding of the interaction requires knowledge of the sensory mechanisms involved in predator avoidance.

10.4.2 Predator diet

Infochemicals can carry remarkably detailed information regarding the identity of the prey recently consumed by the predator, and recent studies show that trait shifts are dependent on this information. Prey trait shifts are generally induced most strongly when predators are feeding on closely related or conspecific prey, and more weakly or not at all when predators are feeding on distantly related prey (e.g. Mathis and Smith 1993; Chivers and Smith 1998; Chivers et al. 2002; Schoeppner and Relyea 2005). For example, Brönmark and Pettersson (1994) found

that the induction of the deep-bodied morph in Crucian Carp is induced by piscivores feeding on other Crucian Carp, but not by the same fish feeding on a macroinvertebrate diet. Similar diet-dependent effects have been observed when fish induce helmets in *Daphnia* (Laforsch et al. 2006) and when crayfish forage on snails (Turner 2008).

The ecological consequences of changing predator diet have not yet been explored, but could have consequences for how we interpret food web interactions involving predator diet shifts. It is well documented that a density-dependent change in predator diet (e.g. switching) can stabilize prey population dynamics or influence the outcome of competition. What has not been appreciated is that the success of prey dropped from the predator diet can be due in part to their release from the costs of trait shifts imposed by the predators, as opposed to the lower death rate imposed by predators. Thus, the indirect effect of diet shifts, like the indirect effects of changes in overall feeding rates, has both a density-mediated and a trait-mediated component.

Diet-specific phenotypic shifts and their ecological consequences are possible only with chemical cues that convey accurate information regarding predator diet. A remote assessment of predator diet would be much more difficult, if not impossible, for prey employing only visual or auditory cues. Thus, the class of indirect effects mediated by trait shifts induced by a change in predator diet is made possible by the existence of infochemicals.

10.4.3 Predator identity

Based on conventional ecological theory, a shift in predator taxonomic identity will have no effect on prey populations or the larger food web if predator feeding rate and diet choice remains unchanged. Prey often respond, however, to a shift in predator identity by employing a different suite of anti-predator traits, especially when the two predators have contrasting modes of foraging (Soluk 1993; McIntosh and Townsend 1996; Laforsch and Tollrian 2004; Hoverman et al. 2005). Chemical information transfer provides a mechanism for predator recognition because each predator species emanates odours with a unique chemical signature. The existence of predator-specific inducible defenses can provide a means by which a change in the identity of a predator can alter food web interactions, even without any change in the feeding patterns of the predator. For example, the pulmonate snail *Physa acuta* responds to chemical cues from sunfish by moving under cover, an adaptive response to a water-column forager. When exposed to crayfish, a benthic forager, snails move to the water's surface. In both cases the snails depleted their periphyton resources within refugia relative to no-risk controls (Fig. 10.5). Thus, the change in predator identity cascaded through the grazers to affect the abundance and spatial distribution of primary producers, even

though the two predators had exactly the same diet (i.e. the same consumptive effect). The complete decoupling of predator diet and predator function in the food web challenges traditional, diet-based approaches to the study of food webs, and is made possible by the rich information content of chemical cues. To our knowledge, all known examples of TMIEs triggered by shifts in predator identity involve, at least in part, chemically based predator-detection mechanisms, and it seems unlikely that visual or mechanical cues could by themselves offer enough information to reliably identify predators.

10.4.4 Chemical cues as an experimental tool

Aquatic ecologists began to design field experiments aimed specifically at investigating trait mediated indirect effects in the 1980s (e.g. Power *et al.* 1985; Turner and Mittelbach 1990; Huang and Sih 1991; Leibold 1991), but devising a method to partition the consumptive effects of predators from the non-consumptive effects proved to be difficult, and these early studies often failed to disentangle the two. At about the same time, there was much excitement generated by the discovery that the trait shifts were often induced by chemical cues associated with predation (e.g. Krueger and Dodson 1981; Petranka *et al.* 1987; Dodson 1988; Crowl and Covich 1990). Community ecologists soon realized that the chemical cues provided a neat solution to the difficult conundrum of partitioning the consumptive and non-consumptive effects of predators. Predators can be fed prey while confined to enclosures that exchange water with the environment, thereby exposing focal animals in the environment to chemical cues associated with predation without altering density. Essentially, chemical cues are used to deceive animals into adopting anti-predator phenotypes. The 'caged predator' approach has proven to be a valuable experimental tool and has become a standard method in aquatic ecology.

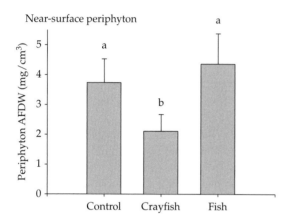

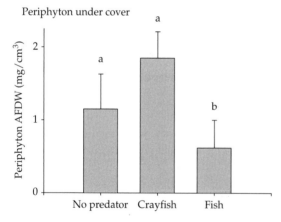

Figure 10.5 Patterns of periphyton growth at the conclusion of an experiment testing the effects of predator identity on the interaction between pulmonate snails and their periphyton resources (data from Bernot and Turner 2001). Bottom: periphyton accumulation on tiles placed under covered substrates. Top: periphyton on tiles placed near the surface of the water. Bars show means ± 1 SE. Different letters indicate significant treatment effects ($P < 0.05$, Tukey's multiple comparison).

10.4.5 Environmental constraints on sensory perception

Constraints on information flow are likely key to community organization, and may have large applied implications, but have received little empirical attention. Environmental factors can compromise the function of animal sensory systems and impair their ability to avoid predators, find food, or locate mates. For example, an emerging body of work shows that hydrodynamic forces such as flow velocity and turbulence can influence chemical odour plumes and how organisms perceive chemical information (Ferner and Weissburg 2005; Smee and Weissburg 2006, Chapter 7). If trait shifts have ecological consequences, then abiotic constraints on perceptual ability will influence the outcome of species interactions. Smee and colleagues have shown that as turbulence and flow increases hard clams (*Mercenaria mercenaria*) are less responsive to crab predators (Smee *et al.* 2008). The overall effect of shifts in the efficacy of chemical cues will depend on the relative impairment of predator and prey. In the case of crabs and clams, the perceptual ability of crabs is also degraded as flow and turbulence increases, but not in a co-linear manner with clams. The clams have a relative advantage at low and high flows, whereas the advantage tilts towards crabs at intermediate flow velocities (Smee *et al.* 2008, 2010). Thus, the relative strength of trait-mediated and density-mediated indirect effects varies in a non-monotonic manner across the hydrodynamic gradients, illustrating the potential complexity of environmental effects on species interactions.

A small but growing number of studies have shown that subtle alterations to the chemical environment can also disrupt chemosensory perception (Chapter 17). Sub-toxic concentrations of heavy metals, pesticides, and surfactants have been shown to impair chemoreception in fish, snails, crustaceans, and amphibians (e.g. 'infodisruption', Blaxter and Hallers-Tjabbes 1992; Lefcort *et al.* 1999; Lurling and Scheffer 2007). Humans have also had a pervasive effect on the chemistry of aquatic systems by altering hydrogen ion concentration (pH). Ocean acidification associated with increasing atmospheric concentration of carbon dioxide, acidification of lakes and streams by sulphur and nitrogen deposition, and photosynthetic elevation of pH associated with eutrophication are well known examples. Marine fish reared in slightly acidified seawater fail to respond appropriately to predator cues (Dixson *et al.* 2009). Similarly, chemoreception of fish (Smith and Lawrence 1988; Brown *et al.* 2002, Leduc *et al.*

2004) and crayfish (Allison *et al.* 1992) is impaired in weakly acidified streams. Eutrophication, which can elevate pH in poorly buffered waters, is a widespread water quality concern, and there is evidence that pulmonate snails fail to respond to predator cues where pH is elevated by eutrophication (Turner and Chislock 2010).

Given the broad extent of acidification and eutrophication of our surface waters, and the prevalence of 'infodisruptors', it is possible that sensory impairment may be quite common. Because studies of chemoreception are heavily biased towards mesocosm and laboratory studies, we have little knowledge of the extent to which chemoreception actually is impaired. Clearly, field assays aimed at assessing the perceptual ability of animals across water quality gradients are essential in order to answer this important question. If infochemicals play an important role in organizing aquatic communities, we expect that disruption of chemical information transfer will have a cascading effect on community structure.

The recognition that environmental factors often constrain chemical information transfer has important implications for the choice of experimental venue and the interpretation of studies showing strong trait-mediated indirect effects. The vast majority of TMIE studies are conducted in small, low flow and low turbulence environments, and use a controlled water source that likely presents a simplified olfactory backdrop (e.g. well water). These conditions will generally maximize perceptual ability of prey and favour strong trait shifts. Animals in a field setting are often faced with the challenge of detecting changes in the concentration of chemical cues embedded within a complex olfactory background (Derby and Sorensen 2008) and delivered via turbulent flow. However, there is a growing list of studies that show strong induction of anti-predator traits in a field setting (reviewed in Werner and Peacor 2003), showing that perceptual ability can remain high and that TMIEs are not confined to laboratory and mesocosm experiments. Inferences regarding NCEs and TMIEs, drawn from laboratory and mesocosm studies, need to consider how environmental factors may enhance or impair chemosensory perception (Peacor 2006; Turner and Chislock 2010).

10.5 Conclusion

Non-consumptive effects and trait-mediated indirect effects are ubiquitous and important determinants of population-, community-, and ecosystem-level processes in both freshwater and marine systems. The majority of NCEs and TMIEs involve infochemicals. Chemical ecologists have developed a detailed understanding of how chemical cues are generated, transmitted, and received. Although chemical signals lie at the heart of most NCEs and TMIEs, little attention has been devoted to examining how the unique informational qualities of chemical signals affects species interactions in aquatic communities. It is likely that an aquatic community in which the available sensory modes did not include infochemicals would look and function very differently than one in which organisms relied heavily on infochemicals. The current challenge is to integrate sensory biology and community ecology, thereby achieving a fuller and more mechanistic understanding of species interactions in food webs.

Acknowledgements

We thank Lars-Anders and Christer for inviting us to participate in the workshop and contribute to this volume. Sharon Montgomery and Matt Gordon provided useful comments on early drafts of the manuscript. AMT acknowledges the support of the National Science Foundation. SDP acknowledges support from National Science Foundation (OCE-0826020) and the Michigan Agricultural Experiment Station.

References

Allison, V., Dunham, D.W., and Harvey, H.H. (1992) Low pH alters response to food in the crayfish *Cambarus bartoni*. *Canadian Journal of Zoology*, **70**: 2416–20.

Abrams, P.A. (1995) Implications of dynamically variable traits for identifying, classifying, and measuring direct and indirect effects in ecological communities. *American Naturalist*, **146**: 112–34.

Abrams, P.A. (2010) Implications of flexible foraging for interspecific interactions: lessons from simple models. *Functional Ecology*, **24**: 7–17.

Bernot, R.J. and Turner, A.M. (2001) Predator identity and trait-mediated indirect effects in a littoral food web. *Oecologia*, **129**: 139–46.

Blaxter, J.H.S. and Hallers-Tjabbes, C.C.T. (1992) The effect of pollutants on sensory systems and behavior of aquatic animals. *Netherlands Journal of Aquatic Ecology*, **26**: 43–58.

Boeing, W. J., Wissel, B. and Ramcharan C.W. (2005) Costs and benefits of *Daphnia* defense against *Chaoborus* in nature. *Canadian Journal of Fisheries and Aquatic Sciences*, **62**: 1286–94.

Bolker, B., Holyoak, M., Krivan, V., Rowe, L., and Schmitz, O. (2003) Connecting theoretical and empirical studies of trait-mediated interactions. *Ecology*, **84**: 1101–14.

Bollens, S.M., Rollwagen-Bollens, G., Quenette, J.A., and Bochdansky, A.B. (2011) Cascading migrations and implications for vertical fluxes in pelagic ecosystems. *Journal of Plankton* Research, **33**: 349–55.

Brönmark, C. and Pettersson, L. B. (1994) Chemical cues from piscivores induce a change in morphology in crucian carp. *Oikos*, **70**: 396–402.

Brönmark, C. and Hansson, L.A. (2000) Chemical communication in aquatic systems: an introduction. *Oikos*, **88**: 103–9.

Brown, G. E., Adrian Jr, J.C., Lewis, M.G. and Tower, J.M. (2002) The effects of reduced pH on chemical alarm signaling in ostariophysan fishes. *Canadian Journal of Fisheries and Aquatic Sciences*, **59**: 1331–8.

Byrnes, J., Stachowicz, J.J., Hultgren, K.M., Hughes, A.R., Olyarnik, S.V., and Thornber, C.S. (2006) Predator diversity strengthens trophic cascades in kelp forests by modifying herbivore behaviour. *Ecology Letters*, **9**: 61–71.

Carpenter, S.R., Kitchell, J.F., and Hodgson, J.R., *et al.* (1987) Regulation of lake primary productivity by food-web structure. *Ecology*, **68**: 1863–76.

Chivers, D. P. and Smith, R. J. F. (1998) Chemical alarm signaling in aquatic predator-prey systems: a review and prospectus. *Ecoscience*, **5**: 338–52.

Chivers, D. P., Mirza, R. S., and Johnston, J. G. (2002) Learned recognition of heterospecific alarm cues enhances survival during encounters with predators. *Behaviour*, **139**: 929–38.

Crowl, T. A. and Covich, A. P. (1990) Predator-induced life history shifts in a freshwater snail: a chemically mediated and phenotypically plastic response. *Science*, **247**: 949–51.

Derby, C.D. and Sorensen, P.W. (2008) Neural processing, perception, and behavioral responses to natural chemical stimuli by fish and crustaceans. *Journal of Chemical Ecology*, **34**: 898–914.

Dicke, M. (2006) Chemical ecology from genes to ecosystems: Integrating 'omics' with community ecology.

In M. Dicke and W. Takken, (eds.) *Chemical Ecology: From Gene to Ecosystem.* Springer-Verlag, New York, pp. 175–89.

Dini, M.L., O'Donnell, J., Carpenter, S.R., Elser, M.M., Elser, J.J., and Bergquist, A.M. (1987) *Daphnia* size structure, vertical migration, and phosphorus redistribution. *Hydrobiologia,* **150**: 185–91.

Dixson, D. L., Munday, P.L., and Jones, G.P. (2009) Ocean acidification disrupts the innate ability of fish to detect predator olfactory cues. *Ecology Letters,* **12**: 1–8.

Dodson, S.I. (1988) The ecological role of chemical stimuli for the zooplankton: Predator-avoidance behavior in *Daphnia. Limnology and Oceanography,* **33**: 1431–9.

Edeline, E., Haugen, T.O., Weltzien, F.-A. *et al.* (2010) Body downsizing caused by non-consumptive social stress severely depresses population growth rate. *Proceedings of the Royal Society B,* **277**: 843–51.

Eklöv, P. and VanKooten, T. (2001) Facilitation among piscivorous predators: effects of prey habitat use. *Ecology,* **82**: 2486–94.

Ferner, M.C. and Weissburg, M.J. (2005) Slow-moving predatory gastropods track prey odors in fast and turbulent flow. *Journal of Experimental Biology,* **208**: 809–19.

Fraker, M.E., Hua, F., Cuddapaha, V. *et al.* (2009) Characterization of an alarm pheromone secreted by amphibian tadpoles that induces behavioral inhibition and suppression of the neuroendocrine stress axis. *Hormones and Behavior,* **55**: 520–9.

Grabowski, J. H. and Kimbro, D.L. (2005) Predator-avoidance behavior extends trophic cascades to refuge habitats. *Ecology,* **86**: 1312–19.

Hall, D.J., Werner, E.E., and Gilliam, J.F. *et al.* (1979) Diel foraging behavior and prey selection in the golden shiner (*Notemigonus crysoleucas*). *Canadian Journal of Fisheries and Aquatic Sciences,* **42**: 1608–13.

Haupt, F., Stockenreiter, M., and Reichwaldt, E.S. *et al.* (2010) Upward phosphorus transport by *Daphnia* diel vertical migration. *Limnology and Oceanography,* **55**: 529–34.

Hay, M.E. and Kubanek, J. (2002) Community and ecosystem level consequences of chemical cues in the plankton. *Journal of Chemical Ecology,* **28**: 2001–16.

He, X. and Kitchell, J.F. (1990) Direct and indirect effects of predation on a fish community: a whole-lake experiment. *Transactions of the American Fisheries Society,* **119**: 825–35.

Hoverman, J.T., Auld, J.R., and Relyea, R.A. (2005) Putting prey back together again: integrating predator-induced behavior, morphology, and life history. *Oecologia,* **144**: 481–91.

Huang, C. and Sih, A. (1991) Experimental studies on direct and indirect interactions in a three trophic-level system. *Oecologia,* **85**: 530–6.

Kitchell, J.F., O'Neill, R.V., Webb, D. *et al.* (1979) Consumer regulation of nutrient cycling. *BioScience,* **29**: 28–34.

Krueger, D. A. and Dodson, S. I. (1981) Embryological induction and predation ecology in *Daphnia pulex. Limnology and Oceanography,* **26**: 212–23.

Laforsch, C. and Tollrian, R. (2004) Inducible defenses in multipredator environments: cyclomorphosis in *Daphnia cucullata. Ecology,* **85**: 2302–11.

Laforsch, C., Beccara, L., and Tollrian, R. (2006) Inducible defenses: the relevance of chemical alarm cues in *Daphnia. Limnology and Oceanography,* **51**: 1466–72.

Leduc, A. O. H. C., J. M. Kelly, and G. E. Brown. (2004) Detection of conspecific alarm cues by juvenile salmonids under neutral and weakly acidic conditions: laboratory and field tests. *Oecologia,* **139**: 318–24.

Lefcort, H., Thomson, S.M. and Cowles, E.E. *et al.* (1999) Ramifications of predator avoidance: predator and heavy-metal mediated competition between tadpoles and snails. *Ecological Applications,* **9**: 1477–89.

Leibold, M.A. (1991) Trophic interactions and habitat segregation between competing *Daphnia* species. *Oecologia,* **86**: 510–20.

Leibold, M. A. and Wilbur, H.M. (1992) Interactions between food web structure and nutrients on pond organisms. *Nature,* **360**: 341–3.

Lima, S. L. (1998) Nonlethal effects in the ecology of predator-prey interactions. *Bioscience,* **48**: 25–34.

Lurling, M. and Scheffer, M. (2007) Info-disruption: pollution and the transfer of chemical information between organisms. *Trends in Ecology and Evolution,* **22**: 374–9.

Mathis, A. and Smith, R. J. F. (1993) Intraspecific and cross-superorder responses to chemical alarm signals by brook stickleback. *Ecology,* **74**: 2395–404.

McCollum, E. W., Crowder, L.B. and McCollum, S.A. (1998) Complex interactions of fish, snails, and littoral zone periphyton. *Ecology,* **79**:1980–94.

McIntosh, A.R. and Townsend, C.R. (1996) Interactions between fish, grazing invertebrates and algae in a New Zealand stream: a trophic cascade mediated by fish induced changes to grazer behavior? *Oecologia,* **108**: 174–81.

McIntosh, A. R., Peckarsky, B. L., and Taylor, B. W. (2004) Predator-induced resource heterogeneity in a stream food web. *Ecology,* **85**:2279–90.

McPeek, M.A. and Peckarsky, B.L. (1998) Life histories and the strengths of species interactions: combining mortality, growth, and fecundity effects. *Ecology,* **79**: 867–79.

Mittelbach, G.G. (1988) Competition among refuging sunfishes and effects of fish density on littoral zone invertebrates. *Ecology,* **69**: 614–23.

Mittelbach, G.G. and Chesson, P.L. (1987) Predation risk: indirect effects on fish populations. In W.C. Kerfoot and

A. Sih, (eds.) *Predation: Direct and Indirect Impacts on Aquatic Communities.*. University Press of New England, Hanover, NH, USA, pp. 315–32.

Pangle, K.L., Peacor, S.D., and Johannsson, O.E. (2007) Large nonlethal effects of an invasive invertebrate predator on zooplankton population growth rate. *Ecology*, **88**: 402–12.

Peacor, S. D. (2002) Positive effect of predators on prey growth rate through induced modifications of prey behavior. *Ecology Letters,* **5**: 77–85.

Peacor, S D. (2006) Behavioral response of bullfrog tadpoles to chemical cues of predation risk is affected by cue age and water source. *Hydrobiologia*, **573**: 39–44.

Peacor, S.D. and Werner, E.E. (2000) The effects of a predator on an assemblage of consumers through induced changes in consumer foraging behavior. *Ecology,* **81**: 1998–2010.

Peacor, S.D. and Werner, E.E. (2001) The contribution of trait-mediated indirect effects to the net effects of a predator. *Proceedings of the National Academy of Sciences (USA)*, **98**: 3904–8.

Peacor, S. D. and Werner, E.E. (2004) Context dependence of nonlethal effects of a predator on prey growth. *Israel Journal of Zoology,* **50**: 139–67.

Peacor, S. D., Schiesari, L., and Werner, E.E. (2007) Mechanisms of nonlethal predator effect on cohort size variation: ecological and evolutionary implications. *Ecology,* **88**: 1536–47.

Peacor, S. D. and Cressler. C. E. In Press. The implications of adaptive prey behavior for ecological communities: a review of current theory. In O. Schmitz, T. Ohgushi, and R. D. Holt, (eds.) *Evolution and Ecology of Trait-Mediated Indirect Interactions: Linking Evolution, Community, and Ecosystem.* Cambridge University Press, Cambrige, UK.

Peckarsky, B. L., Taylor, B.W., McIntosh, A.R., McPeek, M.A., and Lytle, D.A. (2001) Variation in mayfly size at metamorphosis as a developmental response to risk of predation. Ecology, **82**: 740–57.

Peckarsky, B. L., Bolnick, D.I., Dill, L.M. *et al.* (2008) Rewriting the textbooks: Considering non-consumptive effects in classic studies of predator-prey interactions. *Ecology*, **89**: 2416–25.

Petranka, J.W., Kats, L.B. and Sih, A. (1987) Predator-prey interactions among fish and amphibians: the use of chemical cues to detect predatory fish. *Animal Behaviour*, **35**: 420–5.

Preisser, E. (2007) When, and how much, does fear matter? Quantitatively assessing the impact of predator intimidation of prey on community dynamics, National Center for Ecological Analysis and Synthesis Data Repository, http://knb.ecoinformatics.org/knb/metacat/nceas.873.24/nceas [accessed 21 September 2011]

Power, M.E., Matthews, W.J., and Stewart, A.J. (1985) Grazing minnows, piscivorous bass, and stream algae: dynamics of a strong interaction. *Ecology*, **66**: 1448–56.

Relyea R.A. (2000) Trait-mediated indirect effects in larval anurans: reversing competition with the threat of predation. *Ecology*, **81**: 2278–89.

Relyea, R.A. and Werner, E.E. (2000) Morphological plasticity in four larval anurans distributed along an environmental gradient. *Copeia*, **2000**: 178–90.

Romare, P. and Hansson, L.A. (2003) A behavioral cascade: Top-predator induced behavioral shifts in planktivorous fish and zooplankton. *Limnology and Oceanography*, **48**: 1956–64.

Schindler, D.E., Carpenter, S.R., Cottingham, K.L. *et al.* (1996) Foodweb structure and littoral zone coupling to pelagic trophic cascades. In G.A. Polis and K.O. Winemiller, (eds.) *Food Webs: Integration of Pattern and Dynamics.* Chapman & Hall, Inc., New York, pp.96–105.

Schindler, D.E., Scheuerell, M.D., Moore, J.W., Gende, S.M., Francis, T.B., and Palen, W.J. (2003) Pacific salmon and the ecology of coastal ecosystems. *Frontiers in Ecology and the Environment*, **1**: 31–7.

Schmitz, O.J. (2008) Effects of predator hunting mode on grassland ecosystem function. *Science*, **319**: 952–4.

Schmitz, O.J., Krivan, V., and Ovadia, O. (2004) Trophic cascades: the primacy of trait-mediated indirect interactions. *Ecology Letters*, **7**: 153–63.

Schoeppner, N. M. and Relyea, R. A. (2005) Damage, digestion, and defense: the roles of alarm cues and kairomones for inducing prey defenses. *Ecology Letters*, **8**: 505–12.

Sih, A., Englund, G., and Wooster, D. (1998) Emergent impacts of multiple predators on prey. *Trends in Ecology and Evolution*, **13**: 350–5.

Simon, K.S., Townsend, C.R., Biggs, B.J.F., Bowden, W.B., and Frew, R.D. (2004) Habitat-specific nitrogen dynamics in New Zealand streams containing native or invasive fish. *Ecosystems*, **7**: 777–92.

Skelly, D.K. and Werner, E.E. (1990) Behavioral and lifehistorical responses of larval American toads to an odonate predator. *Ecology*, **71**: 2313–22.

Slos, S. and Stoks, R. (2008) Predation risk induces stress proteins and reduces antioxidant defense. *Functional Ecology*, **22**: 637–42.

Smee, D. L., and Weissburg, M.J. (2006) Clamming up: environmental forces diminish the perceptive ability of bivalve prey. *Ecology*, **87**: 1587–98.

Smee, D.L., Ferner, M.C., and Weissburg, M.J. (2008) Environmental conditions alter prey reactions to risk and the scales of nonlethal predator effects in natural systems. *Oecologia*, **156**: 399–409.

Smee, D.L., Ferner, M.C., and Weissburg, M.J. (2010) Hydrodynamic sensory stressors produce nonlinear predation patterns. *Ecology*, **91**: 1391–400.

Smith, R.J.F. (1992) Alarm signals in fishes. *Reviews in Fish Biology and Fisheries*, **2**: 33–63.

Smith, R. J. F., and Lawrence, B.J. (1988) Effects of acute exposure to acidified water on the behavioral response of fathead minnows, *Pimephales promelas*, to alarm substance (Schreckstoff). *Environmental Toxicology and Chemistry*, **7**: 329–35.

Soluk, D.A. (1993) Mutiple predator effects: predicting the combined functional response of stream fish and invertebrate predators. *Ecology*, **74**: 219–25.

Stief, P. and Holker, F. (2006) Trait-mediated indirect effects of predatory fish on microbial mineralization in aquatic sediments. *Ecology*, **87**: 3152–9.

Tollrian, R. and Harvell, C. D. (1999) *The Ecology and Evolution of Inducible Defenses*. Princeton Univ. Press. New Jersey.

Trussell, G.C., P.J. Ewanchuk, M.D. Bertness, and B.R. Silliman. (2004) Trophic cascades in rocky shore tide pools: distinguishing lethal and nonlethal effects. *Oecologia*, **139**: 427–32.

Trussell, G.C., Ewanchuk, P.J., and Matassa, C.M. (2006a) The fear of being eaten reduces energy transfer in a simple food chain. *Ecology*, **87**: 2979–84.

Trussell, G.C., Ewanchuk, P.J., and Matassa, C.M. (2006b) Habitat effects on the relative importance of trait- and density-mediated indirect interactions. *Ecology Letters*, **9**: 1245–52.

Turner, A.M. (1997) Contrasting short-term and long-term effects of predation risk on consumer habitat use and resources. *Behavioral Ecology*, **8**: 120–5.

Turner, A.M. (2004) Nonlethal effects of predators on prey growth depend on environmental context. *Oikos*, **104**: 561–9.

Turner, A.M. (2008) Predator diet and prey behaviour: freshwater snails discriminate among closely related prey in a predator's diet. *Animal Behaviour*, **76**: 1211–17.

Turner, A.M., and Mittelbach, G.G. (1990) Predator avoidance and community structure: interactions among piscivores, planktivores, and plankton. *Ecology*, **71**: 2241–54.

Turner, A.M., Bernot, R.J., and Boes, C.M. (2000) Chemical cues modify species interactions: the ecological consequences of predator avoidance by freshwater snails. *Oikos*, **88**: 148–58.

Turner, A.M. and Montgomery, S.L. (2003) Spatial and temporal scales of predator avoidance: experiments with fish and snails. *Ecology*, **84**: 616–22.

Turner, A.M. and Chislock, M.J. (2010) Blinded by the stink: nutrient enrichment impairs the perception of predation risk by freshwater snails. *Ecological Applications*, **20**: 2089–95.

Van Buskirk, J. and Yurewicz, K.L. (1998) Effects of predators on prey growth rate: relative contributions of thinning and reduced activity. *Oikos*, **82**: 20–8.

Van Buskirk, J. and Arioli, M. (2002) Doseage response of an induced defense: how sensitive are tadpoles to predation risk? *Ecology*, **83**: 1580–5.

Vanni, M. J. (2002) Nutrient cycling by animals in freshwater ecosystems. *Annual Review of Ecology and Systematics*, **33**: 341–70.

Walters C.J. and Juanes F. (1993) Recruitment limitation as a consequence of optimal risk-taking behaviour by juvenile fish. *Canadian Journal of Fisheries and Aquatic Sciences*, **50**: 2058–70.

Werner, E.E. (1991) Non-lethal effects of a predator on competitive interactions between two anuran larvae. *Ecology*, **72**: 1709–20.

Werner E.E. and Anholt, B.R. (1996) Predator-induced indirect effects: consequences to competitive interactions in anuran larvae. *Ecology*, **77**: 157–69.

Werner, E. E. and Peacor, S.D. (2003) A review of trait-mediated indirect interactions. *Ecology*, **84**: 1083–100.

Werner, E. E. and Peacor, S.D. (2006) Interaction of lethal and non-lethal effects of a predator on a guild of herbivores mediated by system productivity. *Ecology*, **87**: 347–61.

Wissinger, S.A., Whiteman, H.H., Denoël, M., Mumford, M., and Aubee, C. (2010) Consumptive and non-consumptive effects of cannibalism in fluctuating age-structured populations. *Ecology*, **91**: 549–59.

CHAPTER 11

Neuroecology of predator–prey interactions

Charles D. Derby and Richard K. Zimmer

Many investigations to date have focused on the physiology of nervous systems, the behaviour that they precipitate, and the effects of organismal interactions on the structure and function of communities. However, few studies have connected the dots between neurobiological mechanisms of chemosensory systems and the roles of chemistry in mediating ecological interactions. By unifying principles from disciplines spanning several levels of biological organization, the emerging field of neuroecology can make these connections (Zimmer and Derby 2007). It encourages investigators to cross, and re-cross, traditional disciplinary boundary lines. Although challenging, this approach is critical for understanding the full ecological impact of nervous systems. The ability to identify environmentally meaningful compounds and to predict their effects on native and non-native species is fundamentally important for developing both basic ecological theory and strategies of conservation management. Thus, a neuroecological approach that links physiology with behaviour, ecology, and evolution would fill a lingering void at the intersection of all four disciplines (Fig. 11.1).

Every living organism has a chemical sense to exploit valuable resources and/or ward off danger. Determination of molecular structures is required to distinguish unique natural products from more common metabolites that serve a variety of roles. Molecules used in attraction and courtship of lepidopterans are an especially useful example (Roelofs and Rooney 2003). These insects have a wide range of novel compounds and blends of mixtures that confer species-specific recognition

(de Bruyne and Baker 2008). Investigation of the molecular basis for biosynthesis of attractants has revealed much about the neuroecology of the systems of chemical communication and the explosive radiation of lepidopterans. As an example, a single point mutation can cause activation of a nonfunctional desaturase gene transcript, alter attractant structure, and lead to insipient speciation within a single generation of moths (Roelofs and Rooney 2003).

Ordinary molecules can also have extraordinary effects. Histamine, for instance, initiates local immune responses, regulates receptor-mediated physiological processes in the gut, acts as a neurotransmitter, and stimulates habitat colonization by organisms of highly divergent species (Swanson *et al*. 2004; Jutel *et al*. 2006; Stuart *et al*. 2007). Studies on the comparative biology of signalling systems across taxa, therefore, can lead to improved appreciation for neuroecological mechanisms that create and maintain biodiversity.

From the perspective of a predator, carnivory is influenced by many different types of chemical cues and signals. Attractants mediate searches for distant prey (e.g. Zimmer-Faust 1993). Stimulants and suppressants increase or decrease localized appetitive and consummatory phases of feeding, respectively (e.g. Carr 1988). Deterrents and defences discourage or act in ways harmful to predators (e.g. Hay 2009). An act of predation thus requires integration of complex chemosensory information as a function of ecological context and proximity of prey.

In the following sections, we discuss the neuroecology of predation while highlighting results for

Neuroecology

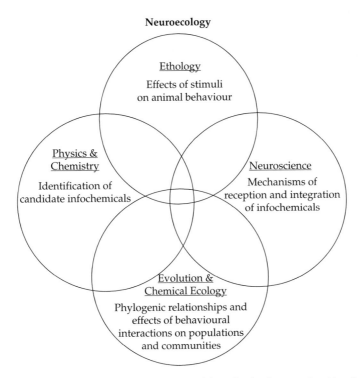

Figure 11.1   A conceptualization of the neuroecological approach to the study of chemical ecology is presented as a Venn diagram with disciplinary approaches shown in the circles.

spiny lobsters (primarily *Panulirus interruptus* and *P. argus*) and blue crabs (primarily *Callinectes sapidus* and *C. similis*). Arguably, there is more known about the integrative biology of these predatory crustaceans than any other aquatic species. With a few notable exceptions (as discussed in later sections), structural identities have not been established for molecules released from prey which serve as attractants, stimulants, deterrents, and suppressants for spiny lobsters and crabs. This review first examines natural histories and trophic ecologies, and then it describes behavioural aspects of predatory search, appetitive feeding, and food consumption, as well as chemical deterence and defence by prey. Next, it considers the physiological basis for chemosensory-mediated behavioural processes in association with predation and carnivorous feeding. Finally, it argues for stronger integration across disciplines towards clarifying basic relationships between neural processes and ecological functions.

11.1 Natural histories and trophic ecologies of spiny lobsters and blue crabs

Spiny lobsters and blue crabs are dominant predators and play seminal roles in structuring species distributions and prey populations within soft sedimentary and rocky reef habitats (e.g. Virstein 1977; Robles *et al.* 1990). Spiny lobsters, for example, feed voraciously in shallow, subtidal environments on sea urchins and other prey species (Tegner and Levin 1983; Shears and Babcock 2002). Urchins are recognized as keystone species because of their seminal roles through herbivory in regulating the dynamics of giant kelp forests (Harrold and Reed 1985). In the presence of spiny lobsters, populations of sea urchins are reduced and impacts of their grazing on kelp forest communities are minimized (Tegner and Levin 1983; Shears and Babcock 2002). Moreover, spiny lobsters move at night on flood tides from offshore reefs to ravage intertidal mussel beds when hydrodynamic forcing (waves and currents) is relatively weak (Robles *et al.* 1990). Thus, spiny lobsters can create significant biotic disturbances to communities dominated by foundation and keystone species, and they can affect considerably the organization, material, and energy flow within trophic webs.

Blue crabs are known as 'invertebrate sharks' and locally wreak havoc on a wide range of phylogenetically diverse prey, exerting strong impacts on communities (Virstein 1977). Perhaps the best understood predator–prey interaction is between blue crabs and hard clams (*Mercenaria mercenaria*). These clams are numerically-dominant suspension feeders of considerable ecological and economic consequence within both muddy and sandy habitats (Mann *et al.* 2005). Rates of predation by blue crabs on hard clams are extremely high under a variety of conditions, including different sediment type, plant cover, and water quality (Laughlin 1982; Orth *et al.* 1984). Experimental caging or removal of predators convincingly demonstrates the seminal role played by blue crabs in structuring populations of hard clams, as well as in determining the composition of species within a community (Virstein 1977; Silliman and Bertness 2002).

The process by which spiny lobsters and blue crabs find and eat prey consists of a series of tasks, each controlled sequentially by different subsets of chemosensors (Box 11.1). The first challenge is to detect and identify chemicals from a distant prey source. This phase is controlled by receptor cells on the first pair of antennae (antennules), through activation of either an olfactory or non-olfactory pathway that leads from the peripheral to the central nervous system, including the brain. Following chemical detection and initiation of search, directional navigation towards a prey is controlled by neural input from the antennules in spiny lobsters and blue crabs, as well as from the legs in crabs (Derby *et al.* 2001; Keller *et al.* 2003; Schmidt and Mellon 2011). Detection of proximal prey involves locating this chemical source using leg chemoreceptors. After contact is made, local grasping reflexes are initiated that lead to prey being picked up and brought to the mouth. In the mouth, a decision to swallow or eject is mediated by chemoreceptors on the mouthparts and esophagus. Multifaceted relationships among neural, behavioural, and ecological processes thus act at each step in chemosensory-mediated predation.

Box 11.1 Crustacean chemical sensors

Chemical sensors are present on most body surfaces of crustaceans, as shown below for spiny lobsters. These include appendages such as the first antennae (= antennules), second antennae, legs, and mouthparts, but also body regions including the cephalothorax, abdomen, and telson (a). The chemosensors are organized as sensilla, which are cuticular extensions of the body surface that are innervated by the dendrites of chemosensory neurons, and are shown by scanning electron microscopy in panels b–e. The aesthetasc sensilla (in panel B) are a prominent type of chemosensor that are located only on the distal half of the antennular lateral flagella. They are the only known unimodal chemosensory sensilla and are the receptors for the olfactory pathway. All of the other known chemosensilla on *P. argus* are bimodal, being innervated by both chemosensory neurons and mechanosensory neurons, and they represent the non-olfactory chemosensory pathways. Examples of the non-olfactory sensilla are shown in panels c–e. Besides the aesthetascs, the distal region of the antennular lateral flagellum contains the bimodal guard setae, companion setae, and asymmetric setae, as shown in panel b.

Hooded sensilla are found on most body surfaces; examples shown here are from the antennular lateral flagellum (c). Simple sensilla are also found on most body surfaces; examples shown here are a medium simple sensillum (d) and a long simple sensillum (e) on the antennule. Modified from figures in Cate and Derby (2001), both with kind permission from Springer Science and Business Media.

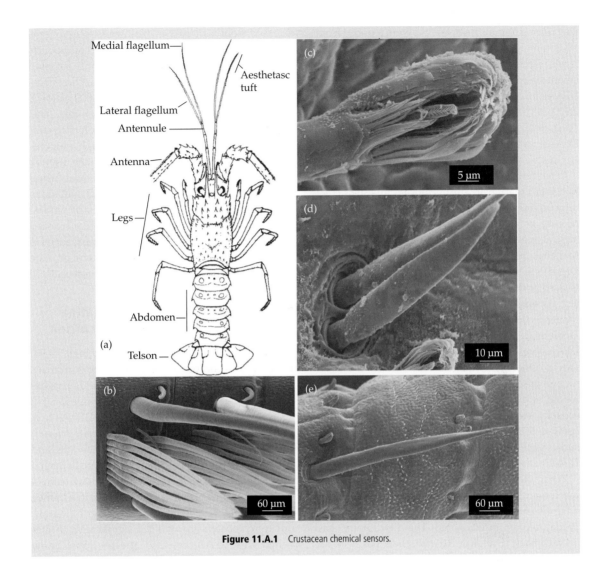

Figure 11.A.1 Crustacean chemical sensors.

11.2 Chemical attraction mediating search for live, intact prey

For many years, dissolved free amino acids (DFAAs) have been considered the principal molecules that attract marine predators, including blue crabs and spiny lobsters, to their prey (Carr 1988). These compounds are highly soluble and diffusible in seawater, stimulate receptors, and cause oriented locomotory responses in laboratory and field experiments (e.g. Derby *et al.* 2001). Previous studies by sensory biologists, however, did not measure release rates of DFAAs from live, intact prey. Consequently, investigations failed to scale appropriately the concentrations of DFAAs in behavioural and physiological tests.

This long-standing paradigm is now under revision. Fluorescent dye studies determined the lag time for water to move in a single pass from the incurrent siphon, through a live, intact clam (*Mercenaria mercenaria*), and into the excurrent siphon. Hence, the same parcel of water could be sampled before and immediately after it passed through a clam (Zimmer *et al.* 1999). HPLC analyses showed that, during a single pass, DFAAs were reduced from 10–20 nmol L^{-1} in natural seawater to undetectable levels (Table 11.1). In contrast, ammonium and ethanolamine, which are breakdown products of DFAAs, increased respectively from 800–900 to 2,500–3,000 nmol L^{-1}, and from 1–5 to 10–20 nmol L^{-1}. Thus, live intact clams take up and metabolize DFAAs rather than release them into the marine environment.

Extending beyond hard clams, nearly all other live, intact marine organisms thus far investigated— representing species in 18 classes and 13 phyla— take-up, or fail to emit, DFAAs as trace organics from natural seawater (e.g. Stephens 1988; Wright and Manahan 1989; Zimmer *et al.* 1999). In cases where DFAAs are leaked by live prey, such as in some crustaceans and teleost fishes, rates are extremely low (e.g. 0.1 nmol min^{-1} g (wet mass tissue)$^{-1}$) and do not attract predators from a distance (Zimmer *et al.* 1999; Finelli *et al.* 2000; Krug and Zimmer, unpubl. data). DFAAs are valuable commodities since they function as osmolytes to maintain cell volume in seawater and as substrates for protein synthesis. Their loss would be expensive energetically, limit homeostatic capacity and scope for growth, and expose prey to being detected by predators. Strong selective pressures therefore favour uptake and retention of DFAAs, as opposed to their release.

What, then, are the natural cues attracting spiny lobsters and blue crabs to live, intact prey? Moreover, what (if any) roles do DFAAs play in determining predator–prey interactions? The first query has been difficult to answer. To date, there has been no complete structural elucidation for any attractant

released from a marine or freshwater prey organism. Suspension-feeding clams release compounds into overlying waters that are attractants for blue crabs. These compounds have been partially characterized as low molecular weight (< 1,000 g mol^{-1}), both ethanol (80%) soluble and insoluble, and degraded through hydrolysis by reaction with concentrated (12 N) hydrochloric acid at high temperature (160 °C) (Zimmer *et al.* 1999; Zimmer unpubl. data). Use of selective enzymic degradations (using endo- and exo-peptidases) and chemical oxidations have so far not illuminated properties of natural attractants. Future research using more advanced chromatographic and spectroscopic approaches is needed to complete chemical isolations and provide full structural elucidations of bioactive molecules.

11.3 Chemical attraction mediating search for a 'free lunch'—fresh carrion

Concentrations of DFAAs are extremely high (1–500 mmol L^{-1}) in the tissues of most marine animals. Consequently, these compounds could provide meaningful cues announcing the availability of recently injured prey or fresh carrion. DFAAs leak profusely (0.3–7.0 μmol min^{-1} g (wet tissue mass)$^{-1}$) when integuments or shells of invertebrates and fish are compromised and their tissues are damaged by a single cut from a crab claw or a blow using a blunt object (Zimmer *et al.* 1999; Krug and Zimmer, unpubl. data). Still, in field trials, neither blue crabs nor spiny lobsters are attracted significantly to sites baited with DFAAs released at rates approaching those described above (Zimmer-Faust and Case 1982, 1983; Finelli *et al.* 2000; Zimmer unpubl. data).

Whereas DFAAs released at ecologically-relevant rates fail to attract crabs and spiny lobsters, they do significantly attract scavenging mud snails (*Ilyanassa obsoleta*) and isopods (*Cirolana diminuta*) (Zimmer *et al.* 1999; Krug and Zimmer, unpubl. data). One potential explanation for this dissimilarity is differences in trophic ecologies of the taxa. Predators, including blue crabs and spiny lobsters, might be more highly discriminating of the food that they eat. They could require more complex mixtures of attractant molecules that include, but are not limited to, DFAAs. For example, adenine nucleotides,

Table 11.1 Concentrations (mean ± 1 SEM) of total dissolved free amino acids (DFAAs) and amino acid breakdown products (ammonium and ethanolamine) in incurrent and excurrent waters of hard clams, *Merceneria mercenaria* (data from Zimmer *et al.* 1999). Each water sample (n = 12 clams) was analysed in triplicate by HPLC, with 18 individual amino acids quantified spectrofluorometrically using modifications of AccQTag (Waters) labelling and separation chemistry

Compound	Concentration (nmol L^{-1})	
	Incurrent	Excurrent
DFAAs	26.5 ± 0.6	None detected
Ammonium	861.4 ± 67.3	2,943.2 ± 885.1
Ethanolamine	3.1 ± 1.0	12.2 ± 3.1

betaine, and other small nitrogenous molecules and organic acids are released from the tissues of damaged prey (Carr 1988), are attractive to spiny lobsters and other crustacean predators, and add to the activity of DFAAs when mixed with them (Carr 1988; Derby 2000). In contrast, high concentrations of DFAAs leaching from fresh carrion are necessary and sufficient to attract mud snails and isopods in natural field habitats (Zimmer *et al.* 1999; Krug and Zimmer, unpubl. data). The attraction of scavenging mud snails and isopods is more tightly coupled to the magnitude of overall release of DFAAs, and less to the specific molecular identity of any single amino acid.

There is, however, a possible alternative explanation. Attraction may be dependent on the density of animals and their proximity to the food source. Experiments on scavengers were performed in habitats having large population densities of mud snails and isopods (Zimmer *et al.* 1999; Krug and Zimmer, unpubl. data). Following release from damaged tissues, DFAAs did not need to travel far to induce attraction. In comparison, population densities of

blue crabs and spiny lobsters were low and each animal had to travel several metres, or farther, to find emission sites. Seawater samples collected 3 cm, but not 32 or 1,500 cm, distant from prey (abalone) flesh were characterized by significantly elevated levels of DFAAs relative to ambient seawater (Fig. 11.2). As a consequence of turbulent mixing, levels of DFAAs may have been diluted rapidly to concentrations below those causing attraction of spiny lobsters or blue crabs. Future experiments will be needed to distinguish among competing hypotheses.

11.4 Chemical stimulation and suppression of feeding

Behavioural investigations of appetitive and consummatory phases of feeding have usually shown that crustaceans are sensitive to low molecular weight ($< 1,000$ g mol^{-1}) compounds. As stated above, select organic nitrogenous substances (including DFAAs), organic acids (including succinate and oxalate), and adenine nucleotides (espe-

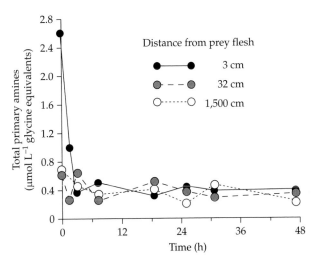

Figure 11.2 Total primary amines (dissolved free amino acids) for water sampled near prey flesh, over a 47-h period. Concurrent with field trapping experiments on spiny lobsters (*Panulirus interruptus*), 180 g of excised abalone foot muscle (prey flesh, *Haliotis rufescens*) was placed in a large mesh bag, centrally positioned in a trap with entry closed, and put in a field study area (3 m ocean depth, More Mesa Reef, Santa Barbara, CA, USA). Concentrations were determined by fluorescence after reaction with ortho-phthaldehyde, using a continuous, flow-injection system. Glycine was used as a standard; for clarity, data are expressed as mean concentrations without standard errors. Time = 0 is defined as the field placement of prey flesh. The concentration of amines fell dramatically with time of leaching and distance from the source, such that the amine level in the leachate was significantly above that in seawater only within 3 h of beginning of leaching and only at 3 cm distance from the flesh. Data are from Zimmer-Faust and Case (1982) and Zimmer (unpubl. data).

cially ATP) are stimulatory to spiny lobsters and blue crabs (Zimmer-Faust et al. 1984; Zimmer-Faust 1987; Carr 1988; Derby 2000). Relatively high concentrations (1–10 µmol L^{-1}) initiate localized search and feeding behaviours such as raising and lowering of the claws (if present), increased rates of antennule flicking, labiating with mouthparts, grooming, grasping and clutching with legs, and finally, ingestion of food.

Single compounds can have significant effects, but often it is the interaction among compounds that exerts the largest impact. Prey tissues are naturally complex chemical milieus. Mixture interactions, therefore, can be either positive (synergistic) or negative (suppressant) (Fig. 11.3). They may inhibit or potentiate behavioural activities to levels far higher or lower than those evoked by isolated compounds (Carr 1988; Derby 2000). For California spiny lobsters, feeding suppression and synergism were caused by substances that resided in a low molecular weight fraction of prey (abalone) flesh (Zimmer-Faust et al. 1984). Using substances (ammonium, glycine, succinate) occurring in the flesh, mixture interactions were shown to affect the chemosensory processes. Ammonium inhibited the behaviour caused by glycine and succinate, without binding these substances (Zimmer-Faust et al. 1984; Zimmer-Faust 1987; Zimmer unpubl. data). Moreover, synergism in the stimulation of spiny lobsters was observed upon combining monocarboxylic amino acids, including glycine, with dicarboxylic acids, including succinate. This mixture was significantly more effective in eliciting behaviour than either of the constituent amino acids or organic acids tested alone, at the same overall concentration (Fig. 11.3).

Synergism likely arises, at least in part, from the simultaneous stimulation of different chemoreceptor sites, each varying in chemical specificity. Behavioural results suggest that glycine and succinate may be stimulating different receptor sites and possibly different cell populations. Because ammonium was inhibitory to glycine and succinate, it is further suggested that (1) ammonium may act nonspecifically to suppress behaviour caused by sensory input from these two different receptor sites, and (2) competitive interactions between suppressants and stimulants seem unlikely to occur, at both

receptor sites for glycine and for succinate, due to major differences in the molecular structures and charges of these substances. Electrophysiological experiments have revealed, at the level of the peripheral and central nervous systems, some of the mechanisms behind mixture interactions (see Section 11.6). However, integrated behavioural and electrophysiological experiments are now needed to test the predictions from behavioural findings.

Relationships have been partially established between mechanisms of mixture interactions and their ecological consequences. Theoretically, suppression and synergism could act to maximize the net rate of nutrient or energy gain, or to minimize the time needed in acquiring a set food reward (Zimmer-Faust 1987, 1993). ATP, for example, is nearly universal as a carrier of chemical energy in metabolic pathways. Not only does ATP in food (at concentrations typical of prey flesh) induce feeding, but free ATP in seawater acts as a localized chemoattractant to spiny lobsters (Zimmer-Faust 1993). The effect of ATP as a behavioural stimulant depends on the sensory information that it conveys, not on its value as food, because once ingested, this molecule is hydrolyzed before uptake across the consumer's gut wall. Food selection by spiny lobsters depends both on ATP concentration and on the mixture blend of adenine nucleotides. When animal flesh begins to degrade, ATP turns over rapidly to adenosine 5′-monophosphate (AMP) because the intermediate, adenosine 5′-diphosphate (ADP), is unstable and quickly dephosphorylates. The observed suppressant effect of AMP on attraction of spiny lobster to ATP therefore focuses foraging on recently killed or injured prey (Fig. 11.3, and Zimmer-Faust 1993). Combined, the profusion of ATP in metabolically active tissues and its rapid decay following cell death make it a reliable chemosensory cue of live (injured) prey and fresh meat for predatory organisms.

DFAAs and ammonium, like adenine nucleotides, could signal the quality of food. DFAAs are a critical source of nitrogen for heterotrophic protein synthesis and comprise as much as 8–10% dry mass of live muscle tissue. Ammonium, on the other hand, is a nitrogenous waste product of transamination in protein catabolism and non-nutritive to heterotrophic consumers. Biodegradatory bacteria

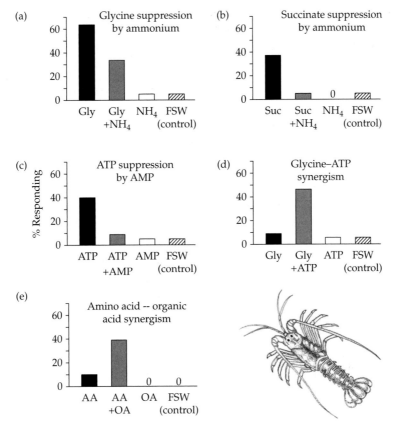

Figure 11.3 Percentages of California spiny lobsters (*Panulirus interruptus*) responding with appetitive feeding behaviour (i.e. a combined increase in antennule flicking, pereiopod (leg) probing, and antennule wiping against mouthparts) to single compounds or simple mixtures. (a) Responses to 10^{-5} mol L^{-1} glycine, 10^{-5} mol L^{-1} ammonium, an equimolar mixture of 10^{-5} mol L^{-1} glycine + 10^{-5} mol L^{-1} ammonium, and seawater control. (b) Responses to 10^{-5} mol L^{-1} succinate, 10^{-5} mol L^{-1} ammonium, an equimolar mixture of 10^{-5} mol L^{-1} succinate + 10^{-5} mol L^{-1} ammonium, and seawater control. (c) Responses to 10^{-5} mol L^{-1} adeonsine 5′-triphosphate (ATP), 10^{-5} mol L^{-1} adenosine monophosphate (AMP), an equimolar mixture of 10^{-5} mol L^{-1} ATP + 10^{-5} mol L^{-1} AMP, and seawater control. (d) Responses to 10^{-7} mol L^{-1} glycine, 10^{-7} mol L^{-1} ATP, an equimolar mixture of 0.5×10^{-7} mol L^{-1} glycine + 0.5×10^{-7} mol L^{-1} ATP, and seawater control. (e) Responses to a monocarboxylic amino acid mixture (glycine, alanine, and serine, each at 3.33×10^{-8} mol L^{-1}), a dicarboxylic organic acid mixture (succinate and oxalate, each at 5×10^{-8} mol L^{-1}), an equimolar solution of amino and organic acids (glycine, alanine, serine, succinate, and oxalate, each at 2×10^{-8} mol L^{-1}), and seawater control. Data are from Zimmer-Faust *et al.* (1984), Zimmer-Faust (1987), and Zimmer (unpubl. data).

assimilate DFAAs (Ogura 1975) while releasing copious amounts of ammonium (Kjosbakken *et al.* 1983). Consequently, the ratio of amino acids to ammonium decreases with increasing age and decomposition of carrion (Kjosbakken *et al.* 1983; Zimmer unpubl. data). For five carnivorous crusta-ceans (spiny lobster, *Panulirus interruptus*; yellow crab, *Cancer anthonyi*; lobster krill, *Pleuroncodes planipes*; deep sea mysid, *Gnathophausia ingens*; and majid crab, *Podochela hemiphilla*), the strength of a feeding response correlates positively with the

DFAA:ammonium ratio of a test solution (Zimmer-Faust 1987). These results are consistent with the premise of energy/nutrient maximization.

11.5 Chemical defences and their effect on ingestion

Chemical cues that affect the search for and selec-tion of food by predators include not only attract-ants, stimulants, and suppressants, but also chemical defences. These defensive chemicals may be constitutively present in their tissues or integument. Many organisms also respond to predatory attacks using activated chemical defences in their tissues or by actively releasing chemicals to the external environment around them to deter predators.

Chemical defences influence the selection and ingestion of food by blue crabs and spiny lobsters. Those defences are often multifunctional and multicomponent. Examples of multifunctional chemical defences in terrestrial animals include the venom of wasps, which can act in defence against enemies and microbes, and in intraspecific interactions as alarm cues, sex pheromones, and identifiers of nestmates and castes (Turillazzi 2006). Another example is the venom of cobras, which is used both in capture of prey by envenomation and as an anti-predator defence by spitting venom into the face of predators (Westhoff *et al.* 2010). The multifunctionality reflects the many challenges facing these organisms. One obvious function is to act directly on consumers to disrupt the attack, through a variety of mechanisms. One mechanism is that the chemicals are aversive by virtue of being distasteful, toxic, irritant, or some other way unpleasant. This is also true of many prey of omnivorous spiny lobsters and blue crabs. Distasteful compounds in a variety of prey, including sea hares (*Aplysia* spp.) and polychaete worms, have been shown to be aversive and to reduce food consumption by blue crabs and its closely related species, the lesser blue crab (*Callinectes similis*) (Kamio *et al.* 2010b; Chapter 16). Defensive chemicals need not be aversive to be effective; they might instead function through phagomimicry or sensory disruption. An example of this is presence of high (mmol L^{-1}) concentrations of DFAAs in defensive ink secretions of sea hares and other molluscs, which affects predatory attacks by spiny lobsters (Kicklighter *et al.* 2005; Derby *et al.* 2007). This DFAA fraction of the secretions is highly attractive to predatory spiny lobsters and can cause the ink secretion to stimulate appetitive feeding responses of spiny lobsters, a process called phagomimicry (Kicklighter *et al.* 2005). This is an example of evolution of a chemical defence through a sensory trap, since spiny lobsters and many other aquatic predators use reception of DFAAs to detect injured prey. The ink secretion might also function

as a sensory disruptor, since the sticky secretion laden with DFAAs may act in either of two ways: physically blocking reception of chemical cues, and/or causing high-amplitude, long-lasting chemo-mechanosensory stimulation that leads to sensory adaptation (Kicklighter *et al.* 2005).

A second general mechanism of chemical defence against predators is the use of alarm signals and cues, which act on conspecifics causing them to flee the area or hide and thus avoid potential predators (Chapter 9). Molecules with alarm functions can be secreted from specialized glands specifically for defensive purposes and thus are considered alarm signals. Alarm signals are especially prevalent in insects. In aquatic environments, many examples exist of chemicals with alarm functions that are passively released from damaged tissue and thus are alarm cues. How alarm cues have evolved in animals that are not highly social or genetically related and thus do not benefit the releaser has received some interest. One scenario is that alarm cues originally protected the integument of the releaser against pathogens, parasites, UV radiation, or some other function. This has been argued for alarm cues released from the skin of injured fish, but neither the molecular identity of the alarm cues nor whether the same molecules serve both protective and alarm effects is known (Chapter 9). Additional support for this idea comes from sea hares, which use mycosporine like amino acids (MAAs) as alarm cues in their ink secretion (Kamio *et al.* 2011; Kicklighter *et al.* 2011). MAAs are common in aquatic invertebrates, which probably acquire MAAs through diet or symbionts rather than through *de novo* synthesis (Shick *et al.* 2002; Carreto and Carignan 2011). MAAs can protect animals from UV radiation by acting as sun screens or antioxidants (Shick *et al.* 2002), and they might also prevent fouling by fungi or algae (Karentz 2001). Sea hares have high concentrations of MAAs in their skin, where they probably function as sun screens (Carefoot *et al.* 2000), but they have even higher concentrations of MAAs in their ink secretion, where they are used as alarm cues (Kamio *et al.* 2011; Kicklighter *et al.* 2011). Thus, the same molecules that act as alarm cues also act as sun screens or anti-foulants. This evidence for alarm cues, and also work on chemical defences (e.g.

Karentz 2001; Kubanek *et al.* 2002), supports the idea that multiple roles of chemicals are a key feature in their elaboration.

Multiple molecules contribute to the efficacy of ink-based alarm cues of sea hares, including MAAs, nucleosides, and nucleic acids (Kicklighter *et al.* 2007). The MAAs in ink of sea hares are diet-dependent, since the type and amount of MAAs varies across the algae available to sea hares (Kicklighter *et al.* 2011). On the other hand, nucleo-sides and nucleic acids are synthesized *de novo* and thus their presence in ink is not directly dependent on diet. Thus, having both types of alarm cues allows for a basal level of protection that is inde-pendent of diet as well as the potential for a diet-dependent enhancement of the levels of alarm cues. The multifunctionality of defensive secretions may derive from the fact that they are multicomponent mixtures. Each component can serve different func-tions, some being aversive to predators, others being alarm cues, and so on. For each function, there may be multiple components. There is a rich body of literature in the field of chemical ecology to support these points (Kubanek *et al.* 2002; Arnold and Targett 2002; Paul *et al.* 2007; Hay 2009). Our own work supports the idea that for active chemi-cal defences, animals that are preyed upon by dif-ferent predators may require a variety of defensive chemicals because each does not work effectively against all predators. Aplysioviolin, which gives the ink of sea hares its purple colour, is highly effec-tive against blue crabs and some fish, but largely ineffective against spiny lobsters (Kamio *et al.* 2010b; Nusnbaum, Aggio, and Derby unpubl. data). There may be multiple effective components against a single species of predator. Such redun-dancy can be a way of having a secretion that always has effective components. For example, ani-mals may protect themselves with diet-derived chemicals when the source material is available, but also have a guaranteed source by having addi-tional deterrents that it produces *de novo*. Ink of sea hares contains aplysioviolin as chemical deterrent, which is diet-dependent (Kamio *et al.* 2010a), and components of the escapin pathway, such as hydro-gen peroxide (Aggio and Derby 2008), that are not diet dependent.

11.6 Neural mechanisms underlying the detection and recognition of feeding stimulants and deterrents

Neural processing of feeding stimulants has been especially well studied in peripheral chemosensory systems of spiny lobsters. The feeding stimuli are detected by chemoreceptor neurons (CRNs) distrib-uted over their body, including the antennules, legs, mouthparts, and body surface (Box 11.1). These CRNs are differentially sensitive to biologically rel-evant chemicals, including select DFAAs, adenine nucleotides, ammonium, and other compounds (Derby 2000; Derby *et al.* 2001; Schmidt and Mellon 2011). The most effective stimulus for a given CRN is typically several orders of magnitude more stim-ulatory than other stimuli. For example, taurine-best cells on antennules of spiny lobsters may also be sensitive to other DFAAs, AMP, and ammonium, but the threshold for taurine is 100–10,000 lower than for other stimuli (Cromarty and Derby 1997).

Knowing how mixtures are processed is important since natural, biologically relevant chemical stimuli are mixtures, and mixtures are generally more effec-tive than individual components in stimulating behaviours (e.g. Carr 1988). Electrophysiological studies on spiny lobsters show that each behaviour-ally discriminable component of a chemically-defined food mixture generates a unique across-neuron pat-tern (Steullet and Derby 1997). Thus, stimulus quan-tity and quality appear to be coded by the frequency of action potentials and the across-neuron pattern of activity respectively (Derby 2000; Schmidt and Mellon 2011), as is generally the case across phyla. Furthermore, the response of a CRN to food mixtures is predictable if one knows several other features of that CRN, including the intensity of its responses to the mixture's components, the receptor types and transduction cascades for that CRN, the coupling of receptor types to those cascades, and which chemi-cals interact with each receptor type and transduction cascade (Cromarty and Derby 1997). The challenge of course is to build a detailed understanding of all of these molecular pathways, which can be quite com-plex because of the diversity of transduction cascades within single CRNs and the diversity of types of CRNs (Ache and Young 2005).

As described in previous sections, the compounds comprising a food stimulus can contribute to behaviour in a non-additive way—i.e. mixture interactions—and these mixture interactions can be either positive (synergistic) or negative (suppressant). Electrophysiological experiments on spiny lobsters show that some mixture interactions can be generated at the level of peripheral receptor neurons, including competitive and non-competitive binding interactions as receptor sites, and some mixture interactions can be generated in the central nervous system due to synaptic integration (reviewed in Derby 2000). To understand the neuroecology of feeding in crustaceans, it will be important in the future to link behavioural and electrophysiological studies using natural and dynamically-scaled stimuli to examine such mixture effects for single species.

Compared to what is known about neural processing of food stimuli by crustaceans, much less is known about neural processing of chemical defences, though progress is being made. Chemical defences in some cases can cause predators to avoid the source of an otherwise attractive object, such as food, from a distance and without the predator contacting or ingesting the source of the chemicals. For example, the ink of sea hares can act via antennular chemoreceptors sensitive to DFAAs, which mediate orientation responses from a distance and contribute to phagomimicry in spiny lobsters (Kicklighter et al. 2005). Other chemical defences act at short range by stimulating mouthparts or esophageal chemoreceptors that control consumption (spiny lobsters: Kicklighter et al. 2005; blue crabs: Aggio and Derby 2010).

The aversive response to defensive chemicals might be mediated by different mechanisms. One mechanism is based on specific types of CRNs that mediate aversive responses by virtue of their central connections, as has been best demonstrated in insects (e.g. de Brito Sanchez et al. 2005). Work on Drosophila has been particularly useful in characterizing mechanisms of this type. For example, gustatory receptors Gr66a and Gr93 expressed in specific gustatory receptor neurons mediate the deterrency of caffeine and related compounds (Lee et al. 2009), and TRPA1 (which is also an irritant sensor in other animals including humans) mediates aversion to noxious compounds such as the reactive electrophiles allyl isothiocyanate and cinnamaldehyde (Kang et al. 2010). A second mechanism is that deterrent chemicals suppress or inhibit the activity of CRNs that are excited by food, thus turning off central neural pathways that stimulate feeding. This is demonstrated in several insect species (e.g. Mitchell and Sutcliffe 1984; de Brito Sanchez et al. 2005; Jørgensen et al. 2007).

The study of cellular mechanisms mediating chemical deterrence in crustaceans is in its infancy. Cells on legs of crayfish and spiny lobsters are known to be selectively activated by pyridyls (Schmiedel-Jakob et al. 1988; Kem and Soti 2001), but the behavioural context of these physiological effects is unknown. We are currently examining properties of CRNs in blue crabs that are responsive to the deterrents in the ink of sea hares such as aplysioviolin and hydrogen peroxide. Electrophysiological studies of crustacean chemoreception have repeatedly shown that individual CRNs have two opposing transduction cascades—one excitatory and one inhibitory—each activated by different chemicals (Ache and Young 2005). Known feeding deterrent compounds have not been tested on these cells, but this cellular organization raises the distinct possibility that deterrents might function by activating the suppressive or inhibitory transduction pathways.

11.7 Conclusions and future directions

Neuroecology requires considerable foresight in selection of appropriate study systems. Genomically enabled species, such as C. elegans, Drosophila, zebrafish, and mice, have been extremely useful in linking molecular mechanisms with physiological functions. For example, these species have been critically important in pushing our knowledge of sensory transduction of chemoreceptor cells, how this information is processed by the CNS, and how these physiological processes are translated into behavioural output (e.g. Ache and Young 2005; Bargmann 2006). However, genomically-enabled species sometimes have their limitations. They have often proven difficult or unhelpful to study in field habitats, and by and large they do not play pivotal roles in structuring native communities. In contrast, many species of considerable ecological relevance—including foundation species (e.g. mussels) and keystone predators (e.g., seastars, whelks, sea otters)—either lack a sequenced genome or are difficult subjects for physiological inquiries. Therefore, inclusion of more ecologically relevant species in genome sequencing projects will help spur investigations spanning genes to ecosystems.

Few investigations to date have identified the complete structures of ecologically important natural products, and even fewer studies have reported experiments in which chemical stimuli are presented dynamically scaled. Consequently, the ecological relevance of many published works is in question. The intellectual integration of insights gained from many findings across many levels of biological organization ultimately reflects the neuroecological approach.

The purpose of our review is to begin creating a common fabric of knowledge about predator–prey interactions and carnivory, woven from the threads of chemosensory physiology, behaviour, and population and community ecology. Our focus is on the value of using marine crustaceans, in particular spiny lobsters and blue crabs, as animals for experimental analyses of the relationships between chemical cues, chemosensitivity, and predation. We encourage an application of the neuroecological approaches to the investigation of other research problems.

Acknowledgements

The authors express their sincere gratitude to the outstanding students and collaborators who have helped us work over the years on neuroecology and predator–prey interactions. Our research has been supported by awards from the National Science Foundation, National Institutes of Health, Office of Naval Research, National Geographic Society, Cold Seas International, Inc., UCLA Council on Research, and Georgia State University.

References

Ache, B.W. and Young, J.M (2005) Olfaction: diverse species, conserved principles. *Neuron*, **48**: 417–30.

Aggio, J.F. and Derby, C.D. (2008) Hydrogen peroxide and other components in the ink of sea hares are chemical defenses against predatory spiny lobsters acting through non-antennular chemoreceptors. *Journal of Experimental Marine Biology and Ecology*, **363**: 28–34.

Aggio, J.F. and Derby, C.D. (2010) Sensory mechanisms of chemical deterrence by sea hare ink against predatory blue crabs. *Chemical Senses*, **31**: A74.

Arnold, T.M., and Targett, N.M. (2002) Marine tannins: the importance of a mechanistic framework for predicting ecological roles. *Journal of Chemical Ecology*, **28**: 1919–34.

Bargmann, C.I. (2006) Comparative chemosensation from receptors to ecology. *Nature*, **444**: 295–301.

Carefoot, T.H., Karentz, D., Pennings, S.C., and Young, C.L. (2000) Distribution of mycosporine-like amino acids in the sea hare *Aplysia dactylomela*: effect of diet on amounts and types sequestered over time in tissues and spawn. *Comparative Biochemistry and Physiology*, **126C**: 91–104.

Carr, W.E.S. (1988) The molecular nature of chemical stimuli in the aquatic environment. In J. Atema, R.R. Fay, A.N. Popper, and W.N. Tavolga, (eds.) *Sensory Biology of Aquatic Animals*. Springer-Verlag, New York, pp. 3–56.

Carreto, J.E. and Carignan M.O. (2011) Mycosporine-like amino acids: relevant secondary metabolites. Chemical and ecological aspects. *Marine Drugs*, **9**: 387–446.

Cate, H.S. and Derby, C.D. (2001) Morphology and distribution of setae on the antennules of the Caribbean spiny lobster *Panulirus argus* reveal new types of bimodal chemo-mechanosensilla. *Cell and Tissue Research*, **304**: 439–54.

Cromarty, S.I. and Derby, C.D. (1997) Multiple excitatory receptor types on individual olfactory neurons: implications for coding of mixtures in the spiny lobster. *Journal of Comparative Physiology A*, **180**: 481–91.

de Brito Sanchez, M.G., Giurfa, M., de Paula Mota, T.R., and Gauthier, M. (2005) Electrophysiological and behavioural characterization of gustatory responses to antennal 'bitter' taste in honeybees. *European Journal of Neuroscience*, **22**: 3161–70.

de Bruyne, M. and Baker, T.C. (2008) Odor detection in insects: volatile codes. *Journal of Chemical Ecology*, **34**: 882–97.

Derby, C.D. (2000) Learning from spiny lobsters about chemosensory coding of mixtures. *Physiology & Behavior*, **69**: 203–9.

Derby, C.D., Steullet, P., Horner, A.J., and Cate, H.S. (2001) The sensory basis to feeding behavior in the Caribbean spiny lobster *Panulirus argus*. *Marine and Freshwater Research*, **52**: 1339–50.

Derby, C.D., Kicklighter, C.E., Johnson, P.M., and Zhang, X. (2007) Chemical composition of inks of diverse marine molluscs suggests convergent chemical defenses. *Journal of Chemical Ecology*, **33**: 1105–13.

Finelli, C.M., Pentcheff, N.D., Zimmer, R.K., and Wethey, D.S. (2000) Physical constraints on ecological processes: a field test of odor-mediated foraging. *Ecology*, **81**: 784–97.

Harrold, C and Reed, D.C. (1985) Food availability, sea urchin grazing, and kelp forest community structure. *Ecology*, **66**: 1160–9.

Hay, M.E. (2009) Chemical signals and cues structure marine populations, communities, and ecosystems. *Annual Review of Marine Science*, **1**: 193–212.

Jørgensen, K., Almaas, T.J., Marion-Poll, F., and Mustaparta, H. (2007) Electrophysiological characterization of gustatory responses of *sensilla chaetica* in the moth *Heliothis virescens*. *Chemical Senses*, **32**: 863–79.

Jutel, M., Blaser, K., and Akdis, C.A. (2006) Histamine receptors in immune regulation and allergen-specific immunotherapy. *Immunology*, **26**: 245–57.

Kamio, M., Nguyen, L., Yaldiz, S., and Derby, C.D. (2010a) How to produce a chemical defense: structural elucidation and anatomical distribution of aplysioviolin and phycoerythrobilin in the sea hare *Aplysia californica*. *Chemistry & Biodiversity*, **7**: 1183–97.

Kamio, M., Grimes, T.V., Hutchins, M.H., van Dam, R., and Derby, C.D. (2010b) The purple pigment aplysioviolin in sea hare ink deters predatory blue crabs through their chemical senses. *Animal Behaviour*, **80**: 89–100.

Kamio, K., Kicklighter, C.E., Nguyen, L., Germann, M.W., and Derby, C.D. (2011) Isolation and structural elucidation of novel mycosporine-like amino acids as alarm cues in the defensive ink secretion of the sea hare *Aplysia californica*. *Helvetica Chimica Acta* , **94**: 1012–8.

Kang, K., Pulver, S.R., Panzano, V.C., Chang, E.C., Griffith, L.C., Theobald, D.L., and Garrity, P.A. (2010) Analysis of *Drosophila* TRPA1 reveals an ancient origin for human chemical nociception. *Nature*, **464**: 597–600.

Karentz, D. (2001) Chemical defenses of marine organisms against solar radiation exposure: UV-absorbing mycosporine-like amino acids and scytonemin. In J. McClintock and W. Baker, (eds.) *Marine Chemical Ecology*. CRC Press, Boca Raton, Florida, pp. 481–520.

Keller, T.A., Powell, I., and Weissburg, M.J. (2003) The identity and role of olfactory appendages in chemically mediated orientation in blue crabs, *Callinectes sapidus*. *Marine Ecology Progress Series*, **261**: 217–31.

Kem, W.R. and Soti, F. (2001) Amphiporus alkaloid multiplicity implies functional diversity: initial studies on crustacean pyridyl receptors. *Hydrobiologia*, **456**: 221–31.

Kicklighter, C.E., Shabani, S., Johnson, P.M., and Derby, C.D. (2005) Sea hares use novel antipredatory chemical defenses. *Current Biology*, **15**: 549–54.

Kicklighter, C.E., Germann, M.W., Kamio, M., and Derby, C.D. (2007) Molecular identification of alarm cues in the defensive secretions of the sea hare *Aplysia californica*. *Animal Behaviour*, **74**: 1481–92.

Kicklighter, C.E., Kamio, M., Nguyen, L., Germann, M.W., and Derby, C.D. (2011) Mycosporine-like amino acids are multifunctional molecules in sea hares and their marine community. *Proceedings of the National Academy of Science of the United States of America*, **108**: 11494–9.

Kjosbakken, J., Storro, L., and Larsen, H. (1983) Bacteria decomposing amino acids in bulk-stored capelin (*Mallotus villosus*). *Canadian Journal of Fisheries and Aquatic Sciences*, **40**: 2092–7.

Kubanek, J., Whalen, K.E., Engel, S., Kelly, S.R., Henkel, T.P., Fenical, W., and Pawlik, J.R. (2002) Multiple defensive roles for triterpene glycosides from two Caribbean sponges. *Oecologia*, **131**: 125–36.

Laughlin, R.A. (1982) Feeding habits of the blue crab, *Callinectes sapidus* Rathbun, in the Apalachicola estuary, Florida. *Bulletin of Marine Science*, **32**: 807–22.

Lee, Y., Moon, S.J., and Montell, C. (2009) Multiple gustatory receptors required for the caffeine response in *Drosophila*. *Proceedings of the National Academy of Science of the United States of America*, **106**: 4495–500.

Mann, R., Harding, J.M., Southward, M.J. and Wesson, J.A. (2005) Northern quahog (hard clam) *Mercenaria mercenaria* abundance and habitat use in Chesapeake Bay. *Journal of Shellfish Research*, **24**: 509–16.

Mitchell, B.K. and Sutcliffe, J.F. (1984) Sensory inhibition as a mechanism of feeding deterrence: effects of three alkaloids on leaf beetle feeding. *Physiological Entomology*, **9**: 57–64.

Ogura, N. (1975) Further studies on decomposition of organic matter in coastal seawater. *Marine Biology*, **31**: 101–11.

Orth, R.J., Heck, K.L., and von Montfrans, J. (1984) Faunal communities in seagrass beds: a review of the influence of plant structure and prey characteristics on predator-prey interactions. *Estuaries*, **7**: 339–50.

Paul, V.J., Arthur, K.E., Ritson-Williams, R., Ross, C., and Sharp, K. (2007) Chemical defenses: from compounds to communities. *Biological Bulletin*, **213**: 226–51.

Robles, C., Sweetnam, D., and Eminike, J. (1990) Lobster predation on mussels: shore-level differences in prey vulnerability. *Ecology*, **71**: 1564–77.

Roelofs, W.L. and Rooney, A.P. (2003) Molecular genetics and evolution of pheromone biosynthesis in Lepidoptera. *Proceedings of the National Academy of Science of the United States of America*, **100**: 9179–84.

Schmidt, M. and Mellon, D. Jr (2011) Neuronal processing of chemical information in crustaceans. In T. Breithaupt and M. Thiel, (eds.) *Chemical Communication in Crustaceans*. Springer, New York, pp. 123–47.

Schmiedel-Jakob, I., Brenninger, V., and Hatt, H. (1988) Electrophysiological studies of chemoreceptors sensitive to pyridine on the crayfish walking leg. II. Characteristics of inhibitory molecules. *Chemical Senses*, **13**: 619–32.

Shears, N.T. and Babcock, R.C. (2002) Marine reserves demonstrate top-down control of community structure on temperate reefs. *Oecologia*, **132**: 131–42.

Shick, J.M. and Dunlap, W.C. (2002) Mycosporine-like amino acids and related gadusols: biosynthesis, accumulation, and UV-protective functions in aquatic organisms. *Annual Review of Physiology*, **64**: 223–62.

Silliman, B.R. and Bertness, M.D. (2002) A trophic cascade regulates salt marsh primary production. *Proceedings of the National Academy of Science of the United States of America*, **99**: 10500–5.

Stephens, G.C. (1988) Epidermal amino acid transport in marine invertebrates. *Biochimica et Biophysica Acta*, **947**: 113–38.

Steullet, P. and Derby, C.D. (1997) Coding of blend ratios of binary mixtures by olfactory neurons in the Florida spiny lobster, *Panulirus argus*. *Journal of Comparative Physiology A*, **180**: 123–35.

Stuart, A.E., Borycz, J., and Meinertzhagen, I.A. (2007) The dynamics of signaling at the histaminergic photoreceptor synapse of arthropods. *Progress in Neurobiology*, **82**: 202–27.

Swanson, R.L., Williamson, J.E., DeNys, R., Kumar, N., Bucknell, M.P., and Steinberg, P.D. (2004) Induction of settlement of larvae of the sea urchin *Holopneustes purpurascens* by histamine from a host alga. *Biological Bulletin*, **206**: 161–72.

Tegner, M.J. and Levin, L.A. (1983) Spiny lobsters and sea urchins: analysis of a predator-prey interaction. *Journal of Experimental Marine Biology and Ecology*, **73**: 125–50.

Turillazzi, S. (2006) *Polistes* venom: a multifunctional secretion. *Annales Zoologici Fennici*, **43**: 488–99.

Virstein, R.W. (1977) The importance of predation by crabs and fishes on benthic infauna in Chesapeake Bay. *Ecology*, **58**: 1199–217.

Westhoff, G., Boetig, M., Bleckmann, H., and Young, B.A. (2010) Target tracking during venom 'spitting' by cobras. *Journal of Experimental Biology*, **213**: 1797–802.

Wright, S.H. and Manahan, D.T. (1989) Integumental nutrient uptake by aquatic organisms. *Annual Review of Physiology*, **51**: 585–600.

Zimmer, R.K. and Derby, C.D. (2007) *Biological Bulletin* virtual symposium: the neuroecology of chemical defence. *Biological Bulletin*, **213**: 205–7.

Zimmer, R.K., Commins, J.E., and Browne, K.A. (1999) The regulatory effects of environmental chemical signals on search behavior and foraging success. *Ecology*, **80**: 1432–46.

Zimmer-Faust, R.K. (1987) Crustacean chemical perception: towards a theory on optimal chemoreception. *Biological Bulletin*, **172**: 10–29.

Zimmer-Faust, R.K. (1993) ATP: a potent prey attractant evoking carnivory. *Limnology and Oceanography*, **38**: 1271–5.

Zimmer-Faust, R.K. and Case, J.F. (1982) Odors influencing foraging behaviour of the California spiny lobster, *Panulirus interruptus*, and other decapod Crustacea. *Marine Behaviour and Physiology*, **9**: 35–58.

Zimmer-Faust, R.K. and Case, J.F. (1983) A proposed dual role of odor in foraging by the California spiny lobster, *Panulirus interruptus*. *Biological Bulletin*, **164**: 341–53.

Zimmer-Faust, R.K., Tyre, J.E., Michel, W.C., and Case, J.F. (1984) Chemical mediation of appetitive feeding in a marine decapod crustacean: the importance of suppression and synergism. *Biological Bulletin*, **167**: 339–53.

CHAPTER 12

Why is the jack of all trades a master of none? Studying the evolution of inducible defences in aquatic systems

Ulrich K. Steiner and Josh R. Auld

Predators come and go; they differ in their peril to prey, their distribution, and their density. How prey cope with this variation is the long, ongoing, and fascinating story of the evolution of induced defences. Evolutionary theory predicts that organisms are often expected to evolve adaptive inducible defences to match predation risk and track changes in it (van Tienderen 1991; Tollrian and Harvell 1999; see Box 12.1 for definitions). Given that temporal and spatial variability in predation risk are essentially ubiquitous, prey in an ideal world should be able to alter their phenotypes to always express an optimal defence. In the real world, however, this is not true. We see a substantial amount of variation in the induction of defences among genotypes within populations, which demonstrates that many plastic genotypes do not match their optimum. At the same time, not all constitutive genotypes are adapted to all environments they experience. The discrepancies between the ideal world and the real world have many sources and, at least from a biodiversity perspective, we can be glad that the jack of all trades is a master of none (or at least only a few). Without limits to the evolution of plasticity, we would expect a single universal perfectly plastic genotype to evolve.

The evolution of inducible defences needs to be considered within the framework of the costs and benefits of defences. It is regulated through natural selection and is limited by physiological and developmental pathways as well as random (evolutionary) processes such as drift, hitchhiking, and mutation (Schlichting and Pigliucci 1998). While environments are constantly changing, such evolutionary processes are simultaneously changing the demographic structure and genotypic frequencies within a population which can influence the arms race between prey and their predators (Kopp and Tollrian 2003). The evolution of the ability to track environmental changes with appropriate induced responses (i.e. adaptive phenotypic plasticity) is not limited to predator–prey interactions. Hence, the arguments we present for the evolution of inducible defences (or the lack thereof) apply to the much more general case of environmental induction of any phenotypic trait.

When the environment changes in a predictable way, and particularly if the change is slow relative to generation time, populations can adapt to the environment by evolving constitutively expressed phenotypes; that is, by fixed genetic adaptation. However, most environmental variation, be it temporal or spatial, occurs at a scale that does not allow mutation or other changes in gene frequencies to occur rapidly enough to facilitate the expression of optimal (adaptive) phenotypes. Thus, the optimal solution is the ability of a single genotype to express various phenotypes according to the environment, which brings us back to phenotypic plasticity. We emphasize that both fixed genetic adaptation and the induced plastic tracking of the environment by phenotypic plasticity have a genetic basis and have been shaped by evolutionary processes. However, inducible defences are not simply the outcome of an adaptive process shaped by natural selection; genetic variation (which forms the basis for the phenotypic variation) is affected by genetic drift, recombination, hitchhiking (genetic draft), and mutations, and we have to keep the contributions of these random processes in mind—something evolutionary ecologists (including ourselves) often neglect. Unfortunately, even though these random evolutionary processes are not predicted to be different for plastic traits and non-plastic traits, our understanding of these random processes in the evolution of induced traits is less specific and less explored compared to the selective forces of natural selection (Via *et al.* 1995). Understanding the phenotype–genotype space for constitutive traits is challenging, and exploring the space for induced traits adds further complexity. For these reasons we will mostly focus on adaptive evolution in the context of natural selection, but try to keep the random processes of evolution in mind.

Evolutionary theory predicts that induced phenotypes should evolve when (1) organisms are exposed to heterogeneous, variable environments with opposing selective forces; (2) reliable environmental cues exist that indicate current or future environmental conditions; and (3) expressing and switching between several adaptive phenotypes is costly (van Tienderen 1991) (Fig. 12.1). Otherwise, natural selection is expected to favour fixed (i.e. non-plastic or constitutive) phenotypes. For the specific case of induced defences, defences are expected to be inducible only if (1) predation risk fluctuates; (2) prey can reliably detect predators; and (3) defences are costly to produce or maintain, but the benefits outweigh the costs. If these conditions are not met, we expect constitutive defences (including no defence) to evolve (Chapter 16). Generally speaking, the main advantage of inducible defences is that they allow individuals to avoid paying the costs of expressing a defence when predators are absent.

We start by describing extrinsic evolutionary drivers for induced defences such as fluctuations in the environment, environmental cues, and coevolution between induced defences and induced offences. Then, we characterize different induced defences in the light of evolutionary processes acting within the organism, such as induction time, resource allocation trade-offs, cost–benefit ratios of defences, passive and adaptive responses to predation risk, and how specific or universal defences might be. Finally, we move to the genetic level and illustrate some quantitative genetic approaches and the genetic basis of induced defences. Evolution unifies all these mechanisms and therefore we cannot isolate single aspects. Arguments about selective forces that promote the evolution of defences can almost always be flipped and then turned into arguments for limits of the evolution of defences. Hence, the limits are as manifold as the promoting factors. One of the most challenging aspects is the time allowed for evolution to find an optimum

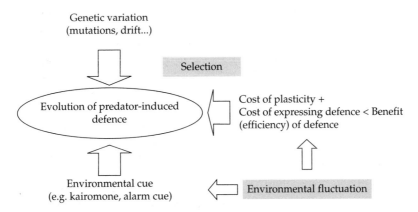

Figure 12.1   Main factors for the evolution of predator-induced defences. The strength of selection depends on the environment. The cost of plasticity is environmentally independent but the expression cost and the benefits are not.

Box 12.1 Definitions

Induced defence: A defence that is only expressed under risk from a predator or other enemy (e.g. a herbivore); it contrasts with *constitutive defences* that are expressed independently of risk (see for instance Chapter 7). Tollrian & Harvell 1999

   Induced offence or reciprocal phenotypic plasticity: An inducible defence of a prey causes the predator to express an inducible counter-offence. Kopp & Tollrian 2003

   Phenotypic plasticity: The ability of a single genotype to express various phenotypes in different environments. Callahan *et al.* 2008

   Generation time: The average age at which a female gives birth to her offspring, or the average time for a population to increase by a factor equal to the average number of daughters a female is expected to produce during her lifetime. Gillespie, J.H. (2004) Population genetics: a concise guide. John Hopkins University Press, Baltimore

   Genetic drift: A change in the relative frequency of a genotype within a population due to random sampling and chance. Gillespie 2004

   Hitchhiking (Genetic draft): The process by which an allele or mutation that is not favoured by natural selection spreads through the gene pool by being linked to a gene that is positively selected. Gillespie 2004

---

expected from theory. In any empirical system evolution is hardly ever at equilibrium, it is a continuous and ongoing dynamic process, and the optimum is constantly shifting. Therefore, it will always remain challenging to differentiate if time has not been sufficient for mutations to occur and selection acting on these mutations to find the appropriate solution, or if the optimal solution is constitutive.

12.1 Extrinsic factors and the evolution of inducible defences

Ecological factors play a major role in the evolution of induced defences. They determine the selective forces acting on the plasticity of a defence.

Environmental fluctuations in the abiotic and biotic environment, including demographic shifts of conspecifics, heterospecifics, and multiple predator species, need to be integrated and provide a fascinating and complex framework for the evolution of predator-induced defences.

12.1.1 Environmental fluctuation

The first requirement for the evolution of inducible defences is that predation risk must vary (Fig. 12.1). This environmental fluctuation can be temporal or spatial, and under most natural circumstances both factors contribute to the variation. Demographic fluctuations within a predator population generate temporal variability in the environment for local prey, and spatial variation in predator densities generates variability for prey that migrates once in a while. The frequency at which predators (or no-predators) are experienced can be rare (Sultan and Spencer 2002). Among-site migration of a few individuals per generation is sufficient variation in predation risk to favour the evolution of inducible defences, though most individuals remain at their local site and never experience a change in their environment throughout their lives. This expectation that induced defences can evolve even in populations where hardly any individual experiences different environments might be surprising at first considering that selection acts on the defence expressed by the individual, but the change in the genotype frequency is what determines the (evolutionary) fitness in the long run. Hence, we need to integrate across generations where environmental change matters; the advantage of inheriting the ability to induce a defence (or the flipside, the ability to save the costs of expressing the defensive phenotype) has to pay-off across generations, not necessarily within a generation. Therefore, migration relaxes the requirements of both temporal environmental heterogeneity and strong selection on inducibility.

These predictions about migration and the frequency of environmental variation are directly related to the spatial distribution of predation risk. If the environment is coarse-grained with respect to predation risk (e.g. no predators in a large lake) and migration is low, we expect local adaptation in the form of constitutive defences (e.g. no defence) against predation risk because fluctuations in predation risk are lacking. Only at the borders of this environment where predation risk changes (e.g. a river with fish connecting to a lake without fish) do we expect the evolution of inducible defences. The other extreme is when predation risk changes at a very fine scale (e.g. every pond has some sit-and-wait predators) and every individual encounters similar levels of predation risk throughout their life and across generations. Here, one also expects the evolution of constitutive defences because predation risk is essentially constant. In most environments predation risk will fluctuate because of varying environmental conditions or migration that influence predator population dynamics, prey population dynamics, or some other species' population dynamics that influence the prey or predator density directly or indirectly (Chapter 10). Surprisingly little empirical information is available on actual variation in predation risk experienced by individuals, populations, or genotypes to test these hypotheses (Van Buskirk 2002). However, the lack of empirical information might be less surprising when we consider what needs to be taken into account: migration and temporal patterns for both predator and prey species across multiple generations as well as their correlation with induced defences. Among-species comparisons frequently agree with the expectations that plasticity will be more frequent in species experiencing greater variability in their environment. As examples, marine invertebrate species with high dispersal rates show more phenotypic plasticity (Hollander 2008); amphibian species occurring in ephemeral or permanent ponds show limited developmental plasticity whereas species occurring in temporary ponds show high levels of plasticity (Richter-Boix et al. 2006); and predator-induced morphological plasticity has increased in amphibian species exposed to greater variability in predators (Van Buskirk 2002). However, within-species and among-population comparisons are often less clear (Van Buskirk 2002, Kishida et al. 2007), but these comparisons do not necessarily have to disagree with the theory: gene flow among populations might be too large, or the time series of variation in predation risk might not be long enough to allow accurate evolutionary predictions. At a regional scale without major physio-

logical barriers, evidence that plasticity evolves in populations exposed to more variable environments was not detected in some cases (Van Buskirk and Arioli 2005), but when major barriers to migration exist (e.g. life on islands) the hypothesis that induced defences are more pronounced in the more variable environments still holds even within species (Bryant et al. 1989; Vourc'h et al. 2001; Kishida et al. 2007; Lind and Johansson 2007).

12.1.2 Environmental cues

If there is enough heterogeneity in predation risk, the second requirement for the evolution of induced defences is that the prey organisms can perceive this information to make accurate predictions about current or future environmental conditions (Fig. 12.1). In many predator–prey systems chemical cues play a major part in this process, especially in aquatic systems. A bird flying over a lake and catching a fish without leaving any trace of the predation event will provide no information to other potential prey (other than some minor change in prey density; though visual cues might be important) (Fig. 12.2). Perceiving information about predation risk is multifold, and it can be variable in its accuracy over diverse time periods. Visual or tactile cues will last for only a short time, but the frequencies of such cues can allow an accurate assessment of predation risk and any change therein. The concentration of alarm cues, kairomones, and other chemical cues will be reliable over a few hours or days but will decay and disperse over multiple days (Peacor 2006). Such rates of cue decay hold important information on the time that the predation event happened and the frequencies of events. Chemical cues will be of limited use when they disperse quickly, such as in streams (but see Chapter 1). Other events that might be associated with specific cues or the lack thereof, like the freezing of a pond to the ground that kills the entire population of large predatory fish—such as pike, might provide information on predation risk across multiple seasons (for general arguments about the evolution of alarm cues see also Chapter 9).

Under many circumstances chemical cues not only indicate the presence or absence of predators, but they also allow prey to differentiate predator

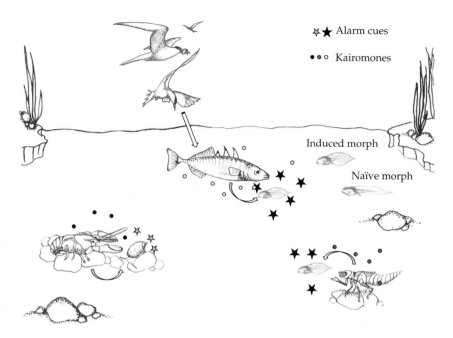

Figure 12.2 The importance of chemical cues in aquatic predator–prey systems. Some predation events, such as a bird depredating a fish, will not leave any chemical cue other prey can use to assess predation risk (visual cues might be relevant). Other predation events, such as fish, crayfish, or dragonfly larvae depredating on tadpoles and snails, leave chemical cues prey can use to induce a response. Alarm cues, cues emitted by prey organisms, do not reveal what predator depredated on the prey, whereas kairomones, cues emitted by the predator, reveal what predator was feeding on which prey. Different predators can induce a multitude of defences. Snails respond to crayfish kairomones by inducing thicker shells and movement towards the surface; while fish induce snails to move under cover. Tadpoles reduce their activity when exposed to alarm cues or kariomones of fish and dragonfly larvae; they seek shelter against fish but not dragonfly larvae. They can induce bulgy morphs against fish, while dragonfly larvae induce deeper tail fins and more pigmentation.

species; the prey that predators are feeding upon; the amount of predation that is occurring; and whether the prey is being consumed or only injured (Van Buskirk 2001; Schoeppner and Relyea 2005; Turner 2008; see also Chapter 8). Furthermore, prey have evolved the ability to utilize these cues in many diverse ways that can change across ontogeny. For example, small *Rana temporaria* tadpoles seek refuge only in the presence of an actively hunting predator but not against a sit-and-wait predator (though under both circumstances they reduce the time spent moving by similar amounts) while large tadpoles show the strongest behavioural response against sit-and-wait predators but no response against a gape-limited predator (Van Buskirk 2001). Both tadpoles and freshwater snails showed differentiation in their anti-predator responses depending on whether the predators were fed various prey species; the magnitude of the prey's response decreased with increasing phylogenetic distance of

the predator's diet (Schoeppner and Relyea 2005; Turner 2008). Tadpoles showed graded responses in reaction to graded risk (dosage of chemical cues) and could differentiate between the number of predators present and the amount they were fed. Behavioural traits were most responsive to the number of tadpoles killed by the predators, whereas morphological traits responded to the number of predators independent of the predators' diet (Van Buskirk and Arioli 2002). In tadpoles and snails, alarm cues (i.e. cues released by injured (crushed) prey) only induced a subset of defences compared to defences induced by kairomones (chemical cues emitted by predators feeding on prey conspecifics; Schoeppner and Relyea 2005; Turner 2008) (see also Fig. 12.2). The complexity of the chemical cues and the highly adaptive specific responses of prey to particular conditions indicate that such information is accurately integrated to assess the predation risk, which in turn relates to the force of selection that shapes the specific induced defences. The actual predation risk for an individual will not only depend on the cues related to predation but also on the density of the prey population. Chemical cues can allow an accurate estimation of conspecific density (Peacor 2003). However, little empirical information is available for understanding if and how individuals integrate the information about densities of other prey, including conspecifics, to regulate their own defences.

So far we have assumed that the environmental cues are accurate, but if they are unreliable the induced response will not be at its optimum or might even be maladaptive, and selective forces will be weakened. Inaccurate, unreliable cues will trigger additional costs because the expression of unnecessary defences or maladaptive phenotypes will be selected against. Unreliable cues might be associated with novel invasive predators (Hagman *et al.* 2009), and such novel cues might trigger evolutionary and ecological traps (Schlaepfer *et al.* 2002). Time may not have allowed evolution to find an adaptive solution to such novel predation risk; that is mutations have not yet arisen that allow the detection and use of reliable cues of the novel predator and potentially a novel induced defence (Freeman and Byers 2006).

12.1.3   Evolutionary arms race: inducible defences and inducible offences

From a predator's perspective it would be of great interest to manipulate kairomones to make them as vague and uninformative as possible. That is, predators that can avoid 'warning' potential prey of their presence should have an advantage (i.e. relatively undefended prey). As an example, pikes avoid contaminating their feeding area with chemical cues that deter their prey by defecating away from areas where they forage; such avoidance was only triggered when pike consumed prey that contained alarm cues (Brown *et al.* 1996). Such results indicate that the evolutionary arms race between predators and prey (often described as inducible defences and inducible offences) is not only going on in terms of efficiency of predation and defence, but also at the information level. Indentifying such arms races in empirical systems is hard, though an

increasing number of empirical examples indicate the existence of such coevolutionary interactions (Kishida *et al.* 2006). Under predation risk from a predatory ciliate, a prey ciliate develops protective lateral wings as an effective defence; this inducible defence then triggers a plastic response of increased cell size and gape size in the predatory ciliate (Kopp and Tollrian 2003). Tadpoles induce a bulgy phenotype as a defence against predation risk from gape-limited salamander larvae; these salamander larvae in turn induce an increase in gape size when bulgy tadpoles were abundant (Kishida *et al.* 2006). More discussions of coevolutionary arms races are presented elsewhere in this book (Chapters 1, 13, 14, and 15).

12.2   Internal factors and the evolution of inducible defences

In addition to the external, environmental factors described above that can influence the evolution of inducible defences, there are several additional factors that can be considered 'internal' (i.e. within-individual) and can play an equally important role in the evolution of inducible defences. Several of these have been referred to as costs or limits of phenotypic plasticity (DeWitt *et al.* 1998).

12.2.1   Reversibility and time lag of induced defence

Once environmental information (e.g. a chemical cue) is recognized by the prey organism, and if this information accurately describes current or future predation risk, the prey organism needs to upregulate its pathways that induce the defence (assuming the adaptive pathway has already evolved). Some behavioural and physiological inducible defences are almost instantaneous (Steiner and Van Buskirk 2009) and highly reversible, which demonstrates that sensing a change in the environment (i.e., a change in cues) can be instantaneous (see also Chapters 1 and 7) and that such behaviour or physiology is costly (otherwise one would not expect an inducible response). The expression of other inducible defences, such as morphological defences, might have substantial lag times of up to a generation; some may be irreversible once a developmental path is set while others are reversible (Tollrian and Harvell 1999, Hoverman and Relyea 2007). These findings indicate not only that building such morphological defences is costly, but also that their maintenance holds costs. Thus, patterns of resource allocation can directly affect the expression of inducible defences. Lag times change the selective forces on induced defences and might therefore play an important role in the evolution of defences because lag times prevent an optimal tracking of the environment in a similar way as having unreliable cues (Gabriel *et al*. 2005). Furthermore, trade-offs between investment in alternative life history strategies may play a role in the evolution of inducible defences.

12.2.2 Benefits and costs of induced defence

A major requirement for the evolution of inducible defences is that the benefits of maintaining them outweigh the alternatives (i.e. both in terms of benefits outweighing costs and in terms of inducibility outweighing constitutive expression of defensive traits) (Fig. 12.1). The strength of selection on an inducible defence depends on the fitness benefit of the defence under predation risk, that is, the efficiency of the defence and the fitness cost of expressing the defence that can be saved in the absence of predators (Tollrian and Harvell 1999; Steiner and Pfeiffer 2007). Hence, the net benefit depends on the frequency of being exposed to predation risk and the frequency of saving costs when predators are absent. This leads to the conclusion that the higher the costs of expressing a defence and the higher the effectiveness of the defence, the stronger the selection pressure must be for an inducible defence to evolve. This point can be somewhat confusing because the net benefit is the difference between total benefits (effectiveness) and costs. We elaborate on this point by describing two extremes. First, if there are only weak (no) costs, a constitutive defence will have similarly strong selective forces acting on it as an induced defence. Imagine a toxic defence is cheaply produced from secondary compounds; individuals that will constitutively be toxic will not have reduced fitness compared to individuals that will only be toxic in the presence of predators. Second, if the defence is less efficient, a constitutive phenotype without any defence has similar selec-

tive forces acting on it compared to an induced phenotype. If the toxicity does not help much to fend off predators, a constitutive non-toxic individual will have similar fitness as the induced toxic individual. It is therefore important to assess the efficiency of the defence, the cost that can be saved by not expressing a defence, and the frequency at which the costs and benefits come into play (van Tienderen 1991; Via *et al*. 1995; Sultan and Spencer 2002).

Decomposing the efficiency and the cost can be challenging empirically but some experimental setups with non-lethal predators emitting chemical cues allow such separation of benefits and costs, which is why they are frequently used in studying induced responses. However, these experiments give little information on the frequency of predation events under natural conditions (Van Buskirk 2000). Only a few studies provide data on predator densities and correlate those to induced trait expression. Predator densities by themselves do not necessarily unveil realized predation risk or predation. Few if any studies on induced defences have quantified the relationship between plasticity and actual predation in natural settings.

Empiricist's are also trapped between accurately assessing costs and fitness under natural conditions, and trying to identify the underlying mechanistic basis for trade-offs in resource or time allocation between defence, maintenance, and reproduction, each of which contribute to fitness (Chapter 15). Measuring trade-offs in natural systems has been challenging. Even widely assumed trade-offs, such as a reduction in activity due to predation risk leading to a reduced resource uptake and therefore to a reduction in growth (Werner and Anholt 1993; McPeek 2004) is not as obvious once the underlying physiology is taken into account (McPeek 2004; Stoks *et al*. 2005; Steiner 2007; Steiner and Van Buskirk 2009). It is thus crucial to get a better mechanistic understanding of the costs and benefits of induced defences to understand their evolution, which can then be used to improve the theory.

12.2.3 Adaptive and passive induced defence

Induced responses might not automatically be adaptive (see above for ecological traps), they might not alter the risk of being depredated (e.g. novel invasive predators), or they might be a passive response (also called indirect response) that accompanies some other response that is selected for. For instance, a classic example of an inducible defence is the deeper body developed by Crucian carps exposed to cues from predatory pike (Brönmark and Miner 1992). However, the adaptively induced deeper body might be a consequence of reduced activity rather than a direct induced response (Johansson and Andersson 2009). Reduced activity is adaptive against predation risk (Werner and Anholt 1993), and such a reduction in activity can cause energetic or biomechanical changes leading to growth patterns that have previously been associated with induced defences, that is less activity leads to increased body depth and a nice byproduct of this induction is reduced predation risk. Such passive responses might not be costly (Chapter 13). They do not have to be linked energetically or biochemically, but they might be genetically linked to other induced responses where induction is selected for (e.g. hitchhiking). The lack of costs may explain the low selective pressure to overcome the genetic linkage. Particularly in field situations, but as the carp example shows also under experimental conditions, one needs to be cautious when separating passive from active responses as defences. As in the carp example, we will not necessarily come to a final conclusion if the defence is adaptive or passive as long as we do not know the underlying genetics.

12.2.4 Multiple benefits and antagonistic induced defence

A defence against one predator might also be efficient against another predator (Van Buskirk 2001). Here, the benefits but not the costs would increase. Similarly, a defence against predation, such as the vertical distribution of *Daphnia*, also provides protection against UV-light (Rhode *et al.* 2001). Under such circumstances we could explain a situation where the benefits related to one predator (one environment) do not outweigh the costs, even though selection acts in favor of the defence, because the net benefits with respect to all predators (environments) need to be considered (Steiner and

Pfeiffer 2007). Antagonistic selection between two predators decreases the benefits of the defence when both predators are present (Mikolajewski *et al.* 2006; Kishida *et al.* 2009), but does not alter the costs of expressing the defence, and therefore will lower the selective forces if there are environments where both predators co-occur. An example of such an antagonistic defence is the spines of dragonfly larvae that are an efficient defence against fish but are selected against by other invertebrate predators (Mikolajewski 2006). In both cases, when benefits are additive, or benefits are reduced, the selective force depends on the frequency of encountering each environment (i.e. predator).

12.2.5 Cost of plasticity and evolutionary time limits

Induced defences are not expected to come without costs, even in a predator-free environment where the induced defence is not expressed. Theory predicts extra costs of plastic genotypes (van Tienderen 1991)—such costs or limits are referred to as 'costs of plasticity' to distinguish them from costs of expressing particular phenotypes (Callahan *et al.* 2008). One can think of this cost as the cost of carrying genes that allow the individual to respond plastically to predation risk, or the cost of maintaining a pathway even though it might not be used (DeWitt *et al.* 1998). Quantitative-genetic approaches have been used to investigate if such costs exist (DeWitt *et al.* 1998), however empirical evidence is scarce and inconsistent (Van Buskirk and Steiner 2009). Only if the costs of plasticity are large, and if they are global rather than environment-specific, will they limit the evolution of induced responses (Sultan and Spencer 2002). In no system have costs of plasticity been unambiguously demonstrated, whereas there is one system where such costs have been ruled out, because the rate at which an induced response was lost did not differ from the expectation based on the measured mutation rates, that is, there was no selection for the loss of the induced response (Masel *et al.* 2007). The latter finding suggests that the main limit of plasticity is time—evolution may not yet have found the optimal adaptive solution to cope with the environmental fluctuations. Further, several studies have examined evidence for a trade-off between the level of trait expression and the magnitude of plasticity to test for 'limits of plasticity'; these studies have generally found weak-to-no correlation between trait expression and the magnitude of plasticity (e.g. Lind and Johansson 2009, Auld et al. 2010).

In systems where multiple biotic components drive the selective processes, such as in predator–prey systems, one expects that we will never approach an evolutionary stable state. This is because such interactions are constantly changing due to evolutionary arms races, red-queen dynamics, or coevolutionary interactions (e.g. induced defences and offenses; Kishida et al. 2009). It is understandable that the adaptive process in a constant environment might be a step ahead of the process in a complex situation where stochastic environmental changes are the norm. Such a perspective may take us a long way towards explaining why specialists often have greater fitness than generalists in the environment that the specialist is adapted to (though see Lind and Johansson 2009). However, this does not imply that the process in the constant environment has already found its optimum. Even after more than 30,000 generations of exposure to the same environment, beneficial mutations can still occur (Blount et al. 2008) and evolution is an ongoing story. Recent environmental changes, such as global climate change, might also give plastic genotypes an adaptive advantage over constitutive ones by either extending the time window for fixed genetic changes to occur to track the novel environment (Lande 2009), or by having a range of trait expressions at hand that might be adaptive to novel predators that colonize new areas previously out of their tolerance zone, in particular if such novel predators require antagonistic defences compared to native predators.

12.3 Genetic variation in and genetic basis of induced defences

As for any trait, an important 'limit' or 'constraint' on the evolution of inducible defences is the amount of genetic variation present as well as the pattern of covariation between the genetic architecture of inducible defences and other traits (Fig. 12.1). Several studies have used quantitative-genetic methods to examine the presence of genetic variation in inducible defences, how inducible traits covary with other traits, and how these patterns change across environments. These studies have been recently reviewed (Callahan et al. 2008; Van Buskirk and Steiner 2009; Auld et al. 2010), but it can be generally stated that significant genetic variation for predator-induced defences is common and the level of genetic variation is often higher for a given trait in the presence of predators (e.g. Parejko and Dodson 1991; Harvell 1998). A subsequent, open, and important evolutionary question is whether this cross-environment change in genetic variation results from a fundamentally different genetic basis to a given trait in predator/no-predator conditions compared to a change in gene expression via regulatory effects (e.g. DNA methylation; Angers et al. 2010). Further, the extent to which pleiotropy and epistasis are involved in the expression and evolution of inducible defences has not been resolved to any extent. Crossing experiments indicate evidence for purely additive effects of autosomal alleles on induced defences (Kishida et al. 2007), however other studies find large amounts of non-additive effects or heritabilities to be highly environment-dependent (Parejko and Dodson 1991; Spitze 1992, Relyea 2005). Recent studies revealing the role of generalized stress-response mechanisms (e.g. heat-shock proteins) in the activation of induced defences suggest that there may be fundamentally similar mechanisms through which defences are expressed across a broad spectrum of organisms (e.g. plants and *Daphnia*; Pijanowska and Kloc 2004; Pauwels et al. 2005; Sangster and Queitsch 2005; Pauwels et al. 2007). The molecular revolution has not halted before the field of phenotypic plasticity, though investigating the molecular basis of plasticity has been difficult despite substantial effort in model systems such as *Arabidopsis* (Pigliucci 2005). Recently, studies have made exciting advances in understanding the specific molecular-genetic basis of an inducible morphological defence in *Rana pirica* tadpoles (Mori et al. 2009). Here, researchers have identified and begun to characterize a specific gene that is involved in the production of a 'bulgy' morphology in the presence of a gape-limited predator. Clearly, future studies that continue to investigate the genetic basis of inducible defences at both the specific molecular-genetic level, as well as the broader (e.g. generalized stress-response) level, will provide the answer to how similar or specific the genetic structure of induced responses is and, hence, their evolution has been.

12.4   Conclusions

Induced defences are widespread, which indicates that evolution has overcome initial limits to their evolution and that selection for induced defences is common. However, none of the observed inducible defences are assumed to be at their optimum because evolutionary arms races likely prevent such evolutionary stable states. The underlying mechanisms for these defences may be physiological, biochemical, or molecular-genetic, and are understudied compared to the higher-level behavioural- or morphological-level responses. However, exciting advances that shed light on the evolution of induced defences await us once more mechanisms are revealed (Chapters 2 and 8). At the other extreme, links between our often rather qualitative knowledge–gained through lab and mesocosm experiments—to the actual selective forces acting in nature through demographics, population structure, and population dynamics, remain relatively unexplored. The diversity of induced defences that have evolved, from highly predator-specific responses to fairly general induced defences that are effective across a range of predators, make the study of the evolution of induced responses an exciting topic.

The expression and evolution of induced defences frequently relies on chemical cues, because such cues mediate predator–prey interactions by providing prey with crucial information about predation risk across time and space. These chemical cues are complex, reliable, and persistent across some time. Without such cues one would expect constitutive (i.e. fixed) defences to evolve. The chemical cues are the reason why predator- (or herbivore-) induced defences became model systems for the study of phenotypic plasticity. The cues provide ideal tools to manipulate predation threat independently of the actual predation, a key feature that is used to approach questions on the ecology and evolution of induced defences from the intra-cellular level to the entire ecological community level (Chapter 10). Therefore, the chemical cues involved in many predator–prey systems make induced defences an ideal system to link evolutionary, population, and ecological processes and should continue to attract attention from students starting their careers and established researchers—each picking their level of complexity and interest.

References

Angers, B., Castonguay, E., and Massicotte, R. (2010) Environmentally induced phenotypes and DNA methylation: how to deal with unpredictable conditions until the next generation and after. *Molecular Ecology,* **19**: 1283–95.

Auld, J.R., Agrawal, A.A., and Relyea, R.A. (2010) Re-evaluating the costs and limits of adaptive phenotypic plasticity. *Proceedings of the Royal Society B-Biological Sciences,* **277**: 503–11.

Blount, Z.D., Borland, C.Z., and Lenski, R.E. (2008) Historical contingency and the evolution of a key innovation in an experimental population of *Escherichia coli. Proceedings of the National Academy of Sciences of the United States of America,* **105**: 7899–906.

Brönmark, C. and Miner, J.G. (1992) Predator-induced phenotypical change in body morphology in Crucian Carp. *Science,* **258**: 1348–50.

Brown, G.E., Chivers, D.P., and Smith, R.J.F. (1996) Effects of diet on localized defecation by northern pike, *Esox lucius. Journal of Chemical Ecology,* **22**: 467–75.

Bryant, J.P., Tahvanainen, J., Sulkinoja, M., Julkunentiitto, R., Reichardt, P., and Green, T. (1989) Biogeographic evidence for the evolution of chemical defense by boreal birch and willow against mammalian browsing. *American Naturalist,* **134**: 20–34.

Callahan, H.S., Maughan, H., and Steiner, U.K. (2008) Phenotypic plasticity, costs of phenotypes, and costs of plasticity toward an integrative view. *Year in Evolutionary Biology 2008,* **1133**: 44–66.

Chivers, D.P., Brown, G.E., and Ferrari, M.C.O. (2011) Evolution of alarm substances. In Brönmark, C. and Hansson, L.-A. (eds.) *Chemically Mediated Interactions in Aquatic Habitats.* Oxford University Press, Oxford.

DeWitt, T.J., Sih, A., and Wilson, D.S. (1998) Costs and limits of phenotypic plasticity. *Trends in Ecology & Evolution,* **13**: 77–81.

Freeman, A.S. and Byers, J.E. (2006) Divergent induced responses to an invasive predator in marine mussel populations. *Science,* **313**: 831–3.

Gabriel, W., Luttbeg, B., Sih, A., and Tollrian, R. (2005) Environmental tolerance, heterogeneity, and the evolution of reversible plastic responses. *American Naturalist*, **166**: 339–53.

Gillespie, J.H. (2004) Population genetics: a concise guide. John Hopkins University Press, Baltimore.

Hagman, M., Phillips, B.L., and Shine, R. (2009) Fatal attraction: adaptations to prey on native frogs imperil snakes after invasion of toxic toads. *Proceedings of the Royal Society B-Biological Sciences*, **276**: 2813–18.

Harvell, C.D. (1998) Genetic variation and polymorphism in the inducible spines of a marine bryozoan. *Evolution*, **52**: 80–6.

Hollander, J. (2008) Testing the grain-size model for the evolution of phenotypic plasticity. *Evolution*, **62**: 1381–9.

Hoverman, J.T. and Relyea, R.A. (2007) How flexible is phenotypic plasticity? Developmental windows for trait induction and reversal. *Ecology*, **88**: 693–705.

Johansson, F. and Andersson, J. (2009) Scared fish get lazy, and lazy fish get fat. *Journal of Animal Ecology*, **78**: 772–7.

Kishida, O., Mizuta, Y., and Nishimura, K. (2006) Reciprocal phenotypic plasticity in a predator-prey interaction between larval amphibians. *Ecology*, **87**: 1599–604.

Kishida, O., Trussell, G.C., and Nishimura, K. (2007) Geographic variation in a predator-induced defense and its genetic basis. *Ecology*, **88**: 1948–54.

Kishida, O., Trussell, G.C., and Nishimura, K. (2009) Top-down effects on antagonistic inducible defense and offense. *Ecology*, **90**: 1217–26.

Kopp, M. and Tollrian, R. (2003) Trophic size polyphenism in *Lembadion bullinum*: Costs and benefits of an inducible offense. *Ecology*, **84**: 641–51.

Lande, R. (2009) Adaptation to an extraordinary environment by evolution of phenotypic plasticity and genetic assimilation. *Journal of Evolutionary Biology*, **22**: 1435–46.

Lind, M.I. and Johansson, F. (2007) The degree of adaptive phenotypic plasticity is correlated with the spatial environmental heterogeneity experienced by island populations of *Rana temporaria*. *Journal of Evolutionary Biology*, **20**: 1288–97.

Lind, M.I. and Johansson, F. (2009) Costs and limits of phenotypic plasticity in island populations of the common frog *Rana Temporaria* under divergent selection pressures. *Evolution*, **63**: 1508–18.

Masel, J., King, O.D., and Maughan, H. (2007) The loss of adaptive plasticity during long periods of environmental stasis. *American Naturalist*, **169**: 38–46.

McPeek, M.A. (2004) The growth/predation risk trade-off: So what is the mechanism? *American Naturalist*, **163**: E88–E111.

Mikolajewski, D.J., Johansson, F., Wohlfahrt, B., and Stoks, R. (2006) Invertebrate predation selects for the loss of a morphological antipredator trait. *Evolution*, **60**: 1306–10.

Mori, T., Kawachi, H., Imai, C., *et al.* (2009) Identification of a Novel Uromodulin-Like Gene Related to Predator-Induced Bulgy Morph in Anuran Tadpoles by Functional Microarray Analysis. *Plos One*, **4**, e5936.

Parejko, K. and Dodson, S.I. (1991) e5936 The evolutionary ecology of an antipredator reaction norm - *Daphnia-pulex and Chaoborus - americanus*. *Evolution*, **45**: 1665–74.

Pauwels, K., Stoks, R., and De Meester, L. (2005) Coping with predator stress: interclonal differences in induction of heat-shock proteins in the water flea Daphnia magna. *Journal of Evolutionary Biology*, **18**: 867–72.

Pauwels, K., Stoks, R., Decaestecker, E., and De Meester, L. (2007) Evolution of heat shock protein expression in a natural population of *Daphnia magna*. *American Naturalist*, **170**: 800–5.

Peacor, S.D. (2003) Phenotypic modifications to conspecific density arising from predation risk assessment. *Oikos*, **100**: 409–15.

Peacor, S.D. (2006) Behavioural response of bullfrog tadpoles to chemical cues of predation risk are affected by cue age and water source. *Hydrobiologia*, **573**: 39–44.

Pigliucci, M. (2005) Evolution of phenotypic plasticity: where are we going now? *Trends in Ecology & Evolution*, **20**: 481–6.

Pijanowska, J. and Kloc, M. (2004) Daphnia response to predation threat involves heat-shock proteins and the actin and tubulin cytoskeleton. *Genesis*, **38**: 81–6.

Relyea, R.A. (2005) The heritability of inducible defenses in tadpoles. *Journal of Evolutionary Biology*, **18**: 856–66.

Rhode, S.C., Pawlowski, M., and Tollrian, R. (2001) The impact of ultraviolet radiation on the vertical distribution of zooplankton of the genus *Daphnia*. *Nature*, **412**: 69–72.

Richter-Boix, A., Llorente, G.A., and Montori, A. (2006) A comparative analysis of the adaptive developmental plasticity hypothesis in six Mediterranean anuran species along a pond permanency gradient. *Evolutionary Ecology Research*, **8**: 1139–54.

Sangster, T.A. and Queitsch, C. (2005) The HSP90 chaperone complex, an emerging force in plant development and phenotypic plasticity. *Current Opinion in Plant Biology*, **8**: 86–92.

Schlaepfer, M.A., Runge, M.C., and Sherman, P.W. (2002) Ecological and evolutionary traps. *Trends in Ecology & Evolution*, **17**: 474–80.

Schlichting, C. and Pigliucci, M. (1998) *Phenotypic Evolution: A Reaction Norm Perspective,* Sunderland, Sinauer Associates.

Schoeppner, N.M. and Relyea, R.A. (2005) Damage, digestion, and defence: the roles of alarm cues and kairomones for inducing prey defences. *Ecology Letters,* **8**: 505–12.

Spitze, K. (1992) Predator-mediated plasticity of prey life-history and morphology - *Chaoborus-americanus* predation on *Daphnia-pulex. American Naturalist,* **139**: 229–47.

Steiner, U.K. (2007) Linking antipredator behaviour, ingestion, gut evacuation and costs of predator-induced responses in tadpoles. *Animal Behaviour,* **74**: 1473–9.

Steiner, U.K. and Pfeiffer, T. (2007) Optimizing time and resource allocation trade-offs for investment into morphological and behavioral defense. *American Naturalist,* **169**: 118–29.

Steiner, U.K. and Van Buskirk, J. (2009) Predator-induced changes in metabolism cannot explain the growth/predation risk tradeoff. *Plos One,* **4**. e6160.

Stoks, R., De Block, M., Van de Meutter, F., and Johansson, F. (2005) Predation cost of rapid growth: behavioural coupling and physiological decoupling. *Journal of Animal Ecology,* **74**: 708–15.

Sultan, S.E. and Spencer, H.G. (2002) Metapopulation structure favors plasticity over local adaptation. *American Naturalist,* **160**: 271–83.

Tollrian, R. and Harvell, C.D. (1999) *The Ecology and Evolution of Inducible Defenses.* Princeton University Press, Princeton.

Turner, A.M. (2011) and Peacor, S.D. Scaling up infochemicals: Ecological consequences of chemosensory assessment of predation risk. In Brönmark, C. and Hansson, L.-A. (eds.) *Chemically Mediated Interactions in Aquatic Habitats.* Oxford University Press, Oxford.

Turner, A.M. (2008) Predator diet and prey behaviour: freshwater snails discriminate among closely related prey in a predator's diet. *Animal Behaviour,* **76**: 1211–17.

Van Buskirk, J. (2000) The costs of an inducible defense in anuran larvae. *Ecology,* **81**: 2813–21.

Van Buskirk, J. (2001) Specific induced responses to different predator species in anuran larvae. *Journal of Evolutionary Biology,* **14**: 482–9.

Van Buskirk, J. (2002) A comparative test of the adaptive plasticity hypothesis: Relationships between habitat and phenotype in anuran larvae. *American Naturalist,* **160**: 87–102.

Van Buskirk, J. and Arioli, M. (2002) Dosage response of an induced defense: How sensitive are tadpoles to predation risk? *Ecology,* **83**: 1580–5.

Van Buskirk, J. and Arioli, M. (2005) Habitat specialization and adaptive phenotypic divergence of anuran populations. *Journal of Evolutionary Biology,* **18**: 596–608.

Van Buskirk, J. and Steiner, U.K. (2009) The fitness costs of developmental canalization and plasticity. *Journal of Evolutionary Biology,* **22**: 852–60.

van Tienderen, P.H. (1991) Evolution of generalists and specialists in spatially heterogeneous environments. *Evolution,* **45**: 1317–31.

Via, S., Gomulkiewicz, R., de Jong, G., Scheiner, S.M., Schlichting, C.D. and Van Tienderen, P.H. (1995) Adaptive phenotypic plasticity - consensus and controversy. *Trends in Ecology & Evolution,* **10**: 212–17.

Vourc'h, G., Martin, J.L., Duncan, P., Escarre, J., and Clausen, T.P. (2001) Defensive adaptations of *Thuja plicata* to ungulate browsing: a comparative study between mainland and island populations. *Oecologia,* **126**: 84–93.

Werner, E.E. and Anholt, B.R. (1993) Ecological consequences of the trade-off between growth and mortality-rates mediated by foraging activity. *American Naturalist,* **142**: 242–72.

CHAPTER 13

How to explore the sometimes unusual chemistry of aquatic defence chemicals

Georg Pohnert

The structural and biosynthetic diversity of marine natural products is overwhelming. From a biosynthetic perspective all major pathways to secondary metabolites have been identified in marine organisms and several metabolic pathways are unique to marine resources (Blunt *et al.* 2005; Moore 2006). Traditionally, drug discovery was the driver for the search of bioactive marine natural products. Bioprospecting has yielded several lead structures, drugs, and drug candidates derived from marine resources and the search is still on (Haefner 2003; Simmons *et al.* 2005). It is, however, well accepted that the driving forces for the evolution of natural product diversity are the environmental and ecological challenges that the producers have to face (Firn and Jones 2003). Thus natural products are involved in very fundamental processes like osmoregulation or adaptation to light stress that ensure the survival of the organism itself (Shick and Dunlap 2002; Stefels 2000). But they also commonly have a function in chemical communication and as a chemical defence and thereby provide a selective advantage in the interaction of an organism with its biotic environment (Paul and Ritson-Williams 2008).

Early studies of defence metabolites focused on compounds that could be readily extracted from the tissue of a well defended organism. A bioassay-guided chromatographic separation then led to pure fractions that could be submitted to further structure elucidation by standard analytical techniques. If the compounds were active against herbivores,

pathogens, or epiphytes in a concentration range that was also found at the site of interaction in the producing organisms, it could be concluded that these so called constitutive chemical defences protected the producer. From marine sources, hundreds of such bioactive metabolites that have an attributed role in chemical defence are known. Since this chapter cannot provide a comprehensive survey of such strategies only a selection of relevant reviews is given (Amsler and Fairhead 2006; Cimino *et al.* 1999; Hay and Fenical 1988; Pawlik 1993; Proksch 1994). Further, induced or activated defence mechanisms will not be discussed here but are rather a focus of other chapters of this book. I will rather only introduce two not very well known defence strategies that do not fall within the categories of constitutive, activated, or induced defences. One relies on very simple primary metabolites or their transformation to achieve a chemical defence, the other on symbionts producing elaborate metabolites. These examples serve the purpose to illustrate that even unpredicted but highly efficient strategies have evolved in aquatic chemical interactions and that we have to be open to every type of metabolic interaction if we want to screen for chemical defence of aquatic organisms. To introduce the diversity of chemical defence metabolites I will provide an insight into the general chemical aspects of aquatic (secondary) metabolites and their biosynthesis. Then I will illustrate the challenges that the diversity of defence strategies, the diversity of physicochemical properties, and the diversity of biosynthetic pathways represent for the structure elucidation of bioactive aquatic natural products. I will introduce techniques that are currently employed in the search for new bioactive compounds. This introduction on general concepts in defence chemistry and methodological approaches provides the chemical background for other chapters in this book.

13.1   The role of primary metabolites in chemical defence

In recent years several studies have proved that not only elaborate secondary metabolites produced by complex biosynthetic pathways are essential for chemical defence. In bioassays activities were identified that could rather be attributed to natural products from the primary metabolism. In a bioassay guided fractionation approach, Fu *et al.* aimed to elucidate the structure of haemolytic compounds from the ichthyotoxic microalga *Fibrocapsa japonica* (Fu *et al.* 2004). This alga produces red tide blooms that induce massive mortality in fish stocks and causes significant damage to coastal fisheries. *F. japonica* is a producer of the complex polyketide brevetoxin, a compound that is a highly active neurotoxin in mammals (Khan *et al.* 1996). Brevetoxins nevertheless can accumulate in live fish by dietary transfer from primary producers such as red tide algae or their consumers and are not haemolytically active (Naar *et al.* 2007). Interestingly, no elaborate polyketide structure showed up in the screening for ichtyotoxic activity but rather very simple fatty acids that are widely distributed in most marine organisms. Between 7% and 89% of the haemolytic activity could be attributed to four polyunsaturated fatty acids, including the common eicosapentaenoic acid and arachidonic acid (De Boer *et al.* 2009). These fatty acids are usually found bound to lipids and levels of free polyunsaturated fatty acids are commonly low. *F. Japonica*, however, transforms storage lipids rapidly after wounding to release the active free fatty acids, thereby mobilizing primary metabolites for their chemical defence (Fig. 13.1) (Fu *et al.* 2004). Bioassay guided structure elucidation showed that polyunsaturated fatty acids that are released during a wound activated process are

also used as grazer defence in freshwater epilithic diatom biofilms (Juttner 2001).

Other primary metabolites can also be involved in chemical defence reactions. Thus, the widely distributed osmolyte dimethylsulfoniopropionate and its transformation products dimethylsulfide and acrylic acid are suggested to be involved in the chemical defence of both micro- and macroalgae (Fredrickson and Strom 2009; Strom *et al.* 2003; Wiesemeier *et al.* 2007; Wolfe *et al.* 1997). These examples demonstrate that chemical defence can be mediated by comparably simple molecules that are also essential for growth and survival of the defended organism in other contexts. This has major consequences if the costs of defence are concerned. Strategies relying on primary metabolites might not be under the same selective pressure compared to those involving elaborate secondary metabolites. It can thus be envisaged that even minor beneficial effects of primary metabolite-based defences can be cost effective and viable.

Another successful defence strategy that involves primary metabolites is reported from damaged micro- and macroalgae. Some of these organisms are capable of actively reducing their food quality upon injury. For example, the wound-activated transformation of fatty acids to polyunsaturated aldehydes that is often observed in marine and freshwater diatoms can result in a massive depletion of unsaturated fatty acids (Fig. 13.1). Since some of these fatty acids are essential for herbivores this results in a reduction of the food quality of the prey algae that no longer represent an optimum nutrition (Wichard *et al.* 2007). A wound-activated depletion of essential amino acids can also be observed in macroalgae. Here elaborate secondary metabolites can serve as chemical tools that react with primary metabolites that are an essential food component for herbivores. The dominant secondary metabolite caulerpenyne that is produced by green algae *Caulerpa spp.* is such an example. It is transformed upon wounding to an effective protein cross-linker termed oxytoxin 2. The reaction of the activated metabolite oxytoxin 2 with amino acids from algal proteins results in the depletion of available proteins. This process goes ahead with a reduction of feeding activity of herbivores. External addition of non-cross-linked protein

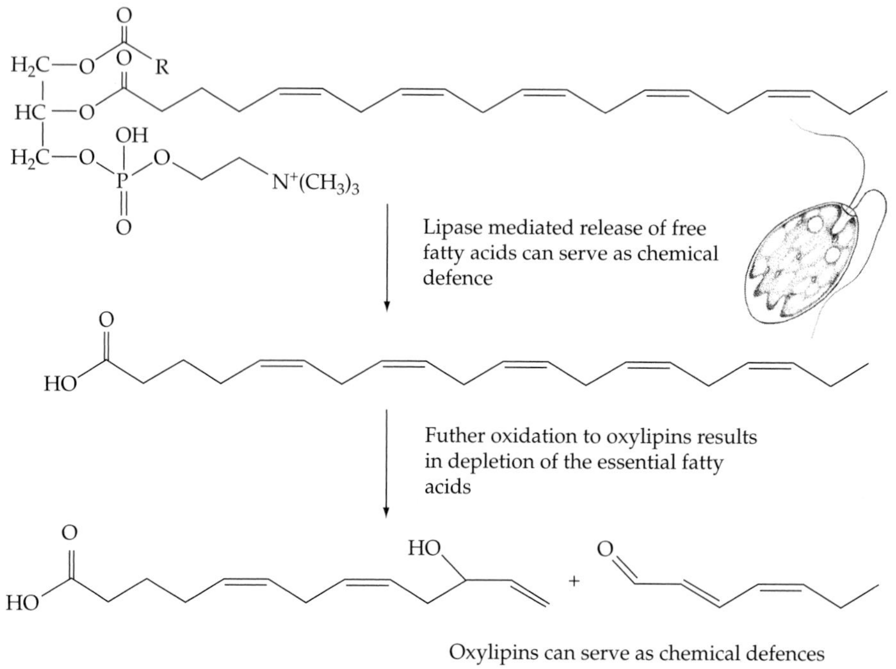

Figure 13.1 Three ways to use one simple enzymatic cascade for a chemical defence exemplified for *Fibrocapsa japonica* and certain diatoms: (1) Primary metabolites, such as the depicted free fatty acid can act as defensive metabolites. The fatty acid with haemolytic activity can be readily released from storage lipids by enzymes. (2) This fatty acid, which is also a limiting nutrient for grazers, can also be actively depleted by further enzymatic reactions thereby reducing the food quality. (3) The resulting end products can then act as defensive metabolites.

restores at least partially the initial food quality (Weissflog *et al.* 2008).

A similar function can be envisaged for protein cross-linking tunichromes from the tunicate *Ascidia nigra* (Fig. 13.2). These small peptides contain one or more dehydrodopa-derived units which can be oxidized enzymatically and then react with proteins or peptides. This process that results in a cross-linking and the depletion of the available proteins is also discussed as a defence strategy against invading microorganisms and a way to assist formation of the thick tunic cover of the adult organisms (Cai *et al.* 2008).

13.1.1 Natural product symbiosis

A strategy employed by several genera of aquatic organisms is the so called natural product symbiosis. Here a macroscopic organism is not producing active secondary metabolites itself but rather relies on associated symbiotic bacteria that provide highly active natural products (Piel 2004; Piel 2009). These natural products can then defend both the host and its symbionts against pathogens and predators and thus provide a cost efficient means of chemical defence for the host. Symbiotic interactions can be highly specific and often the symbiont cannot be cultured without its host. Such natural product symbioses are frequently associated with secondary metabolites from marine sponges as hosts of highly complex symbiont communities (Schmitt *et al.* 2007). Molecular techniques in particular were successful in providing evidence that associated bacteria were indeed the true producers of bioactive secondary metabolites of sponges (Piel *et al.* 2004). Good evidence for natural product symbioses is given in cases where gene clusters coding for enzymes involved in the biosynthesis of the secondary metabolite can be located within the genome of symbiotic bacteria (Schmidt 2008). Besides sponges,

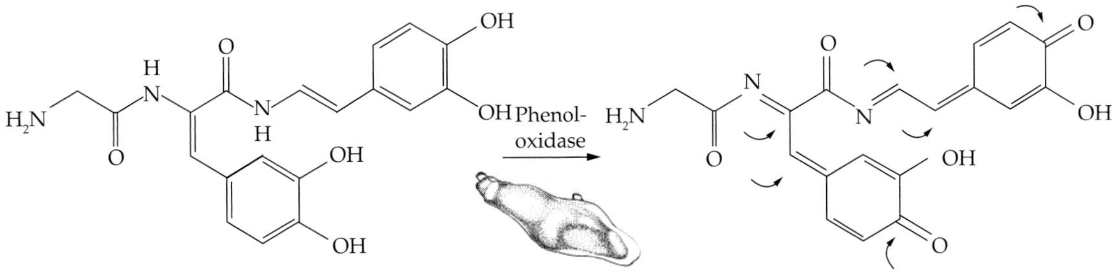

Figure 13.2   Enzyme-mediated oxidation of the tunichrome Mm-1 in *Ascidia nigra* results in the corresponding quinone methides that can form adducts and cross-links with proteins and other macromolecular side chains at the sites indicated by arrows. This process leads to a polymerization of the proteins (Cai *et al.* 2008).

several other marine organisms including corals or bryozoans rely on symbiotic interactions with bacteria that provide them with natural products for defence and other physiologically and ecologically important functions (Engel *et al.* 2002; Lim-Fong *et al.* 2008; Lopanik *et al.* 2004).

In contrast to these examples, natural product symbioses from algae are rather rare, but a remarkable exception has recently been reported from the marine brown alga *Lobophora variegata*. General considerations about complex metabolites that can be isolated from the brown alga, that are structurally nearly identical with metabolites previously found in cyanobacteria, provided a first idea about their potential symbiotic origin. Brown macroalgae are usually not known for their production of highly active secondary metabolites. Algal defence often relies on rather unspecific metabolites, such as phlorotannins, that are present in higher amounts (Amsler and Fairhead 2006; Toth and Pavia 2000). Relatively simple natural products including terpenes, oxylipins, or simple halogenated hydrocarbons have also been frequently reported as defence metabolites (Cronin and Hay 1996; La Barre *et al.*; Schnitzler *et al.* 2001). It was thus surprising that the seaweed *L. variegata*, which is chemically defended against pathogenic and saprophytic marine fungi, relies on a highly complex defensive metabolite. The active 22-membered cyclic lactone, lobophorolide, is of presumed polyketide origin but this class of metabolites has not been previously reported from brown algae (Kubanek *et al.* 2003) (Fig. 13.3). Strikingly similar metabolites are found in cyanobacteria. Tolytoxin, for example, a metabolite from

cyanobacteria of the genera *Tolypothrix* and *Scytonema* shares structural similarities with lobophorolide from the brown alga. It is thus likely that lobophorolide is not a brown algal metabolite but rather the product of a microbial symbiont (Kubanek *et al.* 2003). Indeed, epiphytic bacteria including cyanobacteria are often found on algal surfaces and could explain the observed activity. Since it is often the case that identical or very similar elaborate metabolites can be found in different organisms a closer inspection for the verification of potential natural product symbioses is warranted. Further experiments can then reveal if the true producers are not the terrestrial or aquatic plants, insects, or animals themselves but rather associated bacteria (Daly 2004; Piel 2002).

It becomes clear that the biosynthetic repertoire of an organism can be significantly extended by metabolites from associated microorganisms. Instead of the chemical defence of an aquatic species we should thus rather consider the defence of the respective organism with its associated microbial community. A focus on axenic organisms might sometimes be useful to unravel specific functions, but an ecological perspective will surely often involve the investigation of a natural assemblage of organisms.

These few first considerations already illustrate that no general motives for the biosynthesis of defensive metabolites can be identified in marine organisms. All biogenetic metabolites—including the primary metabolites—from one species and its associated microbial community have to be considered if a comprehensive survey of defence potential in the ecological context is desired.

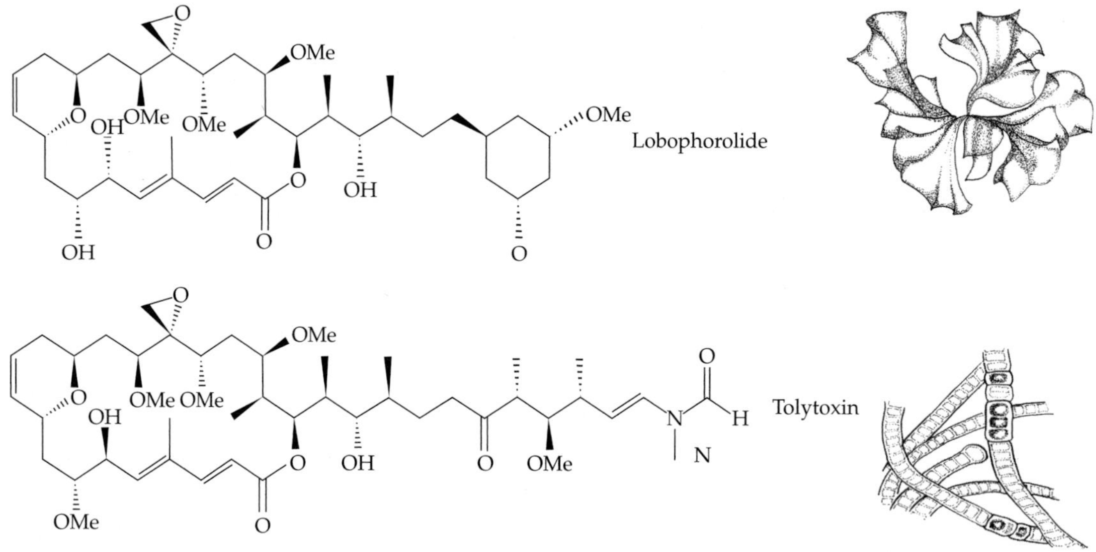

Lobophorolide

Tolytoxin

Figure 13.3 The unusual polyketide lobophorolide from *Lobophora variegata* (top) and tolytoxin, a structurally related polyketide from cyanobacteria of the genus *Tolypothrix* (below).

13.2 General chemical properties and biosynthesis of bioactive metabolites

Marine and freshwater organisms have not developed a chemistry that would be specific for this environment. A broad range of examples demonstrates that the molecules involved in chemically mediated interactions have no common structural or physicochemical properties that would classify them as prototypic aquatic signal or defence compounds. Thus small volatile gases like dimethylsulfide or halomethanes (Barreiro *et al.* 2006; Pohnert *et al.* 2007) as well as large polysaccharides and proteins (Matsumura *et al.* 1998) and a plethora of intermediately sized metabolites are known to play important roles in chemical interactions. Even simple inorganic molecules can serve as an effective chemical defence. A well documented example is the production of high amounts of sulphuric acid by the brown alga *Desmarestia spp.* that deters feeding by sea urchins (Pelletreau and Muller-Parker 2002). Good solubility in water seems to be no issue if the general quality of a natural product as a signal or defence is what is concerned. Biological activity has been reported from non-polar hydrocarbon pheromones of brown algae (Pohnert and Boland

2002) to highly functionalized small organic acids that are readily soluble in water (Fusetani and Kem 2009). It can be argued that highly hydrophobic structures are enriched in hydrophobic receptor pockets thereby leading to a higher signal-to-noise ratio in the aqueous environment (Maier *et al.* 1994). But also highly polar soluble metabolites could bear an advantage in certain contexts since they are readily solubilized and transported by diffusion mediated processes. Aquatic organisms also rely on different reactivity of their secondary metabolites. This is illustrated by strategies that involve the production of highly reactive metabolites, such as hydrogen peroxide and other reactive oxygen species, or the storage or release of very stable aromatic compounds that are persistent over a long time (Kubanek *et al.* 2004; Weinberger *et al.* 2002). It is fascinating to observe that numerous ways to respond to the multiple challenges in the aquatic environment have evolved that are nearly unrestricted in 'chemical creativity'.

The physicochemical diversity of aquatic bioactive natural products is also reflected in the diversity of biosynthetic pathways that are utilized for their production. It is central to understand these pathways towards natural products since this can be a key to the understanding of metabolic costs and regulative principles for their production. Once the complete enzymatic assembly line for a given metabolite is understood we can address metabolic fluxes but also manipulate production rates or even the structures of metabolites. This can be done by simple targeted inhibition approaches or even by elaborate metabolic engineering, which can, for example, involve the use of altered starter units that are incorporated instead of the natural precursor into the metabolite in question (Moore and Hertweck 2002). Comprehensive surveys have been published that classify marine natural products according to their biosynthesis (Moore 2005; Moore 2006 and earlier reviews in this series) and more than 100 specialized reviews focusing on specific biosynthetic pathways of marine natural products are available (e.g. Paul and Ritson-Williams 2008; Pohnert 2004). Major compound classes include the often highly active polyketides (Hertweck 2009) and alkaloids (Facchini 2001), the structurally diverse terpenes (Wise 2003) and phlorotannins (La Barre *et al.* 2010), or oxylipins that frequently play roles in signalling and defence (Pohnert 2005). But also metabolites with mixed origin involving two or more biosynthetic pathways are very commonly found in marine organisms.

13.2.1 Structure elucidation

This huge metabolic diversity also requires a creative usage of a broad spectrum of chemical methods to discover novel active metabolites. A bioassay guided or metabolomics enabled elucidation of active compounds has initially to screen for any metabolite without bias for an analytical protocol. Also the mode of action, distribution in the water, stability, and potential biotic or abiotic transformation has to be considered for every novel metabolite or class of metabolites with potential activity. Nevertheless, standard methods can be used for the characterization of a multitude of active metabolites and it is advised to initially test these before applying elaborate additional techniques.

Common extraction techniques include the head space methods or solid phase microextrac-

tion for the analytics of volatiles (Stashenko and Martinez 2007; Zhang and Li 2010). Diverse liquid or solid phase extraction methods for lipophilic metabolites are available that can be used for an initial test of the extractability of the active compounds (Hennion 1999; Jian *et al.* 2010). If these procedures are unsuccessful or if the extraction yields are poor, novel hydrophilic interaction liquid chromatographic protocols offer a promising alternative for the extraction of more polar metabolites from an aqueous matrix (Jian *et al.* 2010; Sticher 2008). The extracts obtained with these methods can be directly submitted to chromatographic separation to isolate pure fractions of the active metabolites. Usually both extraction and chromatography should be monitored with bioassays to demonstrate that no substantial amount of activity is lost during the process. Another approach that does not rely on bioassay guided fractionation involves novel metabolomic techniques for the search for active compounds. This approach can be used if a certain species can be found in a defended and non-defended state or if the defence is only switched on during certain stages. In these cases quantitative data of as many as possible metabolites from the organism will be recorded in the active and inactive state. These complex data sets can be compared using chemoinformatic methods and metabolites that are only present in the defended stages are candidates for the defence metabolites. This approach is reviewed in more detail in Prince and Pohnert (2010).

Once a given compound is purified, structure elucidation usually relies primarily on nuclear magnetic resonance and mass spectrometry. While these approaches give a good overview of most natural products, specialized methods have to be applied where the structure elucidation of certain more unusual marine metabolites is concerned. Thus, for example, the analysis of iodinated species and iodine is not possible with the routine procedures described above. These species play, however, a central role in the physiology and ecology of brown algae. To circumvent analytical limitations of the above mentioned routine methods, the unusual technique of non-destructive x-ray absorption spectroscopy was conducted at a synchrotron facility. Küpper *et al.* (2008) selected Laminariales (kelps) that are the strongest accumulators of iodine among living organisms to test the chemical state of iodinated species in the algae. They suggested that the accumulated species is iodide, which is massively enriched by the algae. Iodide might serve as a scavenger for reactive oxygen species since it is effluxed by the thallus upon oxidative stress and detoxifies aqueous oxidants (Küpper *et al.* 2008). Other studies used, for example, EPR (electron paramagnetic resonance) spectroscopy to investigate the role of metal mediated cross-linking in mussel secretions (Sever *et al.* 2004). This spectroscopy allowed conclusion of how iron is incorporated into a polymer mediating mussel adhesion. The highly unusual natural product arsenicin A, from the sponge *Echinochalina bargibanti*, is another example of a natural product that required high chemical inventiveness for its structural elucidation (Fig. 13.4) (Mancini *et al.* 2006).

Motivated by its high antibacterial and antifungal activity, arsenicin A was purified by a bioassay guided approach. The proton and carbon NMR spectra of the pure compound showed only very few signals that were insufficient for structure elucidation. Neither could mass spectrometry provide a final clue about the structure of the metabolites. Mancini *et al.* had thus revert to chemical synthesis, and comparison of the synthetic compound with the natural product finally revealed the unusual structure of the metabolite, which is exceptionally rich in arsenic (Mancini *et al.* 2006). These examples illustrate how chemical ecologists should be open to analytical methods that are not commonly in the repertoire of natural product chemists. Novel activities and principles still await discovery and one should not be discouraged if routine analytics does not immediately provide the structure of the compound in the active fraction.

13.2.2 Determination of ecologically relevant concentrations and localization of defence metabolites

Besides the use of more elaborative methods to investigate novel metabolites there is an urgent need in aquatic chemical ecology for the determination of ecologically relevant concentrations of metabolites. It is obvious that any metabolite can only act as a defence if sufficiently high concentrations are available for the ecological interaction. Pure tests of activity cannot serve as valid approaches in chemical ecology without the quantification of the active principle. But bioassays conducted with concentrations that are similar to the concentrations in crude extracts can also be misleading. Seasonality (Amade and Lemee 1998), possible induction of defence metabolites (Toth and Pavia 2000), or enrichment in specific tissues of the producer (Iken *et al.* 2007) can lead to massive variations of the (local) concentrations of defence metabolites. This variability represents different ways to effectively allocate resources to chemical defence and/or to reduce self toxicity and poses challenges for chemical ecologists. In the comparably simple cases of seasonal variations or variable contents of defence metabolites in different tissue types of the producer, a series of quantitative analytical measurements are required to obtain a picture of the variability. Case to case considerations are, however, required when it has to be decided how frequent the sampling or how high the resolution has to be. In some cases simple dissection of, for example, an algal blade is sufficient; in others localization of defensive metabolites in specific glands or mucus might exhibit surprisingly high local concentrations that might explain an observed activity of defence metabolites (Salgado *et al.* 2008). Also associations of a host with microorganisms that produce active metabolites might result in high local variations of defence metabolites. Such accumulations could be clearly visualized by matrix laser desorption ionization mass spectrometry imaging (MALDI MS) of the sponge *Dysidea herbacea*. Some metabolites

Figure 13.4 The unusual sponge metabolite arsenicin A.

were localized on the outer edges, while others have a more complete and uniform distribution, and others appear to have distinct, internal localization. The implication for chemical ecology of such different chemical microenvironments within a sponge is obvious. Chemical interactions will be very localized and even specific tissues could be differentially defended (Fig. 13.5) (Esquenazi *et al.* 2008). Putting the sponge in a blender with some solvent and determining the concentration of the metabolites would thus clearly not provide the ecologically relevant concentrations of the metabolites. MALDI-MS imaging can also be used to locate specific cyanobacteria in complex assemblages based on specific ions of their natural products (Esquenazi *et al.* 2008).

Clearly more method development and also the transfer of methods that are already established for medicinal applications or material science are required to get further insight into the truly relevant concentrations of active principles. The investigation of inhibitors of surface colonization of marine organisms is also posing a problem for analytical chemists. Almost no experimental information is available on the release of metabolites into the diffusion limited layer surrounding biological surfaces in an aqueous environment. Only a few reliable studies address the release of relevant metabolites, as well as their effects on associated microbial communities, taking into account the true surface concentrations. The investigation of defence metabolites on algal surfaces using a brief hexane dipping extraction, which leaves the cells intact, is a useful technique (De Nys *et al.*

1998). During the short contact of the algal surface unpolar metabolites are transferred into the solvent. If recovery rates of metabolites from neutral surfaces can be determined in separate experiments this method gives at least a semi-quantitative impression of the presence of hydrophobic compounds on surfaces of marine organisms. With this information concentrations used in bioassays can be justified and conclusions about surface interactions are warranted. These techniques were successfully used for the elucidation of the defensive role of halogenated metabolites in the red alga *Bonnemaisonia hamifera* (Nylund *et al.* 2008). Based on the observation that this alga is less fouled by bacteria relative to co-occurring seaweeds, antibacterial metabolites could be isolated by bioassay-guided fractionation. 1,1,3,3-tetrabromo-2-heptanone proved to be an effective and specific inhibitor of bacterial growth. This compound could be quantified on the surface of *B. hamifera* without destroying algal cells by hexane dipping, and natural concentrations of this secondary metabolite on artificial surfaces were effective against gram-positive bacteria and flavobacteria (Nylund *et al.* 2008). Recently, a dipping procedure employing a mixture of methanol and hexane, which extracts significantly more polar metabolites, has been suggested for the mechanically more robust brown alga *Fucus vesiculosus* (Lachnit *et al.* 2010). This method picks up even sugars and amino acids but care has to be taken that algal cells are not destroyed due to the increased stress caused by the organic solvents. *F. vesiculosus* surface extracts obtained by this method influenced the community composition

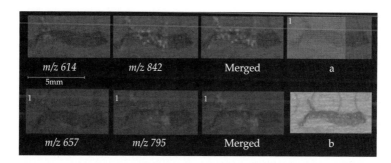

Figure 13.5   Differential localizations of ion masses associated with the sponge *Dysidea herbacea*; (a) shows the raster on the image; (b) shows the photomicrograph of the sponge-section itself (Esquenazi *et al.* 2008). See also colour Plate 6.

of bacterial colonizers and inhibited barnacle cyprid settlement.

Another tool to access surface chemistry of aquatic organisms is the rapidly emerging technique of desorption electrospray mass spectrometry (DESI-MS) (Ifa *et al.* 2010). By guiding a stream of charged solvent microdroplets over surfaces, imaging of metabolites with a reasonable resolution (< 200 mm) is possible (Takats *et al.* 2004). This technique was used for the imaging of bromophycolides and callophycoic acids that are potent antifungal chemical defences from the red macroalga *Callophycus serratus* (Lane *et al.* 2009). DESI-MS imaging revealed that surface-associated fungicides are not homogenously distributed across algal surfaces but are instead associated with distinct surface patches (Fig. 13.6).

The survey suggests that they reach high enough concentrations to act as efficient antifungal agents in the algal defence. Even if this study represents a large step forward it still involved extensive preparation of the algal material and does not allow actual concentration determination in the diffusion-limited boundary layer between surface and water.

13.3 Conclusions

From the selected examples above it will be clear that the structures of the majority of aquatic defensive metabolites are in principle accessible using chemical analytical methods, and that the results from these approaches can be used to design ecologically realistic bioassays. Chemical analytics is, however, still a major challenge and by no means a routine procedure. This is mainly due to the fact that there is no general structural motive for defensive metabolites: active compounds can for example be highly volatile, of intermediate size, or even macromolecules; they can be polar salts, unpolar lipids, or everything in-between; they can be stable during extraction or highly reactive, and so on. All these variable properties lead to the fact that for each compound or compound class a special analytical procedure has to be developed and validated. Even if the methods are in place it has thus to be carefully considered if the endeavour to search for the active principle shall be undertaken. If bioassays are rather slow and the compound does not fall into standard categories the search for an active principle can easily

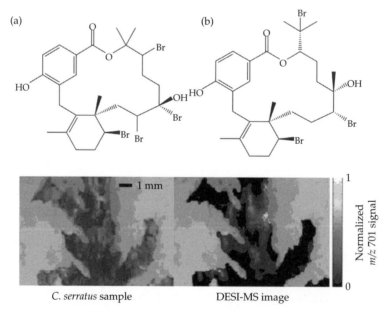

(a)   (b)

1 mm

Normalized
m/z 701 signal

C. serratus sample   DESI-MS image

Figure 13.6   The relative concentrations of the bromophycolides (a) (above left) and (b) (above right) can be monitored on the surface of the red alga *C. serratus*. The left picture shows a photography of the algal surface, the right one a distribution map of the metabolites that can be monitored by their [M + Cl-H]⁻ ions (Lane *et al.* 2009). See also colour Plate 7.

require some years. In logical consequence significant resources have to be invested in such a quest. Several compounds or compound classes, however, still pose major problems that cannot be solved today. This includes, for example, the high molecular weight amphiphilic lytic metabolites from the red tide dinoflagellate *Alexandrium tamarense* (Ma *et al.* 2009) or metabolites that are active in very low concentrations that cannot be detected or extracted in sufficient quantities (in the micro to milligram range) for direct analytical investigations. Also, very labile or reactive metabolites can still at best be elucidated using indirect methods that often involve chemical synthesis or derivatization. Analytical approaches that require extraction and fractionation would lose these labile metabolites during the purification process. But even if the work itself is not trivial, the chemical structure, the pure compound in hand, and the information about the biosynthesis, opens novel fields in ecological investigations that are definitely worth pursuing.

References

Amade, P. and Lemee, R. (1998) Chemical defence of the Mediterranean alga *Caulerpa taxifolia*: variations in caulerpenyne production. *Aquatic Toxicology*, **43**: 287–300.

Amsler, C.D. and Fairhead, V.A. (2006) Defensive and sensory chemical ecology of brown algae. *Advances in Botanical Research, Vol 43*, **43**: 1–91.

Barreiro, A., Guisande, C., Frangopulos, M., Gonzalez-Fernandez, A., Munoz, S., Perez, D., Magadan, S., Maneiro, I., Riveiro, I., and Iglesias, P. (2006) Feeding strategies of the copepod *Acartia clausi* on single and mixed diets of toxic and non-toxic strains of the dinoflagellate *Alexandrium minutum*. *Marine Ecology-Progress Series*, **316**: 115–25.

Blunt, J.W., Copp, B.R., Munro, M.H.G., Northcote, P.T., and Prinsep, M.R. (2005) Marine natural products. *Natural Product Reports*, **22**: 15–61.

Cai, M.M., Sugumaran, M., and Robinson, W.E. (2008) The crosslinking and antimicrobial properties of tunichrome. *Comparative Biochemistry and Physiology B-Biochemistry & Molecular Biology*, **151**: 110–17.

Cimino, G., Fontana, A., and Gavagnin, M. (1999) Marine opisthobranch molluscs: Chemistry and ecology in sacoglossans and dorids. *Current Organic Chemistry*, **3**: 327–72.

Cronin, G. and Hay, M.E. (1996) Induction of seaweed chemical defenses by amphipod grazing. *Ecology*, **77**: 2287–301.

Daly, J.W. (2004) Marine toxins and nonmarine toxins: Convergence or symbiotic organisms. *Journal of Natural Products*, **67**: 1211–15.

De Boer, M.K., Tyl, M.R., Fu, M., Kulk, G., Liebezeit, G., Tomas, C.R., Lenzi, A., Naar, J., Vrieling, E.G., and Van Rijssel, M. (2009) Haemolytic activity within the species *Fibrocapsa japonica* (Raphidophyceae). *Harmful Algae*, **8**: 699–705.

De Nys, R., Dworjanyn, S.A., and Steinberg, P.D. (1998) A new method for determining surface concentrations of marine natural products on seaweeds. *Marine Ecology-Progress Series*, **162**: 79–87.

Engel, S., Jensen, P.R., and Fenical, W. (2002) Chemical ecology of marine microbial defense. *Journal of Chemical Ecology*, **28**: 1971–85.

Esquenazi, E., Coates, C., Simmons, L., Gonzalez, D., Gerwick, W.H., and Dorrestein, P.C. (2008) Visualizing the spatial distribution of secondary metabolites produced by marine cyanobacteria and sponges via MALDI-TOF imaging. *Molecular BioSystems*, **4**: 562–70.

Facchini, P.J. (2001) Alkaloid biosynthesis in plants: Biochemistry, cell biology, molecular regulation, and metabolic engineering applications. *Annual Review of Plant Physiology and Plant Molecular Biology*, **52**: 29–66.

Firn, R.D. and Jones, C.G. (2003) Natural products - a simple model to explain chemical diversity. *Natural Product Reports*, **20**: 382–91.

Fredrickson, K.A. and Strom, S.L. (2009) The algal osmolyte DMSP as a microzooplankton grazing deterrent in laboratory and field studies. *Journal of Plankton Research*, **31**: 135–52.

Fu, M., Koulman, A., Van Rijssel, M., Lutzen, A., M.K., D.B., Tyl, M.R., and Liebezeit, G. (2004) Chemical characterisation of three haemolytic compounds from the microalgal species *Fibrocapsa japonica* (Raphidophyceae). *Toxicon*, **43**: 355–63.

Fusetani, N. and Kem, W. (2009) Marine toxins: an overview. *Progress in Molecular and Subcellular Biology*, **46**: 1–44.

Haefner, B. (2003) Drugs from the deep: marine natural products as drug candidates. *Drug Discovery Today*, **8**: 536–44.

Hay, M.E. and Fenical, W. (1988) Marine plant-herbivore interactions - the ecology of chemical defense. *Annual Review of Ecology and Systematics*, **19**: 111–45.

Hennion, M.C. (1999) Solid-phase extraction: method development, sorbents, and coupling with liquid chromatography. *Journal of Chromatography A*, **856**: 3–54.

Hertweck, C. (2009) The biosynthetic logic of polyketide diversity. *Angewandte Chemie-International Edition*, **48**: 4688–716.

Ifa, D.R., Wu, C.P., Ouyang, Z., and Cooks, R.G. (2010) Desorption electrospray ionization and other ambient ionization methods: current progress and preview. *Analyst*, **135**: 669–81.

Iken, K., Amsler, C.D., Hubbard, J.M., Mcclintock, J.B., and Baker, B.J. (2007) Allocation patterns of phlorotannins in Antarctic brown algae. *Phycologia*, **46**: 386–95.

Jian, W.Y., Edom, R.W., Xu, Y.D., and Weng, N.D. (2010) Recent advances in application of hydrophilic interaction chromatography for quantitative bioanalysis. *Journal of Separation Science*, **33**: 681–97.

Juttner, F. (2001) Liberation of 5,8,11,14,17-eicosapentaenoic acid and other polyunsaturated fatty acids from lipids as a grazer defense reaction in epilithic diatom biofilms. *Journal of Phycology*, **37**: 744–55.

Khan, S., Arakawa, O., and Onoue, Y. (1996) Neurotoxin production by a chloromonad *Fibrocapsa japonica* (Raphidophyceae). *Journal of the World Aquaculture Society*, **27**: 254–63.

Kubanek, J., Jensen, P.R., Keifer, P.A., Sullards, M.C., Collins, D.O., and Fenical, W. (2003) Seaweed resistance to microbial attack: A targeted chemical defense against marine fungi. *Proceedings of the National Academy of Sciences of the United States of America*, **100**: 6916–21.

Kubanek, J., Lester, S.E., Fenical, W., and Hay, M.E. (2004) Ambiguous role of phlorotannins as chemical defenses in the brown alga *Fucus vesiculosus*. *Marine Ecology-Progress Series*, **277**: 79–93.

Kupper, F.C., Carpenter, L.J., Mcfiggans, G.B., Palmer, C.J., Waite, T.J., Boneberg, E.M., Woitsch, S., Weiller, M., Abela, R., Grolimund, D., Potin, P., Butler, A., Luther, G.W., Kroneck, P.M.H., Meyer-Klaucke, W., and Feiters, M.C. (2008) Iodide accumulation provides kelp with an inorganic antioxidant impacting atmospheric chemistry. *Proceedings of the National Academy of Sciences of the United States of America*, **105**: 6954–8.

La Barre, S., Potin, P., Leblanc, C., and Delage, L. (2010) The halogenated metabolism of brown algae (Phaeophyta), its biological importance and its environmental significance. *Marine Drugs*, **8**: 988–1010.

Lachnit, T., Wahl, M., and Harder, T. (2010) Isolated thallus-associated compounds from the macroalga *Fucus vesiculosus* mediate bacterial surface colonization in the field similar to that on the natural alga. *Biofouling*, **26**: 247–55.

Lane, A.L., Nyadong, L., Galhena, A.S., Shearer, T.L., Stout, E.P., Parry, R.M., Kwasnik, M., Wang, M.D., Hay, M.E., Fernandez, F.M., and Kubanek, J. (2009) Desorption electrospray ionization mass spectrometry reveals surface-mediated antifungal chemical defense of a tropical seaweed. *Proceedings of the National Academy of Sciences of the United States of America*, **106**: 7314–19.

Lim-Fong, G.E., Regali, L.A., and Haygood, M.G. (2008) Evolutionary relationships of 'Candidatus Endobugula' bacterial symbionts and their Bugula bryozoan hosts. *Applied and Environmental Microbiology*, **74**: 3605–9.

Lopanik, N., Lindquist, N., and Targett, N. (2004) Potent cytotoxins produced by a microbial symbiont protect host larvae from predation. *Oecologia*, **139**: 131–9.

Ma, H.Y., Krock, B., Tillmann, U., and Cembella, A. (2009) Preliminary characterization of extracellular allelochemicals of the toxic marine dinoflagellate *Alexandrium tamarense* using a *Rhodomonas salina* bioassay. *Marine Drugs*, **7**: 497–522.

Maier, I., Muller, D.G., and Boland, W. (1994) Spermatozoid vhemotaxis in *Laminaria digitata* (Phaeophyceae) 3. Pheromone receptor specificity and threshold concentrations. *Zeitschrift Fur Naturforschung C-a Journal of Biosciences*, **49**: 601–6.

Mancini, I., Guella, G., Frostin, M., Hnawia, E., Laurent, D., Debitus, C., and Pietra, F. (2006) On the first polyarsenic organic compound from nature: Arsenicin A from the New Caledonian marine sponge *Echinochalina bargibanti*. *Chemistry-a European Journal*, **12**: 8989–94.

Matsumura, K., Nagano, M., and Fusetani, N. (1998) Purification of a larval settlement-inducing protein complex (SIPC) of the barnacle *Balanus amphitrite*. *Journal of Experimental Zoology*, **281**: 12–20.

(2005) Biosynthesis of marine natural products: microorganisms (Part A). *Natural Product Reports*, **22**: 580–93.

Moore, B.S. (2006) Biosynthesis of marine natural products: macroorganisms (Part B). *Natural Product Reports*, **23**: 615–29.

Moore, B.S. and Hertweck, C. (2002) Biosynthesis and attachment of novel bacterial polyketide synthase starter units. *Natural Product Reports*, **19**: 70–99.

Naar, J.P., Flewelling, L.J., Lenzi, A., Abbott, J.P., Granholm, A., Jacocks, H.M., Gannon, D., Henry, M., Pierce, R., Baden, D.G., Wolny, J., and Landsberg, J.H. (2007) Brevetoxins, like ciguatoxins, are potent ichthyotoxic neurotoxins that accumulate in fish. *Toxicon*, **50**: 707–23.

Nylund, G.M., Cervin, G., Persson, F., Hermansson, M., Steinberg, P.D., and Pavia, H. (2008) Seaweed defence against bacteria: a poly-brominated 2-heptanone from the red alga Bonne maisonia hamifera inhibits bacterial colonisation. *Marine Ecology-Progress Series*, **369**: 39–50.

Paul, V.J. and Ritson-Williams, R. (2008) Marine chemical ecology. *Natural Product Reports*, **25**: 662–95.

Pawlik, J.R. (1993) Marine invertebrate chemical defenses. *Chemical Reviews (Washington, DC, United States)*, **93**: 1911–22.

Pelletreau, K.N. and Muller-Parker, G. (2002) Sulfuric acid in the phaeophyte alga *Desmarestia munda* deters feeding by the sea urchin *Strongylocentrotus droebachiensis*. *Marine Biology*, **141**: 1–9.

Piel, J. (2002) A polyketide synthase-peptide synthetase gene cluster from an uncultured bacterial symbiont of

Paederus beetles. *Proceedings of the National Academy of Sciences of the United States of America*, **99**: 14002–7.

Piel, J. (2004) Metabolites from symbiotic bacteria. *Natural Product Reports*, **21**: 519–38.

Piel, J. (2009) Metabolites from symbiotic bacteria. *Natural Product Reports*, **26**: 338–62.

Piel, J., Hui, D.Q., Wen, G.P., Butzke, D., Platzer, M., Fusetani, N., and Matsunaga, S. (2004) Antitumor polyketide biosynthesis by an uncultivated bacterial symbiont of the marine sponge *Theonella swinhoei*. *Proceedings of the National Academy of Sciences of the United States of America*, **101**: 16222–7.

(2004) Chemical defense strategies of marine organisms. *Topics in Current Chemistry*, **239**: 179–219.

Pohnert, G. (2005) Diatom/Copepod interactions in plankton: The indirect chemical defense of unicellular algae. *ChemBioChem*, **6**: 946–59.

Pohnert, G. and Boland, W. (2002) The oxylipin chemistry of attraction and defense in brown algae and diatoms. *Natural Product Reports*, **19**: 108–22.

Pohnert, G., Steinke, M., and Tollrian, R. (2007) Chemical cues, defence metabolites and the shaping of pelagic interspecific interactions. *Trends in Ecology & Evolution*, **22**: 198–204.

Prince, E.K. and Pohnert, G. (2010) Searching for signals in the noise: metabolomics in chemical ecology. *Analytical and Bioanalytical Chemistry*, **396**: 193–7.

Proksch, P. (1994) Defensive roles for secondary metabolites from marine sponges and sponge-feeding nudibranchs. *Toxicon*, **32**: 639–55.

Salgado, L.T., Viana, N.B., Andrade, L.R., Leal, R.N., Da Gama, B.A.P., Attias, M., Pereira, R.C., and Filho, G.M.A. (2008) Intra-cellular storage, transport and exocytosis of halogenated compounds in marine red alga *Laurencia obtusa*. *Journal of Structural Biology*, **162**: 345–55.

Schmidt, E.W. (2008) Trading molecules and tracking targets in symbiotic interactions. *Nature Chemical Biology*, **4**: 466–73.

Schmitt, S., Wehrl, M., Bayer, K., Siegl, A., and Hentschel, U. (2007) Marine sponges as models for commensal microbe-host interactions. *Symbiosis*, **44**: 43–50.

Schnitzler, I., Pohnert, G., Hay, M., and Boland, W. (2001) Chemical defense of brown algae (*Dictyopteris* spp.) against the herbivorous amphipod *Ampithoe longimana*. *Oecologia*, **126**: 515–21.

Sever, M.J., Weisser, J.T., Monahan, J., Srinivasan, S., and Wilker, J.J. (2004) Metal-mediated cross-linking in the generation of a marine-mussel adhesive. *Angewandte Chemie-International Edition*, **43**: 448–50.

Shick, J.M. and Dunlap, W.C. (2002) Mycosporine-like amino acids and related gadusols: Biosynthesis, accumulation, and UV-protective functions in aquatic organisms. *Annual Review of Physiology*, **64**: 223–62.

Simmons, T.L., Andrianasolo, E., Mcphail, K., Flatt, P., and Gerwick, W.H. (2005) Marine natural products as anticancer drugs. *Molecular Cancer Therapeutics*, **4**: 333–42.

Stashenko, E.E. and Martinez, J.R. (2007) Sampling volatile compounds from natural products with headspace/solid-phase micro-extraction. *Journal of Biochemical and Biophysical Methods*, **70**: 235–42.

Stefels, J. (2000) Physiological aspects of the production and conversion of DMSP in marine algae and higher plants. *Journal of Sea Research*, **43**: 183–97.

Sticher, O. (2008) Natural product isolation. *Natural Product Reports*, **25**: 517–54.

Strom, S., Wolfe, G., Slajer, A., Lambert, S., and Clough, J. (2003) Chemical defense in the microplankton II: Inhibition of protist feeding by beta-dimethylsulfoniopropionate (DMSP). *Limnology and Oceanography*, **48**: 230–7.

Takats, Z., Wiseman, J.M., Gologan, B., and Cooks, R.G. (2004) Mass spectrometry sampling under ambient conditions with desorption electrospray ionization. *Science*, **306**: 471–3.

Toth, G.B. and Pavia, H. (2000) Water-borne cues induce chemical defense in a marine alga (*Ascophyllum nodosum*). *Proceedings of the National Academy of Sciences of the United States of America*, **97**: 14418–20.

Weinberger, F., Pohnert, G., Kloareg, B., and Potin, P. (2002) A signal released by an enclophytic attacker acts as a substrate for a rapid defensive reaction of the red alga *Chondrus crispus*. *ChemBioChem*, **3**: 1260–3.

Weissflog, J., Adolph, S., Wiesemeier, T., and Pohnert, G. (2008) Reduction of herbivory through wound-activated protein cross-linking by the invasive macroalga *Caulerpa taxifolia*. *ChemBioChem*, **9**: 29–32.

Wichard, T., Gerecht, A., Boersma, M., Poulet, S.A., Wiltshire, K., and Pohnert, G. (2007) Lipid and fatty acid composition of diatoms revisited: Rapid wound-activated change of food quality parameters influences herbivorous copepod reproductive success. *ChemBioChem*, **8**: 1146–53.

Wiesemeier, T., Hay, M., and Pohnert, G. (2007) The potential role of wound-activated volatile release in the chemical defence of the brown alga *Dictyota dichotoma*: Blend recognition by marine herbivores. *Aquatic Sciences*, **69**: 403–12.

Wise, M.L. (2003) Monoterpene biosynthesis in marine algae. *Phycologia*, **42**: 370–7.

Wolfe, G.V., Steinke, M., and Kirst, G.O. (1997) Grazing-activated chemical defence in a unicellular marine alga. *Nature*, **387**: 894–7.

Zhang, Z.M. and Li, G.K. (2010) A review of advances and new developments in the analysis of biological volatile organic compounds. *Microchemical Journal*, **95**: 127–39.

CHAPTER 14

Allelochemical interactions among aquatic primary producers

Elisabeth M. Gross, Catherine Legrand, Karin Rengefors, and Urban Tillmann

14.1 The quagmires of allelopathy— what defines allelopathic interactions?

Allelopathy in its original definition by the Viennese botanist Hans Molisch (1937) describes any inhibitory or stimulatory effect of one plant on another plant, mediated by the release of some chemical factor(s). This concept has been widely used in terrestrial plant communities, but also in marine and freshwater systems to explain patterns of dominance or changes in community structure that apparently cannot be explained by either competitive or trophic interactions.

Allelopathy is often used synonymously with allelochemistry by many authors. The latter term was coined by Whittaker and Feeny (1971) to describe organismic interactions based on chemicals not used as nutrition, that are affecting development, fitness, behaviour, or population biology of organisms of other species. Molisch included microorganisms besides plants as producers and target organisms in his original classification. In our chapter, we will focus on all aquatic 'primary producers' as allelopathically active species and target organisms, to explicitly include all photosynthetic protists and prokaryotes, which often dominate in aquatic systems but do not belong to 'plants'. Potential anti-grazer roles of allelochemicals will be considered by Pavia *et al.* (Chapter 15).

Allelopathy occurs among almost all types of aquatic primary producers (Gross 2003), both in pelagic and benthic habitats. It may protect sessile higher plants and macroalgae from dense colonization by epiphyton, provide an advantage for mat-forming cyanobacteria and algae in the competition for space, or affect phytoplankton succession, including the development of harmful algal blooms (HABs). Algae are most important in aquatic systems, which cover up to 70% of the surface area on Earth; moreover, planktonic algae are responsible for about 50% of carbon dioxide fixation and oxygen production (Falkowski *et al.* 1998). Larger efforts and investments in aquatic allelopathy may benefit water quality, maintenance, and provision of ecosystem services (e.g. algal blooms), as well as bioprospecting for bioactive natural products. Many allelopathy studies in water systems focus on 'weed' management, as several higher aquatic plants and macroalgae are nuisance invasive species, often introduced via ship ballast water or aquarium trade.

In allelopathic interactions, we expect a producing primary producer to be releasing a bioactive compound that reaches a target organism and affects its performance (Fig. 14.1). Experimental proof for the release and transport of active compounds is often very difficult, especially if the intracellular precursor in the producing organism differs from the compound(s) set free. How active compounds reach intracellular targets in afflicted organisms is also largely unknown. Both membrane transfer by passage through defined uptake systems (e.g. ATP-binding cassette (ABC) transporters) or passive transport across the lipid bilayers are possible. While actively transported compounds can be large (i.e. molecular weight > 1000 Da.) and polar (hydrophilic), substances capable of passively crossing membrane(s) should be lipo- or at least

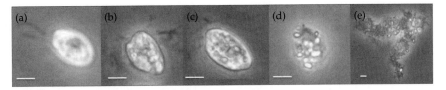

Figure 14.1 Lysis of the cryptophyte *Rhodomonas salina* after addition of algal allelochemicals. Scale bar: 5 μm. (a) control cell, (b) and (c) cells in osmotic shock, (d) cell membrane ruptured and cell content is released in the surrounding medium, (e) bacteria colonize algal cell remains. Photographs: P. Uronen. See also colour Plate 8.

amphiphilic (having both polar and non-polar molecule elements). Most of the identified allelopathically active compounds, not only in aquatic habitats, are of low molecular weight and amphi- to lipophilic (e.g. Whittaker and Feeny 1971; Gross 2003, 2009). In eukaryotic organisms, compounds affecting photosynthesis need not only to pass the outer cell membrane but also that of the chloroplast. In general, this aspect is sidestepped in many studies by providing evidence that active compounds affect intact target organisms. In this chapter we will highlight our current knowledge in this field. We will also discuss the direct proof for the involvement of allelopathically active compounds in interactions among primary producers, and that resource competition is not solely responsible for species declines or replacement (Legrand *et al*. 2003; Gross 2003).

A range of techniques is used to explore allelopathic interactions. Co-culturing exposure to exudates and a bioassay-directed identification of active compound(s) identify allelopathic interactions between species and verify the chemical nature of the interaction. The direct transfer of laboratory results in allelopathy to field conditions is difficult. Yet, we now have plentiful experimental evidence on active compounds, mode of action, and effects on target species, all of which suggest that allelopathy as a mode of interference competition is common and widespread. Even if we still observe a trade-off between the ecological relevance and the definite experimental proof for allelopathic effects (Legrand *et al*. 2003; Hilt and Gross 2008), this should not lead to discussions on the existence of allelopathy at all. Instead of asking whether allelopathy exists or not, we should better ask how primary producers allelochemically interfere with other primary producers. Current research further

indicates that allelopathy might be an evolutionary stable trait (Legrand *et al*. 2003; Roy 2009).

A critical aspect in the discussion of allelopathy is whether producing and target organisms have coevolved. It is possible that co-occurring primary producers have developed resistance against allelopathically active compounds, thus effects are more pronounced against organisms from different habitats. On the other hand, the classical allelopathy study in freshwater phytoplankton (Keating 1978) indicates effects within one community in one habitat. We therefore discuss these issues in detail below. In this chapter we will also compare aquatic and terrestrial habitats to see what we can learn from the other system and, finally, we discuss the potential of up to date molecular methods, and if and how they may help solving the many riddles allelopathy still poses.

14.2 Out and gone—what is an efficient way to distribute allelochemicals?

In order to elicit a response from other species, allelochemicals simply have to come into contact with the target cells. Whereas intracellular compounds may contact their targets after ingestion and/or after complete lysis/destruction of the producing cell only, release of compounds into the extracellular realm is an efficient way for living cells to distribute allelochemicals. Release of organic matter generally might be due to an unspecific loss ('exudation, leakage'), for example by a continuous passive permeation of low molecular weight organic molecules through the cell membrane; or a more specific transport or active release ('excretion'), for example the active release of excess photosynthates under nutrient stress. In any case, compounds released to the environment might either be still more or less tightly connected to the cells (attached, bound, etc.) or freely released into the surrounding water. In the first case, cell-to-cell contact between allelopathic species and their targets can be assumed to play an important role in allelochemical interactions. Cell-to-cell contact could be expected to play a major role in benthic and biofilm communities but a role for the plankton has also been suggested (Uchida 2001; Gross *et al.* 1991). Compounds released into the water phase might be expected to be more hydrophilic, but a number of marine allelochemicals have been shown to be largely amphiphilic (Ervin and Wetzel 2003), while lipophilic allelochemicals might be expected when epiphytes are the targets (Gross 2003). Nevertheless, when compounds are released in the water phase, they are inevitably subject to dilution.

The world of planktonic microorganisms with a size generally well below 1 mm is shaped by low Reynolds numbers and dominated by viscous forces (see also Chapters 1 and 7). Extracellular water soluble compounds are thus distributed by simple diffusion and advective laminar flow. Molecular diffusion is slow and limits transport to the region of a few to hundred micrometres; however, it allows exuded chemicals to accumulate in a diffusion limited boundary layer around plankton cells. In the phycosphere around cells, local concentrations may be orders of magnitude higher than bulk concentrations, especially when cell concentration is low (Jonsson *et al.* 2009). Microzone gradients, however, are distorted and dissipated by all kind of water movement relative to the cell, like swimming, sinking, and turbulence at the millimetre scale (Wolfe 2000). In addition to dilution and advection, the efficient concentration of allelochemicals is also influenced by a wide array of different factors (Fig. 14.2). Although parameterization and modelling of all these processes is available (Jonsson *et al.* 2009), the realm of small scale gradients remains largely unexplored experimentally due to profound technical difficulties in observation (Wolfe 2000). In spatially structured populations, the outcome of competition depends on how effective allelopathy is in relation to its cost, while in homogeneous environments the initial conditions determine the outcome of competition in populations of *Escherichia*

coli (Durrett and Levin 1997). This indicates that distinct differences in allelopathic effects might occur between benthic and pelagic habitats, and in both environments the microscale also needs to be considered.

The rate of release is of fundamental importance in determining the effective concentration of extracellular compounds, but this has been little studied. If compounds are produced at a high rate but stored inside the cells, effective concentrations for target cells may be low. In contrast, extracellular concentration might increase drastically if stored compounds are released in sudden bursts. Moreover, environmental factors might affect both the intracellular production and the fate of compounds after release. For example, light was essential for toxin production in *Prymnesium parvum*, with higher irradiance augmenting toxin production, but once the toxins were extracellular, they were rapidly inactivated by exposure to both visible light and UV radiation (Shilo 1981). The bioactivity of allelochemicals can also be modulated by water chemistry. For instance, the allelopathic effect of the haptophyte *Chrysochromulina polylepis* is affected by pH (Schmidt and Hansen 2001). In addition to chemical modifications, biological transformation of dissolved organic compounds has to be considered. Bacteria are generally capable of metabolizing all kinds of algal exudates, however, little is known specifically on bacterial degradation of allelochemicals. For a few phycotoxins, it is well established that the bacterial metabolism may yield either known congeners or even novel chemically modified derivatives with potentially modified bioactivity. For allelochemicals of the dinoflagellate *Alexandrium tamarense*, long term stability indicates that these compounds are rather refractory to bacterial degradation, at least to that of the bacterial consortium present in the culture supernatant (Ma *et al.* 2009). It should be noted that bacteria might be involved in allelochemical interactions in various ways, ranging from modulators of algal production of bioactive compounds to mediating the release of intracellular substances through lysis of algal cells.

In summary, extracellular compounds are prone to lose biological activity by dilution and chemical or biological modifications. At the community or ecosystem level this means that maintaining

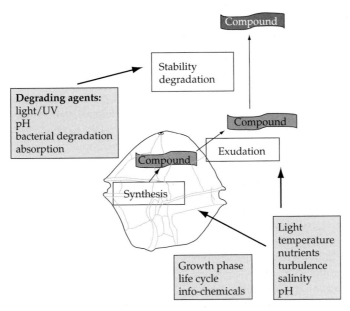

Figure 14.2   Processes (white boxes) and regulatory factors (greyboxes) related to synthesis, release, and degradation of allelochemicals.

biologically active threshold concentrations of allelochemicals requires a certain minimum level of stability and a regular/continuous replacement of compounds.

14.3   Exploitative versus interference competition

Whenever two species interact, their interests are seldom aligned. They may impede the growth of one another through both competition and interference. Exploitative competition denotes competition for one or more resources, for example two protists competing for nutrients or two macroalgae competing for space, and can act both intra- and interspecifically. Interference competition refers to an antagonistic action of one species against another, and in the case of allelopathy as chemical interference implies the production and release of inhibitory bioactive substances, which may be essential for community structure and function. We suggest that allelopathy could theoretically explain succession. Traditionally though, most research into aquatic ecology and ocean biogeochemistry focuses on how resources and abiotic parameters (bottom-up) and predation (top-down) constrain the distribution and function of viruses, bacteria, protists, and algae. However, as pointed out by Strom (2008), these factors are insufficient to explain the dominance of certain functional groups in the sea. For example, models predicting the diatom spring bloom are often based on the superiority of diatoms over photosynthetic flagellates in resource competition, while mounting evidence suggests that grazing and chemical interactions, including allelopathy, may explain diatom distribution/spring bloom patterns (Strom 2008; Ianora *et al*. 2006; Pohnert 2005; Legrand *et al*. 2003). Other examples are the differential impact of cyanobacteria on predecessors and successors in freshwater (Keating 1978) or the interaction between *Peridinium* and *Microcystis* in Lake Kinneret ((Sukenik *et al*. 2002); see also Section 14.4.3). Most photosynthetic flagellates have lower nutrient uptake affinities and maximum growth rates than diatoms, thus flagellates are inferior competitors for inorganic nutrients; vertical migration and chemical interference could make up for this ecological disadvantage (Smayda 1997). However, it is difficult to distinguish between exploitative competition and allelopathy. In laboratory experiments, nutrient competition can be alleviated through the addition of inorganic nutrients. But is this experimental procedure satisfying when considering nutrient dynamics and phytoplankton patchiness in natural environments?

In the case of biofilms, microbial aggregates of microalgae and bacteria, physical constraints are different as microbes are bound to a polysaccharide matrix. In a biofilm, microbes compete both for nutrients and space/light, which adds to the selection pressure and could favor allelopathic interactions. To test whether allelopathic interactions are favoured under a resource-competition scenario (i.e. reducing exploitative competition), it would be necessary to show how allelopathy contributes to niche or resource partitioning. Several modelling studies indicate that spatially structured communities are more likely to exhibit allelopathic interactions than communities in non-structured habitats (Durrett and Levin 1997; Czárán et al. 2002; Szabo et al. 2007).

The release of allelochemicals may be an adaptive strategy for protists weak in exploitative competition under nutrient limitation. Studies on the impact of the resource availability on the production of allelochemicals are scarce, as most research focuses on the targets and only occasionally on the donor. Allelopathic interference may differ depending on the physiological status of both donor and target. Allelopathic effects that are not detectable in a favourable environment may become evident under nutrient limitation or when exposed to other stressors. Under stress conditions, allelopathic effects can increase while the target becomes more sensitive (Fistarol et al. 2005). The cyanobacterium *Trichormus doliulum*, the dinoflagellate *Peridinium aciculiferum*, and the haptophyte *Prymnesium parvum* can increase their allelopathic effect under nutrient stress (Rengefors and Legrand 2001; Granéli and Johansson 2003; von Elert and Jüttner 1997). Different target species can exhibit different responses to allelochemicals, suggesting that both the sensitivity and physiological status of competing species can determine their lasting coexistence. The death of the target can in turn increase nutrient availability (DOC leaking, Uronen et al. 2005) and can reduce exploitative competition for the donor and secondary beneficiaries (bacteria, less sensitive

targets). This positive feedback may also maintain species diversity and regulate community function together with other sources of mortality because new interactions are created and competitive interactions will be altered.

Even if a clear distinction between exploitative and interference competition is not possible under natural or semi-natural conditions, we suggest that allelopathy should be investigated at the community level to assess its impact on organismic and trophic interactions.

14.4 Evolution of allelopathy

14.4.1 Benefits and costs of bioactivity against primary producers

'*Umsonst ist nur der Tod, und der kostet das Leben*' ('Just death is free but it costs your life') - This little piece of worldly wisdom nicely expresses our intuitive assumption that there is no benefit without associated costs or risks. In fact, cost–benefit calculations and models are common and widely used in a variety of areas ranging from economics to evolution. The benefit for aquatic organisms to produce bioactive compounds against other primary producers can be manifold. Reducing the concentration of competitors, either by directly causing cell death of surrounding cells or by reducing competitor population growth rate, represents a relief mechanism for the donor from resource competition. Compared to this more indirect benefit, allelopathic species might also profit directly from damaging accompanying cells. For marine haptophytes, Estep and MacIntyre (1989) proposed the concept of dasmotrophy. They speculated that extracellular compounds produced by certain haptophytes simply punch holes in the cell membrane of other accompanying protists, producing a transient nutrient leakage, and that this organic material then benefits the donor organism via osmotrophic uptake. This fascinating hypothesis of 'tax collection' (the Greek word 'dasmos' means 'tribute, tax'), however, still needs to be experimentally tested. In any case, mixotrophy, defined as the combination of photosynthesis and uptake of organic material in the same individual, either as osmotrophic uptake of dissolved organic matter and/or ingestion of particulate matter (phagotrophy), seems to be widespread among planktonic microorganisms. The phago-trophic uptake of other algae, a strategy of 'eating your competitor' is particularly intriguing as it involves killing two birds with one stone: in other words, the number of competitors is reduced and the mixotrophic alga receives all kinds of macro- and micronutrients in concentrated parcels. The important point to stress here is that there is good evidence that among mixotrophs, bioactive com-pounds are involved in prey capture. A chemical means to paralyse or trap prey during prey capture has, for example, been shown for a haptophyte (Tillmann 2003) and a dinoflagellate (Sheng *et al.* 2010).

As stated above, evolutionary theory assumes that these benefits are somehow related to costs. If the finite resources of an organism are invested in allelochemicals, they cannot be used for other pur-poses. Bioactive secondary metabolites are gener-ally complex in structure and thus require complex synthesis machineries, which in turn are expected to be metabolically expensive. However, the exact nature and amount of such costs have rarely, if ever, been specified in the plankton. Surprisingly, many plankton organisms producing extremely elaborate metabolites do not seem to suffer slower growth or other obvious reductions in physiological processes. In a study on clonal variability in allelochemical potency of the dinoflagellate *Alexandrium tamarense*, no correlation between growth rate and lytic activ-ity could be found (Tillmann *et al.* 2009). This indi-cates that at least under non-limiting conditions no obvious costs in terms of growth rate reduction seem to be associated with the production of these allelochemicals.

In addition to potential costs in terms of energy and/or metabolites, there are certain risks associ-ated with producing biologically active compounds. 'Autotoxicity' poses the risk that any poisonous compound may damage the producing cell and may involve costs to gain immunity. One general strat-egy for a reduction of autotoxicity is compartmen-talization. Whereas multicellular organisms may compartimentalize into specialized tissues or organs, unicellular plankton may use compartmented organelles. Phycotoxins of dinoflagellates may be localized in chloroplasts (Anderson and Cheng

1988), whereas non-toxic precursors may be stored inside the cell and are only after excretion enzymati-cally converted to the potent end product. Another way to gain immunity is if specific target molecules are absent, replaced and/or modified in the produc-ing species. The marine dinoflagellate *Karlodinium* spp. produces allelochemically active karlotoxins, which interact with membrane sterols of the target cell, whereas *Karlodinium* itself is protected by its specific membrane sterol composition (Deeds and Place 2006). Protection against own compounds may also help in other situations; the allelochemically active haptophyte *Prymnesium parvum* is immune against lytic compounds produced by the dinoflag-ellate *Alexandrium* spp. (Tillmann *et al.* 2007). As both species rarely, if ever, co-occur, the insensitivity of *Prymnesium* against *Alexandrium* lysins is not likely to be the result of a selective adaptation but probably results from auto-immunity of *Prymnesium* against its own lytic compounds.

Autotoxicity, however, is not the only risk associ-ated with producing bioactive compounds. Most planktonic grazers use chemical cues to locate their prey, so release of any kind of organic compound might implicate the unintentional attraction of graz-ers. This may explain why a number of marine algal allelochemicals have been shown to affect both competitors and potential protist grazers (Tillmann *et al.* 2008).

If we assume production costs of allelochemicals to be substantial, simple cost–benefit optimization would suggest the need for induced/activated sys-tems compared to constitutively produced com-pounds. Although rarely explicitly studied for allelopathic compounds, most available examples from marine plankton indicate allelochemicals, like all known phycotoxins, to be constitutively pro-duced. This can be deduced from the finding that high allelochemical activity occurs in the cell-free fraction of donor monocultures, but total toxicity might be modulated by environmental factors like nutrient limitation. Cost–benefit considerations also lead to the argument that natural selection will favour those metabolites that possess multiple func-tions. Polyunsaturated aldehydes (PUAs) in dia-toms, which were first described in connection with anti-predatory function, and may also act as allelo-pathic agents, can affect growth of some bacterial strains, and possibly also have a stress signalling function (Vardi *et al.* 2006, Ribalet *et al.* 2008).

14.4.2 Does allelopathy benefit single cells or the many individuals in a genetically diverse population?

A recurring question in discussing the existence of allelopathy in planktonic primary producers is if there is any benefit or adaptive advantage for the single cell, the population, or even the species. In order to address this question we must backtrack and consider the very basics of natural selection. Selection occurs on the level of the individual gene or genotype (or actually heritable phenotype), thus, in these systems the units of selection are the individual cells that compete with other cells. Traits evolve in a population because they increase the producer's fitness relative to competing individuals (Freeman and Herron 2007). We hope we can leave behind the misconception that allelopathy may be a trait that has evolved because it is beneficial for the species.

A complication with planktonic protists, macrophytes, and cyanobacteria is that they entirely, or for the most part, reproduce asexually, which is then rendered more complicated by occasional sexual reproduction (most protists). Thus, the question is whether the cells in a population can be considered to be different individuals as in higher animals, or to belong to a few clonal lineages. This has implications for both genetic variation in allelopathy and selection. In the case of selection, the unit being selected would be all the cells within a clone. Until recently, it was believed that phytoplankton populations were composed of only a few clonal lineages. However, during the past few years it has been shown by use of DNA fingerprinting methods (AFLP) and specific microsatellite markers that individual cells sampled in a population are genetically distinct (see Rynearson and Armbrust 2004). These findings, which show that every single isolate is genetically distinct, suggest that most cells are actually individuals. This empirical evidence is in line with the conclusions of Mes (2008), who suggested that because microbial population sizes are so large, the diversity is huge. Consequently, our working hypothesis is that a population of phytoplankton consists of a huge number of individual cells on which selection acts.

How did allelopathy evolve? An adaptive trait is a characteristic that increases the fitness of the individual compared to an individual without the trait. A possible scenario is that the ability to produce allelochemicals conveys a competitive advantage for individual cells by providing a means to obtain limiting nutrients, thereby allowing them to reproduce quicker than other individuals. Since most phytoplankton reproduce mainly asexually, a genotype with the allelopathy trait can quickly proliferate. The allelochemical trait, if adaptive, should be inherited during sexual reproduction, that is at least once per growth season, although the frequency of sexual events in phytoplankton populations could be much higher or lower.

On the other hand, Lewis (1986) argued that the evolution of allelopathy as an adaptive trait is in conflict with the basic principles of natural selection. Lewis's reasoning was that the release of allelochemicals not only benefits the donor but also competitors, and that the allelochemical is quickly diluted as the cell moves away, so it is questionable if the donor cell is benefited at all. If other cells benefit without paying the metabolic costs involved (i.e. 'cheaters'), Lewis reasoned that allelopathy is not an evolutionarily stable strategy. As for empirical evidence, it is generally lacking. However, in recent years there have been some modelling studies that suggest that allelopathy in microorganisms can be an evolutionary stable strategy. Most importantly, Czárán *et al.* (2002) and Kirkup and Riley (2004) show in a bacterial allelopathic system that 'cheaters' or resistant cells that do not produce toxins actually stabilize the system, both theoretically and experimentally (see section 14.4.3 on coevolution). Jonsson *et al.* (2009) further argue that an allelopathic effect is only effective in cell-to-cell interactions due to the dilution and swimming away-effect. They suggest that allelopathy may be a non-adaptive trait, and could be a side effect of predator–prey or parasitic interactions, since there is lack of evidence of allelopathic interactions at below bloom-densities. Thus, it is under debate if allelopathy is adaptive and/or if allelopathy is an evolutionary stable strategy. Clearly, further research, both modelling and empirical, is needed.

14.4.3 Coevolution of donor and target—an arms race?

Coevolution refers to selection that occurs when interactions between species (such as predation, parasitism, mutualism) produces adaptations in both species. Coevolution may take place in allelopathic interactions as well. Species subjected to allelochemicals from a competitor may either evolve resistance or tolerance in response to the allelochemicals they are exposed to. As a consequence, there may be selection for individuals of the allelochemical producer agent to produce more potent allelochemicals, and an evolutionary arms race will take place. In studies of bacteria, it has been shown that *E. coli* produce a multitude of different toxins used to kill closely related strains, and molecular studies of the genes behind these toxins indicate that there has been a strong selection for innovation and change (Riley 1998). However, so far there have been no experiments to specifically test coevolution in allelopathic phytoplankton and protists. The first step is to look for evidence of a variation in the allelochemical effect of an allelopathic species on a number of other co-occurring target species. If the same allelochemical producer can have negative, neutral, as well as positive effects this suggests that target organisms have developed tolerance, very much like hosts develop tolerance to parasites. Several laboratory experiments show that the allelochemical effect from one single strain varies among target species from high mortality to no effect, or even a stimulatory effect. Kubanek *et al.* (2005) found that the marine dinoflagellate *Karenia brevis* had a negative effect on the growth rate on most sympatric and temporally co-occurring competitors, while it enhanced growth for some other species. Kubanek *et al.* suggested that the latter may have evolved resistance against the allelochemicals of their competitors, much in the same way that auto-resistance has evolved. It should be noted though that the strains tested were not all isolated from the same location as the *K. brevis* strain. Fistarol *et al.* (2004) showed that *Alexandrium* spp. (from one location) had both negative and neutral effects on different phytoplankton species within a natural community from another location, and they further showed that the

dinoflagellate *Scrippsiella trochoidea* formed temporary cysts as a response to filtrate from three allelopathic species. They suggested that encysting and sinking out of the allelopathic zone was a defense mechanism that has evolved to avoid being killed. Suikkanen *et al.* (2005) worked with producer and target species from the same location and showed that three cyanobacterial species from the Baltic Sea had mostly positive effects on concurrent phytoplankton from the natural community. Likewise, the freshwater cyanobacterium *Gloeotrichia echinulata* ranges from having a slightly negative to a huge stimulatory effect on other microalgae (Carey and Rengefors 2010). Possibly, some of these species have coevolved and become resistant to allelopathy.

The most compelling evidence of coevolution is that found by Keating (1978). She showed that the dominant cyanobacteria in a freshwater pond had a negative or neutral effect on species preceding them in the seasonal succession, but had none or a positive effect on their immediate successor. These findings are indirect evidence that coevolution has helped shape seasonal succession. Similarly, the freshwater cyanobacterium *Microcystis* and the dinoflagellate *Peridinium gatunense*, which alternate in dominating Lake Kinneret, have an inhibitory effect on each other (Sukenik *et al.* 2002). The authors suggest that allelopathy could explain the strong negative correlation between the abundance of these organisms. Coevolutionary mechanisms may underlie these observations. To test whether coevolution has actually taken place it would be necessary to compare the effect of an allelopathic species on strains that it has coevolved with compared to strains of the same species that they have not coevolved with.

Another aspect of coevolution is the coexistence of phytoplankton species. In the past, the contradiction of the high biodiversity of phytoplankton found in an apparently uniform environment (i.e. with a limited number of resources) was referred to as the *Paradox of the Plankton* (Hutchinson 1961). Since then, the presence of density-dependent predation along with fluctuations in the environment and resources due to physical forcing has been used to explain the paradox. Coexistence of allelochemical and non-allelochemical species/strains has been explored mainly in the toxin (colicin)-producing *E. coli* system. A toxin-producing microorganism is generally immune to its own allelochemicals, and thus there may be a cost in maintaining that immunity as well as a cost in producing the allelochemical. Durret and Levin (1997) demonstrated that spatial localization is necessary for the evolution of allelopathy in a bacterial system with a sensitive and a killer strain. In other words, they showed that coexistence does not happen in a stirred liquid solution. They also showed that a 'cheater', that is a resistant but not toxin-producing strain, could invade the system. The idea of chemical warfare in microorganisms promoting biodiversity was shown theoretically by Czárán *et al.* (2002). By modelling killer (K), resistant (R), and sensitive (S) bacterial strains, Czárán *et al.* (2002) could show that the interplay between resource and interference competition results in a cyclical dominance structure. The bacterial strains can be renamed to an allelopathic, a resistant, and a sensitive target phytoplankton strain or species. S outcompete R due to R's cost of maintaining resistance. R outcompete K in resource competition since they do not have the metabolic cost of toxin production and autoimmunity. However, K outcompete S by using their toxin (interference competition). The system can be applied to both different species and to strains within a species.

The question is whether these results from structured (2-D) space can be transferred to allelopathic protists and cyanobacteria in a mixed 3-D environment? Recent modelling efforts show that allelopathy can act as a stabilizing factor that boosts coevolution of two phytoplankton species competing for a single limiting nutrient (Roy 2009). By killing the competitor the availability of nutrients increases in the system, both by reduced uptake and recycling of the dead cells. Theoretically this leads to a reduction in competition pressure and stabilizes co-existence. Empirical evidence is to date absent.

Another issue that must be addressed in this context is that both benthic and pelagic aquatic ecosystems are not homogenous environments. Phytoplankton patchiness is known to vary from millimetre to kilometre scale and from seconds to years. This is caused by mobility of species (and

many of the allelopathic species are in fact motile), zooplankton predation, nutrients and light, as well as physical forcing. Both horizontal and vertical patchiness driven by wind and vertical migration may persist despite alternations of mixed and calm periods. Given the patchiness of phytoplankton, it may be incorrect to model allelopathy in mixed conditions.

At least in freshwater systems, the sensitivity of possible target species towards allelopathically active compounds varies. Cyanobacteria and diatoms exposed to allelochemicals from freshwater macrophytes are more strongly inhibited than chlorophytes, and epiphytes are less susceptible than planktonic species (Hilt and Gross 2008). Even within the group of planktonic cyanobacteria, the sensitivity towards *Myriophyllum spicatum* inhibitors varied manifold (Körner and Nicklisch 2002). This raises the question of how strong the effects of allelopathy on community processes are, if some species resist negative effects. Or in other words: Do we expect changes in community structure depending on the presence or absence of allelopathic species? Rigorous studies to clarify these questions are needed. Some authors already considered that co-occurring species are less susceptible to allelopathy than species from foreign habitats, and that the coevolutionary history of secondary metabolite producing plants with other organisms (herbivores, pathogens, other plants) needs to be considered (Macias *et al.* 2007, see also 14.5).

14.5   Are aquatic systems different from terrestrial habitats?

Differences in the dominant primary producers in aquatic and terrestrial biomes are evident, with more 'microphytes' in freshwater and marine systems. Allelopathic interactions in terrestrial higher plants rely mainly on leaf and root leachates or on volatile compounds. Allelopathic interactions involving thallophytes such as lichens are rare. While page limits forbid an extensive comparison of environmental and organism-based factors in both habitats, some facets might be important to better understand critical processes also among aquatic primary producers.

In contrast to primary producing microorganisms (cyanobacteria, protists, microalgae), higher aquatic plants are all secondarily aquatic, derived from terrestrial ancestors (Cook 1999). Secondary metabolites and their ecological functions in plants are well-studied in the terrestrial biome, but little is known for their freshwater and marine relatives. Either these plants have lost some potential to produce bioactive secondary metabolites, or we might expect a similar array and ecological function of secondary metabolites in aquatic plants. Microphytic primary producers may have very distinct allelopathically active compounds, following different biosynthetic pathways. Almost nothing is known about the evolution of such pathways. It might even be that bacteria intimately associated with some aquatic organisms are the actual producers of inhibitory compounds. Solving the evolution of secondary metabolites in different classes of primary producers might help us to understand which types of secondary metabolites to expect from which producing organism.

Recently, Bais and coworkers (2003) highlighted the apparent stronger allelopathic activity of root-exuded catechin by the invasive terrestrial plant *Centaurea* spp. against native grasses and shrubs in North America compared to its European congeners. Catechin is a widespread plant secondary metabolite present in angiosperms and also some phaeophytes, which has amphiphilic characteristics. However, several studies disputed the *Centaurea* findings, as the active concentrations of catechin in the soil strongly depended on humidity and substrate composition, that is the proportion of clay or humic compounds (Blair *et al.* 2006). Indeed, the authors of the *Centaurea* study later had to clarify that catechin concentrations in the field were low or not detectable, and the exudation by roots could not be confirmed (Bais *et al.* 2010), but they confirmed that catechin has the signalling and phytotoxic activities indicated in the original paper (Bais *et al.* 2003). This terrestrial example shows the complexity of allelopathy studies, especially when tracing down the active compounds in the environment. Thus, the crucial step is the transport and accumulation of the active compound in the medium connecting producing and target organisms. Similar problems occur in aquatic environments, where individual compounds are difficult to trace and quantify due to dilution effects and possible binding to other compounds. The *Centaurea* example indicates clearly that varying environmental conditions will strongly affect the active concentration of allelochemicals, and that we need to include sophisticated analytical techniques to trace the pathways of exudation and uptake.

Differences in active concentrations of allelochemicals might be caused by phenotypic or genotypic differences of the producing species. Specifically we should ask: (1) Are invasive plants more allelopathically active in their new habitat? This would reflect potential coevolutionary relations between producing and target organisms (see also 14.4.3), and (2) do different cultivars (subspecies, genotypes) exhibit different levels of allelopathic activity? Regarding (1), we strongly recommend more studies comparing the allelopathic activity of aquatic primary producers against target species from their home range and invaded areas. Epiphytic cyanobacteria and algae associated with different freshwater macrophytes were less sensitive towards their 'home' plants than foreign plants (Erhard and Gross 2006). Regarding (2), we already discussed genotypic differences in allelopathic strength within a population (Section 14.4.2), yet this topic has not been sufficiently addressed in the past. Rice, the world's most important agricultural crop, is known to allelopathically affect associated weeds. However, not all cultivars are the same, and both breeding systems and genetic engineering have been used to develop allelopathically active strains that produce a high yield despite possible metabolic costs to keep up the allelopathic activity (Jensen *et al.* 2008).

14.6 Will disentangling the molecular mechanisms solve the riddles of allelopathy?

Proof for allelopathy involves the design of appropriate bioassay systems to establish the allelopathic interaction between two species. It further requires analytical techniques in elucidating the structure of the active compound(s) and tracing them from the producing to the target organism. Last, but not least, the mode of action of certain active compounds might allow us to distinguish clearly between allelopathy and other biotic or abiotic interactions. To enhance the understanding of allelopathic interactions, controlled bioassays with sensitive target strains are valuable. However, to increase our perception of the impact of allelopathy at the community or ecosystem level, we additionally need bioassay systems which adequately reflect the natural settings, meaning that some target organisms should be from the natural environment, or include organisms that co-occur with the producing species. Different target cells might be selectively affected by allelochemicals. This is not adequately reflected by most assays. Most assays use relatively simple 'end points' such as abundance, cell density, or growth.

Common modes of action in allelopathy are interference with photosynthesis, membrane integrity, and the inhibition of enzymes (Gross 2003). Apparently, some active compounds exert more than one mode of action on target cells, and it is very likely that we do not know the full range of activity for several allelochemicals. QSAR— Quantitative-Structure-Activity-Relationships—is a technique widely used in ecotoxicology, which allows deducing modes of action from structural traits of a compound. This technique might also provide information on changes in allelopathic activity once compounds are released and subjected to microbial degradation or photochemical changes. Adapting molecular methods from cell biology to determine more specific interactions is challenging but already showing good results. Programmed cell death (apoptosis) has been investigated in some cases as a population response towards allelochemicals (Vardi *et al.* 2002). We should also expand the search for possible modes of action by using, for example, fluorescent marked cell processes.

Expressed sequence tags (EST) show changes in expression profiles of treated versus control cells, and they may provide answers on the induction of defence reactions or the biosynthesis of active compounds in the producing cell. Although this method has been widely used in cell biology, few results have been reported for allelopathy. The production of ESTs might not be the difficult part,

but the critical issue is to link the results to expressed proteins and ecologically relevant functions. Other similar methods related to expression are cDNA-AFLP (Amplified Fragment Length Polymorphism) followed by cloning and sequencing. Otherwise the most promising method probably involves transcriptome analyses using next generation sequencing techniques such as Solexa or 454-sequencing. The crucial issue with all these methods is an appropriate experimental design that allows for detection of differential expression between strains (or treatments) that exhibit differences in allelopathy.

Metabolic profiling might aid tracing allelochemicals or their metabolites in target cells. Possibly, the use of stable isotopes could help the narrowing down of possible candidates, but only if we know something about the biosynthesis of the active compounds, and if we can apply suitable labelled precursor molecules. We should certainly intensify the chemical analysis of so far unknown active metabolites, since this is the best base to proceed from with studies on exudation and modes of action.

Will molecular mechanisms be able to (better) explain allelopathic interactions? Not unambiguously, but they can strongly aid in increasing our mechanistic understanding of the physiological processes, that is biosynthesis and cellular transport of allelochemicals, uptake by target species, and mode of action. Yet new approaches have to provide answers not only under laboratory conditions, but should also prove compound-specific cell responses *in situ*. This would allow the estimation of their relevance under different environmental conditions (e.g. nutrient availability, limitations for either producer or target species) and at the community or even ecosystem level. Recently, some studies included microbial communities in their analysis of allelopathic interactions. The role of heterotrophic bacteria associated with producing and target cells or present in the water has long been neglected, although they may both enhance or decrease the allelopathic activity. A special focus could be on the induction of degrading enzymes in associated bacteria, for example the degradation of allelopathically active polyphenols by heterotrophic bacteria associated with *Myriophyllum spicatum* (Mueller *et al.* 2007).

14.7   Conclusions

Allelopathy is a complex process involved in the interaction of primary producers with other photosynthetic target organisms. This interference competition may help explain species successions and interactions where other factors such as exploitative competition or grazing fail to predict the outcome. Nevertheless, a full proof for the involvement of allelopathy in such interactions has been and probably will continue to be difficult. While laboratory studies allow the bioassay-directed fractionation of active compounds, and their testing against various target organisms, the field situation is generally much more complex and less predictable. Several studies have taken into account that environmental factors and the community composition can strongly affect the outcome of allelopathic interactions. We now have a good picture of which species are involved in allelopathic interactions (especially (benthic) cyanobacteria, HAB-forming algae, and dominant/invasive aquatic plants), and we understand that the spatial structure of the (micro-) - environment is important to predict how allelochemicals affect target species. A remaining controversial issue is whether the producer and target organisms have coevolved, and if allelopathy is more effective against species from other habitats. A comparison of allelopathic interactions in different biomes (terrestrial, marine, freshwater) shows some similarities, for example in the chemical structure of released compounds, and indicates common research themes, such as the efficiency by which allelochemicals affect neighbouring primary producers. Nevertheless, major differences in chemical and biological modification of released compounds can be expected in different media (air, soil, and water). Future allelopathy research will include advanced molecular methods, and also methods commonly used in other fields (ecotoxicology, cell biology, pharmacology) to answer some of the open questions in the field of allelopathy in aquatic systems. However, the foundation of any allelopathy study requires detailed field observations and the asking of relevant questions.

Acknowledgements

This work was supported by grants to CL from the European Commission (ALGBACT-IEF-220732) and the Swedish Research Council (Formas).

References

Anderson, D.M. and Cheng, T.P.O. (1988) Intracellular localisation of saxitoxins in the dinoflagellate *Gonyaulax tamerensis*. *Journal of Phycology*, **24**: 17–22.

Bais, H.P. (2010) Allelopathy and exotic plant invasion: From molecules and genes to species interactions (Correction) *Science*, **327**: 781.

Bais, H.P., Vepachedu, R., Gilroy, S., Callaway, R.M., and Vivanco, J.M. (2003) Allelopathy and exotic plant invasion: From molecules and genes to species interactions. *Science*, **301**: 1377–80.

Blair, A.C., Nissen, S.J., Brunk, G.R., and Hufbauer, R.A. (2006) A lack of evidence for an ecological role of the putative allelochemical (±)-catechin in spotted knapweed invasion success. *Journal of Chemical Ecology*, **32**: 2327–31.

Carey, C.C. and Rengefors, K. (2010) The cyanobacterium *Gloeotrichia echinulata* stimulates the growth of other phytoplankton. *Journal of Plankton Research*, **32**: 1349–54.

Cook, C.D.K. (1999) The number and kinds of embryo-bearing plants which have become aquatic: a survey. *Perspectives in Plant Ecology, Evolution and Systematics*, **2**: 79–102.

Czárán, T.L., Hoekstra, R.F., and Pagie, L. (2002) Chemical warfare between microbes promotes biodiversity. *Proceedings of the National Academy of Science of the United States of America*, **99**: 786–90.

Deeds, J.R. and Place, A.R. (2006) Sterol-specific membrane interactions with the toxin from *Karlodinium micrum* (Dinophyceae) - s strategy for self-protection? *African Journal of Marine Science*, **28**: 421–5.

Durrett, R. and Levin, S. (1997) Allelopathy in spatially distributed populations. *Journal of Theoretical Biology*, **185**: 165–71.

Erhard, D. and Gross, E.M. (2006) Allelopathic activity of *Elodea canadensis* and *Elodea nuttallii* against epiphytes and phytoplankton. *Aquatic Botany*, **85**: 203–11.

Ervin, G.N. and Wetzel, R.G. (2003) An ecological perspective of allelochemical interference in land-water interface communities. *Plant and Soil*, **256**: 13–28.

Estep, K.W. and MacIntyre, F. (1989) Taxonomy, life cycle, distribution and dasmotrophy of *Chrysochromulina*: a theory accounting for scales, haptonema, muciferous bodies and toxicity. *Marine Ecology Progress Series*, **57**: 11–21.

Falkowski, P.G., Barber, R.T., and Smetacek, V. (1998) Biogeochemical controls and feedbacks on ocean primary production. *Science*, **281**: 200–6.

Fistarol, G.O., Legrand, C., and Granéli, E. (2005) Allelopathic effect on a nutrient-limited phytoplankton species. *Aquatic Microbial Ecology*, **41**: 153–61.

Fistarol, G.O., Legrand, C., Selander, E., Hummert, C., Stolte, W., and Granéli, E. (2004) Allelopathy in *Alexandrium* spp.: effect on a natural plankton community and on algal monocultures. *Aquatic Microbial Ecology*, **35**: 45–56.

Freeman, S. and Herron, J.C. (2007) *Evolutionary Analysis*. Pearson Education Inc., Upper Saddle River, NJ.

Granéli, E. and Johansson, N. (2003) Increase in the production of allelolpathic substances by *Prymnesium parvum* cells grown under N- or P-deficient conditions. *Harmful Algae*, **2**: 135–45.

Gross, E.M. (2003) Allelopathy of aquatic autotrophs. *Critical Reviews in Plant Sciences*, **22**: 313–39.

Gross, E.M. (2009) Allelochemical reactions. In G. E. Likens, *Encyclopedia of Inland Waters*. Elsevier, Oxford, Vol. 2, pp. 715–26.

Gross, E.M., Wolk, C.P., and Jüttner, F. (1991) Fischerellin, a new allelochemical from the freshwater cyanobacterium *Fischerella muscicola*. *Journal of Phycology*, **27**: 686–92.

Hilt, S. and Gross, E.M. (2008) Can allelopathically active submerged macrophytes stabilise clear-water states in shallow lakes? *Basic and Applied Ecology*, **9**: 422–32.

Hutchinson, G.E. (1961) The paradox of the plankton. *American Naturalist*, **65**: 137–44.

Ianora, A., Boersma, M., Casotti, R., Fontana, A., Harder, J., Hoffmann, F., Pavia, H., Potin, P., Poulet, S., and Toth, G. (2006) New trends in marine chemical ecology. *Estuaries and Coasts*, **29**: 531–51.

Jensen, L.B., Courtois, B., and Olofsdotter, M. (2008) Quantitative trait loci analysis of allelopathy in rice. *Crop Science*, **48**: 1459–69.

Jonsson, P.R., Pavia, H., and Toth, G. (2009) Formation of harmful algal blooms cannot be explained by allelopathic interactions. *Proceedings of the National Academy of Science of the United States of America*, **106**: 11177–82.

Keating, K.I. (1978) Blue-green algal inhibition of diatom growth: transition from mesotrophic to eutrophic community structure. *Science*, **199**: 971–3.

Kirkup, B.C. and Riley, M.A. (2004) Antibiotic-mediated antagonism leads to a bacterial game of rock-paper-scissors in *vivo*. *Nature*, **428**: 412–14.

Körner, S. and A. Nicklisch (2002) Allelopathic growth inhibition of selected phytoplankton species by submerged macrophytes. *Journal of Phycology*, **38**: 862–71.

Kubanek, J., Hicks, K.M., Naar, J., and Villareal, T.A. (2005) Does the red tide dinoflagellate *Karenia brevis* use allelopathy to outcompete other phytoplankton? *Limnology and Oceanography*, **50**: 883–95.

Legrand, C., Rengefors, K., Fistarol, G.O., and Graneli, E. (2003) Allelopathy in phytoplankton: Biochemical, ecological and evolutionary aspects. *Phycologia*, **42**: 406–19.

Lewis, W.M.J. (1986) Evolutionary interpretations of allelochemical interactions in phytoplankton algae. *The American Naturalist*, **127**: 184–94.

Ma, H., Krock, B., Tillmann, U., and Cembella, A. (2009) Preliminary characterization of extracellular allelochemicals of the toxic marine dinoflagellate *Alexandrium tamarense* using a *Rhodomonas salina* bioassay. *Marine Drugs*, **7**: 497–522.

Macias, F.A., Galindo, J.L.G., and Galindo, J.C.G. (2007) Evolution and current status of ecological phytochemistry. *Phytochemistry*, **68**: 2917–36.

Mes, T.H.M. (2008) Microbial diversity - insights from population genetics. *Environmental Microbiology*, **10**: 251–64.

Molisch, H. (1937) *Der Einfluss einer Pflanze auf die andere – Allelopathie*, Fischer, Jena, Germany.

Mueller, N., Hempel, M., Philipp, B., and Gross, E.M. (2007) Degradation of gallic acid and hydrolysable polyphenols is constitutively activated in the freshwater plant-associated bacterium *Matsuebacter* sp. FB25. *Aquatic Microbial Ecology*, **47**: 83–90.

Pohnert, G. (2005) Diatom/copepod interactions in plankton: the indirect chemical defense of unicellular algae. *ChemBioChem*, **6**: 946–59.

Rengefors, K. and Legrand, C. (2001) Toxicity in *Peridinium aciculiferum* - an adaptive strategy to outcompete other winter phytoplankton? *Limnology and Oceanography*, **46**: 1990–7.

Ribalet, F., Intertaglia, L., Lebaron, P., and Casotti, R. (2008) Differential effect of three polyansaturated aldehydes on marine bacterial isolates. *Aquatic Toxicology*, **86**: 249–55.

Riley, M.A. (1998) Molecular mechanisms of bacteriocyn evolution. *Annual Review of Genetics*, **32**: 255–78.

Roy, S. (2009) The coevolution of two phytoplankton species on a single resource: Allelopathy as a pseudo-mixotrophy. *Theoretical Population Biology*, **75**: 68–75.

Rynearson, T.A. and Armbrust, E.V. (2004) Genetic differentiation among populations of the planktonic marine diatom *Ditylym brightwellii* (Bacillariophyceae). *Journal of Phycology*, **40**: 34–43.

Schmidt, L.E. and Hansen, P.J. (2001) Allelopathy in the prymnesiophyte *Chrysochromulina polylepis*: effect of cell concentration, growth phase and pH. *Marine Ecology Progress Series*, **216**: 67–81.

Sheng, J., Malkiel, E., Katz, J., Adolf, J., and Place, A.R. (2010) A dinoflagellate exploits toxins to immobilize prey prior to ingestion. *Proceedings of the National Academy of Science of the United States of America*, **107**: 2082–7.

Shilo, M. (1981) The toxic principle of *Prymnesium parvum*. In Carmichael, W. W. (ed.) *The Water Environment. Algal Toxins and Health*. Plenum Press, New York.

Smayda, T.J. (1997) Harmful algal blooms: Their ecophysiology and general relevance to phytoplankton blooms in the sea. *Limnology and Oceanography,* **42**: 1137–53.

Strom, S.L. (2008) Microbial ecology of ocean biogeochemistry: A community perspective. *Science,* **320**: 1043–5.

Suikkanen, S., Fistarol, G.O., and Granéli, E. (2005) Effects of cyanobacterial allelochemicals on a natural plankton community. *Marine Ecology Progress Series,* **287**: 1–9.

Sukenik, A., Eshkol, R., Livne, A., Hadas, O., Rom, M., Tchernov, D., Vardi, A., and Kaplan, A. (2002) Inhibition of growth and photosynthesis of the dinoflagellate *Peridinium gatunense* by *Microcystis* sp (cyanobacteria): A novel allelopathic mechanism. *Limnology and Oceanography,* **47**: 1656–63.

Szabo, P., Czárán, T., and Szabo, G. (2007) Competing associations in bacterial warfare with two toxins. *Journal of Theoretical Biology,* **248**: 736–44.

Tillmann, U. (2003) Kill and eat your predator: a winning strategy of the planktonic flagellate *Prymnesium parvum*. *Aquatic Microbial Ecology,* **32**: 73–84.

Tillmann, U., Alpermann, T., Purificacao, R., Krock, B., and Cembella, A. (2009) Intra-population clonal variability in allelochemical potency of the toxigenic dinoflagellate *Alexandrium tamarense*. *Harmful Algae,* **8**: 759–69.

Tillmann, U., John, U., and Cembella, A. (2007) On the allelochemical potency of the marine dinoflagellate *Alexandrium ostenfeldii* against heterotrophic and autotrophic protists. *Journal of Plankton Research,* **29**: 527–43.

Tillmann, U., John, U., Krock, B., and Cembella, A. (2008) Allelopathic effects of bioactive compounds produced by harmful algae. In Moestrup, O. (Ed.) *Proceedings of the 12th International Conference on Harmful Algae.* 2008 Kopenhagen, International Society for the Study of Harmful Algae and Intergovernmental Oceanographic Commission of UNESCO.

Uchida, T. (2001) The role of cell contact in the life cycle of some dinoflagellate species. *Journal of Plankton Research,* **23**: 889–91.

Uronen, P., Lehtinen, S., Legrand, C., Kuuppo, P., and Tamminen, T. (2005) Haemolytic activity and allelopathy of the haptophyte *Prymnesium parvum* in nutrient-limited and balanced growth conditions. *Marine Ecology Progress Series,* **299**: 137–48.

Vardi, A., Formiggini, F., Casotti, R., De Martino, A., Ribalet, F., Miralto, A., and Bowler, C. (2006) A stress surveillance system based on calcium and nitric oxide in marine diatoms. *Public Library of Science (PLoS) Biology,* **4**: 411–19.

Vardi, A., Schatz, D., Beeri, K., Motro, U., Sukenik, A., Levine, A., and Kaplan, A. (2002) Dinoflagellate-cyanobacterium communication may determine the composition of phytoplankton assemblage in a mesotrophic lake. *Current Biology,* **12**: 1767–72.

von Elert, E. and Jüttner, F. (1997) Phosphorus limitation and not light controls the extracellular release of allelopathic compounds by *Trichormus doliolum* (cyanobacteria). *Limnology and Oceanography,* **42**: 1796–802.

Whittaker, R.H. and Feeny, P.P. (1971) Allelochemics: Chemical interactions between species. *Science,* **171**: 757–70.

Wolfe, G.V. (2000) The chemical defense ecology of marine unicellular plankton: constrains, mechanisms, and impacts. *Biological Bulletin,* **198**: 225–44.

CHAPTER 15

Chemical defences against herbivores

Henrik Pavia, Finn Baumgartner, Gunnar Cervin, Swantje Enge,
Julia Kubanek, Göran M. Nylund, Erik Selander, J. Robin Svensson,
and Gunilla B. Toth

This chapter will highlight recent and emerging research involving chemical defences against herbivory in aquatic primary producers. In aquatic systems herbivory is intense, with up to three times more primary production consumed by grazers compared to terrestrial systems (Cyr and Pace 1993). Grazing pressure impacts significantly on aquatic communities across many spatial and temporal scales (Lubchenco and Gaines 1981; Carpenter 1986; Lewis 1986; Cyr and Pace 1992) and is considered a strong selective agent for defensive mechanisms (Duffy and Hay 1990).

Aquatic primary producers are a taxonomically and functionally diverse group of organisms that includes macroalgae, microalgae, and vascular plants (Fig. 15.1). These organisms differ widely in traits such as size, morphology, life history, habitat use, and so on. Despite the fact that aquatic primary producers constitute a large and diverse group of organisms that vary considerably in their evolutionary histories (Dawes 1998; Lee 1999), selection for chemical defences (in the form of secondary metabolites) to resist or reduce grazing are commonplace across the phylogenetic boundaries (e.g. Hay 1996 and references therein; Kubanek et al. 2007; Pohnert et al. 2007). A plethora of different secondary metabolites have been identified from macroalgae that are derived from a number of different metabolic pathways (Harper et al. 2001). Phytoplankton produce a wide variety of secondary metabolites that include potent neurotoxins, and recently inroads have been made into understanding the ecological functions of these compounds with regards to their action against consumers (Ianora et al. 2006; Selander et al. 2006; Paul et al. 2007; Pohnert et al. 2007). Conversely, a substantial number of macroalgal secondary metabolites have been tested as chemical defences (see Paul and Puglisi 2004 and references therein) with variable effects against an assemblage of consumers that are, with few exceptions, generalists that vary greatly in size, mobility, and feeding mode (Fig. 15.1) (Hay and Steinberg 1992).

Because herbivore assemblages can vary so dramatically across different temporal and spatial gradients, so does the production of chemical defences. For the most part, algal chemical defences, similar to terrestrial plant chemical defences, are produced either constitutively (constantly produced at effective concentrations) or induced in response to grazing pressure (Hay 1996; Karban and Baldwin 1997; Toth and Pavia 2007). However, a few species of macroalgae possess activated defences whereby a constantly produced chemical defence is enzymatically transformed to a compound with greater deterrence upon wounding (e.g. Jung et al. 2002 and references therein).

Selection for defence production strategy is considered to be largely dependent on the type of grazers and grazing times (Hay 1996; Karban and Baldwin 1997). Induction of chemical defences is postulated to be common when herbivore pressure is variable in both space and time, whereas

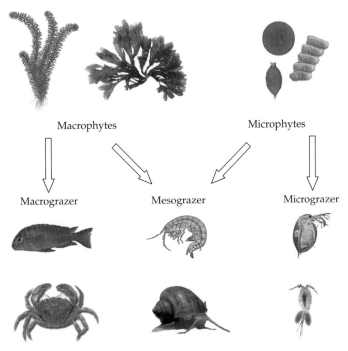

Figure 15.1 Examples of aquatic primary producers and the herbivorous grazers that feed on them. Macrophytes generally include macroalgae and vascular plants, while microphytes are typically unicellular microalgae. Macrograzers tend to be larger, (>2.5 cm), mobile herbivores, e.g. fish and crabs, which consume macrophytes. Mesograzers are generally smaller, (< 2.5 cm) less-mobile herbivores, e.g. amphipods and small gastropods, which may exploit microphytes or macrophytes. Micrograzers are typically small (< 0.5 cm) herbivorous crustaceans, e.g. copepods and cladocerans, which feed on microphytes.

constitutive production tends to be prominent when herbivory is constant and predictable (Karban and Baldwin 1997; Chapter 12). As an example, constitutive production of chemical defences are presumed to be more common in tropical macroalgae, as grazer assemblages consist primarily of mobile macroherbivores (fish, urchins, crabs etc.) that exert a constant pressure and can inflict a mortal blow on a seaweed individual in a single bite (Hay 1996). In contrast, induction of chemical defences is hypothesized to be more widespread in temperate macroalgae as herbivory is temporally variable and mesograzers (isopods, amphipods, gastropods etc.; Fig. 15.1), which exploit macroalgae for both food and shelter, are dominant and associate with their host for sufficiently long periods for induced responses to be effective (Hay 1996). Meta-analysis (Toth and Pavia 2007) has shown that induction of chemical defences is common and widespread in macroalgae, in contrast to earlier notions (see Pavia

and Toth 2000a). Moreover, induced responses to zooplankton grazers in phytoplankton have recently been described (Selander *et al.* 2006; Bergkvist *et al.* 2008).

Most theories pertaining to production of chemical defences in algae are tested in a framework derived from terrestrial plant defence theories (e.g. Cronin and Hay 1996a; Pavia *et al.* 2002; Toth *et al.* 2005; Selander *et al.* 2008). To what degree and in what way these theories are relevant to aquatic systems still remains a point of conjecture. It has generally been suggested that largely different herbivore assemblages and grazing intensities drive selection of chemical defences in terrestrial plants compared to aquatic primary producers, that is specialist, low intensity grazing in terrestrial systems versus generalist, high intensity grazing in aquatic systems (Hay and Steinberg 1992). However, a recent database approach suggests generalists also have substantial impacts on terrestrial plant communities (Parker *et al.* 2006). Indeed the paradigm of specialists, predominantly insects, being primary selective agents of chemical defences in terrestrial plants may be an artefact of a greater research focus in response to their significant impacts on plant agriculture. Whether future studies uncover a need for alterations of terrestrial plant defence theories to better suit aquatic systems will be interesting to see. Here we provide an overview of plant chemical defence theories and highlight recent research on aquatic primary producers that has addressed a number of aspects of these theories, concluding with new chemical approaches to tackle these questions and suggestions for future research directions.

15.1 Theories of chemical defences

During the 1950s and 1960s the focus of research on natural products gradually changed from mere discovery and description of new unique compounds into an increased interest in the function and ecological roles of these compounds (e.g. Fraenkel 1959). The discovery and characterization of tens of thousands of natural products in plants, algae, and other organisms called for theories that could provide explanations and categorizations for this ample form of biodiversity. This led ecologists and evolutionary biologists to formulate a series of general models with an emphasis on the defensive role of secondary metabolites in terrestrial plants. These defence models have allowed natural product chemists and chemical ecologists to put their specific work into a broader context, and they have become a significant driving force for further studies of secondary metabolites and their ecological roles. The publications introducing or extending the primary defence models represent some of the most widely cited papers within the field of plant ecology (e.g. Feeny 1976; Rhoades 1979; Bryant et al. 1983; Coley et al. 1985; Herms and Mattson 1992), and the models have been reviewed and discussed repeatedly (e.g. Hay and Steinberg 1992; Karban and Baldwin 1997; Cronin 2001; Stamp 2003). Here, we give a brief overview of the models that have been most influential in explaining secondary metabolite variation of aquatic primary producers (Table 15.1).

Chemical defence theories can be categorized in different ways based on the patterns they are intended to explain, for example inter- or intra-specific variation, or on the type of explanations they offer, such as proximate physiological mechanisms or ultimate evolutionary rationales. The most prominent theories for interspecific variation in plant defences are the *Resource Availability Model* (RAM; Coley et al. 1985) and the *Plant Apparency Model* (PAM; Feeny 1976). The RAM focuses on differences in resource availability among habitats to explain variation in defences among species over evolutionary time-scales. It assumes that loss of biomass is more difficult and costly to replace in resource-poor environments, where plant growth rates tend to be relatively low. Plants living in oligotrophic environments are therefore predicted to have high levels of immobile defences with broad effects and little turnover, such as polyphenolic compounds. In resource-rich environments, where tissue is more readily replaced and competition among plants is strong, the RAM predicts that plants should invest in low concentrations of potent, mobile defences that are easily catabolized, such as terpenes or alkaloids. The PAM also addresses interspecific differences over evolutionary time scales, but in contrast to the RAM it is a demand-based model that focuses primarily on the need of plants to defend themselves. It assumes that large, common, and long-lived plants, that is, apparent plants, are easily found by herbivores, while short-lived, fast-growing, and opportunistic species have a higher probability of escaping their consumers. Consequently, the PAM predicts that apparent plants should invest in high levels of so-called quantitative defences, which are defences that are supposed to be effective in a dose-dependent and generalized way that is hard for herbivores to overcome, for example digestibility reducers. In contrast, more opportunistic plant species should contain low concentrations of qualitative defences, that is compounds that are highly toxic or deterrent to most consumers, except for a few specialists.

The most influential theories to explain intraspecific variation in defences are the *Optimal Defence Model* (ODM; Rhoades 1979), the *Carbon Nutrient Balance Model* (CNBM; Bryant et al. 1983), and the *Growth-Differentiation Balance Model* (GDBM; Loomis 1932; Herms and Mattson 1992). These three models have all significantly influenced studies of chemical

Table 15.1 Summary of the general features of the most influential models developed to explain variation in secondary metabolites of primary producers.

Model	Level of variation	Temporal scale	Assumptions	Predictions
Resource Availability Model (RAM)	Interspecific	Evolutionary	Plant growth rate is low and biomass loss is costly in resource-poor environments	• Plants have high levels of broad-effect, immobile (quantitative) defences in low-resource environments (e.g. pholyphenols) • Plants have low levels of potent, mobile (qualitative) defences in high-resource environments (e.g. alkaloids and terpenes)
Plant Apparency Model (PAM)	Interspecific	Evolutionary	Large, common, and long-lived plants are apparent to herbivores	• Large, common, long-lived plants have high levels of quantitative defences (e.g. tannins) • Small, short-lived, fast-growing plants have low levels of qualitative defences (e.g. alkaloids)
Optimal Defence Model (ODM)	Intraspecific Intra-individual	Ecological/ Evolutionary	Chemical defences are produced in direct proportion to the risk and the negative fitness consequences of attack, and in inverse relation to the cost of their production	• Plants optimally allocate resources in space and time, e.g. through inducible defences
Carbon Nutrient Balance Model (CNBM)	Intraspecific	Ecological	The production of defence chemicals is determined by the relative availability of carbon and nutrients	• Plants have high levels of 'carbon-based' compounds (e.g. polyphenols) when growth is nutrient-limited • Plants have high levels of 'nitrogen-based' compounds (e.g. alkaloids, peptides) under low carbon:nitrogen ratios.
Growth-Differentiation Balance Model (GDBM)	Interspecific Intraspecific Intra-individual	Ecological	There is a physiological trade-off between growth and differentiation processes (including defence production)	• Plants have high levels of defences when growth is more inhibited than photosynthesis by any factor

defences in terrestrial plants, although there has been considerable confusion about the scopes, assumptions, and predictions of the models (Stamp 2003). The ODM is a demand-based model focusing on the plants' need for defences. In short, the ODM states that defence chemicals are produced in direct proportion to the risk and the negative fitness consequences of attacks, and in inverse relationship to the costs of their production (Zangerl and Bazzaz 1992). The assumption that chemical defences are costly, that is that they can be produced at the direct expense of other functions such as growth and reproduction, is an important underlying principle of the ODM. As a consequence, the ODM predicts that there should be selection for optimal allocation of defence chemicals in space and time, for example higher concentration in plant parts that are most vulnerable to attacks and contribute most to plant fitness (Rhoades 1979). One way to optimize the use of costly defence chemicals is to have an inducible increase in the production of these compounds in response to attacks (Karban and Baldwin 1997). This potentially allows plants to produce large amounts of defence chemicals when they are needed, without having to pay the cost of being defended when enemies are absent.

In contrast to the ODM, the CNBM is a resource-based model, stating that the production of defence chemicals is determined by the relative availability of carbon and nutrients (Bryant et al. 1983; Reichardt et al. 1991). It suggests that when acquired resources exceed the necessary levels for primary metabolism (growth), the excess can be used for production of secondary metabolites. For example, when growth is nitrogen-limited, plants will increase their production of 'carbon-based' (i.e. non-nitrogen containing) secondary metabolites, such as tannins. The CNBM, especially in its recent versions (e.g. Reichardt et al. 1991; Herms and Mattson 1992), is in many ways closely related to overflow models of secondary metabolism, where secondary metabolites are seen essentially as waste products (Haslam 1985).

Like the CNBM, the GDBM has been regarded as a resource-based or supply-side model, as it emphasizes that the levels of defences are indirectly determined by the supply of secondary metabolites, rather than actively regulated in order to fulfil defensive demands. The GDBM is based on the premise that there is a physiological trade-off between growth and differentiation processes (in which secondary metabolism is considered a function of highly differentiated tissues), and it focuses on changes in this trade-off with resource availability and ontogenetic shifts (Herms and Mattson 1992). Herms and Mattson (1992) used the framework of the GDBM to discuss and predict how physiological constraints of differentiation processes are related to the evolution of different plant life histories. In conformity with the CNBM, but in a more explicit and general way, the GDBM predicts that any factor that will inhibit growth more than photosynthesis will increase the availability of internal resources for differentiation processes, including chemical defence production. The GDBM is also a more general model than the CNBM in that it can be used to predict intraspecific, including intra-individual, as well as interspecific variation in the production of secondary metabolites (Cronin 2001).

The vast majority of empirical studies addressing defence models concern vascular plants in terrestrial systems. The empirical evidence is mixed for all models and some of the models, as well as their underlying assumptions, have been severely criticized over the years. For example, several authors have advocated that it is time to abandon the CNBM as a viable model (e.g. Berenbaum 1995; Koricheva 2002), while others have argued that it still contains useful aspects and that much of the critique is due to a fundamental misunderstanding of the model and misstatements of its predictions (Lerdau and Coley 2002; Stamp 2003). In fact, all of the defence models presented here have withstood the test of time in that they are still frequently cited, discussed, and addressed in experimental studies. This probably reflects the fact that each of the models has something to contribute to our understanding of chemical defences by emphasizing different aspects and patterns, or simply that there is a lack of better, replacement theories.

Most studies addressing general defence models in aquatic systems deal with seaweeds, while very little work has been done on microalgae or freshwater macrophytes. Research on secondary metabolites (toxins) in microalgae has mainly been fuelled by socioeconomic effects of harmful algal blooms and has only rarely been conducted within the same theoretical framework as studies of chemical defences in macroalgae and vascular plants. Several studies inspired by the PAM and RAM were performed on seaweeds 20–30 years ago, but with disappointing results. In a review, Hay and Steinberg (1992) concluded that neither the PAM nor the RAM appears to be a good predictor of chemical defence patterns in seaweeds. The results from the only recent test of the RAM that we know of were no more encouraging; in contradiction to what the RAM predicts, seaweeds from nutrient-rich Antarctic waters did not contain nitrogenous defence compounds more often than seaweeds from lower latitudes (Amsler *et al.* 2005). During the last 15 years there has been an increased number of studies testing models for intraspecific variation in macroalgal defences, especially the ODM and its corollary Induced Defence Model (reviewed in Cronin 2001; Amsler and Fairhead 2006; Pavia and Toth 2007; Toth and Pavia 2007). Tests of predictions derived from the ODM concerning inducible defences have been summarized and analysed with meta-analysis (Toth and Pavia 2007), but the number of studies and model species used to test hypotheses derived from the CNBM and GDBM are still too few to draw any firm conclusions. To date, the ODM has received most of its support from experiments on inducible defences, while the results from studies investigating intra-plant variation in macroalgal defences within the context of the ODM are more difficult to interpret, mainly due to the lack of objective estimates of fitness values and attack risks among different tissues (but see Pavia *et al.* 2002). Furthermore, predictions from the CNBM have almost exclusively studied variation in brown algal phlorotannins (polyphenolic compounds produced by the polyketide pathway), with variable results. Explanations for these variable results could be that the effect of nutrients on phlorotannins varies between different seaweed species, or that phlorotannins are problematic metabolites to use in tests of the CNBM because they also serve primary (non-defence) functions. Experiments showing that increased light availability can result in higher phlorotannin levels (e.g. Pavia and Toth 2000b) are concordant with predictions of the CNBM, but the effects of light as a

resource could be confounded by induced defence responses to increased UV exposure. The only two studies that have tested predictions from the GDBM show mixed results (Cronin and Hay 1996a, Van Alstyne et al. 1999). Moreover, we know of no previous studies that have formulated hypotheses, based on the GDBM, concerning how the pattern of secondary metabolites should vary between algae or aquatic plants with different life histories. Overall, few aquatic studies have resulted in the formulation of specific, *a priori* hypotheses that are distinctive and contrasting between the different models.

15.2 Cost of chemical defence

Some plant defence theories predict that allocation to herbivore defences imposes a cost on plants in the absence of enemies, manifested as a reduction in growth and reproduction (Herms and Mattson 1992). Resistance costs will determine whether a stable polymorphism of resistant and susceptible genotypes can be maintained in a population, and are considered a significant explanation for the evolution of induced resistance (Heil and Baldwin 2002). Costs can arise directly from the resistance trait itself, or indirectly through interactions with other species (Strauss *et al.* 2002). Examples of direct resistance costs include allocation costs (defence-allocated resources unavailable for other functions), autotoxicity (defence metabolites that are not only toxic to enemies, but also to the producer), or opportunity costs, which occur when small costs early in the life history are compounded with age (Strauss *et al.* 2002). Indirect or ecological costs of resistance require ecological interaction with other species to be accrued. For example, defence compounds can deter mutualists or be used by specialists to locate their plant food (Strauss *et al.* 2002). Inducible defence, which is considered a cost-saving strategy, may also be costly in the absence of enemies. Allocation costs and opportunity costs of inducible defence could occur because plants must have wound-detection pathways, defence precursors, and storage vesicles, which all require constitutive allocation of both energy and resources (Cipollini *et al.* 2003). Induced resistance may also carry ecological costs, such as increased susceptibility to untargeted herbivores (Strauss *et al.* 2002).

Research on terrestrial plants has commonly detected both direct resistance costs and ecological costs of chemical defence (reviewed by Purrington 2000; Heil and Baldwin 2002; Strauss et al. 2002; Cipollini et al. 2003). However, much less is known about the costs of chemical defences in aquatic primary producers. Only a few published studies have addressed this issue for aquatic primary producers and these studies have mainly focused on seaweeds. One of the first studies assessing the costs of chemical defence in seaweeds utilized the kelp, Alaria nana, whereby total growth was negatively correlated with frond concentrations of phlorotannins (Pfister 1992). Many brown algae, such as kelps and especially fucoids, contain high levels of phlorotannins (Ragan and Glombitza 1986), which can act as a defence against herbivory (Amsler and Fairhead 2006). In a study on the brown alga, Fucus vesiculosus, a significant inverse relationship between phlorotannin concentration and growth rate was also found, but only at elevated nitrogen levels (Yates and Peckol 1993). Similarly, a significant inverse relationship between levels of phlorotannins and growth rate for the brown alga Ecklonia radiata was only found at high nutrient levels in spring when both phlorotannin concentrations and growth rates reached their seasonal maximum (Steinberg 1995). That variation in resource availability is important for the cost of defence is further supported by a study on the brown alga, Ascophyllum nodosum, in which it was found that the trade-off between growth and production of phlorotannins was more pronounced in individuals with high nitrogen tissue content (Pavia et al. 1999). These results are in agreement with the theoretical prediction that trade-offs between growth and defence in plants should primarily occur at moderate to high resource levels (Herms and Mattson 1992), a prediction that is supported by a meta-analysis of 70 studies showing that phenotypic correlations between defence and fitness were negative only at high levels of nutrient availability (Koricheva 2002). This is because at high resource levels growth should be fast enough to replace losses from herbivory, so resources can be allocated to growth at expense of defence (Herms and Mattson 1992). However, it has also been predicted that the cost of defence should be greatest when resources are limiting (Bergelson

and Purrington 1996), which is supported by another study on F. vesiculosus. In that study, a cost of phlorotannin production was only found for specimens living in deep water with low light availability (Jormalainen and Ramsay 2009).

As the above examples indicate, most studies on the costs of chemical defences in seaweeds have inferred costs from negative phenotypic correlations between growth and levels of defence metabolites. However, inferring costs from phenotypic correlations is problematic due to the lack of causality, since phenotypic correlations may occur because of co-varying responses to environmental factors and are not direct evidence for trade-offs between traits. Furthermore, it is difficult to make evolutionary inferences in the absence of any genetic information.

Costs of chemical defence in marine plants other than brown macroalgae have been reported for only one species, the red alga Delisea pulchra (Dworjanyn et al. 2006). This alga produces a series of structurally related non-polar secondary metabolites, halogenated furanones, that deter feeding by herbivores and also protect the alga against biofouling (Steinberg et al. 2001). The cost of producing halogenated furanones in D. pulchra was measured by a combination of (1) phenotypic correlations between concentrations of furanones and fecundity; (2) genotypic correlation between concentrations of furanones and growth for clones derived from tetraspores; and (3) manipulative experiments where growth rates of algae, for which furanone production was experimentally inhibited, were compared with growth rates of furanone producing individuals. The observed significant negative relationship between fecundity and levels of furanones, as well as a significantly higher growth rate for algae unable to synthesize furanones, indicated a cost of furanone production in D. pulchra. However, the genotypic correlations showed that the concentrations of furanones were strongly positively correlated with growth (Dworjanyn et al. 2006). The authors argued that the apparently conflicting results are consistent with the consequences of apical growth in this alga and it is probable that the cost of furanone production is only manifested at critical development stages, such as during tissue differentiation.

Even though many phytoplankton species produce toxic metabolites, it is often not known whether these compounds are involved in the defence of the cell, which makes assessments of costs of chemical defences in phytoplankton difficult (Cembella 2003). However, increased production of paralytic shellfish toxin in the dinoflagellate *Alexandrium minutum* in the presence of copepod cues, which is a rare example in which a defensive role of the toxins appears valid, is not associated with a reduced growth rate (Selander *et al.* 2006, 2008). This may indicate that the cost inflicted by grazer-induced toxin production in *A. minutum* is low, at least in terms of growth (Selander *et al.* 2008).

Although feeding specialization among marine herbivores is rare, a few studies suggest that chemical defences in seaweeds can carry indirect or ecological costs. For example, the crab *Caphyra rotundifrons* is a specialist feeder on the tropical green alga *Chlorodesmis fastigiata* and is stimulated to feed by the diterpenoid chlorodesmin, the major defence metabolite of *C. fastigiata* against herbivorous reef fishes (Hay *et al.* 1989). Another example where a specialist herbivore cues to an algal defence compound is the amphipod *Pseudoamphithoides incurvaria*, which lives in a portable domicile that it constructs from the chemically defended *Dictyota bartayresii* (Hay *et al.* 1990). In both choice and no-choice laboratory tests, this amphipod selectively consumed species of *Dictyota* that produced dictyol-class diterpenes that deter feeding by reef fishes. Furthermore, pachydictyol-A, which is the major secondary metabolite of *D. bartayresii*, directly cued domicile building (*Hay et al.* 1990). These studies suggest that chemical defences that are effective against generalist macrograzers carry some ecological costs in terms of attracting specialists.

Being an alternative system to test plant defence theories, further research on seaweeds may add new insights in costs of chemical defences. Unlike terrestrial plants, most have poorly developed internal transport mechanisms for nutrients and photosynthetic products. Instead they have the capacity to photosynthesise and absorb nutrients through most of the thallus. Growth of cells other than at the meristem, either through cell division or enlargement, is also limited. These features may make the

view and magnitude of allocation costs in seaweeds different from terrestrial higher plants, as illustrated by the work on the red alga *Delisea pulchra* (Dworjanyn *et al.* 2006). However, one weakness of working with long-lived organisms like seaweeds and vascular plants is that growth rate may not be the most appropriate measure of fitness, and therefore costs of defence may not be adequately detected in the form of reduced growth. Phytoplankton provides an alternative system in which asexual growth (via mitosis) is the most common form of reproduction, with sexual reproduction occurring only rarely. Thus, simple measures of growth (i.e. in the form of changes in cell concentration) may accurately represent fitness, allowing a fair assessment of the costs of defences. As mentioned above, these studies are currently limited to the few cases in which particular phytoplankton compounds have been shown to act as chemical defences.

15.3 Induced herbivore resistance in aquatic primary producers

Aquatic primary producers can respond to herbivory by inducing morphological or chemical tissue changes (e.g. changes in toughness, concentrations of defence chemicals, or nutritional value) that result in decreased herbivore preference or performance (i.e. induced resistance). The evolution of inducible, as opposed to constitutive, resistance is favoured when herbivory is unpredictable, when the production of resistance traits involves fitness costs, and when reliable environmental cues, such as direct physical contact or water-borne cues, are available (Karban and Baldwin 1997; Tollrian and Harvell 1999, Chapter 12).

In a recent review of herbivore-induced resistance in marine macroalgae, a categorical meta-analysis was used to evaluate overall algal responses to herbivore damage or damage-related cues, as well as factors explaining the observed variation in inducible resistance (Toth and Pavia 2007). There was a highly significant overall effect of damage on induced macroalgal resistance to herbivory. Brown and green, but not red, macroalgae induced significant resistance to herbivory in response to grazing by small crustaceans and gastropods, but not in response to large gastropods and sea urchins (Toth and Pavia 2007). Most of the studies included in the meta-analysis tested the hypothesis that herbivore-induced resistance is present in macroalgae, and the authors called for future studies testing more advanced hypotheses including effects of damage-induced responses on other organisms and trophic levels, genetic variation, costs and limits of induced responses, and the genetic and biochemical mechanisms behind induced resistance in macroalgae (Toth and Pavia 2007).

A series of studies published after the review by Toth and Pavia (2007) show that research questions regarding the presence versus absence of induced resistance in marine macroalgae still is the main focus of many investigations (e.g. Long *et al.* 2007; Yun *et al.* 2007; Macaya and Thiel 2008; Molis *et al.* 2008; Rhode and Wahl 2008a, b). Not surprisingly, these studies find that some algal species induce resistance against some herbivore species under certain conditions, while there is no induced resistance under other conditions (Table 15.2). However, a few recent studies investigate more advanced hypotheses concerning the genetic variation of induced defences (Jormalainen and Ramsay 2009; Haavisto *et al.* 2010), and the impact of induced resistance on herbivore competition (Long *et al.* 2007; Molis *et al.* 2010). Genetic variation of induced herbivore resistance in the brown seaweed *Fucus vesiculosus* was found in response to grazing by the isopod *Idotea baltica* (Haavisto *et al.* 2010). However, no genetic variation of induction of putative defence compounds (phlorotannins) was found between individuals of this species from the same area (Jormalainen and Ramsay 2009), indicating that induced resistance may be conveyed by changes in traits other than phlorotannin concentration. Furthermore, competition between different herbivore species through induced changes in their common algal resource was found for *F. vesiculosus* (Long *et al.* 2007). Grazing by the gastropod *Littorina obtusata* induced responses that reduced the palatability of the alga to two co-occurring herbivore species (*L. littorea* and *I. baltica*, Long *et al.* 2007). In contrast, feeding by the gastropod *Lacuna vincta* on the kelp *Laminaria digitata* did not induce resistance to further grazing, and in fact increased the palatability of the algae to the isopods *I. granulosa* and *I. emarginata* (Molis *et al.* 2010).

In contrast to marine macroalgae, much less is known about herbivore-induced responses in aquatic vascular plants and phytoplankton. To our knowledge, only two previously published studies have investigated herbivore-induced responses in marine vascular plants (Arnold *et al.* 2008; Vergés *et al.* 2008), and only four have studied similar responses in freshwater macrophytes (Jeffries 1990; Bolser and Hay 1998; Lemoine *et al.* 2009; Morrison and Hay 2011), all with mixed results. Marine turtlegrass (*Thalassia testudinum*) induced higher levels of condensed tannins in leaves and roots in response to simulated fish grazing and grazing by a sea urchin, although the deterrent role of tannins was not tested (Arnold *et al.* 2008). In contrast, the concentration of total phenolic compounds decreased or remained unchanged, and no induced resistance was detected in the temperate seagrass *Posidonia oceanica* in response to simulated herbivory (Vergés *et al.* 2008). Furthermore, pondweed (*Pomatogeton coloratus*) induced resistance to grazing by caddis larvae (*Triaenodes bicolor*) after artificial damage (Jeffries 1990), and water lilies (*Nuphar luteum*) induced resistance to a generalist herbivore (the crayfish *Procambarus clarkii*), but not to a specialist herbivore (the beetle *Galerucella nymphaeae*) (Bolser and Hay 1998). In a test with three freshwater macrophytes (*Elodea canadensis*, *E. nuttallii*, and *Myriophyllum spicatum*), induction of total phenolics in response to snail (*Lymnea stagnalis*) grazing was only found in one species (*M. spicatum*, Lemoine *et al.* 2009). Furthermore, the freshwater macrophyte *Camboba caroliniana* induced chemical defences that reduced the fitness of a crayfish (*Procambrus clarkii*) and a snail (*Pomacea canaliculata*). These examples show that it is still too early to draw any firm conclusions about herbivore-induced resistance in aquatic vascular plants. Considering that herbivory can be high in aquatic environments (Cyr and Pace 1993), the vast number of examples of induced herbivore resistance in terrestrial plants (Karban and Baldwin 1997), and the fact that aquatic vascular plants originate from terrestrial habitats, there are reasons to predict that herbivore-induced responses could be important in aquatic vascular plants.

Phytoplankton are important primary producers in aquatic habitats and can be exposed to high grazing pressure by herbivorous zooplankton. Marine

Table 15.2   Summary of results from studies published 2007–09 testing the hypothesis that induced herbivore resistance is present in marine macroalgae in response to direct grazing and damage-related waterborne cues.

Algal species	Herbivore species	Induction (direct grazing/ waterborne cues)	Reference
Ascophyllum nodosum	Littorina obtusata	No/No	Long et al. 2007
Dictyopteris membranacea	Gammarus insensibilis	No/No	Yun et al. 2007
	Gamariella fucicola		
	Cymadusa filosa		
Fucus vesiculosus	Gammarus insensibilis	Yes/Yes	Yun et al. 2007
	Gamariella fucicola		
	Cymadusa filosa		
Gelidium sesquipedale	Gammarus insensibilis	No/No	Yun et al. 2007
	Gamariella fucicola		
	Cymadusa filosa		
Sphaerococcus coronopifolius	Gammarus insensibilis	No/No	Yun et al. 2007
	Gamariella fucicola		
	Cymadusa filosa		
Dictyota kunthii	Parhyalella ruffoi	Yes/No	Macaya and Thiel 2008
Macrocystis integrifolia	Parhyalella ruffoi	No/No	Macaya and Thiel 2008
Ulva pertusa	Littorina brevicula	No/Not applicable	Molis et al. 2008
	Halotis discus	No/Not applicable	
Laminaria japonica	Littorina brevicula	Yes/Not applicable	Molis et al. 2008
	Halotis discus	No/Not applicable	
Furcellaria lumbricalis	Idotea baltica	Yes/No	Rhode and Wahl 2008a
Delesseria sanguinea	Idotea baltica	Yes/No	Rhode and Wahl 2008a
Phyllophora pseudoceranoides	Idotea baltica	Yes/No	Rhode and Wahl 2008a
Fucus serratus	Idotea baltica	Yes/No	Rhode and Wahl 2008a
Fucus evanescence	Idotea baltica	Yes/No	Rhode and Wahl 2008a
Fucus vesiculosus	Idotea baltica	Yes/Not applicable	Rhode and Wahl 2008b

dinoflagellates (*Alexandrium minutum*) have been shown to induce production of paralytic shellfish toxins (PSTs) in response to waterborne cues from different copepod species (Fig. 15.2) (Selander *et al.* 2006; Bergkvist *et al.* 2008). The induction of PST is strongly dependent on the copepod species producing the cue. Waterborne cues from *Centropages typicus* induced more than 20-times higher toxin concentrations compared to negative controls, while no response was found to cues from *Pseudocalanus* sp. (Bergkvist *et al.* 2008). Furthermore, *A. minutum* showed a more pronounced induction of PST in high compared to low nitrate treatments (Selander *et al.* 2008). The cyanobacterium *Microcystis aeruginosa* induce higher toxin (microcystin) content in response to grazing by zooplankton (Jang *et al.* 2003) and fish (Jang *et al.* 2004), although whether micro-

cystins deter these grazers is uncertain. Furthermore, some *M. aeruginosa* strains were highly responsive to herbivore cues while others showed little or no increase in toxin content (van Gremberghe *et al.* 2009), indicating that there are genetic differences in toxin induction in this species.

Research on herbivore-induced defences in primary producers has traditionally focused on plant–herbivore interactions, ignoring the fact that both terrestrial and aquatic ecosystems are characterized by enormous species diversity and a corresponding complexity of interactions among these species. Terrestrial ecologists have become interested in interactions involving three or more trophic levels (Tscharntke and Hawkins 2002). For example, it is now well known that terrestrial plants can release volatile cues that attract predators and/or parasitoids

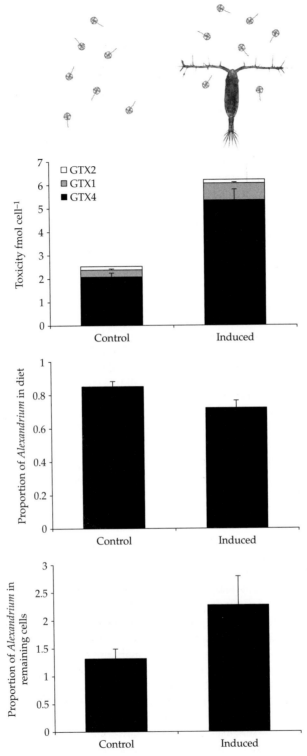

Figure 15.2 Example of grazer-induced chemical defence. (a) the dinoflagellate *Alexandrium minutum* responds to chemical cues from copepod grazers by increased toxicity. (b) The grazer-induced, more toxic cells are ingested at a lower rate compared to non-induced cells. (c) The feeding preference of the grazer results in increased relative abundance of the induced, more toxic, *Alexandrium* cells. The figure is modified from Selander *et al*. (2006).

when wounded by herbivores, and that this response leads to a decreased grazing pressure and an increase in plant fitness (Dicke 2009). Aquatic predators are under selection pressure to use reliable chemical cues to locate their prey because the use of visual cues can be impaired in aqueous environments (Zimmer and Butman 2000). Prey species, on the other hand, are under selection pressure to minimize release of such cues, and the emission of volatile or waterborne compounds from wounded aquatic primary producers could be important foraging cues for aquatic predators.

Volatile organic compounds (VOCs) are common cues in herbivore-induced indirect defences of terrestrial plants (Arimura et al. 2009). Aquatic macroalgae and phytoplankton can also emit VOCs in response to grazing (Van Alstyne 2008). Some aquatic VOCs, such as dimethyl sulphide (DMS), function as foraging cues for predatory birds and fish (Nevitt and Haberman 2003; deBose et al. 2008), which indicates that the release of these compounds may increase algal fitness by attracting predators of their grazers. Although this hypothesis has been proposed previously (e.g. Van Alstyne 2008), it has not been adequately tested in manipulative experiments. To our knowledge, there is only one published study testing potential indirect defences in a marine macroalgae (Coleman et al. 2007). *Ascophyllum nodosum* released signals when grazed by herbivorous snails (*Littorina obtusata*), which attracted predatory crabs and fish in a laboratory setting (Coleman et al. 2007). However, whether the attraction of predators resulted in decreased grazing and increased algal fitness was not determined, and the chemical nature of the cues was not identified. Therefore, more research is needed to investigate herbivore-induced indirect defences in aquatic primary producers.

15.4 Herbivore performance and fitness effects of chemical defences

Chemical defences of aquatic primary producers and their impacts on herbivores have been recently and thoroughly reviewed by several authors (McClintock and Baker 2001; Landsberg 2002; Paul and Puglisi 2004; Paul et al. 2006a, Paul and Ritson-Williams 2008; Hay 2009; Poulson et al. 2009; Sotka et al. 2009). One of the most striking patterns in the

responses of herbivores to chemical defences is their high variability. Slight differences in the structure of defensive metabolites have large implications for their effects on herbivores (Cronin and Hay 1996b). Different species react differently to the same metabolite (Sotka et al. 2009), ranging from acute toxic responses (Bricelj et al. 2005), reduced feeding (e.g. Colin and Dam 2003; Cruz-Rivera and Hay 2003), and decreased fecundity and fertility (e.g. Sotka and Hay 2002) to positive effects in the form of associational defences (Hay et al. 1989). Sequestered secondary metabolites may, for example, protect grazers on defended algae against predation (e.g. Kvitek et al. 1991). Variation also occurs at the population level where local populations of a single species have different sensitivities to defensive metabolites (Colin and Dam 2003; Bricelj et al. 2005; Sotka and Hay 2002). Finally, the effects on herbivore performance vary over time. Initial effects of consuming chemically defended algae may vanish or develop after an acclimatization period (Colin and Dam 2003; Kubanek et al. 2007). This multi-level variation is the main reason why the defensive roles and ecological significance of many secondary metabolites are still controversial and debated (Cembella 2003; Ianora and Miralto 2010). Recent studies have shown that an important part of this variation can be explained by the coevolution of resistance in herbivores exposed to chemical defences in evolutionary time. The best documented example of coevolution is the soft shell clam *Mya arenaria* along the North American east coast. Mussels from areas more frequently exposed to paralytic shellfish toxins (PSTs) were less sensitive to PSTs compared to mussels from areas where PST-producing algae rarely occurred. Resistance was shown to depend on a single amino acid substitution that reduced the affinity of the targeted sodium channel protein to PSTs by three orders of magnitude (Bricelj et al. 2005). Similar differences in resistance have been found for populations of the copepod *Acartia hudsonica* (Colin and Dam 2003). Copepods with little history of encountering PST-containing prey stopped feeding and reduced respiration when fed PST containing dinoflagellates (*Alexandrium fundyense*) whereas copepods that had historical experience of feeding on PST-containing prey did not exhibit any of these effects (Colin and

Dam 2003). Local amphipod populations (*Amphitoe longimana*) that share habitat with defended seaweeds (*Dictyota*) are less affected by *Dictyotas* chemical defence and have higher fitness compared to *A. longimana* from outside *Dictyotas*'s geographical range when fed *Dictyota* (Sotka and Hay 2002). Coevolution of defended aquatic primary producers and herbivores nevertheless appears to be an understudied topic that is likely to contribute to our understanding of the variable response of herbivores to chemical defences.

Reviews have repeatedly identified the fitness effects of secondary metabolites on aquatic herbivores as understudied (e.g. Hay 1996; Paul and Puglisi 2004; Sotka *et al.* 2009). The number of articles describing physiological effects on consumers continues to increase, but the mechanisms by which herbivore performance is affected are still understood in only a few cases (Sotka *et al.* 2009). A more detailed understanding of the physiological responses of herbivores to chemical defences would increase our understanding of the dynamic coevolution of chemical defences and resistance (Sotka *et al.* 2009).

When it comes to studying grazing on marine macroalgae, the major relevant herbivore groups (fish, echinoderms, small crustaceans, and molluscs) are represented in the literature. On the other hand, reported studies of herbivory on phytoplankton emphasize small crustaceans (copepods and cladocerans) and commercially important bivalves, whereas the, at least equally, important heterotrophic protists are under-studied (Calbet and Landry 2004; some examples reviewed by Tillmann 2004). Research on phytoplankton chemical defences has mainly been fuelled by the interest in harmful algal blooms, as phytoplankton toxins cascade through food webs and impact higher trophic levels and entire ecosystems (Hay and Kubanek 2002). As a consequence, the physiological effects on herbivores and higher trophic levels are generally better understood compared to the macroalgal counterpart (Sotka *et al.* 2009), although experimental evidence that harmful algal toxins act as chemical defences is sparse (Hay and Kubanek 2002). The use of small crustaceans with short life cycles has resulted in a higher proportion of studies accounting for fitness correlated traits like egg production (Jonasdottir *et al.* 1998), hatching success, and larval development, in addition to feeding behaviour. The most well investigated example concerns the effects of short chain polyunsaturated aldehydes and oxylipins produced by diatoms on copepod reproduction. Copepod females fed diatoms have been shown to produce abnormal eggs with low hatching success or eggs that hatch into malformed nauplii, although this is still debated (Ianora and Miralto 2010). Effect of macroalgal defences on grazers is more often assessed as changes in biomass (Hay 1996; but see Cruz-Rivera and Hay 2003; Taylor *et al.* 2003; and Toth *et al.* 2005). Research on macroalgae has in return benefited from more precise experimental manipulations, whereby it has been possible to manipulate specific compounds in artificial diets to investigate their effect on grazers (Hay 1996). The physiological effects on herbivores feeding on phytoplankton have often been inferred from experiments using live alga, and are potentially confounded by morphological features (e.g. size, hardness) or nutritional effects (Prince *et al.* 2006). Recently, giant liposomes were successfully used as carriers of test substances, which provide a promising alternative for more precise manipulations of chemical defences in feeding experiments (Buttino *et al.* 2006). There are several examples of grazers sequestering defensive metabolites from macroalgal diets, but we only managed to find a single herbivore sequestering secondary metabolites from phytoplankton, the butter clam (*Saxiodomus giganteus*), which incorporates saxitoxins from dinoflagellates in their diet (Kvitek *et al.* 1991). Similarly, specialist herbivores which associate with, and adapt to, a single host species are more common (but still rare compared to terrestrial herbivores) among macroalgal grazers than zooplankton even though specialist grazers do exist (Tillmann 2004). Whether the higher proportion of sequestered defences and specialized grazers on macroalgae represents true differences or the effects of different research efforts remains to be tested.

15.5 Invasion ecology and chemical defences

Invasive non-indigenous species are, next to habitat fragmentation, the largest threat to biodiversity (Levine *et al.* 2003), and, by any criteria, major agents of current global change (Mack *et al*. 2000). Although problems caused by biological invasions have been increasing lately due to growing international commerce (Pimentel *et al*. 2001), this is a long-standing phenomenon observed already by Darwin (Darwin 1859). A subset of the introduced non-indigenous species become pests in their new geographic ranges, causing damage to natural ecosystems as well as agri- and aquaculture, leading to immense economic costs (Pimentel *et al*. 2001). Pimentel *et al*. (2001) estimated the global cost of these damages as 1.4 trillion US dollars per year, which represents roughly 5% of the world economy. Clearly, there is much to be gained, in both environmental and monetary values, by efficient management.

The earliest attempts to understand and predict biological invasions are the enemy release hypothesis (ERH; Darwin 1859) and the biotic resistance hypothesis (BRH; Elton 1958). The ERH explains the success of invasive exotics by their escape from their regulation of coevolved herbivores and pathogens in their native range (Fig. 15.3), and the BRH predicts communities with low diversity to be more readily invaded because the invader will encounter less competition for resources (Keane and Crawley 2002). Even though both hypotheses have been supported in various environments, and are still the most commonly cited, inconsistencies in outcomes of tests of the two hypotheses are numerous (see Maron and Vila 2001 and references therein). Thus, why species differ in invasion success in exotic ranges is still far from understood.

More recent efforts to understand biological invasions have considered the biochemical properties of species. Invasive non-indigenous species that possess novel chemical defences are often competitively superior to natives, because these compounds not only have strong effects on native consumers, but also on competitors, pathogens, and soil microbes (Callaway and Ridenour 2004). In a literature study on invasive plants in North America, Cappuccino and Arnason (2006) showed that approximately half of the highly invasive exotics possessed prominent secondary metabolites that were not previously known from the native plant communities. They also proposed that by creating an index of phytochemical novelty, where the chem-

istry of possible invaders can be compared to those in probable recipient communities, it would be possible to identify invasive exotics *a priori*, which could aid management decisions in preventing possible invasions (Cappuccino and Arnason 2006).

The basis for this predictive tool is the Novel Weapons Hypothesis (NWH; Callaway and Ridenour 2004). The NWH predicts that successful exotic invaders bring unique compounds to their new ranges, which allow them to establish and proliferate through allelopathy because their new neighbours are not adapted to these novel compounds (Fig. 15.3). An extension of the NWH called the 'Allelopathic Advantage against Resident Species' (AARS), predicts populations of invaders to evolve greater concentrations of allelopathic, antibiotic, and grazer deterrent compounds compared to conspecifics in the native range (Inderjit *et al*. 2006). This is explained by a greater selection pressures for the traits that are advantageous in the new range (Inderjit *et al*. 2006).

However, the evolutionary novelty argument is still highly debated, and the logic of the AARS is in direct contrast to the Evolution of Increased Competitive Ability hypothesis (EICA; Blossey and Nötzold 1995). According to the EICA, exotics are suggested to be competitively superior because they can allocate resources away from defence to growth, since defences carry costs which limit growth or reproduction, when they are released from former enemies. Thus, unlike that predicted by the AARS, invasion success according to the EICA requires that the introduced species does not have to defend itself from new enemies. In yet a different view, contrasting both the AARS and EICA, Parker *et al*. (2006) argued that evolutionary novelty leaves plants at greater risk of attack by newly encountered generalist herbivores since they are not evolved to defend themselves against these new enemies. In their meta-analysis they found plants containing defensive compounds similar to native species to be more resistant, and would therefore gain nothing from biochemical novelty. Hence, both Parker *et al*. (2006) and the NWH, and to some extent even the EICA, suggest linking the success of invasions to the phytochemical properties of plants and the resistance to generalist herbivores, even though their predictions differ widely.

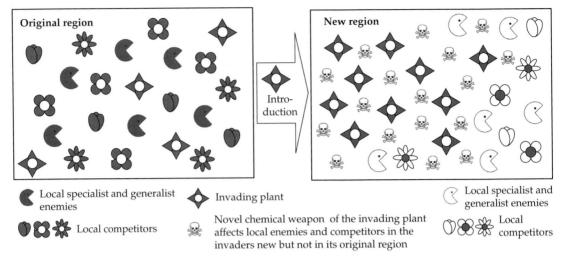

Figure 15.3 Illustration of the Enemy Release Hypothesis (ERH) and the Novel Weapon Hypothesis (NWH). In the native region all plants suffer from their generalist and specialized enemies as well as from interspecifc competition. On introduction to a new region the non-indigenous plant escapes regulation by its enemies, as predicted by ERH. In the new region the introduced plant will be less attacked by local enemies than native plants and experiences less competition due to its effective chemicals as predicted by the NWH. As a consequence, the non-indigenous plant will gain a competitive advantage and become invasive in the new region.

Hypotheses regarding biological invasions are commonly developed from examples of terrestrial plant invasions. Unfortunately, their validity for biological invasions in aquatic systems has rarely been tested (but see Parker and Hay 2005; Wikström *et al.* 2006). The ERH and NWH rely on interspecific competition and herbivory as the most important structuring agents of plant communities. Fundamentally different to terrestrial systems, in aquatic environments specialist enemies are rare and generalist consumers usually dominate (Hay and Steinberg 1992). Thus, the effects of escaping specialist enemies on introduction may be negligible for aquatic plants and algae. However, generalist herbivory has been shown to have pronounced effects on the growth and community composition of seaweeds (Duffy and Hay 2000) and aquatic vascular plants (Lodge *et al.* 1998). In aquatic communities structured by herbivory and competition, new biotic interactions with new enemies will determine the invasion success of non-indigenous aquatic plants. Introduced species that are chemically defended may thus have a competitive advantage over native species and can therefore become successful invaders.

Although there are few formal tests in aquatic systems of the hypotheses on biological invasions and species' phytochemical properties, there are studies that show substantial effects of the chemical compounds of invading aquatic plants on native species in the new range. One example is *Caulerpa taxifolia*, which is probably the most well-known out of the estimated 277 introduced seaweeds worldwide. This siphonous green alga usually grows in tropical waters, but in the early 1980s a cold-water resistant and faster growing strain of *C. taxifolia* escaped from a public aquarium (Jousson *et al.* 2000). Since then, the alga strain has spread to more than 6,000 ha in the Mediterranean, by attaching to rocks, sand, and mud at depths between 0 to 50 m, and has further been introduced to the Californian Coast (Jousson *et al.* 2000). Through invasion of sensitive sea grass meadows of *Posidonia oceanica* and competition with other algae and corals, this alga changed ecosystems and habitats in parts of the Mediterranean (Meinesz 1999). Extensive research has been done on the reasons and consequences of *C. taxifolia*'s uncontrolled proliferation, and the lack of natural enemies seems to be a crucial factor. Compared to its consumption of native macroalgae, the sea urchin *Paracentrotus lividus* (one of the major herbivores in the Mediterranean) consumed only small amounts of *C. taxifolia* and performed badly (spine loss, atrophy of gonads, increased mortality) on a long term diet of *C. taxifolia* (Boudouresque *et al.* 1996). *C. taxifolia* is known to produce toxins, mainly oxytoxins and caulerpenyne (Paul and Fenical 1986; Amade and Lemee 1998), the latter being an important compound in wound healing in this siphonous green alga (Adolph *et al.* 2005). Caulerpenyne, as well as its degradation products, have been shown to negatively affect microorganisms and macroalgae and to deter herbivores, such as *Ampithoe longimana* and *Paracerceis caudata* (Paul and Fenical 1986; Ferrer *et al.* 1997; Weissflog *et al.* 2009). In contrast, an experiment on herbivorous fish showed no deterrence by caulerpenyne (Wylie and Paul 1988). In the native range of *C. taxifolia*, sacoglossans (e.g. *Elysia subornata*, *Oxynoe ayuropunctatas*, *Lobiger viridis*, and *Stiliger smaragdinus*) have morphologically and metabolically specialized on *Caulerpa* species (Jensen 1997), but *C. taxifolia* is the least preferred due to the high concentration of its secondary metabolites (Baumgartner *et al.* 2009). In the Mediterranean, the native sacoglossans *Oxynoe olivacea* and *Lobiger serradilfalci* usually feed on the native *Caulerpa prolifera* but cannot control the spread of *Caulerpa taxifolia* since they do not consume effective amounts of this alga (Thibaut and Meinesz 2000). Caulerpenyne is also present in this native *Caulerpa* species, but in much lower concentrations (Amade and Lemee 1998, but see Jung *et al.* 2002). In conclusion, *C. taxifolia* escapes herbivory due to its high concentrations of defensive compounds and is, thus, an example of the presence of a chemical defence in an invasive species. However, since the chemical defensive compounds are not novel to native species of the Mediterranean, the findings do not support the NWH.

A second marine example where chemical properties are likely to play a role in the invasion process is the fucoid brown alga *Fucus evanescens*. This species is native to the Arctic, but has recently established populations in the North and Baltic Sea. Wikström *et al.* (2006) showed that herbivore richness and abundance on *F. evanescens* was lower in the introduced compared to the native range. The

ecologically relevant herbivores, such as the isopod *Idotea granulosa* and the periwinkles *Littorina littorea* and *L. obtusata*, generally preferred native fucoid species over *F. evanescens* in the introduced ranges (Wikström *et al.* 2006). In the native range, *F. evanescens* is readily consumed and preferred over other fucoids by small amphipods (*Gammarus oceanicus* and *Echinogammarus fumarchicus*) and periwinkles (*L. littorea* and *L. obtusata*) (Barker and Chapman 1990; Denton and Chapman 1991). Fucoids are known to produce phlorotannins and their concentration negatively correlates with consumption rates of herbivores (Toth *et al.* 2005). Introduced populations of *F. evanescens* were demonstrated to have higher phlorotannin levels than other coexisting fucoid species in the introduced range (Wikström *et al.* 2006). Phlorotannin concentrations in *F. evanescens* were lower in native Icelandic populations compared to introduced populations, but seem to be comparable with native Canadian populations (Denton and Chapman 1991; Wikström *et al.* 2006). Herbivores in both the native and introduced ranges of *F. evanescens* share an evolutionary history with fucoids and are accustomed to their chemical defence, which contradicts the novel weapon argument. However, the higher levels of phlorotannins compared to native fucoid species may give *F. evanescens* a competitive advantage over native fucoids and thus explain its success in establishment in its new ranges.

Two additional examples of macroalgal invasions come from the red algal family Bonnemaisoniaceae. Both *Bonnemaisonia hamifera* and *Asparagopsis armata* are successful invaders in the North Atlantic and the Mediterranean, respectively (Maggs and Stegenga 1999). *A. armata* was first described from Western Australia and has been shown to produce several brominated compounds including bromoform and dibromoacrylic acid (Paul *et al.* 2006b). In its native range, *A. armata* is chemically defended against small generalist herbivores and bacterial colonization, but is consumed by sea slugs, such as *Aplysia parvula* (Paul *et al.* 2006b). In the Mediterranean, the sea urchin *Paracentrotus lividus*, showed low consumption and growth when subjected to diets of *A. armata* (Frantzis and Gremare 1993), which may be an effect of the secondary metabolites produced by *A. armata*. *B. hamifera* has been introduced from the Northwest Pacific, whereafter the tetrasporo-phytic phase became a common filamentous alga on many temperate coasts in the North Atlantic. The impacts of herbivores in its native range are not known, but *B. hamifera* produces a polybrominated heptanone which inhibits bacterial colonization and has allelopathic effects on algal spores and micro-algae in its introduced ranges (Nylund *et al.* 2008; Svensson *et al.* unpublished). Due to its chemical defence, the alga is also avoided by different gener-alist herbivores in the introduced region (Enge *et al.* unpublished). The chemical defences of both *A. armata* and *B. hamifera* are suggested to contribute to their invasion success.

There are few examples of invasions by phyto-plankton that have been directly linked to their sec-ondary metabolites, but *Alexandrium minutum* is known to be an invasive exotic and has been show to produce secondary metabolites with allelopathic and grazer deterrent properties (Selander *et al.* 2006). Bolch and de Salas (2007) have also shown that the saxitoxin-producing dinoflagellates *Gymnodinium catenatum* and *Alexandrium catenella* were introduced to Australia and New Zealand from Asia 30–100 years ago, most likely via ship ballast water. In Australia, *A. minutum* and *G. catenatum* are also identified to be among the top ten most damaging invasive species, based on human, environmental, and economic impact, as well as among the ten potential domestic target species most likely to be spread to uninfected regions by shipping (Hayes *et al.* 2005).

In conclusion, even though more research has to be done to evaluate the role of chemical defences in biological invasions of aquatic plants and algae, both direct and indirect evidence of invasion facili-tated by secondary metabolites of invaders to date are quite convincing. The examples above suggest that the chemistry of non-indigenous aquatic pri-mary producers affects interactions with both native competitors and consumers. The alteration of such interactions by chemical means is likely to be of larger importance than previously thought, especially since top-down control is a vital and common regulation mechanism in aquatic environ-ments. However, since there are directly conflicting hypotheses, as well as evidence, on the effects of biochemical properties of invaders on invasion

success, both the general applicability and the underlying mechanisms of such patterns must be investigated more thoroughly if any form of con-sensus is to be reached. Nonetheless, these differ-ent hypotheses all similarly predict that the chemistry of exotics and natives will influence the success of biological invasions, albeit in different ways. Consequently, the number of studies on the importance of chemically mediated interactions in biological invasions is likely to increase, and thus our knowledge, as this question gets more atten-tion in the coming years.

15.6   New approaches and future directions

In order to make progress, we need to systemati-cally develop and adapt general ecological theories to aquatic systems. With the differences that exist in herbivore assemblages and grazing intensities between terrestrial and aquatic systems, these theo-ries need more input from chemically-mediated interactions between algae and consumers. Chemical ecologists have already unravelled the role of a plethora of exciting marine secondary metabolites. It has, however, been problematic to identify deterrent molecules that are difficult to purify or characterize against a background of many other primary and secondary metabolites. This has been especially problematic for unstable, water soluble, structurally complex, and low con-centration compounds. The traditional bioassay-guided fractionation of chemical defences has failed in many of these cases, and new approaches are needed. Today, techniques exist that make it possi-ble to unravel previously hidden specific molecules that are active in the algae–consumer interactions, which gives us the new tools needed to develop and adapt the ecological theories.

15.6.1   Metabolomics

Among the promising newer techniques is metabo-lomics, the comprehensive study of metabolites from an organism. Metabolomics should allow chemical ecologists to identify candidate chemical defences from a complex mixture and correlate the presence of certain compounds with their ecological effects, without purification of individual components. Because all compounds are measured simultaneously, additive or synergistic effects are also more likely to be detected. In addition to the indirect nature of this type of investigation, metabolomics carries some methodological weaknesses, such as inadequate sensitivity for some compounds if much more abundant (but non-ecologically relevant) molecules co-vary in concentration and the need for intense data handling and statistical analysis.

To date, metabolomics has been applied sparingly in aquatic chemical ecology and not at all in studies of anti-herbivore chemical defences from marine or freshwater systems. Metabolomics could be used to pinpoint previously unknown chemical defences or to analyse differences in compounds across populations, individuals, or tissues. It could also be applied to study changes in compound production when aquatic primary producers are exposed to cues from herbivores, as has been done in terrestrial plants, and to compare the induction of defences when exposed to herbivory versus mechanical wounding (see Prince and Pohnert (2010) for a sampling of this literature and Wiesemeier *et al.* (2008) for a preliminary metabolomic study of this type using brown macroalgae). Because metabolomics involves the simultaneous analysis of thousands of small molecules from an organism or tissue, metabolic pathways involved in induction of chemical defences could also be suggested based upon groups of compounds whose concentrations fluctuate when exposed to herbivory. Finally, we suggest that metabolomics could enable the detection of metabolic pathways that are suppressed when chemical defences are up-regulated, providing testable hypotheses regarding allocation costs of chemical defences. Some of these applications for metabolomics have been considered by chemical ecologists working with terrestrial plants (e.g. Bezemer and van Dam 2005).

To date, most metabolomic studies have utilized the whole organism or tissue extracts. However, identification of waterborne cues from herbivores which lead primary producers to induce chemical defences may also be excellent targets for metabolomic analysis (see Paul *et al.* (2009) for an example involving phytoplankton-phytoplankton interac-

tions). This approach has been termed 'metabolic fingerprinting' in other literature.

15.6.2 Proteomics

Secondary metabolites that function as chemical defences are synthesized by enzymes, so the analysis of proteins (and their fluctuating concentrations depending on environmental conditions) can provide insights into the physiology of chemically defended organisms. Particularly for aquatic primary producers, whose defences are induced in response to herbivores or other environmental cues, proteomic analysis could lead us to identify pathways responsible for biosynthesis and induction of chemical defences. Proteomic analysis is achieved by mass spectrometry (MS) typically using a whole tissue or cell extract whose proteins have been digested into small peptides. The MS fragmentation pattern of these peptides enables determination of their amino acid sequence, with reference to a database of known peptide fragments. Such a database is only available for organisms whose genomes or transcriptomes have been reported or for well-studied proteins whose sequences are highly conserved. It is possible to sequence peptides *de novo* without a reference database, but this is still a laborious and expensive process involving considerable expertise, and is therefore not feasible for the comparison of many samples in an ecologically rigorous study. The lack of full genome sequence data for most aquatic organisms of interest to chemical ecologists probably accounts for the underutilization of this technology so far. Nevertheless, Chan *et al.* (2004) used proteomic analysis to pin-point phytoplankton proteins whose expression varies with growth phase and environmental conditions. Terrestrial chemical ecologists have used a selective proteomics approach (focusing on a few known enzymes) to understand induction of secondary metabolite (terpene) biosynthesis (Zulak *et al.* 2009).

15.6.3 Transcriptomics

Whereas metabolomics studies the full complement of small molecules in an organism, and proteomics focuses on proteins including biosynthetic enzymes, transcriptomics allows the comprehensive analysis of gene expression, particularly focusing on up- or down-regulated genes in response to environmental perturbation. Historically, this has been achieved in a semi-quantitative way using microarrays of selected gene sequences in a chip format, but recent innovations in sequencing technology allow a more quantitative and comprehensive analysis of expression levels. Transcriptomics has rarely been applied to questions in aquatic chemical ecology, but we see great potential for this approach in elucidating genes responsible for biosynthesis of chemical defences in cases where defences are known to be up-regulated (e.g. when damaged or exposed to cues from consumers). Other genes of interest that may be expected to be up-regulated in such a system include those involved in induction of biosynthesis and genes involved in sensory pathways for recognizing specific grazers. Moustafa *et al.* (2010) detected substantially altered gene expression patterns of the toxic dinoflagellate *Alexandrium tamarense* when exposed to marine bacteria, compared to bacteria-free cultures, suggesting that interactions between these organisms are worthy of future attention. More closely related to herbivore defences, Van Oosten *et al.* (2008) identified four genes, using expression profiling, whose expression co-varied with microbially induced chemical defences of the terrestrial plant *Arabidopsis*. In a marine system, but involving defences against predators, Hoover *et al.* (2008) showed that induction of chemical defences upon damage to the soft coral *Sinularia polydactyla* is accompanied by up-regulation of a small number of unidentified genes. As with proteomics, the lack of whole genome data has impaired the application of transcriptomic technology to aquatic chemical ecology.

15.6.4 Consumer aspect

How does consumption of chemically defended primary producers affect the physiology of herbivores? Transcriptomic and proteomic approaches are expected to be valuable in exploring the responses of herbivores to chemical defences. One would expect that stress and detoxification pathways may be up-regulated, and that the physiological basis for fitness costs might be illuminated by transcriptional analysis of herbivores fed chemically defended

foods. As with prey potentially up-regulating sensory genes when exposed to consumers, expression of genes and proteins involved in chemoreception in consumers may also be altered. Transcriptomic and proteomic analyses could provide valuable inroads, given that we know very little about how aquatic herbivores sense and assimilate chemical defences in their diets. There likely needs to be a stronger focus on model species in order to take advantage of these approaches.

Research on chemical defences has primarily focused on defended plants and prey, and less on the behavioural and physiological impacts on consumers. We suggest that future studies should place additional focus on the fitness consequences of consuming chemically defended food, since herbivores are known to sample deterrent plants and therefore ingest potentially toxic compounds. Recent work with predatory snails points to the importance of detoxification enzymes in allowing generalists to tolerate chemically defended coral diets (Whalen *et al.* 2010); similar research with herbivores could help us understand how some animals, especially specialists and small grazers, tolerate these defences. The sensory perception of chemical defences by animals has been little studied, except in insects (although see Jordt and Julius (2002) for how birds and mammals differently experience the hot chilli compound capsaicin). A recent study identified a novel transmembrane receptor in fish that allows them to discriminate against marine sponges defended with triterpene glycosides (Cohen *et al.* 2008; Cohen *et al.* 2010). To date, we are unaware of studies that explore the chemoreception process in aquatic herbivores.

15.6.5 A changing world

Most studies of chemical defences assume a relatively unchanged biosphere over recent evolutionary history, which does not take into account recent and ongoing environmental changes caused by humans. Major herbivores and predators have become extinct in many ecosystems, while in other ecosystems species have been introduced. Chemical defences that evolved to protect a primary producer against one enemy may function differently with new enemies and altered abiotic conditions.

We are only starting to address the ecological and evolutionary implications of anthropogenic environmental change to chemically mediated interactions. Changes such as pH, UV, temperature, and so on, are all likely to be important and to interact with regard to chemical interactions. Changing ecosystems in concert with changing anthropogenic environments are challenges for future theories.

Related to global warming, the following questions seem relevant: how will increased seawater and freshwater temperatures affect the production of chemical defences in aquatic primary producers? How will herbivores' ability to detect and react to chemically defended food be affected by increased temperatures? How might physiological stress caused by increased temperature affect herbivores' tolerance to toxins in their diet? Which species or populations are most likely to thrive in a warmer climate—chemically defended or undefended ones? If populations of herbivores decline due to altered climate, will that lead to increased competition among primary producers, favouring those that do not pay the costs of defending themselves, driving chemically defended species locally extinct? The coupling of altered abiotic conditions and new biotic interactions as a result of the expanding ranges of some species (while others' ranges shrink) will lead to interesting and likely unexpected outcomes related to chemical defences.

Increasing atmospheric carbon dioxide is acidifying the world's oceans, which will likely have direct and indirect effects on chemical defences. How will decreased pH affect the stability and ionization state of chemical defences, the biosynthesis of chemical defences, and the receptor–ligand interactions that allow herbivores to recognize and avoid defended primary producers? We hypothesize that the expected loss of calcium carbonate skeletons for some macroalgae that should occur with decreasing pH may weaken their resistance to herbivory, in cases in which chemical defences and calcium carbonate structural defences act synergistically (Hay et al. 1994). The potential loss of stony corals as a result of ocean acidification is expected to benefit non-calcified macroalgae, many of which are chemically defended against herbivorous fishes. The disruption of the innate ability for settling fish larvae

to recognize cues from predators due to lowered pH (Dixson et al. 2010), could further exacerbate the loss of tropical coral reefs.

In concert with increased temperature and decreased pH, the world's oceans are continuing to lose herbivores as a result of overfishing (and occasionally disease), which has led to coral overgrowth by macroalgae. Although we would expect undefended macroalgae to benefit the most by removal of herbivores, coral overgrowth by both undefended and chemically defended macroalgae occurs widely in the Caribbean and tropical Pacific Ocean. Since the effects of chemical defences on herbivores are species-specific, restoration of reefs by the addition of herbivores needs to take into account which algae the reintroduced fish will consume (Burkepile and Hay 2008). Some macroalgae that are chemically defended against herbivores are also allelopathic to corals, further stressing corals on algal dominated reefs (Rasher and Hay 2010). Equivalent food web disruptions in marine and freshwater plankton may also be expected to affect chemically mediated interactions among phytoplankton and grazers, although this has yet not been studied.

References

Adolph, S., Jung, V., Rattke, J., and Pohnert, G. (2005) Wound closure in the invasive green alga *Caulerpa taxifolia* by enzymatic activation of a protein cross-linker 13. *Angewandte Chemie International Edition*, **44**: 2806–8.

Amade, P. and Lemee, R. (1998) Chemical defence of the Mediterranean alga *Caulerpa taxifolia*: variations in caulerpenyne production. *Aquatic Toxicology*, **43**: 287–300.

Amsler, C.D. and Fairhead, V.A. (2006) Defensive and sensory chemical ecology of brown algae. *Advances in Botanical Research*, **43**: 1–91.

Amsler, C.D., Iken, K., McClintock, J.B. et al. (2005) Comprehensive evaluation of the palatability and chemical defenses of subtidal macroalgae from the Antarctic Peninsula. *Marine Ecology Progress Series*, **294**: 141–59.

Arimura, G., Matsui, K., and Takabayashi, J. (2009) Chemical and molecular ecology of herbivore-induced plant volatiles: proximate factors and their ultimate functions. *Plant Cell Physiology*, **50**: 911–23.

Arnold, T.M., Tanner, C.E., Rothen, M., and Bullington, J. (2008) Wound-induced accumulations of condensed tannins in turtlegrass, *Thalassia testudinum*. *Aquatic Botany*, **89**: 27–33.

Barker, K.M. and Chapman, A.R.O. (1990) Feeding preferences of periwinkles among 4 species of *Fucus*. *Marine Biology*, **106**: 113–18.

Baumgartner, F.A., Motti, C.A., de Nys, R., and Paul, N.A. (2009) Feeding preferences and host associations of specialist marine herbivores align with quantitative variation in seaweed secondary metabolites. *Marine Ecology Progress Series*, **396**: 1–12.

Berenbaum, M.R. (1995) The chemistry of defense: theory and practice. *Proceedings of the National Academy of Science USA*, **92**, 2–8.

Bergelson, J. and Purrington, C.B. (1996) Surveying patterns in the cost of resistance in plants. *American Naturalist*, **148**: 536–58.

Bergkvist, J., Selander, E., and Pavia, H. (2008) Induction of toxin production in dinoflagellates: the grazer makes a difference. *Oecologia*, **156**: 147–54.

Bezemer, T.M. and van Dam, N.M. (2005) Linking aboveground and belowground interactions via induced plant defenses. *Trends in Ecology & Evolution*, **20**: 617–24.

Blossey, B. and Nötzold, R. (1995) Evolution of increased competitive ability in invasive non-indigenous plants - a hypothesis. *Journal of Ecology*, **83**: 887–9.

Bolch, C.J.S and de Salas, M.F. (2007) A review of the molecular evidence for ballast water introduction of the toxic dinoflagellates *Gymnodinium catenatum* and the *Alexandrium 'tamarensis* complex' to Australasia. *Harmful Algae*, **6**: 465–85.

Bolser, R.C. and Hay, M.E. (1998) A field test of inducible resistance to specialist and generalist herbivores using the water lily *Nuphar luteum*. *Oecologia*, **116**: 143–53.

Boudouresque, C.F., Lemée, R., Mari, X., and Meinesz, A. (1996) The invasive alga *Caulerpa taxifolia* is not a suitable diet for the sea urchin *Paracentrotus lividus*. *Aquatic Botany*, **53**: 245–50.

Bricelj, V.M., Connell, L., Konoki, K., *et al.* (2005) Sodium channel mutation leading to saxitoxin resistance in clams increases risk of PSP. *Nature*, **434**: 763–7.

Bryant, J.P., Chapin, F.S. III, and Klein, D.R. (1983) Carbon/nutrient balance of boreal plants in relation to vertebrate herbivory. *Oikos*, **40**: 357–68.

Burkepile D.E. and Hay, M.E. (2008) Herbivore species richness and feeding complementarity affect community structure and function: the case for Caribbean reefs. *Proceedings of the National Academy of Science USA*, **105**: 16201–6.

Buttino, I., De Rosa, G., Carotenuto, Y. *et al.* (2006) Giant liposomes as delivery system for ecophysiological studies in copepods. *Journal of Experimental Biology*, **209**: 801–9.

Calbet, A. and Landry, M.R. (2004) Phytoplankton growth, microzooplankton grazing, and carbon cycling in marine systems. *Limnology and Oceanography*, **49**: 51–7.

Callaway, R.M. and Ridenour, W.M. (2004) Novel weapons: invasive success and the evolution of increased competitive ability. *Frontiers in Ecology and the Environment*, **2**: 436–43.

Cappuccino, N. and Arnason, J.T. (2006) Novel chemistry of invasive exotic plants. *Biology Letters*, **2**: 189–93.

Carpenter, R.C. (1986) Partitioning herbivory and its effects in coral-reef alagal communities. *Ecological Monographs*, **56**: 345–63.

Cembella, A.D. (2003) Chemical ecology of eukaryotic microalgae in marine ecosystems. *Phycologia*, **42**: 420–47.

Chan, L.L., Hodgkiss, I.J., Wan, J.M. *et al.* (2004) Proteomic study of a model causative agent of harmful algal blooms, *Prorocentrum triestinum* II: the use of differentially expressed protein profiles under different growth phases and growth conditions for bloom prediction. *Proteomics*, **4**: 3214–26.

Cipollini, D., Purrington, C.B., and Bergelson, J. (2003) Costs of induced responses in plants. *Basic and Applied Ecology*, **4**: 79.

Cohen, S.A.P., Hatt, H., Kubanek, J., and McCarty, N.A. (2008) Reconstitution of a chemical defense signaling pathway in a heterologous system. *Journal of Experimental Biology*, **211**: 599–605.

Cohen, S.P., Haack, K.K., Halstead-Nussloch, G.E. *et al.* (2010) Identification of RL-TGR, a coreceptor involved in aversive chemical signaling. *Proceedings of the National Academy of Science USA*, **107**: 12339–44

Coleman, R.A., Ramshunder, S.J., Davies, K.M., Moody, A.J., and Foggo, A. (2007) Herbivore-induced infochemicals influence foraging behaviour in two intertidal predators. *Oecologia*, **151**: 454–63.

Coley, P.D., Bryant, J.P., and Chapin, F.S. (1985) Resource availability and plant antiherbivore defense. *Science*, **230**: 895–9.

Colin, S.P. and Dam, H.G. (2003) Effects of the toxic dinoflagellate *Alexandrium fundyense* on the copepod *Acartia hudsonica*: a test of the mechanisms that reduce ingestion rates. *Marine Ecology Progress Series*, **248**: 55–65.

Cronin, G. (2001) Resource allocation in seaweeds and marine invertebrates: chemical defense patterns in relation to defense theories. In J.B. McClintock and B.J. Baker, (eds.) *Marine Chemical Ecology*. CRC Press, Boca Raton, pp. 325–53.

Cronin, G. and Hay, M. E. (1996b) Induction of seaweed chemical defenses by amphipod grazing. *Ecology*, **77**: 2287–301.

Cronin, G. and Hay, M.E. (1996a) Within-plant variation in seaweed palatability and chemical defenses: optimal defense theory versus the growth-differentiation balance hypothesis. *Oecologia*, **195**: 361–8.

Cruz-Rivera, E. and Hay, M.E. (2003) Prey nutritional quality interacts with chemical defenses to affect consumer feeding and fitness. *Ecological Monographs,* **73**: 483–506.

Cyr, H. and Pace, M.L. (1992) Grazing by zooplankton and its relationship to community structure. *Canadian Journal of Fisheries and Aquatic Science,* **49**: 1455–65.

Cyr, H. and Pace, M.L. (1993) Magnitude and patterns of herbivory in aquatic and terrestrial ecosystems. *Nature,* **361**: 148–50.

Darwin, C. (1859) *On the Origin of Species by Means of Natural Selection.* John Murray, London.

Dawes, C.J. (1998) *Marine Botany, 3rd edition.* John Wiley & Sons, New York.

deBose, J.L., Lema, S.C., and Nevitt, G.A. (2008) Dimethylsulfoniapropionate as a foraging cue for reef fishes. *Science,* **319**: 1356.

Denton, A.B. and Chapman, A.R.O. (1991) Feeding preferences of gammarid amphipods among 4 species of *Fucus. Marine Biology,* **109**: 503–6.

Dicke, M. (2009) Behavioural and community ecology of plants that cry for help. *Plant, Cell and Environment,* **32**: 654–65.

Dixson, D.L., Munday, P.L., and Jones, G.P. (2010) Ocean acidification disrupts the innate ability of fish to detect predator olfactory cues. *Ecology Letters,* **13**: 68–75.

Duffy, J.E. and Hay, M.E. (1990) Seaweed adaptations to herbivory: Chemical, structural, and morphological defenses are adjusted to spatial or temporal patterns of attack. *Bioscience,* **40**: 368–75.

Duffy, J.E. and Hay, M.E. (2000) Strong impacts of grazing amphipods on the organization of a benthic community. *Ecological Monographs,* **70**: 237–63.

Dworjanyn, S.A., Wright, J.T., Paul, N.A., de Nys, R., and Steinberg, P.D. (2006) Cost of chemical defence in the red alga *Delisea pulchra. Oikos,* **113**: 13–22.

Elton, C.S. (1958) *The Ecology of Invasion by Animals and Plants.* Chapman and Hall, London.

Feeny, P.P. (1976) Plant apparency and chemical defense. *Recent Advances in Phytochemistry,* **10**: 1–40.

Ferrer, E., Garreta, A.G., and Ribera, M.A. (1997) Effect of *Caulerpa taxifolia* on the productivity of two Mediterranean macrophytes. *Marine Ecology Progress Series,* **149**: 279–87.

Fraenkel, G.S. (1959) The raison d'être of secondary plant substances. *Science,* **129**: 1466–70.

Frantzis, A. and Gremare, A. (1993) Ingestion, absorption, and growth-rates of *Paracentrotus lividus* (Echinodermata, Echinoidea) fed different macrophytes. *Marine Ecology Progress Series,* **95**: 169–83.

Haavisto, F., Välikangas, T., and Jormalainen, V. (2010) Induced resistance in a brown alga: phlorotannins, genotypic variation and fitness costs for the crustacean herbivore. *Oecologia,* **162**: 685–95.

Harper, M.K., Bugni, T.S., Copp, B.R, *et al.* (2001) Introduction to the chemical ecology of marine natural products. In J.B. McClintock, and B.J. Baker, (eds.) *Marine Chemical Ecology.* CRC Press, Boca Raton, pp. 3–69.

Haslam, E. (1985) *Metabolites and Metabolism.* Clarendon Press, Oxford.

Hay, M.E. (1996) Marine chemical ecology: What's known and what's next? *Journal of Experimental Marine Biology and Ecology,* **200**: 103–34.

Hay, M. E. (2009) Marine chemical ecology: chemical signals and cues structure marine populations, communities, and ecosystems. *Annual Review of Marine Science,* **1**: 193–212.

Hay, M.E. and Kubanek, J. (2002) Community and ecosystem level consequences of chemical cues in the plankton. *Journal of Chemical Ecology,* **28**(10).

Hay, M.E., Duffy, J.E., and Fenical, W. (1990) Host-plant specialization decreases predation on a marine amphipod - an herbivore in plants clothing. *Ecology,* **71**: 733–43.

Hay, M.E., Kappel, Q.E., and Fenical, W. (1994) Synergisms in plant defenses against herbivores: interactions of chemistry, calcification, and plant quality. *Ecology,* **75**: 1714–26.

Hay, M.E., Pawlik, J.R., Duffy, J.E., and Fenical, W. (1989) Seaweed-herbivore-predator interactions—host-plant specialization reduces predation on small herbivores. *Oecologia,* **81**: 418–27.

Hay, M.E. and Steinberg, P.D. (1992) The chemical ecology of plant-herbivore interactions in marine versus terrestrial communities. In G.A. Rosenthal and M.R. Berenbaum, (eds.) *Herbivores: Their Interactions with Secondary Plant Metabolites.* Academic Press, New York, pp. 371–413.

Hayes, K., Sliwa, C., Migus, S., McEnnulty, F., and Dunstan, P. (2005) National priority pests—part II: ranking of Australian marine pests, *CSIRO Marine Research for the Australian Department of Environment and Heritage,* Parkes, Australia. Available at http://www.environment.gov.au/coasts/publications/imps/priority2.html [accessed 5 October 2011].

Heil, M. and Baldwin, I.T. (2002) Fitness costs of induced resistance: emerging experimental support for a slippery concept. *Trends In Plant Science,* **7**: 61–7.

Herms, D.A. and Mattson, W.J. (1992) The dilemma of plants: to grow or to defend. *The Quarterly Review of Biology,* **67**: 283–335.

Hoover, C.A., Slattery, M., Targett, N.M., and Marsh, A.G. (2008) Transcriptome and metabolite responses to predation in a South Pacific soft coral. *Biological Bulletin*, **214**: 319–28.

Ianora, A., Boersma, M., Cassoti, R. *et al.* (2006) New trends in marine chemical ecology. *Estuaries and Coasts*, **29**: 531–51.

Ianora, A. and Miralto, A. (2010) Toxigenic effects of diatoms on grazers, phytoplankton and other microbes: a review. *Ecotoxicology*, **19**: 493–511.

Inderjit, Callaway, R.M., and Vivanco, J.M. (2006) Can plant biochemistry contribute to understanding of invasion ecology? *Trends in Plant Science*, **11**: 574–80.

Jang, M.H., Ha, K., Joo, G.J., and Takamura, N. (2003) Toxin production of cyanobacteria is increased by exposure to zooplankton. *Freshwater Biology*, **48**: 1540–50.

Jang, M.H., Ha, K., Lucas, M.C., Joo, G.-J., and Takamura, N. (2004) Changes in microcystin production by *Microcystis aeruginosa* exposed to phytoplanktivorous and omnivorous fish. *Aquatic Toxicology*, **68**: 51–9.

Jeffries, M. (1990) Evidence of induced plant defences in a pondweed. *Freshwater Biology*, **23**: 265–9.

Jensen, K. (1997) Evolution of the sacoglossa (Mollusca, Opisthobranchia) and the ecological associations with their food plants. *Evolutionary Ecology*, **11**:301–35.

Jonasdottir, S.H., Kiorboe, T., Tang, K.W. *et al.* (1998) Role of diatoms in copepod production: good, harmless or toxic? *Marine Ecology Progress Series*, **172**: 305–8.

Jordt, S.E. and Julius, D. (2002) Molecular basis for species-specific sensitivity to 'hot' chili peppers. *Cell*, **108**: 421–30.

Jormalainen, V. and Ramsay, T. (2009) Resistance of the brown alga *Fucus vesiculosus* to herbivory. *Oikos*, **118**: 713–22.

Jousson, O., Pawlowski, J., Zaninetti, L. *et al.* (2000) Invasive alga reaches California. *Nature*, **408**: 157–8.

Jung, V., Thibaut, T., Meinesz, A., and Pohnert, G. (2002) Comparison of the wound-activated transformation of caulerpenyne by invasive and noninvasive *Caulerpa* species of the Mediterranean. *Journal of Chemical Ecology*, **28**: 2091–105.

Karban, R. and Baldwin, I.T. (1997) *Induced Responses to Herbivory*. University of Chicago Press, Chicago.

Keane, R.M. and Crawley, M.J. (2002) Exotic plant invasions and the enemy release hypothesis. *Trends in Ecology and Evolution*, **17**: 164–70.

Koricheva, J. (2002) Meta-analysis of sources of variation in fitness costs of plant antiherbivore defenses. *Ecology*, **83**: 176–90.

Kubanek, J., Snell, T.W., and Pirkle, C. (2007) Chemical defense of the red tide dinoflagellate *Karenia brevis* against rotifer grazing. *Limnology and Oceanography*, **52**: 1026–35.

Kvitek, R.G., Degange, A.R., and Beitler, M.K. (1991) Paralytic shellfish poisoning toxins mediate feeding-behavior of sea otters. *Limnology and Oceanography*, **36**: 393–404.

Landsberg, J.H. (2002) The effects of harmful algal blooms on aquatic organisms. *Reviews in Fisheries Science*, **10**: 113–390.

Lee, R.E. (1999) *Phycology*. Cambridge University Press, Cambridge.

Lemoine, D.G., Barrat-Segretain, M.H., and Roy, A. (2009) Morphological and chemical changes induced by herbivory in three common aquatic macrophytes. *International Review of Hydrobiology*, **94**: 282–9.

Lerdau, M. and Coley, P.D. (2002) Benefits of the carbon-nutrient balance hypothesis. *Oikos*, **98**: 534–6.

Levine, J.M., Vila, M., D'Antonio, C.M., Dukes, J.S., Grigulis, K., and Lavorel, S. (2003) Mechanisms underlying the impacts of exotic plant invasions. *Proceedings of the Royal Society of London Series B-Biological Sciences*, **270**: 775–81.

Lewis, S.M. (1986) The role of herbivorous fishes in the organization of a Caribbean reef community. *Ecological Monographs*, **53**: 183–200.

Lodge, D.M., Cronin, G., Van Donk, E., and Froehlich, A.J. (1998) *Impact of Herbivory on Plant Standing Crop: Comparison Among Biomes, Between Vascular and Nonvascular plants, and Among Freshwater Herbivore Taxa*. Springer Verlag, New York.

Long, J.D., Hamilton, R.S., and Mitchell, J.L. (2007) Asymmetric competition via induced resistance: specialist herbivores indirectly suppress generalist preference and populations. *Ecology*, **88**: 1232–40.

Loomis, W.E. (1932) Growth-differentiation balance vs. carbohydrate-nitrogen ratio. *Proceedings of the American Society for Horticultural Science*, **29**: 240–5.

Lubchenco, J. and Gaines, S.D. (1981) A unified approach to marine plant-herbivore interactions. I. Populations and communities. *Annual Review of Ecology and Systematics*, **12**: 405–37.

Macaya, E. and Thiel, M. (2008) In situ test on inducible defenses in *Dictyota kunthii* and *Macrocystis integrifolia* (Phaeophyceae) from the Chilean coast. *Journal of Experimental Marine Biology and Ecology*, **354**: 28–38.

Mack, R.N., Simberloff, D., Lonsdale, W.M., Evans, H., Clout, M., and Bazzaz, F.A. (2000) Biotic invasions: Causes, epidemiology, global consequences, and control. *Ecological Applications*, **10**: 689–710.

Maggs, C.A. and Stegenga, H. (1999) Red algal exotics on North Sea coasts. *Helgoländer Meeresuntersuchungen*, **52**: 243–58.

Maron, J.L. and Vila, M. (2001) When do herbivores affect plant invasion? Evidence for the natural enemies and biotic resistance hypotheses. *Oikos*, **95**: 361–73.

Meinesz, A. (1999) *Killer Algae*. The University of Chicago Press, Chicago and London.

McClintock, J.B. and Baker, B.J. (2001) *Marine Chemical Ecology*. CRC Press, Washington, D.C.

Molis, M., Enge, A., and Karsten, U. (2010) Grazing impact of, and indirect interactions between mesograzers associated with kelp (*Laminaria digitata*). *Journal of Phycology*, **46**: 76–84.

Molis, M., Körner, J., Ko, Y.W., and Kim, J.H. (2008) Specificity of inducible seaweed anti-herbivory defences depends on identity of macroalgae and herbivores. *Marine Ecology Progress Series*, **354**: 97–105.

Morrison, W.E. and Hay, M.E. (2011) Induced chemical defenses in a freshwater macrophyte suppress herbivore fitness and the growth of associated microbes. *Oecologia*, **165**: 427–36.

Moustafa, A., Evans, A.N., Kulis, D.M. *et al.* (2010) Transcriptome profiling of a toxic dinoflagellate reveals a gene-rich protist and a potential impact on gene expression due to bacterial presence. *PloS ONE*, **5**: e9688.

Nevitt, G.A. and Haberman, K. (2003) Behavioral attraction of Leach's storm-petrels (*Oceanodroma leucorhoa*) to dimethyl sulfide. *Journal of Experimental Biology*, **206**: 1497–501.

Nylund, G.M., Cervin, G., Persson, F., Hermansson, M., Steinberg, P.D., and Pavia, H. (2008) Seaweed defence against bacteria: a poly-brominated 2-heptanone from the red alga *Bonnemaisonia hamifera* inhibits bacterial colonisation. *Marine Ecology Progress Series*, **369**: 39–50.

Parker, J.D., Burkepile, D.E., and Hay, M.E. (2006) Opposing effects of native and exotic herbivores on plant invasions. *Science*, **311**: 1459–61.

Parker, J.D. and Hay, M.E. (2005) Biotic resistance to plant invasions? Native herbivores prefer non-native plants. *Ecology Letters*, **8**: 959–67.

Paul, C., Barofsky, A., Vidoudez, C., and Pohnert, G. (2009) Diatom exudates influence metabolism and cell growth of co-cultured diatom species. *Marine Ecology Progress Series*, **389**: 61–70.

Paul, N.A., de Nys, R., and Steinberg, P.D. (2006b) Chemical defence against bacteria in the red alga *Asparagopsis armata*: linking structure with function. *Marine Ecology Progress Series*, **306**: 87–101.

Paul, V.J., Arthur, K.E., Ritson-Williams, R., Ross, C., and Sharp, K. (2007) Chemical defenses: From compounds to communities. *Biological Bulletin*, **213**: 226–51.

Paul, V.J. and Fenical, W. (1986) Chemical defense in tropical green-algae, order *Caulerpales*. *Marine Ecology Progress Series*, **34** (1–2): 157–69.

Paul, V.J. and Puglisi, M.P. (2004) Chemical mediation of interactions among marine organisms. *Natural Product Reports*, **21**: 189–209.

Paul, V.J., Puglisi, M.P., and Ritson-Williams, R. (2006a) Marine chemical ecology. *Natural Product Reports*, **23**: 153–80.

Paul, V.J. and Ritson-Williams, R. (2008) Marine chemical ecology. *Natural Product Reports*, **25**: 662–95.

Pavia, H. and Toth, G.B. (2000a) Influence of light and nitrogen on the phlorotannin content of the brown seaweeds *Ascophyllum nodosum* and *Fucus vesiculosus*. *Hydrobiologia*, **440**: 299–305.

Pavia, H. and Toth, G.B. (2000b) Inducible chemical resistance to herbivory in the brown seaweed *Ascophyllum nodosum*. *Ecology*, **81**: 3212–25.

Pavia, H. and Toth, G.B (2007) Macroalgal models in testing and extending defense theories. In C.D. Amsler, (ed.) *Algal Chemical Ecology*. Springer, Berlin Heidelberg, pp. 147–72.

Pavia, H., Toth, G., and Åberg, P. (1999) Trade-offs between phlorotannin production and annual growth in natural populations of the brown seaweed *Ascophyllum nodosum*. *Journal of Ecology*, **87**: 761–71.

Pavia, H., Toth, G.B., and Åberg, P. (2002) Optimal defense theory: elasticity analysis as a tool to predict intraplant variation in defenses. *Ecology*, **83**: 891–7.

Pfister, C.A. (1992) Costs of reproduction in an intertidal kelp - patterns of allocation and life-history consequences. *Ecology*, **73**: 1586–96.

Pimentel, D., McNair, S., Janecka, J. *et al.* (2001) Economic and environmental threats of alien plant, animal, and microbe invasions. *Agriculture Ecosystems & Environment*, **84**: 1–20.

Pohnert, G., Steinke, M., and Tollrian, R. (2007) Chemical cues, defence metabolites and the shaping of pelagic interspecific interactions. *Trends in Ecology and Evolution*, **22**: 198–204.

Poulson, K.L., Sieg, R.D., and Kubanek, J. (2009) Chemical ecology of the marine plankton. *Natural Product Reports*, **26**: 729–45.

Prince, E. K., Lettieri, L., Mccurdy, K. J., and Kubanek, J. (2006) Fitness consequences for copepods feeding on a red tide dinoflagellate: deciphering the effects of nutritional value, toxicity, and feeding behavior. *Oecologia*, **147**: 479–88.

Prince, E.K. and Pohnert, G. (2010) Searching for signals in the noise: metabolomics in chemical ecology. *Analytical and Bioanalytical Chemistry*, **396**: 193–7.

Purrington, C.B. (2000) Costs of resistance. *Current Opinion in Plant Biology*, **3**: 305.

Ragan, M. and Glombitza, K. (1986) Phlorotannins, brown algal polyphenols. *Progress in Phycological Research*, **4**: 129–241.

Rasher, D.B. and Hay, M.E. (2010) Chemically rich seaweeds poison corals when not controlled by herbivores.

Proceedings of the National Academy of Science USA, **107**: 9683–8.

Reichardt, P.B., Chapin, F.S. III, Bryant, J.P., Mattes, B.R., and Clausen, T.P. (1991) Carbon/nutrient balance as a predictor of plant defense in Alaskan balsam poplar: potential importance of metabolic turnover. *Oecologia,* **88**: 401–6.

Rhoades, D.F. (1979) Evolution of plant chemical defense against herbivores. In G.A. Rosenthal and D.H. Janzen, (eds.) *Herbivores: Their Interaction with Secondary Plant Metabolites.* Academic Press, New York, pp. 3–54.

Rhode, S. and Wahl, M. (2008a) Antifeeding defense in baltic macroalgae: induction by indirect grazing versus waterborne cues. *Journal of Phycology,* **44**: 85–90.

Rhode, S. and Wahl, M. (2008b) Temporal dynamics of induced resistance in marine macroalgae: Time lag of induction and reduction in *Fucus vesiculosus. Journal of Experimental Marine Biology and Ecology,* **367**: 227–9.

Selander, E., Cervin, G., and Pavia, H. (2008) Effects of nitrate and phosphate on grazer-induced toxin production in *Alexandrium minutum. Limnology and Oceanography,* **53**: 523–30.

Selander, E., Thor, P., Toth, G., and Pavia, H. (2006) Copepods induce paralytic shellfish toxin production in marine dinoflagellates. *Proceedings of the Royal Society of London B,* **273**: 1673–80.

Sotka, E.E., Forbey, J., Horn, M., Poore, A.G.B., Raubenheimer, D., and Whalen, K. E. (2009) The emerging role of pharmacology in understanding consumer-prey interactions in marine and freshwater systems. *Integrative and Comparative Biology,* **49**: 291–313.

Sotka, E. E. and Hay, M. E. (2002) Geographic variation among herbivore populations in tolerance for a chemically rich seaweed. *Ecology,* **83**: 2721–35.

Stamp, N. (2003) Out of the quagmire of plant defense hypotheses. *The Quarterly Review of Biology,* **78**: 23–55.

Steinberg, P.D. (1995) Seasonal variation in the relationship between growth rate and phlorotannin production in the kelp *Ecklonia radiata. Oecologia,* **102**: 169–73.

Steinberg, P.D., de Nys, R., and Kjelleberg, S. (2001) Chemical mediation of surface colonization. In J.B. McClintock and B.J. Baker, (eds.) *Marine Chemical Ecology.* CRC Press, Boca Raton, pp. 355–87.

Strauss, S.Y., Rudgers, J.A., Lau, J.A., and Irwin, R.E. (2002) Direct and ecological costs of resistance to herbivory. *Trends In Ecology and Evolution,* **17**: 278–85.

Taylor, R.B., Lindquist, N., Kubanek, J., and Hay, M.E. (2003) Intraspecific variation in palatability and defensive chemistry of brown seaweeds: effects on herbivore fitness. *Oecologia,* **136**: 412–23.

Thibaut, T. and Meinesz, A. (2000) Are the Mediterranean ascoglossan molluscs *Oxynoe olivacea* and *Lobiger serradifalci* suitable agents for a biological control against

the invading tropical alga *Caulerpa taxifolia? Comptes Rendus de l'Académie des Sciences - Series III - Sciences de la Vie,* **323**: 477–88.

Tillmann, U. (2004) Interactions between planktonic microalgae and protozoan grazers. *Journal of Eukaryotic Microbiology,* **51**: 156–68.

Tollrian, R. and Harvell, C.D. (1999) *The Ecology and Evolution of Inducible Defenses.* Princeton University Press, Princeton.

Toth, G.B., Langhamer, O., and Pavia, H. (2005) Inducible and constitutive defenses of valuable seaweed tissues: Consequences for herbivore fitness. *Ecology,* **86**: 612–18.

Toth, G.B. and Pavia, H. (2007) Induced herbivore resistance in seaweeds: a meta-analysis. *Journal of Ecology,* **95**: 425–34.

Tscharntke, T. and Hawkins, B.A. (2002) *Multitrophic Level Interactions.* Cambridge University Press, Cambridge.

Van Alstyne, K.L., McCarthy, J.J III, Hystead, C.L. and Kearns, L.J. (1999) Phlorotannin allocation among tissues of Northeastern Pacific kelps and rockweeds. *Journal of Phycology,* **35**, 483–492.

Van Alstyne, K.L. (2008) Ecological and physiological roles of dimethyl sulfoniopropionate and its products in marine macroalgae. In C.D. Amsler, (ed.) *Algal Chemical Ecology.* Springer-Verlag, Berlin, Heidelberg, pp. 173–94.

Van Gremberghe, I., Vanormelingen, P., Van der Gucht, K. *et al.* (2009) Influence of *Daphnia* infochemicals on functional traits of *Microcystis* strains (Cyanobacteria). *Hydrobiologia,* **635**: 147–55.

Van Oosten, V.R., Bodenhausen, N., Reymond, P. *et al.* (2008) Differential effectiveness of microbially induced resistance against herbivorous insects in *Arabidopsis. Molecular Plant-Microbe Interactions,* **21**: 919–30.

Weissflog, J., Welling, M., Adolph, S., Jung, V., and Pohnert, G. (2009) Wound healing and chemical defense in siphonous green algae. *Phycologia,* **48**: 307.

Vergés, A., Pérez, M., Alcoverro, T., and Romero, J. (2008) Compensation and resistance to herbivory in seagrasses: induced responses to simulated consumption by fish. *Oecologia,* **155**: 751–60.

Whalen, K.E., Lane. A.L., Kubanek, J., and Hahn, M.E. (2010) Biochemical warfare on the reef: the role of glutathione S-transferases in consumer tolerance of dietary prostaglandins. *PLoS ONE,* **5**: e8537.

Wiesemeier, T., Jahn, K., and Pohnert, G. (2008) No evidence for the induction of brown algal chemical defense by the phytohormones jasmonic acid and methyl jasmonate. *Journal of Chemical Ecology,* **34**: 1523–31.

Wikström, S.A., Steinarsdottir, M.B., Kautsky, L., and Pavia, H. (2006) Increased chemical resistance explains low herbivore colonization of introduced seaweed. *Oecologia,* **148**: 593–601.

Wylie, C.R. and Paul, V.J. (1988) Feeding preferences of the surgeonfish *Zebrasoma flavescens* in relation to chemical defenses of tropical algae. *Marine Ecology Progress Series*, **45**: 23–32.

Yates, J.L. and Peckol, P. (1993) Effects of nutrient availability and herbivory on polyphenolics in the seaweed *Fucus vesiculosus*. *Ecology*, **74**: 1757.

Yun, H.Y., Cruz, J., Treitschke, M., Wahl, M., and Molis, M. (2007) Testing for the induction of anti-herbivory defences in four Portuguese macroalgae by direct and water-borne cues of grazing amphipods. *Helgoland Marine Research*, **61**: 203–9.

Zangerl, A.R. and Bazzaz, F.A. (1992) Theory and pattern in plant defense allocation. In R.S. Fritz and E.L. Simms, (eds.) *Plant Resistance to Herbivores and Pathogens*. Chicago Press, Chicago, pp. 363–91.

Zimmer, R.K. and Butman, C.A. (2000) Chemical signalling processes in the marine environment. *Biological Bulletin*, **198**: 168–87.

Zulak, K.G., Lippert, D.N., Kuzyk, M.A. *et al.* (2009) Targeted proteomics using selected reaction monitoring reveals the induction of specific terpene synthases in a multi-level study of methyl jasmonate-treated Norway spruce (*Picea abies*). *Plant Journal*, **60**: 1015–30.

CHAPTER 16

Chemical defences against predators

Cynthia Kicklighter

16.1 Characteristics of defended organisms

The interactions between predators and their prey can have significant impacts on the structure and composition of communities. The strong influence of predation is illustrated by the 'life–dinner' principle (Dawkins and Krebs 1979): if a predator doesn't capture and consume a prey individual, the predator loses out on a meal, but can likely reproduce in the future; however, if a prey is eaten by a predator, the cost is losing life and any chance of future reproduction. Thus, the selective pressure on prey species to evolve anti-predation strategies can be significant. As a result, numerous aquatic species are under selection to produce, sequester, or associate with other organisms that produce chemical defences that deter predators from consuming them. The majority of chemically defended species are marine invertebrates, with chemical defences demonstrated in sponges, corals, hydroids, bryozoans, ascidians, platyhelminths, nemerteans, annelids, hemichordates, echinoderms, molluscs, arthropods, phoronids, and brachiopods (Paul *et al.* 2006). Defended freshwater invertebrates include beetles and bugs (Scrimshaw and Kerfoot 1987). Some marine fish (e.g. Scorpaenidae, pufferfishes, and gobies) and freshwater amphibians use chemical defences as well (Schubert *et al.* 2003; Williams 2010; Andrich *et al.* 2010).

The majority of defended species are soft-bodied, overt, sessile species, such as sponges, gorgonians, and ascidians. These species can be easily detected by predators and have no means of escape, likely selecting for the production of chemical defences. Some of these taxonomic groups contain a high proportion of defended species and

surveys have demonstrated that 69% of tropical Atlantic sponges, 100% of tropical Pacific sponges, 100% of Mediterranean sponges, 78% of Antarctic sponges, 94% of ascidians, 100% of gorgonians, and 95% of echinoderms are chemically deterrent to at least some predators (reviewed in Kicklighter and Hay 2006; Peters *et al.* 2009).

For some species, aposematic colouring is correlated with a chemical defence. This has been demonstrated in marine worms (Kicklighter and Hay 2006), large larvae of marine invertebrates (Lindquist and Hay 1996), gobies (Schubert *et al.* 2003), and Rittson-Williams and Paul (2007) found that reef fish will avoid brightly-coloured contrasting colour patterns similar to the colouration of some chemically defended nudibranchs.

This chapter is not meant to be an exhaustive review, but attempts to highlight some of the areas of recent study on chemical defences that deter predators (for reviews focused on induced defences and alarm cues, see Chapters 8, 9, and 12). Because the diversity of species and number of studies is greater for marine species, especially invertebrates, the majority of examples come from these groups.

16.2 Source of defences

16.2.1 Sequestration

Because the production of chemical defences can be physiologically costly (reviewed in Leong and Pawlik 2010), one strategy for prey is to sequester defensive metabolites from their diet. This appears to be most common in marine organisms, especially in opisthobranch molluscs, primarily sacoglossans (Becerro 2001) and sea hares (Ginsburg and Paul 2001). These species most often sequester chemicals from sponges, ascidians, cnidarians, seaweeds, and, in some cases, bryozoans (Faulkner 1992). For example, Becerro *et al.* (2006) examined the opisthobranchs *Sagaminopteron nigropunctatum* and *S. psychedelicum*. These species are generally found on the sponge *Dysidea granulosa*, which contains three polybrominated diphenyl ether metabolites, with the major one being 3,5-dibromo-2-(20,40-dibromophenoxy) phenol. Gas chromatography-mass spectrometry (GC-MS) examination of the molluscs revealed that they sequestered this compound in their tissues, some of which (parapodia) contained more than twice the concentration of the sponge itself. 3,5-Dibromo-2-(20,40-dibromophenoxy) phenol has been shown to deter feeding by a wide range of predators, including pufferfish, xanthid crabs, and a sea hare (reviewed in Becerro *et al.* 2006), even at concentrations below those found in *S. nigropunctatum* and *S. psychedelicum*. Becerro *et al.* (2006) found that the pufferfish *Canthigaster solandri* rejected *S. nigropunctatum*, suggesting that 3,5-dibromo-2-(20,40-dibromophenoxy) phenol defends this opisthobranch from predation. In addition, this compound was found in the mucus secretion these two opisthobranchs release when disturbed, which is thought to mediate rejection by fish predators. *Sagaminopteron nigropunctatum* also delivers 3,5-dibromo-2-(20,40-dibromophenoxy) phenol to its egg masses, which may also deter predation (Becerro *et al.* 2006).

One cost of sequestration is that availability of sequestered chemicals may vary qualitatively and/or quantitatively, which may impact defences against predators. For example, Fahey and Garson (2002) examined the nudibranch *Asteronotus cespitosus* from three geographically separate locations in the tropical Pacific. Similar to the study above, they identified in the nudibranchs several secondary metabolites found in the sponge *D. herbacea*, including the predator-deterrent 3,5-dibromo-2-(20, 40-dibromophenoxy) phenol. This was the case for two of the three populations, but the third population did not contain this metabolite and instead contained 3,5-dibromo-2-(30,50-dibromo-20-methoxyphenoxy) phenol (also a known metabolite from *D. herbacea*). Whether this metabolite deters preda-

tion, however, and whether this nudibranch population is deterrent to predators, is unknown.

In many cases, compounds sequestered from prey are not modified, but in some cases consumers may alter the structure of sequestered metabolites, either to enhance deterrence or as a detoxification mechanism. For example, Kamio *et al.* (2010a, b) investigated the sea hares *Aplysia californica* and *A. dactylomela*, which consume red algae and sequester the red photosynthetic pigment phycoerythrin. Phycoerythrin is composed of phycobiliprotein covalently linked to the chromophore, phycoerythrobilin. Sea hares cleave the covalent bond in phycoerythrin first, forming phycoerythrobilin in the digestive gland. Then, this metabolite is methylated in the ink gland to form aplysioviolin. Aplysioviolin and, to a lesser extent, phycoerythrobilin, deter feeding by blue crabs (*Callinectes sapidus*), which is the first demonstration of an animal converting a photosynthetic pigment into a chemical deterrent.

Sequestration of chemical defences can also occur via non-dietary sources. For example, Fattorini *et al.* (2010) found that vanadium is sequestered from seawater by the Antarctic sabellid polychaete *Perkinsiana littoralis* and is found in the highest concentrations in the branchial crown (some of the most exposed tissues of the worm). When portions of the worm body and branchial crown were separately offered to the rock cod *Trematomus bernacchii*, fish always consumed the body, but rejected the branchial crown, suggesting that the vanadium may deter feeding. Similarly, the Mediterranean sabellid *Sabella spallanzanii* sequesters arsenic in high concentrations in its branchial crown and feeding experiments demonstrated that the branchial crown is also rejected by consumers, while the body was palatable (Notti *et al.* 2007).

16.2.2 *De novo* production of defences

Most chemical defences are thought to be produced by the animals that possess them, which alleviates the ecological cost of going without defences when suitable prey cannot be found, as well as issues associated with quantitative and qualitative variation in defences. The production of a chemical defence, however, is thought to be energetically costly (reviewed in Leong and Pawlik 2010). Studies demonstrating costs to chemically defended animals are rare, but two studies with Caribbean sponges lend some support to this hypothesis. Pawlik *et al.* (2008) examined the recruitment of sponges to an artificial reef and found that undefended sponges recruited more quickly than chemically defended sponges. Four years after submergence, the sponge fauna was dominated by undefended, palatable species (6 of 8 species), while adjacent, natural reefs were dominated by chemically defended, unpalatable species (6 of 10 species). The authors suggest that this difference may be due to differences in recruitment, such that chemically defended species have less energy to allocate to reproduction than do palatable, undefended species. In addition, Leong and Pawlik (2010) examined growth of defended versus undefended sponges inside enclosed cages to exclude predators. They found that the growth inside cages of two of three undefended, palatable sponge species was significantly greater than growth of three of four chemically defended, unpalatable species inside cages (Fig. 16.1). Defended and undefended sponges

were also grown in the absence of cages and the growth of chemically defended sponges inside and outside of cages was similar for three of four species, and growth of undefended sponges inside cages was significantly greater than outside cages for all three species investigated, demonstrating that undefended species were consumed by fishes (Fig. 16.1).

In many cases, it has been assumed, rather than experimentally demonstrated, that defences are produced *de novo* by organisms from which chemicals are extracted. This is due to the fact that many defended species do not feed upon chemically defended prey from which they could sequester secondary metabolites. Thus, *de novo* production of chemical defences appears to be common in deposit- and filter-feeding species, such as sponges, gorgonians, ascidians, annelids, and hemichordates, but some studies have demonstrated that defences in these groups can be produced by bacterial symbionts (see below).

While many opisthobranch molluscs sequester chemical defences from their prey, some dorid nudibranch species are thought to produce their

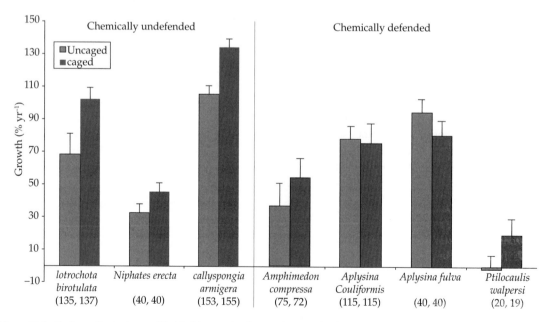

Figure 16.1 Yearly growth rates (mean + SE) of Caribbean coral reef sponges in uncaged and caged treatments on reefs off Key Largo, Florida. Number of replicates (uncaged, caged) are shown below each species name. Figure from Leong and Pawlik (2010), used with permission from Inter-Research.

own metabolites. In one of the few studies to experimentally demonstrate *de novo* production, Kubanek and colleagues (Kubanek *et al.* 1997) investigated secondary metabolite production using stable isotope incorporation experiments in the Northeastern Pacific nudibranch *Cadlina luteomarginata*. While this nudibranch does sequester metabolites from its sponge diet, three terpenoid compounds were found in this species that did not exhibit geographic variation (typical of metabolites that are diet-derived) and were shown to be produced *de novo*. One compound, albicanyl acetate, deters fish feeding, but the other two were not bioassayed for chemical defensive function.

Many *de novo* defences are constitutive (including most of the examples in this chapter), where defensive chemicals are always present in the organism. This is thought to be energetically costly, so, for some species, consumption by predators initiates the production of a chemical defence (= an induced

chemical defence). This has been demonstrated in several species of algae (see Chapter 15), but no examples exist for animals. Another defence strategy is to use an activated defence, in which molecules lacking predator deterrent properties are immediately converted to ones that deter predators upon wounding by a consumer. One advantage to this strategy is that the potential cost of possessing autotoxic compounds is avoided. For example, when the hydroid *Tridentata marginata* is damaged, dithiocarbamates (tridentatols E-H) in nematocysts react via an enzymatic reaction to form tridentatols A-D, which are deterrent to fish consumers (Lindquist 2002a). As another example, in the sponge *Aplysinella rhax*, psammaplin A sulfate is converted to psammaplin A (likely via an enzymatic reaction), which deters feeding by the pufferfish *Canthigaster solandri* (Thoms and Schupp 2008).

The sea hare *Aplysia californica* also utilizes an activated defence, through the mixing of components

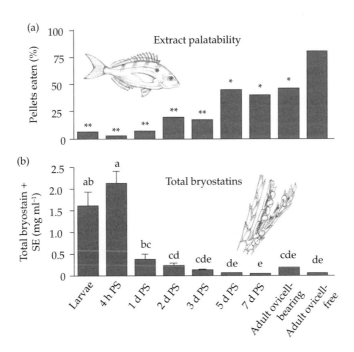

Figure 16.2 Predation rate (percentage pellets eaten) and total bryostatin content of *Bugula neritina* over time. (a) Palatability of crude extracts to the fish *Lagodon rhomboids* (N = 14–18 fish tested). Significance of difference from control (McNemar's test for significance of change; *P < 0.01; **P < 0.001). (b) Concentration of total bryostatins (mg ml^{-1}) measured by analytical HPLC in volumetric extracts (N = 3 replicate samples). Letters indicate significant differences in concentrations (1-way ANOVA, log transformed data, Tukey's HSD, p < 0.05). PS = post-settlement. Figure from Lopanik *et al.* (2006), used with permission from Inter-Research.

in two separate glands. When *A. californica* is attacked, it releases an ink secretion from an ink gland and opaline from the opaline gland. When these two secretions mix, an L-amino acid oxidase protein (escapin) in the ink reacts with L-lysine in opaline to form hydrogen peroxide (among other products), which depresses feeding and handling time by blue crabs *Callinectes sapidus* (Kamio *et al.* 2010a). In addition, when intermediate products resulting from this reaction are mixed with hydrogen peroxide and added to a palatable food, feeding by señorita and bluehead wrasses (*Oxyjulis californica* and *Thalassoma bifasciatum*, respectively) is depressed (Nusnbaum and Derby 2010).

16.2.3 Chemical defences produced by symbionts

While the chemical defences of numerous species are thought to be produced by the organism in which they are found, several studies have demonstrated, however, that defensive chemicals are actually produced by microbial symbionts. For example, Lopanik *et al.* (2006) found that the larvae and juveniles of the bryozoan *Bugula neritina* are chemically

defended against the pinfish *Lagodon rhomboides* by three bryostatins, which are produced by the bacterial symbiont *Candidatus* Endobugula sertula. Predation pressure on the large, overt larval stage of *B. neritina* may have selected for the establishment and maintenance of this symbiotic relationship. After the larvae settle and begin to metamorphose into adults, palatability of extracts increases and bryostatin concentration decreases over time, demonstrating that that adult *B. neritina* are not chemically defended, but may deter predation through their structural aspects (Fig. 16.2).

Microbial symbionts and their hosts may both participate in the production of chemical defences. One of the most common marine endosymbionts is zooxanthellae, which associate with most corals, including gorgonians (octocorals). In some cases, zooxanthellae may produce secondary metabolites, such as sterols. The gorgonian *Pseudoterogoria americana* is defended by 9,11-secogorgosterol and 9,11-secodinosterol (Epifanio *et al.* 2007), and previous work found that enzyme preparations of whole *P. americana* colonies were capable of transforming gorgosterol into labelled 9,11-secogorgosterol (Kerr *et al.* 1996). Because zooxanthellae can produce

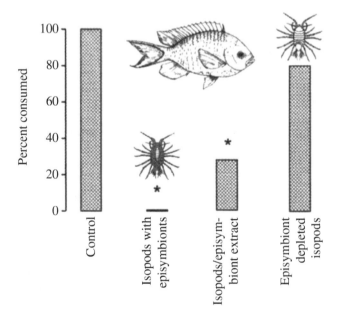

Figure 16.3 Results of *in situ* feeding assays testing the palatability of: (A) Red isopods with their episymbiotic microbial community; (B) A crude extract of the red isopod/microbial association; and (C) Episymbiont depleted red isopods. A single control bar is shown for comparison with each assay result to simplify the graphical presentation of the data because consumption of control pellets was 100% for all assays, n = 10 replicates per assay. Figure from Lindquist *et al.* (2005), used with permission from The Royal Society.

gorgosterol, it is thought that this compound is oxidized by *P. americana* to secosterols (i.e. 9,11-secogorgosterol and 9,11-secodinosterol), which provide protection to both the gorgonian and zoothanthellae.

Symbionts may not only produce chemical defences that protect their host, they may also provide a nutritional benefit. This is the case for the marine isopod *Santia* sp. and its episymbiotic microbes (primarily *Synechococcus*-type cyanobacteria), examined by Lindquist and co-workers (2005). The bright-red symbionts colonize the dorsal carapace of the isopods. Isopods consume the symbionts and facilitate their growth by staying in exposed, sunlit areas. Isopods avoid predation through the protection of defensive, unidentified metabolites by the symbionts. Feeding assays demonstrated that reef fishes rejected 100% of isopods possessing symbionts, versus rejecting only 20% of isopods lacking symbionts (Fig. 16.3). An organic extract of isopods + symbionts resulted in 70% rejection of food pellets laced with extract (Fig. 16.3).

Tetrodotoxin is a potent neurotoxin that blocks sodium channels and can function as a chemical defence, in addition to facilitating prey capture and functioning in intra- and interspecific communication. It is often found on the skin of several species of pufferfishes, blue ringed octopus *Octopus maculosus*, a goby (*Yongeichthys criniger*), a flatworm (*Planocera multitentaculata*), and some salamanders (reviewed in Williams 2010). For some species, tetrodotoxin is thought to be produced by symbiotic bacteria, and in other cases it may be accumulated through diet, or even produced *de novo*, as is the case for the rough skinned newt *Taricha granulosa*, but tetrodotoxin origins are often debated (reviewed in Williams 2010).

16.3 Secreted defences

Most chemical defences function passively, localized in body tissues, likely repelling predators when they taste prey. Some species, however, actively defend themselves against predators by releasing chemical defence secretions when attacked. As described above, sea hares of the genus *Aplysia* release two secretions—ink and opaline—from two separate glands when attacked by predators. Not

only do ink and opaline secretions create a defence when mixed together, they also separately function as chemical defences. When examined separately, ink and opaline from *A. californica*, *A. dactylomela*, *A. kurodai*, and *A. brasiliana* have been shown to deter feeding by birds, fishes, crustaceans, and sea anemones (reviewed in Kicklighter and Derby 2006; Nusnbaum and Derby 2010). Nolen *et al.* (1995) demonstrated that ink facilitated the rejection of *Aplysia californica* when fed to the anemone *Anthopleura xanthogrammica*, but when sea hares were depleted of their ink and opaline stores, they were eaten. In an attempt to elucidate the defensive chemicals that deterred feeding by anemones, Kicklighter and Derby (2006) found that both polar and non-polar unidentified metabolites in ink caused tentacle shriveling and retraction in *A. sola*, which presumably facilitate rejection of sea hares by anemones. Conversely, opaline elicited a feeding response by anemones (Kicklighter and Derby 2006).

Sea hare defensive secretions also arrest attack by the spiny lobster *Panulirus interruptus*. In this case, ink and opaline divert the attention of the predator by mimicking the stimulatory properties of food (phagomimicry) via the presence of amino acids, and by possibly disrupting predator sensory systems (Kicklighter *et al.* 2005; Chapter 13). In addition, opaline contains a polar, non-proteinaceous component that inhibits feeding (Kicklighter *et al.* 2005). Phagomimicry may also be one strategy by which ink protects cephalopods from predators (Derby *et al.* 2007), but not for the Caribbean reef squid *Sepioteuthis sepiodea*, whose ink appears to deter feeding (Wood *et al.* 2010).

Some defensive secretions are less dramatic, but just as effective. Gobies of the genus *Gobiodon* release large amounts of a frothy white mucus from the skin when disturbed and this has been shown to cause loss of equilibrium in other fishes and to deter feeding on a palatable food, and on the gobies themselves (Schubert *et al.* 2003). In addition, some goby species exhibit variation in toxicity of the secretion, as drably coloured gobies that reside deep within coral branches are less toxic (as determined by an equilibrium assay), while gobies that are more overt and brightly coloured possessed a more toxic secretion. Thus, species most susceptible to predation possessed the strongest defence. The chemical(s) responsible for these predator effects were not investigated, but it is possible that the secretion contains tetrodotoxin, as another goby (*Yongeichthys criniger*) is known to contain this toxin in its skin and some pufferfish secrete this toxin in their mucus (Williams 2010).

Many freshwater bugs (Hemiptera) and beetles (Coleoptera) release defensive secretions from a variety of glands, either as an ooze or through forcible discharge (Scrimshaw and Kerfoot 1987). Eisner and Aneshansley (2000) investigated the whirligig beetle *Dineutes hornii* (Fig. 16.4) and found that when attacked by largemouth bass, *Micropterus salmoides*, the beetle releases a secretion, prompting the fish to exhibit an oral flushing behaviour. This activity most likely functions to rid the oral cavity of the secretion, however, even when exhibiting this behaviour, the fish always rejected the beetles, uninjured. The metabolite eliciting flushing behaviour was determined to be the norsesquiterpene gyrinidal, as mealworms coated with this metabolite

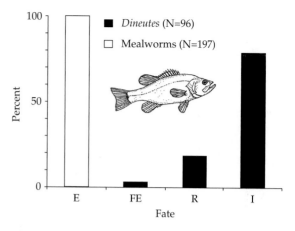

Figure 16.5   Fate of live *Dineutes* and of untreated mealworms offered to bass (data from six fish lumped). E, eaten outright; FE, eaten after flushing; R, rejected outright or after flushing; I, ignored. Figure from Eisner and Aneshansley (2000), used with permission from *Proceedings of the National Academy of Sciences, USA*.

prompted flushing behaviour (Fig. 16.5). Interestingly, the beetle releases the secretion slowly, such that the average duration of release is longer than the average oral flush time of the fish. Härlin (2005) conducted a more recent study of the whirligig beetles and investigated a species that contains only norsesquiterpenes in its defensive secretion (*Gyrinus minutus*) and *G. aeratus*, whose secretion contains norsesquiterpenes plus volatile components, such as 3-methyl-1-butanal and 3-methyl-l-butanol. In feeding assays, *G. minutus*, whose defensive secretion lacked volatile metabolites, was consumed by 65% of rainbow trout (*Onchorhynchus mykiss*), versus *G. aeratus*, which was consumed by only 4.5% of fish. In addition, when experienced fish were offered these beetles, they rejected *G. aeratus* more quickly than *G. minutus*, suggesting that the volatile components of the secretion may make beetles more distasteful or that fish remember the volatiles more readily than the norsesquiterpenes.

16.4   Defences of early developmental stages

The vast majority of chemical defence investigations have been conducted on adult life stages of organisms, despite the fact that many freshwater and marine vertebrates and invertebrates lay eggs that are unprotected and exhibit larval stages that

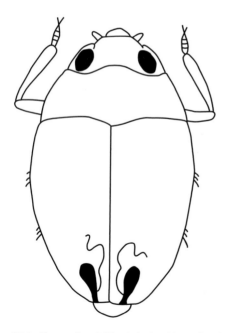

Figure 16.4   Diagram of a gyrinid beetle (such as *Dineutes hornii*) showing defensive glands in rear of abdomen. Glands consist of a pair of sac-like reservoirs in which the secretion is stored. Figure from Eisner and Aneshansley (2000), used with permission from *Proceedings of the National Academy of Sciences*, USA.

differ greatly in mobility and morphology from adult forms. The few studies that have been conducted have mainly focused on marine invertebrates, such as sponges, ascidians, and bryozoans. These investigations show that species that produce large, often brooded, larvae that are released during the day and spend little, if any, time in the plankton before settling are most commonly chemically defended (Lindquist 2002b). For example, Lopanik *et al.* (2006) found that larvae of the bryozoan *Bugula neritina* were chemically defended from generalist fish predators by bryostatins, but adults were not. Despite the fact that a wide variety of terrestrial and freshwater insects and terrestrial egg and larval stages have been shown to be chemically defended, no defended aquatic egg or larval stages have been identified to date.

Recently, McClintock *et al.* (2003) examined chemical defences against the sea star *Odontaster validus* of the larvae of two Antarctic sea stars (*Neosmilaster georgianus* and *Lysasterias perrieri*), both of which produce large, lecithotrophic larvae. In addition, extracts of the brooded juveniles of an isopod (*Glyptonotus antarcticus*) were examined. As has been found for other species with large and/or brooded larvae, chemical extracts of larvae of the sea star *L. perrieri* and juveniles of the isopod *G. antarcticus*, deterred consumption. The extract of the larvae of *N. georgianus* was palatable, however, despite the fact that whole larvae were rejected by the sea star predator. The authors speculate that defensive chemicals may not have been extracted efficiently from larvae or that defences are allocated to the exterior surface, in which case their bioassays would have tested extracts at too low a concentration.

Other studies have also failed to detect deterrent larval extracts. For example, Tarjuelo *et al.* (2002) examined the palatability to fishes and crustaceans of the larval stages of six ascidian species in the Mediterranean Sea. Larvae of four of the species (*Cystodytes dellechiajei*, *Polysysncraton lacazei*, *Diplosoma spongiforme*, and *Pseudodistoma crucigaster*) were large and unpalatable, while the smaller larvae of *Clavelina lepadiformis* and *Ecteinascidia herdmanni* were palatable. When organic extracts of the unpalatable larvae were assayed using crustacean consumers, deterrence was not detected for any species. Thus, some lar-

vae may possess unstable metabolites that decompose upon extraction or are not efficiently extracted by the methods employed by investigators.

While the use of tetrodotoxin as a chemical defence by some adult salamanders and newts has been investigated by several researchers, a study by Hanifin *et al.* (2003) examined defences in eggs of the newt *Taricha granulosa*. Their results demonstrated that there was a direct correlation between the concentration of tetrodotoxin in female skin and the concentration found in her eggs, suggesting that the tetrodotoxin found in the eggs is maternally derived. Tetrodotoxin concentration was not found to vary with egg size, suggesting that the incorporation of this defensive compound in eggs is independent of the deposition of other egg resources, such as lipids or other provisions.

16.5   Integration of a chemical defence with other strategies

When the production of chemical defences is thought to be costly it may be beneficial for species to utilize only one type of defence—one best suited to the most common or dangerous predator. Thus, some species may trade-off the use of a chemical defence with other defensive strategies, such as structural aspects, behaviour, or refuge use. For example, Stachowicz and Lindquist (2000) examined the palatability of six hydroids off the coast of North Carolina and found that species deterred a fish consumer by either stinging nematocysts or chemical deterrents, but not with both.

While most chemically defended species are sessile, some defended species do have some capability for mobility and may exhibit a trade-off between chemical defence and behavioural escape. For example, marine sabellid polychaete worms dwell in tubes and extend their feather-like radioles for feeding and respiration. The body remains inside the tube and the radioles retract when threatened. Kicklighter and Hay (2007) examined eight species and found that all species had unpalatable, often chemically-defended radioles, while only two species had chemically defended bodies. For these two species, *Bispira brunnea* and *B. variegata*, they had weak tubes, they did not retract until disturbances were very close, and *B. brunnea* retracted slowly and incompletely, even when touched. Thus, they relied less on behavioural escape and refuge use than the other species, but were more strongly chemically defended than the other species. Variation in the integration of anti-predation strategies even occurs intraspecifically within this group. A North Carolina population of *Bispira variegata* had a body that was chemically-defended against a temperate and tropical fish (mummichog and bluehead wrasse, respectively), while the body of a Panama population was readily consumed by both these fishes. In addition, the Panama population was induced to retract in response to a distant movement, while the unpalatable North Carolina population did not retract until nearly touched, again, suggesting a trade-off between defence and escape strategies. The use of a defensive strategy appeared to correlate with predation pressure. The North Carolina population was collected from a subtidal jetty, where consumer pressure can be high, while the Panama population was collected around mangroves, where both densities and species richness of predators are often lower than on reefs.

More commonly, however, multiple defensive strategies are integrated, which may be selected for in environments where consumer diversity is high and modes of attack and consumer preferences are diverse. For example, sponge spicules and chemicals can act synergistically to reduce feeding. Hill *et al.* (2005) found that feeding by the hermit crab *Pagurus longicarpus* on *Microciona prolifera* was only deterred when both crude extract and spicules were added to an artificial food, but not when either was tested alone. Jones *et al.* (2005) also detected synergistic interactions between spicules and chemicals in four of eight sponges investigated. In addition, a survey of Western Atlantic ascidians by Pisut and Pawlik (2002) found that 16 of 17 species were defended by chemical extracts deterrent to the blue head wrasse *Thalassoma bifasciatum*. In addition, nine of these species also produced inorganic acids of a pH ≤ 3, which is an acidity level effective at deterring *T. bifasciatum*.

16.6 Variation in defences

The optimal defence theory asserts that within an organism, costly defences should be allocated to tissues in relation to the vulnerability and value of the tissues to the fitness of the organism. Furrow *et al.* (2003) investigated the Antarctic sponge *Latrunculia apicalis* and found that it allocates 52% of the defensive metabolite discorhabdin G to the outermost (2 mm thick) tissues. This tissue caused tube feet of the sea star *Perknaster fuscus* to retract significantly longer than did inner tissues (8–10 mm depth). Because tube feet responses are related to feeding behaviour, it was assumed that retraction varies directly with deterrence. When inner tissues were spiked with a concentration of discorhabdin G similar to the concentration in the outer tissues, tube feet retracted for a similar amount of time as that elicited by outer tissues.

Similarly, the Caribbean sponge, *Ectoplasia ferox*, was found to vary in the concentration of defensive metabolites. When layers 0–2 mm, 2–6 mm, and 6–14mm in thickness were examined for defensive triterpene glycosides, Kubanek *et al.* (2002) demonstrated that defences were highest in the outermost layer and decreased in deeper layers. Thus, defences were allocated to surface layers, where they effectively served as deterrence against fish predators, in addition to allelopathic agents that prevented overgrowth by other sponge species.

Kicklighter and Hay (2006, 2007) found that for most sabellid and terebellid polychaete worms, exposed tissues (radioles and tentacles) were unpalatable and usually chemically defended, while palatable bodies remained inside tubes or in burrows. Defending exposed feeding appendages not only increases fitness through decreasing the likelihood of losing tissue to predation, but it may also increase fitness by allowing species greater freedom in terms of feeding.

Variation in the presence of chemical defences also varies among species and may be correlated with predation pressure. This has been demonstrated with sponges on temperate reefs in the South Atlantic Bight by Ruzika and Gleason (2009) who found that sponges growing on vertical, rocky outcroppings were more often chemically-defended than species living on the reef top, where there were fewer fish predators. When species from the reef top were transplanted to vertical areas, some species were consumed, suggesting that their lack of chemical defences relegated them to this area, where the abundance of spongivorous fishes was lower. At larger scales, some studies have documented variation in the presence or strength of chemical defences among species living in tropical versus temperate latitudes. Because the abundance and diversity of consumers is generally higher in the tropics, prey species are likely under stronger selective pressure to develop defences. Very few studies, however, have addressed this possibility for animals. Beccero *et al.* (2003) examined 20 sponge species and found that chemical extracts of species from the tropics (Guam) deterred the same fish consumers just as frequently as extracts from temperate (Northeast Spanish Coast) sponges, suggesting that temperate species are just as strongly defended as tropical species. Conversely, Ruzika and Gleason (2009) found that tropical sponge species living in a temperate location (South Atlantic Bight) exhibited a lower frequency of chemical deterrence to sympatric consumers than did the same tropical sponges living in tropical habitats. Clearly, more studies are needed to fully understand latitudinal variation in animal chemical defences.

16.7 Environmental effects

The environmental stress theory of resource allocation states that a stressed organism is less able to acquire resources and will allocate a greater proportion of these reduced resources to maintenance, compared to an unstressed organism. Thus, fewer resources are available for the production of defensive metabolites. Climate change and other human disturbances will likely impact the defensive capabilities of organisms, but few studies have addressed this issue. Coral bleaching is an indication of stress on the coral–zooxanthellae symbiosis and has been shown to negatively affect chemical defences. Slattery and Paul (2008) examined the impact of natural and artificially induced bleaching on the soft coral *Sinularia maxima*, off Guam. In both bleaching regimes, both lipid and chemical defence (pukalide) concentration decreased compared to unbleached individuals. When artificial foods containing pukalide and lipid content of bleached versus unbleached corals were offered to the pufferfish *Canthigaster solandri*, fish were not deterred from feeding on food mimicking bleached corals, but

were deterred by foods mimicking unbleached corals. Surprisingly, lower lipid content caused pufferfish to be less attracted to the food. Nevertheless, in the field, the number of bite marks on bleached corals was 4 times higher than the number on adjacent unbleached corals. Thus, bleaching decreases the production of a defensive metabolite, leading to increased predation on this coral.

As another example, chemical defences may facilitate the success of invading species, as consumers in the new habitat have not evolved mechanisms to tolerate invaders' defensive chemistry. This may be the case for two chemically-defended sea slugs (*Haminoea cyanomarginata* and *Melibe viridis*) that have been detected in the Mediterranean Sea (Mollo *et al.* 2008). In addition, an exotic soft coral, *Chromonephthea braziliensis*, has become established off the coast of Brazil and there is concern that it will become invasive. It contains a hemiketal steroid, 23-ketocladiellin A, which deters feeding by reef fishes sympatric with where this coral has invaded (Fleury *et al.* 2008). With little to no consumer regulation, this species may come to dominate Brazilian reefs. Similarly, the ahermatypic coral *Tubastrea coccinea* has invaded the Southwestern Atlantic and is chemically deterrent to native generalist fish (Lages *et al.* 2010).

16.8 Community-level effects

Predator defences may play critical roles in allowing unpalatable species to increase their densities in places or times when consumers are common and have suppressed the densities of more palatable competitors. As an example, the chemically-defended hemichordate *Saccoglossus kowalevskii* not only persists, but increases in abundance on mudflats during seasons when predation pressure is high; palatable worms in these habitats decline dramatically during the same time period (Kicklighter and Hay 2006). As another example, chemically-defended surface-feeding species, such as members of the polychaete family Terebellidae, may differentially affect local community structure by consuming settling larvae because their chemically defended tentacles allow them to forage for extensive periods of time and to extend their appendages far from burrow openings with lessened risk of loss. These species often include diatoms, other unicellular algae, and small invertebrates, including larvae, in their diets and these feeding patterns, coupled with their high densities, could allow their feeding to strongly impact meiofaunal communities.

Chemically defended species can be dominant competitors for resources, such as space, but their presence can facilitate species diversity if palatable species associate with these species and gain a refuge from predation. Thus, species that would be excluded from predator-rich habitats, such as coral reefs, can persist if in association with a deterrent species. A diversity of mesofauna often resides in and around sponges and may gain a refuge from predation through this association. Huang *et al.* (2008) investigated this possibility for the tropical sponge *Amphimedon viridis*. Mesofauna associating with this sponge were mainly crustaceans (82% of all fauna), and consisted mainly of amphipods, most of which resided on the sponge surface. To examine whether this sponge might protect amphipods from predation, small (2 mm) bite-sized pieces of sponge tissue (similar to the size that fish might ingest while attempting to consume an amphipod) were presented to pinfish, *Lagodon rhomboides*. Pinfish rejected 100% of sponge portions. Food pellets containing sponge spicules elicited weak, but significant deterrence, but sponge extract elicited much stronger deterrence. Thus, chemical defences (and, to a lesser extent, structural defences) mediate the unpalatability of this sponge to a fish predator, facilitating the association between mesofauna and *A. viridis*.

16.9   Conclusions

A vast number of studies have investigated the use of chemical defecses in animals against their predators. These studies have provided the basis for our understanding of the life history characteristics associated with the use of defence strategies, which suggest the pressures that select for the evolution of defences. Our knowledge of the use of chemical defences by freshwater, compared to marine species, however, is much less broad. Whether this may be due to the lower diversity of freshwater versus marine species, due to differences in the abundance and/or diversity of predators between these two systems that select for defences, or simply because

fewer studies have investigated freshwater species, is unknown. Thus, whether patterns elucidated for marine species for traits such as behaviour and morphology of defended species, and the integration of multiple defensive strategies, hold for freshwater species is unknown. Further emphasis on this group is required for comparative studies.

Our ability to understand the causes and consequences of variation in chemical defences within and among species is limited by our ability to confidently identify species. For example, it had been assumed that larvae of the bryozoan *Bugula neritina* varied in the presence of bryostatins, which protect them from consumption by fish predators. Larvae of individuals south of Cape Hatteras, North Carolina, USA are chemically defended against fish consumers, while those from more northern latitudes are palatable (Mcgovern *et al.* 2003). Recent research by Mcgovern and Hellberg (2003) using mitochondrial cytochrome oxidase c subunit I (COI) sequences as molecular markers, found that there are at least two cryptic species of *B. neritina*, corresponding to the presence/absence of larval chemical defence. In addition, these two subspecies differ in the microbial symbionts. The chemically defended subspecies associates with *Endobugula sertula*, which is known to produce bryostatins, but the undefended subspecies does not possess *E. sertula* symbionts and instead associates with a γ-proteobacteria. Thus, variation in palatability of larvae is not a result of local adaptation to variation in predation pressure, but due to cryptic species which harbour different symbionts. With increased application of molecular tools, we can expand our understanding of the prevalence of cryptic species and of species' phylogenies and how they may relate to patterns of chemical defence.

Finally, stress on aquatic species will continue to increase, due to impacts stemming from disturbances such as habitat degradation, invasive species, pollution, increased temperature, and ocean acidification. Results from the few studies that have investigated how stress impacts prey defences, suggest that predator–prey interactions will be altered. Thus, our ability to understand how community structure and function may be impacted due to these disturbances is currently limited by our understanding of how stress may influence chemical defences.

References

Andrich, F., J.B.T. Carnielli, J.S. Cassoli, R.Q. Lautner, R.A.S. Santos, A.M.C. Pimenta, M.E. de Lima, and S.G. Figueiredo. (2010) A potent vasoactive cytolysin isolated from *Scorpaena plumieri* scorpionfish venom. *Toxicon*, **56**: 487–96.

Beccerro, M.A., Thacker, R.W., Turon, X., Uriz, M.J., and V.J. Paul (2003) Biogeography of sponge chemical ecology: comparisons of tropical and temperate defenses. *Oecologia*. **135**: 91–101.

Dawkins, R. and Krebs, J.R. (1979) Arms races between and within species. *Proceedings of the Royal Society of London B*, **205**: 489–511.

Derby, C.D., Kicklighter, C.E. , Johnson, P.M., and Xu, Z. X. (2007) Chemical composition of inks of diverse marine molluscs suggests convergent chemical defenses. *Journal of Chemical Ecology*, **33**: 1105–13.

Epifanio, R.A., Maia1, L.F., Pawlik, J.R. and Fenical, W. (2007) Antipredatory secosterols from the octocoral *Pseudopterogorgia Americana*. *Marine Ecology Progress Series*, **329**: 307–10.

Eisner T. and Aneshansley, D.J. (2000) Chemical defense: Aquatic beetle (*Dineutes hornii*) vs. fish (*Micropterus salmoides*). *Proceedings of the National Academy of Science*, **97**: 11313–18.

Fahey, S.F. and Garson, M.J. (2002) Geographic variation of natural products of the tropical nudibranch *Asteronotus cespitosus*. *Journal of Chemical Ecology*, **28**: 1773–85.

Faulkner, D.J. (1992) Chemical defenses in marine molluscs. In V.J. Paul, (ed.) *Ecological Roles of Marine Natural Products*. Cornell University Press, Ithaca. pp. 119–163.

Fattorini, D, Notti, A., Nigro, M., and Regoli, F. (2010) Hyperaccumulation of vanadium in the Antarctic polychaete *Perkinsiana littoralis* as a natural chemical defense against predation. *Environmenal Science Pollution Research*, **17**: 220–8.

Fleury, B.G., Lages, B.G., Barbosa, J.P., Kaiser, C.R., and Pinto, A.C. (2008) New hemiketal steroid from the introduced soft coral *Chromonephthea braziliensis* is a chemical defense against predatory fishes. *Journal of Chemical Ecology*, **34**: 987–993.

Furrow, F.B., Amsler, C.D., McClintock, J.B., and Baker, B.J. (2003) Surface sequestration of chemical feeding deterrents in the Antarctic sponge *Latrunculia apicalis* as an optimal defense against sea star spongivory. *Marine Biology*, **143**: 443–9.

Ginsburg, D.W. and Paul, V.J. (2001) Chemical defenses in the sea hare *Aplysia parvula*: importance of diet and sequestration of algal secondary metabolites. *Marine Ecology Progress Series*, **215**: 261–74.

Hanifin, C.T., Brodie III, E.D., and Brodie, Jr. E.D. (2003) Tetrodotoxin levels in the eggs of the rough-skinned newt, *Taricha granulosa*, are correlated with female toxicity. *Journal of Chemical Ecology*, **29**: 1729–39.

Härlin, C. (2005) To have and have not: volatile secretions make a difference in gyrinid beetle predator defence. *Animal Behavior*, **69**: 579–85.

Hill M.S., Lopez N.A., and Young K.A. (2005) Antipredator defenses in western North Atlantic sponges with evidence of enhanced defense through interactions between spicules and chemicals. *Marine Ecology Progress Series*, **291**: 93–102.

Huang, J.P, McClintock, J.B., Amsler, C.D., and Huang, Y.M. (2008) Mesofauna associated with the marine sponge *Amphimedon viridis*. Do its physical or chemical attributes provide a prospective refuge from fish predation? *Journal of Experimental Marine Biology and Ecology*, **362**: 95–100.

Jones, A.C, Blum, J.E., and Pawlik, J.R. (2005) Testing for defensive synergy in Caribbean sponges: Bad taste or glass spicules? *Journal of Experimental Marine Biology and Ecology*, **322**: 67–81.

Kamio, M., Grimes, T.V., Hutchins, M.H., van Dam, R., and Derby, C.D. (2010a) The purple pigment aplysioviolin in sea hare ink deters predatory blue crabs through their chemical senses, *Animal Behavior*, **80**: 89–100.

Kamio, M., Nguyen, L., Yaldiz, S., and Derby, C.D. (2010b) How to produce a chemical defense: structural elucidation and anatomical distribution of Aplysioviolin and Phycoerythrobilin in the Sea Hare *Aplysia californica*. *Chemical Biodiversity*, **7**: 1183–97.

Kerr, R.G., Rodrigues L.C., and Kellman , J. (1996) A chemoenzymatic synthesis of 9(11)-secosteroids using an enzyme extract of the marine gorgonian *Pseudopterogorgia americana*. *Tetrahedron Letters*, **37**: 8301–4.

Kicklighter, C.E. and Derby, C.D. (2006) Multiple components in ink of the sea hare *Aplysia californica* are aversive to the sea anemone *Anthopleura sola*. *Journal of Experimental Marine Biology and Ecology*, **334**: 256–68.

Kicklighter, C.E. and Hay, M.E. (2006) Integrating prey defensive traits: contrasts of marine worms from temperate and tropical habitats. *Ecol. Monogr.* **76**: 195–215.

Kicklighter, C.E. and Hay, M.E. (2007) To avoid or deter: interactions among defensive and escape strategies in sabellid worms. *Oecologia*, **151**: 161–73.

Kicklighter, C.E., Shabani, S., Johnson, P.M., and Derby, C.D. (2005) Sea hares use novel antipredatory chemical defenses. *Current Biology*, **15**: 549–54.

Kubanek, J., Graziani, E.I., and Andersen, R.J. (1997) Investigations of terpenoid biosynthesis by the dorid nudibranch Cadlina luteomarginata. *Journal of Organic Chemistry*, **62**: 7239–7246.

Kubanek, K., Whalen, K.E., Engel, S., Kelly, S.R., Henkel, T.P., Fenical, W., and Pawlik, J.R. (2002) Multiple defensive roles for triterpene glycosides from two Caribbean sponges. *Oecologia*, **131**: 125–136.

Lages, B.G., Fleury, B.G., Pinto, A.C., Creed, J.C. (2010) Chemical defenses against generalist fish predators and fouling organisms in two invasive ahermatypic corals in the genus *Tubastraea*. *Journal of Chemical Ecology*, **31**: 473–82.

Leong, W. and Pawlik, J.R. (2010) Evidence of a resource trade-off between growth and chemical defenses among Caribbean coral reef sponges, *Marine Ecology Progress Series*, 406, 71–8.

Lindquist, N. (2002a) Tridentatols D–H, nematocyst metabolites and precursors of the activated chemical defense in the marine hydroid *Tridentata marginata* (Kirchenpauer 1864). *Journal of Natural Products*, **65**: 681–4.

Lindquist, N. (2002b) Chemical defense of early life stages of benthic marine invertebrates. *Journal of Chemical Ecology*, **28**: 1987–2000.

Lindquist, N., Barber, P.H., and Weisz, J.B. (2005) Episymbiotic Microbes as Food and Defence for Marine Isopods: Unique Symbioses in a Hostile Environment, *Proceedings of Royal Society B*, **272**: 1209–16.

Lindquist, N., and Hay, M. E. (1996) Palatability and chemical defense of marine larvae. *Ecological Monographs*, **66**: 431–50.

Lopanik, N.B., Targett, N.M., and Lindquist, N. (2006) Ontogeny of a symbiont-produced chemical defense in *Bugula neritina* (Bryozoa), *Marine Ecology Progress Series*, **327**: 183–91

McClintock, J.B., Mahon, A.R., Peters, K.J., Amsler, C.D., and Baker, B.J. (2003) Chemical defences in embryos and juveniles of two common Antarctic sea stars and an isopod, *Antarctic Science*, **15**: 339–44.

McGovern, T.M. and Hellberg, M.E. (2003) Cryptic species, cryptic endosymbionts, and geographical variation in chemical defences in the bryozoans *Bugula neritina*, *Molecular Ecology*, **12**: 1207–15.

Mollo, E., Gavagnin, M., Carbone, M., Castelluccio, F., Pozone, F., Roussis, V., Templado, J., Ghiselin, M.T., and Cimino, G. (2008) New hemiketal steroid from the introduced soft coral *Chromonephthea braziliensis* is a chemical defense against predatory fishes. *Journal of Chemical Ecology*, **34**: 987–93.

Nolen, T.G., Johnson, P.M., Kicklighter, C.E., and Capo, T. (1995) Ink secretion by the marine snail *Aplysia californica* enhances its ability to escape from a natural predator.

Journal of Comparative Physiology, A Sensory and Neural Behavior and Physiology, **176**: 239–54.

Notti, A, Fattorini, D., Razzetti, E.M., and Regoli, F. (2007) Bioaccumulation and biotransformation of arsenic in the Mediterranean polychaete *Sabella spallanzanii*: experimental observations. *Environmental Toxicology and Chemistry*, **26**: 1186–91.

Nusnbaum, M. and Derby, C.D. (2010) Effects of sea hare ink secretion and its escapin-generated components on a variety of predatory fishes. *Biological Bulletin*, **218**: 282–92.

Paul, V.J., Puglisi, M.P., and Rittson-Williams, R. (2006) Marine chemical ecology. *Nat. Prod. Rep.* **23**: 153–80.

Pawlik, J.R., Henkel, T.P., McMurray, S.E., López-Legentil, S., Loh, T., and Rohde, S. (2008) Patterns of sponge recruitment and growth on a shipwreck corroborate chemical defense resource trade-off. *Marine Ecology Progress Series*, **368**: 137–43.

Peters, K.J., Amsler, C.D., McClintock, J.B., van Soest, R.W.M., and Baker, B.J. (2009) Palatability and chemical defenses of sponges from the western Antarctic Peninsula. *Marine Ecology Progress Series*, **385**: 77–85.

Pisut, D.P. and Pawlik, J.R. (2002) Anti-predatory chemical defenses of ascidians: secondary metabolites or inorganic acids? *Journal of Experimental Marine Biology and Ecology*, **270**: 203–14.

Ritson-Williams, R. and Paul, V.J. (2007) Marine benthic invertebrates use multimodal cues for defense against reef fish. *Marine Ecology Progress Series*, **340**: 29–39.

Ruzika, R. and Gleason, D.F. (2009) Sponge community structure and anti-predator defenses on temperate reefs of the South Atlantic Bight. *Journal of Experimental Marine Biology and Ecology*, **380**: 36–46.

Scrimshaw, S. and Kerfoot, W.C. (1987) Chemical defences of freshwater organisms: beetles and bugs. In W.C. Kerfoot and A. Sih, (eds.) *Predation. Direct and Indirect Impacts on Aquatic Communities*. University Press of New England, Lebanon, pp. 240–62.

Schubert, M., Munday, P.L., Caley, M.J., Jones, G.P., and Llewellyn, L.E. (2003) The toxicity of skin secretions from coral-dwelling gobies and their potential role as a predator deterrent. *Environmental Biology of Fishes*, **67**: 359–67.

Stachowicz, J.J. and Lindquist, N. (2000) Hydroid defenses against predators: the importance of secondary metabolites versus nematocysts, *Oecologia*, **124**: 280–8.

Slattery, M. and Paul, V.J. (2008) Indirect effects of bleaching on predator deterrence in the tropical Pacific soft coral *Sinularia maxima, Marine Ecology Progress Series*, **354**: 169–79.

Tarjuelo, I, López-Legentil, S., Codina, M., and Turon, X. (2002) Defence mechanisms of adults and larvae of colonial ascidians: patterns of palatability and toxicity. *Marine Ecology Progress Series*, **235**: 103–15.

Thoms, C. and Schupp, P.J. (2008) Activated chemical defense in marine sponges—a case study on *Aplysinella rhax. Journal of Chemical Ecology*, **34**: 1242–52.

Williams, B.L. (2010) Behavioral and chemical ecology of marine organisms with respect to tetrodotoxin. *Marine Drugs*, **8**: 381–98.

Wood, J.B., Maynard, A.E., Lawlor, A.G., Sawyer, E.K., Simmons, D.M., Pennoyer, K.E., and Derby, C.D. (2010) Caribbean reef squid *Sepioteuthis sepioidea*, use ink as a defense against predatory French grunts, *Haemulon flavolineatum, Journal of Experimental Marine Biology and Ecology*, **388**: 20–7.

CHAPTER 17

Infodisruption: pollutants interfering with the natural chemical information conveyance in aquatic systems

Miquel Lürling

Communication is the one essential property for life (De Loof 1993). Indeed, information-bearing compounds are universally present in the biosphere. The use of chemicals as carriers of information is a common, widespread means of endogenous and exogenous signalling, ranging from well-documented chemical signalling systems inside cells and within organisms that control cell and organ activity and steer development, to chemical communication between organisms (e.g. Birch 1974; Tollrian and Harvell 1999; McLachlan 2001; Ward 2004; Dicke and Takken 2006; Fig. 17.1). In eukaryotic cells, the genetic information flow from DNA to protein assimilation regions in the cytoplasm is directed through mRNA, and many other chemicals, such as Ca^{2+}, cyclic AMP, and G proteins, may serve as intracellular messengers triggering responses to extracellular signals. In algae, even the chloroplast talks (Jarvis 2001, 2003). Intensive chemical communication between cells in multicellular organisms serves to control growth, tune development, and coordinate function and organization into tissues. The intricate web of hormones forms perhaps the most renowned chemical signalling pathways. On a higher organizatorial level, organisms use chemical messengers intensively in tuning their behaviour (Fig. 17.1).

All organisms utilize chemical information about their environment, and for many organisms that lack sophisticated organs it may be the dominant sensory modality. Organisms use chemical messengers intensively to locate food, find a mate, recognize close kin, mark a territory, and detect enemies (e.g. Dicke and Takken 2006). Chemicals that transfer information between organisms may be considered metabolic products that leak to the environment and fortuitously convey information. Such compounds are referred to as *infochemicals*, which are chemicals that, in the natural context, convey information between two organisms, evoking in the receiving organism a behavioural or physiological response that is adaptively favourable to one or both organisms (Dicke and Sabelis 1988). Infochemicals influence the temporal and spatial distribution of organisms, creating a tight connection between the information network and food web functioning (Dicke 2006). In fact, the food web is overlaid with an even more complex infochemical web that includes both chemically-mediated trophic, and indirect non-trophic, interactions (Dicke 2006; Chapter 10). The natural infochemicals form a complex 'smellscape'.

17.1 Infodisrupting chemicals

Anthropogenic activities have resulted in the discharge of many manmade chemicals, of which many, such as heavy metals, pesticides, plasticizers, and surfactants, are omnipresent in the environment. These chemicals have the potential for deregulating natural chemical communication systems (Fig. 17.2). The resulting maladaptive responses can have consequences for the individual organisms, populations, and—through biological interactions—for the community and

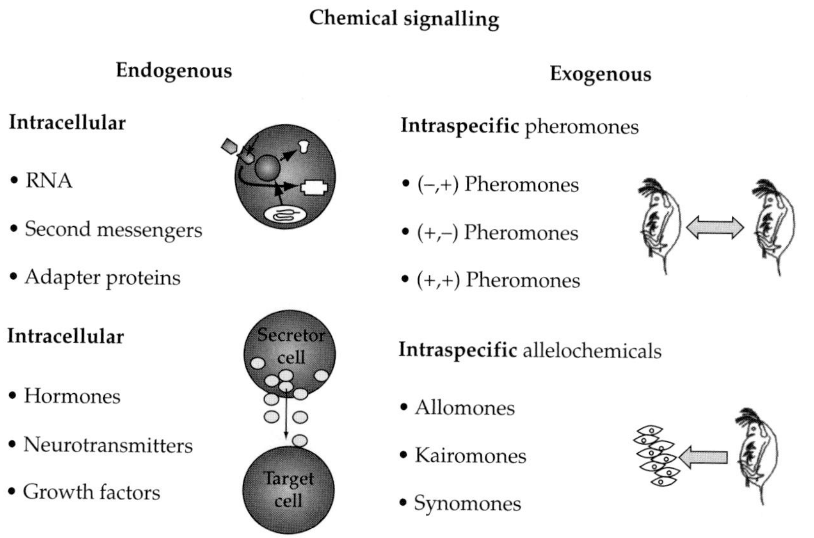

Chemical signalling

Endogenous	Exogenous

Intracellular

- RNA
- Second messengers
- Adapter proteins

Intraspecific pheromones

- (−,+) Pheromones
- (+,−) Pheromones
- (+,+) Pheromones

Intracellular

- Hormones
- Neurotransmitters
- Growth factors

Intraspecific allelochemicals

- Allomones
- Kairomones
- Synomones

Figure 17.1 Chemical signalling is ubiquitous and vital for coordination of functions at different levels within organisms (endogenous signalling) and in communicating between organisms (exogenous signalling). (See text for further explanation).

ecosystem functioning (Hanazato 1999, 2001). Aquatic environments may be especially at risk, as lakes, rivers, and even marine habitats are confronted with noticeable concentrations of numerous chemical substances brought in by humans.

Traditionally, the impact of pollutants is quantified by determining the doses that cause an X% of inhibition or mortality ($IC_x/EC_x/LC_x$) (Newman 2001). The fundamental basis is the sigmoid dose-response model, in which the response of isolated cells (*in vitro*), eggs (*in ovo*), or organisms (*in vivo*) to varying doses of a chemical is used as a measure of toxicity. The derived ecotoxicological data are then extrapolated to estimate ecological effects (Calow and Forbes 2003). However, high-dose exposure studies might be inadequate in predicting low-dose effects (e.g. vom Saal *et al*. 1997). Dose–response relationships might be non-linear (e.g. Calabrese and Baldwin 2003), effects being expressed non-monotonically, being permanent and latent or transgenerational (Colborn and Thayer 2000; French *et al*. 2001). In fact, there is a growing body of evidence on 'unusual' dose-response curves, including U- and inverted U-shaped ones (Calabrese and Baldwin 2003). For example, exposure to low and

environmentally realistic concentrations of the herbicide atrazine resulted in hermaphroditic and demasculinized leopard frogs (*Rana pipiens*), but this did not occur at high doses (Hayes *et al*. 2002). Because the low-dose effects that resulted in developmental and reproductive abnormalities could be mediated by the endocrine systems, the term 'endocrine disruption' was launched (Colborn and Clement 1992; Colborn *et al*. 1993). Endocrine disrupting chemicals are sometimes called environmental estrogens, but their action is much wider than just mimicking estrogens (Clotfelter *et al*. 2004) and broader than causing over- or underproduction of hormones (Tabb and Blumberg 2006). There is a growing body of evidence pointing towards interference with the thyroid system (Brucker-Davis 1998) with far reaching consequences (Colborn and Thayer 2000). Hormone action is only one of the intercellular communication means (Fig. 17.1).

As highlighted before, information networks not only exist inside organisms, but a sea of infochemicals, chemical messenger molecules, form the smellscape in which organisms live (Fig. 17.2). The environmental communication flows link numerous species in this network beyond the classical trophic interactions (Dicke 2006). Because pollutants

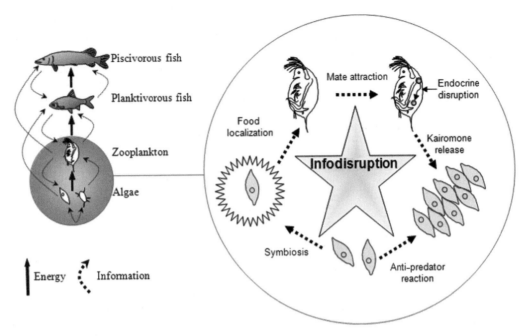

Figure 17.2 A typical pelagic food chain in which energy is transported up the trophic levels from algae to water fleas and fish by trophic interactions, and where chemical information flows between and within organisms, populations, communities and the various trophic levels. This infochemical network forms a 'smell-scape' in which the organisms live. Chemically-mediated interactions, of which examples are given for the plankton community, to the right, are potentially all vulnerable to pollutant distortions referred to as infodisruption.

might disrupt the information flow at several levels in the chemical signalling pathways/networks, the term *infodisrupters* has been proposed recently for this class of pollutants (Lürling and Scheffer 2007). In another review, Klaschka (2008) pointed out that the ecological importance of infodisruption, the 'infochemical effect', is much greater than previously was assumed. This is corroborated by recent comprehensive reviews on olfactory toxicity in fish (Tierney *et al.* 2010), infodisruption by metals (Boyd 2010), and fragrances in the environment (Klaschka 2007). Any chemically-mediated information system seems vulnerable to disruption by anthropogenic chemicals.

How pollutants might influence chemically-mediated biotic interactions is usually beyond the scope of the standard risk assessments. The plethora of information on endocrine disrupting chemicals on one side, and the ubiquity of natural chemical information transfer between organisms on the other side, suggests potential disturbance of the natural chemical information flows to be an obvious

ecological risk. In fact, the risk of pollutants affecting the chemical senses in aquatic animals was pointed out decades ago (Sutterlin 1974).

In this chapter, I will highlight the present understanding of the disruption of information transfer between aquatic organisms. The examples are concentrated around responses and behaviours that are crucial for population dynamics and food web functioning, including mating- and foraging behaviour and anti-predator reactions. For each information system, I will present the infodisruptive effects of four different categories of pollutants or environmental disturbances, namely pesticides, metals, hydrocarbons and, in addition, eutrophication, acidification, and increasing levels of humic acids.

17.2 Infodisruption of sex pheromone systems

Reproductively mature females of many species advertise mating receptivity by the production and release of a blend of species-specific pheromones (see Chapter 3). There is an increasing weight of evidence showing the disruption of the pheromone system on both the female production and the male reception/response side by a wide variety of compounds and factors (Table 17.1).

17.2.1 Pesticides

Pesticides are chemicals that are deliberately released into the environment for controlling pests and they may enter nearby surface waters by run-off during the spraying season and periods of high precipitation through leaching and/or accidental spills (Wauchope 1978). Pesticides are intended to disrupt a primary target in the pest so that it is no longer harmful (Casida 2009). Insecticides especially are designed to disrupt neurotransmission, primarily targeting the cholinergic system, sodium channels, and chloride channels (Casida 2009). Sub-lethal concentrations of the insecticide carbofuran reduced acetylcholinesterase (an enzyme regulating nerve-impulse transmission) in the olfactory organ of the common carp *Cyprinus carpio* (Haubruge and Toutant 1997). It is, therefore, not surprising that insecticides might interfere with pheromonal systems.

The insecticide endosulfan, for example, caused a significantly delayed response of male newts, *Notophthalmus viridescens*, to female odours, which resulted in a strong reduction in mating success (Park and Propper 2002). Further, when female newts were exposed to endosulfan, males exhibited a significantly lower response to their odours compared to odour from control females, and again mating-success was lowered considerably (Park *et al.* 2001). Measurements of the olfactory signal (electro-olfactogram) in mature male Atlantic salmon parr (*Salmo salar*) revealed that low concentrations of the insecticides diazinon, carbofuran, and cypermethrin, and of the herbicides atrazine and simazine, caused significant reductions in the ability to detect the female pheromones (Moore and Waring 1996, 1998, 2001; Waring and Moore 1997; Moore and Lower 2001). The exact mechanism of the herbicide-induced olfactory impairment is not known, but atrazine might inhibit acetylcholinesterase and could also act as a modifier of the endogenous androgen metabolism (Moore and Waring 1998). Herbicides such as

dichlobenil can interfere with the G-proteins that play a crucial role in signal transduction in olfaction through formation of the second messengers adenosine 3'5'-monophosphate (cAMP) and inositol 1,4.5-trisphosphate (InsP$_3$), and thus may affect the ability of exposed animals to respond to chemical cues (Andreini *et al.* 1997).

In the amphipod *Corophium volutator*, females release gender specific pheromones that guide males in their search for receptive burrowed females. However, the anti-fouling compound medetomidine significantly reduced pheromone-induced mate search in male amphipods (Krång and Dahlström 2006).

17.2.2 Metals

Metals can effectively block ion channels at the receptor site, but can also be transported along nerves into more central portions of the nervous system exerting disruptive effects (Sloman 2007; Boyd 2010; Tierney *et al.* 2010). Mud snails, *Nassa obsoleta*, from high imposex sites (indicating severe organotin pollution) no longer responded to gender specific pheromones (Straw and Rittschof 2004). Copper (0.1, 0.5 mg l^{-1}) affected the ability of male shore crabs (*Carcinus maenas*) to detect female pheromones, perform specific mating behaviours, and to form pairs (Krång and Ekerholm 2006).

17.2.3 Hydrocarbons

Pollution of the seawater by petroleum-derived aliphatic and aromatic hydrocarbons and oxygenated derivatives at levels of ng l^{-1} has been observed (Müller *et al.* 2005). A fraction in crude oil, consisting of C$_5$-alkylated benzenes, appeared more potent in inducing sperm release in male polychaetes (*Nereis succinea* and *Platynereis dumerilii*) than the natural sex pheromones (Müller *et al.* 2005). Naphthalenes also had a stimulatory effect on the polychaetes (Müller *et al.* 2005), but naphthalene (0.5–5 µg g^{-1} sediment) significantly reduced pheromone-induced mate search in male amphipods *C. volutator* (Krång 2007). Prudhoe Bay crude oil suppressed the reproductive chemotaxis of the dorid nudibranch *Onchidoris bilamellata* leading to reduced reproduction (cited in Blaxter and Ten Hallers-Tjabbes 1992).

Table 17.1   Chemically-mediated processes and examples of their disruption by pollutants (infodisruption).

Foraging

Pollutant	Disruption	Concentration	Reference
Hydrocarbons	Chemoreception of potential food sources is blocked in motile bacteria	0.6%	Mitchell et al. (1972)
Kerosene	Impaired/reduced chemically-mediated attraction to food extracts in the marine snails	1 µg l^{-1}	Jacobson and Boylan (1973)
Naphthalene	Reduced pheromone-induced mate search in male amphipods	0.5–5 µg g^{-1} sed.	Kräng (2007)
Prudhoe Bay crude oil	Reduced detection of chemical food cues in Dungeness crabs	0.013–0.34 ppm	Pearson et al. (1981)
Lo Rosa Crude oil	Reduced reaction in lobster to food cues	10 ppm	Atema and Stein (1974)
Heavy metals	Impaired response in crayfish to food cues		Steele et al. (1992)
Copper	Crayfish unable to locate food	0.02, 0.2 mg l^{-1}	Sherba et al. (2000)
	Leeches unable to locate food	5, 10 µg l^{-1}	Pyle and Mirza (2007)
	Inhibited chemo-attraction response of ciliates to yeast extract	50–200 µg l^{-1}	Roberts and Berk (1993)
Chromium	Impaired chemoreception of food in snails, enhanced response by zinc	30 mg l^{-1}	Cheung et al. (2002)
Manganese	Reduced chemically-induced food search behaviour in Norway lobster	~5.5–11 mg l^{-1}	Kräng and Rosenqvist (2006)
2,4-D Roundup®	Inhibited chemo-attraction responses of the ciliate to yeast extract	150–200 mg l^{-1} 25–100 ppm	Roberts and Berk (1993)
Irgarol 1051	Inhibition of chemoreception to detect food in snails	0.375–1.5 mg l^{-1}	Finnegan et al. (2009)
Metalochlor	Crayfish unable to locate food	25–75 µg l^{-1}	Wolf and Moore (2002)
Carbofuran	Decreased attraction of goldfish to chironomids filtrate	1–100 µg l^{-1}	Saglio et al. (1996)
pH	Low pH impaired the chemoreception of food in crayfish	pH 4.5	Allison et al. (1992)
pH	Suppressed feeding behaviour in palmate newts and smooth newts	pH 5	Griffiths (1993)
pH	Impaired food search in gold mollies	pH 5	Tembo (2009
Humic acids	Loss of attraction to food odors in female swordtails	20 mg l^{-1}	Fisher et al. (2006)

Mating

Pollutant	Disruption	Concentration	Reference
Diazinon	Impaired detection of female pheromones in male Atlantic salmon parr	1–20 µg l^{-1}	Moore and Waring (1996)
Carbofuran	Impaired detection of female pheromones in male Atlantic salmon parr	1–20 µg l^{-1}	Waring and Moore (1997)
Atrazine	Impaired detection of female pheromones in male Atlantic salmon parr	2–20 µg l^{-1}	Moore and Waring (1998)
Cypermethrin	Impaired detection of female pheromones in male Atlantic salmon parr	<0.004 µg l^{-1}	Moore and Waring (2001)
Simazine	Impaired detection of female pheromones in male Atlantic salmon parr	1 µg l^{-1}	Moore and Lower (2001)
Endosulfan	Impaired pheromone detection in male red-spotted newts and reduced production in females	0.5–5 µg l^{-1} 5, 10 µg l^{-1}	Park and Propper (2002) Park et al. (2001)
Medetomidine	Reduced pheromone-induced mate search in male amphipods	10 µg l^{-1}	Kräng and Dahlström (2006)
C$_5$-alkylated benzenes	Induce release of gametes in male Nereid polychaetes	4 nmol l^{-1}	Müller et al. (2005)
Nitrate	Reduced production of olfactory cues in male palmate newts lowers sexual attractiveness of male newts	75 mg l^{-1}	Secondi et al. (2009)
Copper	Male shore crabs fail to detect female pheromones, perform specific mating behaviours, and to form pairs	0.1, 0.5 mg l^{-1}	Kräng and Ekerholm (2006)
pH	Enhanced use of male cues in mate choice in gravid female threespined sticklebacks	pH 9.5	Heuschele and Candolin (2007)

Humic acids	Reduced olfactory response of goldfish to pheromones	1–1,000 mg l^{-1}	Hubbard et al. (2002)
	Impaired ability of female swordtails to respond to conspecific male chemical cues	20 mg l^{-1}	Fisher et al. (2006)
	Adult zebrafish lost ability to distinguish conspecific and heterospecific urinary chemical cues	200 mg l^{-1}	Fabian et al. (2007)

Induced defence

Pollutant	Disruption	Concentration	Reference
Triton-X	Bitterlings lost reaction to chemical cues from predators	0.5%	Kasumyan and Pashchenko (1985)
BPMC, Carbaryl Temephos Diazinon Fenitrothion	Helmet formation in *Daphnia*	5 µg l^{-1} 5 µg l^{-1} 4,6 µg l^{-1} 4 µg l^{-1} 20 µg l^{-1}	Hanazato (1991)
Carbaryl	Neckteeth prolongation and reduced *Daphnia* population growth		Hanazato and Dodson (1992)
	Neckteeth, high helmet, and tailspine formation in *Daphnia*		Hanazato and Dodson (1993)
	Higher helmets and longer helmeted period in *Daphnia* in presence of *Chaoborus* kairomone	5–20 µg l^{-1}	Hanazato and Dodson (1995)
	Interaction effects with *Anisops* kairomone on *Daphnia* life history	0.1–3.2 µg l^{-1}	Barry (1999)
	Inhibition of defensive morphology in *Bosmina*	1, 2 µg l^{-1}	Sakamoto et al. (2006)
	Inhibition of defensive morphology in *Bosmina*	2 µg l^{-1}	Sakamoto et al. (2009)
Endosulfan	Crest formation in *Daphnia*	0.1–10 µg l^{-1}	Barry (1998)
	Inhibition of defensive neckteeth formation in *Daphnia*	100 µg l^{-1}	Barry (2000)
Amitrole	Impaired chemical recognition of predators	0.1–10 mg l^{-1}	Mandrillon and Saglio (2007)
Pentachlorophenol	Negative phototaxis in the waterflea *Daphnia*	0.2–1.6 mg l^{-1}	Michels et al. (1999)
	Negative phototaxis in the waterflea *Daphnia*	0.4–1.6 mg l^{-1}	Kieu et al. (2001)
	Reversed anti-predator response in the rotifer *Brachionus*	0.19, 0.33 mg l^{-1}	Preston et al. (1999)
Atrazine and Diuron	Decreased anti-predator behaviour in goldfish	0.5–50 µg l^{-1}	Saglio and Trijasse (1998)
Diazinon	Inhibits response of salmon to alarm pheromones	1, 10 µg l^{-1}	Scholz et al. (2000)
Metalochlor	Altered response of crayfish to alarm pheromones	25–75 ppb	Wolf and Moore (2002)
Imidacloprid	Inhibited kairomone induced life history response in *Daphnia*	2.2–8.8 mg l^{-1}	Pestana et al. (2010)
Heavy metals	Reduced fright response in bullfrog		Raimondo et al. (1998)
	No response of mud snails to alarm cues		Lefcort et al. (1999)
	Snails fail to exhibit anti-predator behaviour in response to alarm cues		Lefcort et al. (2000)
Lead and Zinc	Tadpoles fail to take evasive action to chemical cues of fish	1–100 ppm	Lefcort et al. (1998)
Copper	Failure of Colorao pike minnow to react to fright pheromone	10–120 µg l^{-1}	Beyers and Farmer (2001)
	Negative phototaxis in the waterflea *Daphnia*	5–40 µg l^{-1}	Michels et al. (1999)
	Impaired response of fathead minnow to alarm pheromones	10 µg l^{-1}	Carreau and Pyle (2005)
	Impaired induction of protective neckteeth in *Daphnia*	5 µg l^{-1}	Hunter and Pyle (2004)
	Altered morphology, reduced survival in *Daphnia*	10 µg l^{-1}	Mirza and Pyle (2009)
Copper and Nickel	Impaired response of darters to alarm pheromones	10, 15 µg l^{-1} 52, 338 µg l^{-1}	McPherson et al. (2004)
Nickel	Impaired induction of protective neckteeth in *Daphnia*	40 µg l^{-1}	Hunter and Pyle (2004)
Cadmium	Eliminated response to alarm pheromones in rainbow trout	2 µg l^{-1}	Scott et al. (2003)
	Deficits in predator avoidance in juvenile zebrafish	14 mg l^{-1}	Blechinger et al. (2007)
	Impaired anti-predatory behaviour of juvenile silver catfish	4.5, 8 µg l^{-1}	Kochhann et al. (2009)
Organotin	Impaired response of mud snails to alarm substances		Straw and Rittschof (2004)

(continues)

Table 17.1 Continued

Induced defence

Pollutant	Disruption	Concentration	Reference
Surfactant SDS, FFD6	Induced anti-grazing reaction in *Scenedesmus*	10 mg l⁻¹	Lürling and Beekman (2002)
pH	No fish avoidance in freshwater snails	pH > 9.0	Turner and Chislock (2010)
	Reduced anti-predator responses in juvenile pumpkinseed sunfish	pH < 6.0	Leduc *et al*. (2003)
	Failure of detection of conspecific alarm cues in brook charr	pH < 6.0	Leduc *et al*. (2004)
	No response to alarm cues in salmon	pH < 6.0	Leduc *et al*. (2006)
	Reduced response to conspecific alarm cues in juvenile rainbow trout	pH < 6.4	Leduc *et al*. (2008)
	No typical anti-predator behaviour in fathead minnows and finescale dace	pH 6.0	Brown *et al*. (2002)

Others

Pollutant	Disruption	Concentration	Reference
Cadmium	No attraction to adult pheromones in juvenile banded kokopu	0.5, 1 µg l⁻¹	Baker and Montgomery (2001)
Zinc	No attraction to female aggregation pheromone in female zebrafish	5 ppm	Bloom *et al*. (1978)
pH	Impaired reaction to chemical settling cues in orange clown fish	pH 7.8	Munday *et al*. 2009

17.2.4 Eutrophication, pH, humic acids

Eutrophication has become the most important water quality issue in inland and coastal waters worldwide (Smith and Schindler 2009). Key symptoms of eutrophication are algal blooms and turbid water that constrains, for example, visual cues in mate choice and courtship behaviour in fish (Candolin *et al*. 2007). Elevated pH is also often related to eutrophication. Increased pH from 8 to 9.5 enhanced the use of male olfactory cues in mate choice in gravid female threespined sticklebacks, *Gasterosteus aculeatus* (Heuschele and Candolin 2007). The population consequences of the olfactory compensation for impaired visual communication are unknown. Further, eutrophication often leads to increasing nitrogen levels, often in the form of nitrate. Exposure of male palmate newts (*Triturus helveticus*) to high, but realistic, nitrate concentrations affected the production of male olfactory cues and consequently reduced the sexual attractiveness of male newts (Secondi *et al*. 2009).

In recent years there has been an increase in the concentration of humic acids, which are a by-product of degrading organic matter (Thomas 1997), in inland waters. Humic acids are large, complex organic molecules that adsorb relatively water-insoluble organic substances, such as steroidal pheromones, reducing their biological availability (Mesquita *et al*. 2003). Consequently, strongly reduced olfactory response in goldfish (*Carassius auratus*) to synthetic pheromones in the presence of humic acids (1–1,000 mg l⁻¹) has been observed (Hubbard *et al*. 2002). Also in swordtails (*Xiphophorus birchmanni*), realistic concentrations of humic acids (20 mg l⁻¹) impaired the ability of females to respond to conspecific male chemical cues (Fisher *et al*. 2006). Likewise, in high humic acid concentration (200 mg l⁻¹) adult zebrafish (*Danio rerio*) could no longer discriminate between conspecific and heterospecific urinary chemical cues (Fabian *et al*. 2007). In addition to direct interaction with the messenger molecule, humic acid can also convey chemical information by itself (Steinberg *et al*. 2008). For example, exposure of swordtails to humic acid (5–180 mg l⁻¹) caused a dose-dependent feminization (Meinelt *et al*. 2004).

17.3 Infodisruption in foraging

In many species, olfaction is common when foraging for food. Herbivorous insects use plant cues to recognize their food (Visser 1986), while malaria mosquitoes use mammalian body odours to locate blood sources (Takken and Knols 1999). Sophisticated means of communicating a source of food, for example through the wagtail dance in bees, can be disrupted by pollutants (Thompson 2003), and chemical food localization seems to be no exception (Table 17.1).

17.3.1   Pesticides

The crayfish *Orconectus rusticus* showed a significantly weakened response towards food odours after exposure to the herbicide metalochlor (Wolf and Moore 2002). The chemo-attraction responses of the ciliate *Tetrahymena pyriformis* to yeast extract was inhibited significantly after exposure to the herbicides 2,4-Dichlorophenoxyacetic acid (2,4-D) and Roundup® (Roberts and Berk 1993). In the snail *Ilyanassa obsoleta* chemoreception to detect food was inhibited when the snails were exposed to Irgarol 1051, an algistatic compound used in copper-based anti-foulant paints (Finnegan *et al.* 2009). The chemical attraction of goldfish to a filtrate of chironomids was significantly reduced after exposure to the insecticide carbuforan (Saglio *et al.* 1996).

17.3.2   Metals

Copper exposure significantly inhibited the chemo-attraction response of *T. pyriformis* to yeast extract (Roberts and Berk 1993). In leeches (*Nephelopsis obscura*) and the crayfish *C. bartonii*, relatively low waterborne copper concentrations impaired their ability to perform chemosensory food detection (Sherba *et al.* 2000; Pyle and Mirza 2007). While the crayfishes *Procambarus clarkii*, *O. rusticus*, and *C. bartonii* were attracted to a feeding stimulant, this attraction was lost in specimens that had been exposed to a blend of four heavy metals (Steele *et al.* 1992). Further, environmentally realistic manganese concentrations significantly reduced the chemically-induced food search behavior in the Norway lobster, *Nephrops norvegicus*, where only half to one-third reached the chemical food stimulus (Krång and Rosenqvist 2006).

17.3.3   Hydrocarbons

Detection of chemical food cues was significantly reduced in Dungeness crabs (*Cancer magister*) when placed in seawater contaminated with crude oil

(Pearson *et al.* 1981). The response of snails (*Nassarius obsolete*) to chemical cues that normally initiate feeding behaviour was inhibited by water-soluble kerosene (Jacobson and Boylan 1973). In contrast, the whole crude oil, not the water-soluble oil fraction, caused reduced chemosensory detection of food cues in the lobster *Homarus americanus* (Atema and Stein 1974).

17.3.4   Eutrophication, pH, humic acids

No examples were found of chemically-impaired food detection at elevated pH. The pH of an aquatic system, however, can also be lowered by acid rain, discharge of mining waste, and industrial wastewater. Indeed, sub-lethal acid pH of 4.5 critically impaired the chemoreception of food cues in *C. bartonii* (Allison *et al.* 1992). The same pH suppressed feeding behaviour in palmate newts *Triturus helveticus* and smooth newts *T. vulgaris*, probably through an impaired chemosensory system (Griffiths 1993). Gold mollies (*Poecilia sphenops*) had difficulties locating food odours at low pH and spent more time searching for food and at a lower swimming speed than at higher pH (Tembo 2009). Female swordtails had a strong preference for food odour cues in water without humic acids, but showed no preference in water with 20 mg l^{-1} humic acid (Fisher *et al.* 2006).

17.4   Infodisruption of anti-predator reactions

In many prey species, chemicals associated with the threat of being consumed, either released from injured or killed conspecifics (alarm pheromones; Chivers and Smith 1998; Chapter 9) and/or the predator itself (kairomones; Kats and Dill, 1998; Chapters 2 and 8), are involved in the detection of the predator and the onset of anti-predator reactions (Schoeppner and Relyea 2005). There are a growing number of studies highlighting pollutant-disruption of chemically-induced anti-predator reactions (Table 17.1).

17.4.1   Pesticides

Tadpoles of the common toad (*Bufo bufo*) exhibited a clear anti-predator reaction when they were exposed to chemical cues from conspecific-fed crayfish, but the anti-predator behaviour was lost when they were exposed simultaneously to sub-lethal concentrations of the herbicide amitrole (Mandrillon and Saglio 2007). In the presence of conspecific alarm cues, crayfish (*O. rusticus*) exposed to the herbicide metolachlor moved faster than non-exposed animals, but moved in the direction of the chemically-mediated alarm signal instead of away from it as the control animals did (Wolf and Moore 2002). Other herbicides, atrazine and diuron, decreased the alarm cue-induced anti-predator behaviours (grouping, sheltering, swimming orientation) in goldfish (Saglio and Trijasse 1998). However, it is not entirely clear how herbicides affect the information processing in animals.

Weis and Weis (1974) observed impairment of schooling behaviour and of regrouping after a fright stimulus in DDT-exposed fish, which could render them more susceptible to predation. The insecticide diazinon impaired the response of Chinook salmon to alarm pheromones (Scholz *et al.* 2000), while the broad spectrum insecticide pentachlorophenol even reversed the anti-predator response in the rotifer *Brachionus* (Preston *et al.* 1999). In contrast, pentachlorophenol evoked a phototactic behaviour in the water flea *Daphnia* resembling the anti-predator behaviour induced by fish smell (Michels *et al.* 1999; Kieu *et al.* 2001), but whether the animals would also show altered responses to fish kairomones remains to be tested.

The induction of morphological defensive structures in *Daphnia*, such as formation of crests, neck-teeth, high helmets, and tailspine elongation have been reported after exposure to a variety of different pesticides (Hanazato 1991; Hanazato and Dodson 1993; Barry 1998). In the presence of *Chaoborus* kairomone, carbaryl caused higher helmets and a longer helmeted period in *Daphnia* (Hanazato and Dodson 1995). Low carbaryl concentrations (1.2 µg l^{-1}) inhibited the development of a defensive morphology in small cladocera species *Bosmina* (Sakamoto *et al.* 2006, 2009). The mechanism of inducible defenses in *Daphnia* is not completely elucidated yet, but it is believed that chemical cues trigger the nervous system, which causes neurosecretory cells to release growth and reproduction hormones (Barry 2002; see also Chapter 8). Because

some pesticides that inhibit γ-aminobutyric acid (GABA), an inhibitory neurotransmitter, can cause inducible defenses in *Daphnia* it has been proposed that GABA might down-regulate the expression of inducible defenses in these zooplankters (Barry 1998, 2002).

In addition to chemically-induced behavioural and morphological anti-predator reactions, organisms also exhibit changes in life history characteristics, such as age and size at maturity, number and size of offspring, in response to predation (Lass and Spaak 2003). Contaminant disrupted life history response has been found in the water flea *D. magna*, where exposure to the insecticide imidacloprid inhibited the fish-kairomones induced life history response (Pestana *et al.* 2010). Exposure of *D. pulex* to *Chaoborus flavicans* kairomones and carbaryl led to a synergistic interaction on age at maturity (Hanazato and Dodson 1992). However, this could not be confirmed in another study (Coors and DeMeester 2008).

17.4.2 Metals

Low concentrations of heavy metals can cause disruption of the chemosensory function in fish (Beyers and Farmer 2001; Tierney *et al.* 2010) and alter predator avoidance behaviour (Webber and Haines 2003). The probability of a normal fright response to alarm pheromones in Colorado pike-minnows (*Ptychocheilus lucius*) decreased significantly with exposure to increasing concentrations of copper (Beyers and Farmer 2001). Similar effects were seen when fathead minnows (*Pimephales promelas*) were exposed to environmentally relevant concentrations of copper (Carreau and Pyle 2005). Cadmium impaired the ability of juvenile rainbow trout (*Oncorhynchus mykiss*) to respond to alarm cues and thus altered predator-avoidance strategies (Scott *et al.* 2003). Short term exposure (3 h) of juvenile zebrafish to cadmium (~14 mg l^{-1}) during early larval development caused continued deficits in olfactory-dependent predator avoidance behaviours 4–6 weeks after a return to clean water (Blechinger *et al.* 2007).

Metal-induced olfactory impairment in fish has also been observed in wild populations. Copper, nickel, and zinc exposure impaired the response of

Iowa darters (*Etheostoma exile*) to chemical alarm cues in nature (McPherson *et al.* 2004). Olfaction in fish is commonly measured with an electro-olfactogram, which is the external registration of the electrical properties of olfactory sensory neurons, where decreased potentials reflect less information conveyance (Tierney *et al.* 2010). Despite being widely used in research on effects of contaminants on fish, a recent study by Mirza *et al.* (2009) clearly demonstrated the importance of including behavioural assays in ecotoxicological studies. They found that wild juvenile yellow perch from contaminated lakes exhibited significantly larger electrophysiological responses to alarm cues than fish from non-polluted lakes, but they showed no anti-predator behaviour in contrast to clean-lake fish (Mirza *et al.* 2009).

In invertebrates, lead and zinc exposure resulted in the failure of tadpoles of the Columbia spotted frog (*Rana luteiventris*) to take evasive action when presented with chemical cues from fish (Lefcort *et al.* 1998). Heavy metals released from coal ash reduced the fright response of bullfrogs (Raimondo *et al.* 1998), while snails (*Physella columbiana*) from heavy metal-polluted sites failed to exhibit antipredator behaviour in response to alarm cues (Lefcort *et al.* 2000). Further, copper and cadmium mimicked the phototactic anti-predator response in *Daphnia* (Michels *et al.* 1999; Kieu *et al.* 2001).

Heavy metals can also disrupt the chemical information system between midge larvae (*Chaoborus*) and their prey, water fleas (*Daphnia*); both copper (5 µg l^{-1}) and nickel (40 µg l^{-1}) impaired the induction of protective neck teeth in *Daphnia* (Hunter and Pyle 2004). Copper (10 µg l^{-1}) had no effect on life history traits, but clearly influenced morphology in such a way that it significantly reduced survival in *D. pulex* neonates when encountering predators (Mirza and Pyle 2009). Pyle *et al.* (2006) showed that copper did not affect kairomone efficiency in the water, nor outcompeted kairomones at chemosensory surface receptors, but probably impaired kairomone perception in *Daphnia* along the molecular signal transduction pathway.

17.4.3 Hydrocarbons

Where the water soluble fraction of crude oil (0.05 ppm) impaired the defense response of sea urchin (*Strongylocentrotus droebachiensis*) resulting in increased predation by starfish (*Pycnopodia helianthoides*), the water soluble fraction (0.1–0.05 ppm) of No 2 Fuel oil enhanced the alarm response of snails (*N. obsolete*) to damaged conspecifics (cited in Blaxter and Ten Hallers-Tjabbes 1992).

17.4.4 Eutrophication, pH, humic acids

In a recent set of experiments, in which nutrients were added to outdoor mesocosms and mid-afternoon pH values consequently climbed to values between 8.5 and 9.7, the freshwater snails *Physa acuta* and *Helisoma trivolvis* only showed fish avoidance behaviour in water with a pH of less than 9.0, but no avoidance at a higher pH (Turner and Chislock 2010). These authors conclude that chemosensory impairment is likely a common occurrence in nature. Indeed, pH may affect the efficacy of kairomones and alarm pheromones and infochemicals in general, altering their availability, which is corroborated by the finding that under low pH conditions the *Chaoborus* kairomone is less water soluble relative to when the water is more alkaline (Tollrian and von Elert 1994).

Several studies have shown that anti-predator behaviour in fish can be lost under weakly acidic conditions. For example, in weakly acidic water (pH < 6.0), juvenile pumpkinseed sunfish (*Lepomis gibbosus*) had significantly lower anti-predator responses to conspecific alarm cues, or even a total loss of response to a cyprinid alarm cue (Leduc *et al.* 2003). Brook charr (*Salvelinus fontinalis*) in a naturally weakly acidic stream failed to detect conspecific alarm cues, while they responded to these cues in a neutral stream (Leduc *et al.* 2004). Likewise, salmon (*Salmo salar*) did not respond to alarm cues in acidic streams, while those in a neutral stream exhibited typical alarm responses (Leduc *et al.* 2006). In a follow-up study, Leduc *et al.* (2008) showed that juvenile rainbow trout exhibited a reduction in their response to conspecific alarm cues proportional to the ambient acidity and that the response was lost at a pH of less than 6.4 (Leduc *et al.* 2008). Two cyprinid species, fathead minnows (*Pimephales promelas*) and finescale dace (*Phoxinus neogaeus*), did not exhibit typical anti-predator behaviour under weakly acidic conditions (Brown *et al.* 2002). This was not caused by damage to the olfactory receptors, but as a result of chemical alteration of the alarm pheromone. The cyprinid alarm substance hypoxanthine-3-N-oxide was converted to 6,8-dioxypurine under acidic conditions, which implied loss of the 3-N-oxide functional group thereby rendering the alarm cue non-functional (Brown *et al.* 2002).

17.4.4.1 *Surfactants*

One of the most common surfactants in the aquatic environment, 4-Nonylphenol, at a low, environmentally relevant dosage (0.5 µg l⁻¹) influenced the shoaling behaviour in juvenile banded killifish, *Fundulus diaphanus* (Ward *et al.* 2008). At dosage levels of 1 and 2 µg l⁻¹ shoaling killifish actually avoided conspecifics which might imply major fitness implications, as shoaling is an adaptive response to predation (Ward *et al.* 2008).

The grazer *Daphnia*–alga *Scenedesmus* interaction has become a paradigm in the field of inducible defenses in phytoplankton. When *Scenedesmus* unicells are grazed upon, infochemicals released during the grazing process promote the formation of colonies to sizes beyond the ingestion capacity of the zooplankton (Lürling 2003). Two commercially available anionic surfactants (sodium dodecyl sulphate and FFD-6) triggered the formation of colonies in *Scenedesmus* in a similar way (Lürling and Beekman 2002; Lürling 2006). When unicellular *Scenedesmus* populations were initially exposed for 48 h to different concentrations of the surfactant FFD-6 (range 0–10 g l⁻¹), colonies were formed at concentrations of 0.01 g l⁻¹, which was far below the no-observed-effect-concentration (NOEC value) of 1 g l⁻¹ for growth inhibition. Hence, the surfactant FFD-6 induced an anti-grazer response and seemed to mimic the natural information conveying chemicals. Meanwhile, natural active substances extracted from *Daphnia* have been identified as different aliphatic sulfates and sulfamates that show strong structural similarity with the synthetic anionic surfactants (Yasumoto *et al.* 2005, 2006, 2008a,b; Fig. 17.3). Recently, one of the active compounds (8-methylnonyl sulfate) was also detected in *Daphnia* culture medium (Uchida *et al.* 2008).

17.5 Infodisrupted aggregation, migration, and settlement

Juvenile migrating fishes, the banded kokopu *Galaxias fasciatus*, are known for attraction to adult pheromones, but this reaction was impaired by exposure to cadmium (Baker and Montgomery 2001). Adult female zebrafish (*Brachydanio rerio*) that had been maintained in water with a sub-lethal dose of zinc were not attracted to female-aggregating pheromone-containing water, while non-treated females were (Bloom *et al.* 1978).

Ocean acidification could lead to maladapted responses in orange clown fish (*Amphiprion percula*) to chemical cues from parents and tropical trees, which are important for determining suitable settlement sites (Munday *et al.* 2009). Brominated flame retardants can modify olfactory function and thyroid hormones during smoltification of salmon, which might affect important behaviours, such as shoaling, predator avoidance, downstream and upstream migration, but also imprinting to the natal river and affect the ability of the adult salmon to return to the natal river to spawn (Lower and Moore 2007). However, population studies are required to determine the impact of these flame retardants.

17.6 Potentiated toxicity

The recent findings that chemical cues induce substantial potentiating of pesticide toxicity are related to chemical information conveyance systems. The pesticide carbaryl became 2–4 times more lethal to gray treefrog tadpoles when predatory cues were present, killing 60–98% of the tadpoles (Relyea and Mills 2001). Roundup® became twice as lethal to *Rana sylvatica* tadpoles in the presence of chemically advertised predation stress (caged predatory newts) (Relyea 2005). Likewise, a combination of fish-based chemical cues and 2.82 µg l⁻¹ malathion resulted in a 76% reduction in survival of the waterflea *Ceriodaphnia dubia* compared to malathion alone (Maul *et al.* 2006). A recent review on the effects of various natural stressors (temperature, low oxygen concentration) and toxicants revealed that synergistic interactions are not uncommon phenomena (Holmstrup *et al.* 2009).

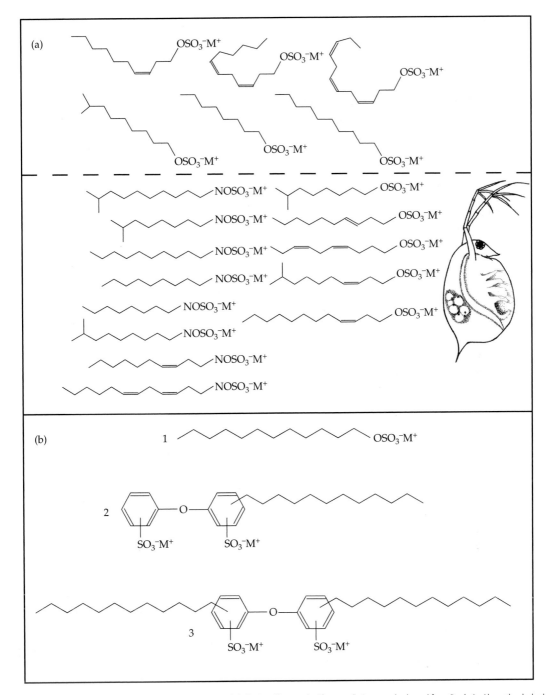

Figure 17.3   Top panel (a) presents the chemical structures of aliphatic sulfates and sulfamates (kairomones) released from *Daphnia*. Above the dashed line six aliphatic sulfates first reported by Yasumoto *et al*. (2005), below the dashed line 13 new kairomones (Yasumoto *et al*. 2008a, b). Bottom panel (b) gives the chemical structures of the synthetic anionic surfactants with morphogenetic effects on the green alga *Scenedesmus*. Numbers indicate: 1 = sodium dodecyl sulfate, 2 = sodium monododecyl disulfonate diphenyloxide, 3 = sodium didodecyl disulfonate diphenyloxide.

17.7 Novel threats

17.7.1 Engineered nanoparticles

Nanotechnology is expanding rapidly as, at the nanoscale, manufactured particles possess properties that make them very useful in a wide range of commercial and industrial applications. Nanoparticles are defined as materials having at least two dimensions between 1 and 100 nm (ASTM 2006). Nanoparticles are being increasingly used as fillers, catalysts, semiconductors, cosmetics, microelectronics, pharmaceuticals, drug carriers, energy storage, and antifriction coatings (Farré *et al.* 2009). Consequently, it is unavoidable that manmade nanoparticles will enter the aquatic environment (Moore 2006; Farré *et al.* 2009). Knowledge of the harmful effects of nanoparticles is very limited, but their properties suggest that environmental contamination from nanotechnologies constitutes one of the most critical challenges for the coming years (Auffan *et al.* 2009).

The physico-chemical properties of being small sized, lipophilic or hydrophilic, with a charged high surface area, might come with unpredictable genotoxicity (Singh *et al.* 2009). Exposure to cerium dioxide caused DNA-strand breakage in *Daphnia* and *Chironomus*, but not when the animals were exposed to titanium dioxide (Lee *et al.* 2009). Nanomaterials such as silver nanoparticles, photosensitive fullerenes, and quantum dots can penetrate the cell causing direct damage to DNA molecules within the cell nucleus, disrupting membrane integrity, and causing protein destabilization and oxidation (Klaine *et al.* 2008; Singh *et al.* 2009); all ingredients for intracellular disruption of chemical information conveyance. However, toxicity of nanomaterials seems not just to be related to size and shape, charge, and so on. (Lee *et al.* 2009); nanoparticles in the environment may perhaps influence chemically-mediated interactions between cells and organisms, as they might bind to cell membrane receptors (Singh *et al.* 2009), aggregate with natural organic compounds (Klaine *et al.* 2008), or react with organic compounds (Fujishima *et al.* 2000). Titanium dioxide is a potent photo-catalyst, which in sunlight breaks down almost any organic compound (Fujishima *et al.* 2000). Black carbon, originating from incomplete combustion of fossil fuels, is another material that

occurs in large quantities at the nanoscale. Black carbon seems to act as a supersorbent for persistent organic pollutants (Koelmans *et al.* 2006), and therefore, could also interact with natural infochemicals. It is obvious that much research is needed on effects, fate, bioavailability, and bioaccumulation of engineered nanomaterials.

17.7.2 Pharmaceuticals and personal care products

Pharmaceuticals are designed to be biologically active at very low concentrations. They end up in surface waters either unchanged, or as active metabolites/polar conjugates, mostly via effluents, agricultural run-off, and aquaculture discharge (Fent *et al.* 2006; Bottoni *et al.* 2010). Concentrations of commonly water-dispersed pharmaceuticals, such as antibiotics, anti-depressants, anti-neoplasics, anti-epileptics, anti-hypertensive drugs, anti-inflammatory drugs, antiseptics, beta-blockers, blood lipid lowering agents, painkillers, sex hormones, X-ray contrast media, and other substances, are in the range ng l^{-1} to µg l^{-1} (Fent *et al.* 2006; Bottoni *et al.* 2010; Santos *et al.* 2010). Generally, effects that are determined in standard assays, with survival, immobilization, inhibition, and growth as the most prominent endpoints, occur at concentrations orders of magnitude higher than those found in surface waters (Fent *et al.* 2006; Kümmerer 2009; Santos *et al.* 2010). However, due to continuous consumption and application of pharmaceuticals, an almost nonstop discharge into surface waters occurs with a consequent chronic, whole life cycle and multigenerational exposure, of aquatic non-target organisms to a rather constant low dose of (a mixture of) pharmaceuticals. In general, effects of long-term exposure are not covered in the standard ecotoxicological assays (Fent *et al.* 2006; Santos *et al.* 2010). At this moment, compounds such as acetylsalicylic acid, caffeine, nicotine, fluoroquinolone antibiotics, carbamazepine, and clofibric acid can be considered ubiquitous and persistent (Bottoni *et al.* 2010). There are, however, only a few indications for possible infodisruption by these compounds. For example, caffeine is an aversive tastant in goldfish (Lamb and Finger 1995), whereas acetylsalicylic acid interfered with the gastropod larval metamorphosis that is triggered by red algal chemical cues (Boettcher and Targett 1998).

Pharmaceuticals are designed to be biologically active and to change mood, behaviour, nerve function, or reduce stress by targeting receptors, biochemical pathways, and enzymes that are conserved in evolutionary terms (Fent *et al.* 2006). Therefore, it seems likely that pharmaceuticals exert effects in non-target organisms through interactions with receptors that are similar to those in humans. For example, low and environmentally realistic concentrations of fluoxetine and ibuprofen ($10-100$ ng l^{-1}) reduced the activity of the amphipod *Gammarus pulex*, as could be expected from their pharmacological purpose in humans (De Lange *et al.* 2006). Since *G. pulex* are well-known for tuning their behaviour in response to chemical cues from predators, injured conspecifics, and food (Åbjörnsson *et al.* 2000; De Lange *et al.* 2005), this organism might be a good candidate for testing the possible interference of pharmaceuticals with natural chemical information conveyance. For a comprehensive risk assessment of pharmaceuticals in surface waters, studies are needed that not only include long-term exposure of relevant mixtures, but also subtle endpoints including natural chemical information conveyance.

17.8 Conclusions

Upon review of the literature, it becomes apparent that infodisruption is a widespread phenomenon in aquatic systems, which can be caused by all kind of anthropogenic activities that change the chemical environment in which the organisms live.

Manmade alterations of the environment may deregulate chemical information conveyance in several ways: (1) Inhibition of chemically-mediated interactions in foraging, mating, and predator avoidance seems prevalent among animals like fish and crayfish. (2) Mimicry, or evoking responses similar to natural infochemical induced reactions, has been observed far less, and mostly in freshwater plankton. (3) Synergism between infochemicals and contaminants has been found in a few cases for both an infochemically-aggravated pollutant effect and a pollutant-enhanced infochemical effect.

Infochemicals might have important effects on food web dynamics, because of their influence on species interactions. In planktonic bi-and tritrophic food chains, infochemicals play a major role in stabilizing dynamics and preventing extinctions at higher trophic levels (Verschoor *et al.* 2004). Infochemicals not only mediate trophic interactions, but underlie a wide variety of interactions, such as mate-finding, food search, and competition, which are crucial in determining food web structure and functioning (Dicke 2006; Van Donk 2006). How anthropogenic stresses influence this intricate aquatic infochemical network is difficult to predict from the studies on single species from a relatively low number of taxa, often also with only one stressor included. Although it remains a key issue to resolve chemically-mediated interactions and possible disruptions on an organism level, more complex studies seem inevitable to obtain a true insight into potential consequences on communities and ecosystems. Models may help in extrapolating experimental findings from the laboratory and might prove powerful in generating hypotheses (Baldwin *et al.* 2009). However, complete understanding of the impact of infodisruption would benefit strongly from studies at the community or even ecosystem level, despite the fact that community and ecosystem studies come with high complexity.

Several decades ago the potential of pollutants in breaking down the communication of aquatic animals with their environment was already recognized (Sutterlin 1974). Chemoreceptive membranes were considered the prime targets for interactions with pollutants, because the chemoreceptors are directly exposed to the environment (Sutterlin 1974; Hara *et al.* 1983). It has since long been well-established that detection of chemical messengers is not restricted to metazoan animals. Infochemistry might be common in freshwater unicellular ciliates (Kusch 1993), in seaweed (Toth and Pavia 2000), in phytoplankton (Hansson 1996; Rengefors *et al.* 1998; Lürling 2003), among bacteria (Waters and Bassler 2005; Keller and Surette 2006; Ryan and Dow 2008), and between plants and bacteria in the rhizosphere (Badri *et al.* 2009). It can be assumed that pollutants also modify all these chemical interactions, as has been found for the phytoestrogen signalling between leguminous plants and symbiotic nitrogen-fixing bacteria that can be disrupted by a variety of organochlorine pesticides, PCBs, PAHs, and plasticizers (Fox 2004; Fox *et al.* 2001, 2004).

Despite a wealth of evidence on chemically-mediated information conveyance between aquatic organisms (dead or alive), in many cases the chemical identity of the natural infochemicals, dose–response relationships, and the mechanisms of action are poorly understood. Resolving the chemical structure of the messenger molecules is key to predicting the fate of infochemicals, especially under altered environmental conditions.

Alterations in environmental conditions (pH, salinity, humic acids, temperature, etc.) could influence the natural chemical information networks through effects on synthesis and excretion of messenger molecules, as well as through effects on the chemical characteristics of the messengers. For example, dimethylsulfoniopropionate (DMSP) is a potent attractant to some planktivorous reef fish (DeBose *et al.* 2008); however, temperature rise of the ocean results in a decrease of the DMSP concentration in the producing algae (van Rijssel and Gieskes 2002) and thus warming of the ocean could influence the chemically-mediated response of the fish. A change in pH might affect basic and acidic functionalities of messenger molecules and could modify their activity (Tollrian and von Elert 1994; Brown *et al.* 2002). Therefore, contrasting data from different studies does not necessarily imply poor quality of the research, but could result from different environmental conditions. For instance, cadmium impaired the anti-predatory behaviour of juvenile silver catfish (*Rhamdia quelen*) at an alkalinity of 63 mg l^{-1} CaCO$_3$, but not at an alkalinity of 92 mg l^{-1} CaCO$_3$ (Kochmann *et al.* 2009). The complexity of the natural systems and the infochemical network, the interactions of infochemicals, producing pathways and receptors with interfering stressors, and the interactions of natural stressors and conditions with anthropogenic stresses, create a major challenge for the deeper understanding of the impact of infodisruption on natural populations and ecosystems. Insight into the potential consequences of infodisruption starts with a systematic approach to identify organisms that react to infochemicals and a mechanistic understanding of how chemical information is perceived and processed by these organisms. Once the chemical structure of the infochemicals involved has been resolved, their environmental fates can be determined and groups of chemicals and stresses identified that could disturb natural information conveyance. The combined efforts of modelling exercises, multi-trophic- and field experiments will be inevitable for obtaining significant insight into the potential consequences of infodisruption.

References

Åbjörnsson, K., Dahl, J., Nyström, P., and Brönmark, C. (2000) Influence of predator and dietary chemical cues on the behaviour and shredding efficiency of *Gammarus pulex. Aquatic Ecology*, **34**: 379–87.

Allison, V., Dunham, D.W., and Harvey, H.H. (1992) Low pH alters response to food in the crayfish *Cambarus bartoni. Canadian Journal of Zoology*, **70**: 2416–20.

Andreini, I., DellaCorte, C., Johnson, L.C., Hughes, S., and Kalinoski D.L. (1997) G-Protein(s), G$_{\alpha q}$/G$_{\alpha 11}$, in the olfactory neuroepithelium of the channel catfish (*Ictalurus punctatus*) is altered by the herbicide, dichlobenil. *Toxicology*, **117**: 111–22.

ASTM, American Society for Testing and Materials. (2006) Standard terminology relating to nanotechnology. E 2456–06. ASTM Internationsl, West Conshohocken, PA.

Atema, J. and Stein, L.S. (1974) Effects of crude oil on the feeding behaviour of the lobster *Homarus americanus. Environmental Pollution*, **6**: 77–86.

Auffan, M., Bottero, J.-Y., Chaneac, C., and Rose, J. (2009) Inorganic manufactured nanoparticles: How their physicochemical properties influence their biological effects in aqueous environments. *Nanomedicine*, **5**: 999–1007.

Badri, D.V., Weir, T.L., van der Lelie, D., and Vivanco, J.M. (2009) Rhizosphere chemical dialogues: plant–microbe interactions. *Current Opinion in Biotechnology*, **20**: 642–50.

Baker, C.F. and Montgomery, J.C. (2001) Sensory deficits induced by cadmium in banded kokopu, *Galaxias fasciatus*, juveniles. *Environmental Biology of Fishes*, **62**: 455–64.

Baldwin, D.H., Spromberg, J.A., Collier, T.K., and Scholz, N.L. (2009) A fish of many scales: extrapolating sublethal pesticide exposures to the productivity of wild salmon populations. *Ecological Applications*, **19**: 2004–15.

Barry, M.J. (1998) Endosulfan-enhanced crest induction in *Daphnia longicephala*: evidence for cholinergic innervation of kairomone receptors. *Journal of Plankton Research*, **20**: 1219–31.

Barry, M.J. (1999) The effects of a pesticide on inducible phenotypic plasticity in *Daphnia*. *Environmental Pollution*, **104**: 217–34.

Barry, M. J. (2000) Effects of endosulfan on *Chaoborus*-induced life-history shifts and morphological defenses in *Daphnia pulex*. *Journal of Plankton Research*, **22**: 1705–18.

Barry, M. J. (2002) Progress toward understanding the neurophysiological basis of predator-induced morphology in *Daphnia pulex*. *Physiological and Biochemical Zoology*, **75**: 179–86.

Beyers, D.W. and Farmer, M.S. (2001) Effects of copper on olfaction of Colorado pikeminnow. *Environmental Toxicology and Chemistry*, **20**: 907–12.

Birch, M.C. (1974) *Pheromones*. Frontiers of Biology: 32. North Holland American Elsevier, Amsterdam.

Blaxter, J.H.S. and Ten Hallers-Tjabbes, C.C. (1992) The effect of pollutants on sensory systems and behaviour of aquatic animals. *Netherlands Journal of Aquatic Ecology*, **26**: 43–58.

Blechinger, S.R., Kusch, R.C., Haugo, K., Matz, C., Chivers, D.P., and Krone, P.H. (2007) Brief embryonic cadmium exposure induces a stress response and cell death in the developing olfactory system followed by long-term olfactory deficits in juvenile zebrafish. *Toxicology and Applied Pharmacology*, **224**: 72–80.

Bloom, H.D., Perlmutter, A., and Seeley R.J. (1978) Effect of a sublethal concentration of zinc on an aggregating pheromone system in the zebrafish, *Brachydanio rerio* (Hamilton-Buchanan). *Environmental Pollution*, **17**: 127–31.

Boettcher, A.A. and Targett, N.M. (1998) Role of chemical inducers in larval metamorphosis of Queen Conch, *Strombus gigas* Linnaeus: Relationship to other marine invertebrate systems. *Biological Bulletin*, **194**: 132–42.

Bottoni, P., Caroli, S., and Barra Caracciolo, A. (2010) Pharmaceuticals as priority water contaminants. *Toxicological and Environmental Chemistry*, **92**: 549–65.

Boyd, R.S. (2010) Heavy metal pollutants and chemical ecology: exploring new frontiers. *Journal of Chemical Ecology*, **36**: 46–58.

Brown, G.E., Adrian, J.C., Lewis, M.G., and Tower, J.M. (2002) The effects of reduced pH on chemical alarm signalling in ostariophysan fishes. *Canadian Journal of Fisheries and Aquatic Sciences*, **59**: 1331–8.

Brucker-Davis, F. (1998) Effects of environmental synthetic chemicals on thyroid function. *Thyroid*, **8**: 827–56.

Calabrese, E.J. and Baldwin, L.A. (2003) Toxicology rethinks its central belief. Hormesis demands a reappraisal of the way risks are assessed. *Nature*, **421**: 691–2.

Calow, P. and Forbes, V.E. (2003) Does ecotoxicology inform ecological risk assessment? *Environmental Science and Technology*, **37**: 146A–51A.

Candolin, U., Salesto, T., and Evers, M. (2007) Changed environmental conditions weaken sexual selection in sticklebacks. *Journal of Evolutionary Biology*, **20**: 233–9.

Carreau, N.D. and Pyle, G.G. (2005) Effect of copper exposure during embryonic development on chemosensory function of juvenile fathead minnows (*Pimephales promelas*). *Ecotoxicology and Environmental Safety*, **61**: 1–6.

Casida, J.E. (2009) Pest toxicology: The primary mechanisms of pesticide action. *Chemical Research in Toxicology*, **22**: 609–19.

Cheung, S.J., Tai, K.K., Leung, C.K., and Siu, Y.M. (2002) Effects of heavy metals on the survival and feeding behaviour of the sandy shore scavenging gastropod *Nassarius festivus* (Powys). *Marine Pollution Bulletin*, **45**: 107–13.

Chivers, D.P. and Smith, R.J.F. (1998) Chemical alarm signalling in aquatic predator-prey systems: A review and prospectus. *Ecoscience*, **5**: 338–52.

Clotfelter, E.D., Bell, A.M., and Levering, K.R. (2004) The role of animal behaviour in the study of endocrine-disrupting chemicals. *Animal Behaviour*, **68**: 665–76.

Colborn, T. and Clement, C. (1992) *Chemically-Induced Alterations in Sexual and Functional Development: The Wildlife-Human Connection*. Princeton Scientific, New Jersey, USA.

Colborn, T. and Thayer, K. (2000) Aquatic ecosystems: harbingers of endocrine disruption. *Ecological Applications*, **10**: 949–57.

Colborn, T., von Saal, F.S., and Soto, A.M. (1993) Developmental effects of endocrine-disrupting chemicals in wildlife and humans. *Environmental Health Perspectives*, **101**: 378–84.

Coors, A. and De Meester, L. (2008) Synergistic, antagonistic and additive effects of multiple stressors: predation threat, parasitism and pesticide exposure in *Daphnia magna*. *Journal of Applied Ecology*, **45**: 1820–8.

DeBose, J.L., Lema, S.C., and Nevitt, G.A. (2008) Dimethylsulfoniopropionate as a foraging cue for reef fishes. *Science*, **319**: 1356.

De Lange, H.J., Noordoven, W., Murk, A.J., Lürling, M., and Peeters, E.T.H.M. (2006) Behavioural responses of *Gammarus pulex* (Crustacea, Amphipoda) to low concentrations of pharmaceuticals. *Aquatic Toxicology*, **78**: 209–16.

De Lange, H.J., Lürling, M., Van den Borne, B. and Peeters, E.T.H.M. (2005) Attraction of the amphipod *Gammarus pulex* to water-borne cues of food. *Hydrobiologia*, **544**: 19–25.

Dicke, M. (2006) Chemical ecology from genes to communities: integrating 'omics' with community ecology. In M. Dicke and W. Takken, (eds.) *Chemical Ecology. From Gene to Ecosystem*. Wageningen UR Frontis Series, Vol. 16, Springer, The Netherlands, pp. 175–89.

Dicke, M. and Sabelis, M.W. (1988) Infochemical terminology: Based on cost-benefit analysis rather than origin of compounds? *Functional Ecology*, **2**: 131–9.

Dicke, M. and Takken, W. (2006) *Chemical Ecology. From Gene to Ecosystem*. Wageningen UR Frontis Series, Vol. 16, Springer, The Netherlands.

De Loof, A. (1993) Schrödinger 50 years ago: 'What is life?' 'the ability to communicate', a plausible reply? *International Journal of Biochemistry*, **25**: 1715–21.

Fabian, N. J., Albright, L.B., Gerlach, G., Fisher, H.S., and Rosenthal, G.G. (2007) Humic acid interferes with species recognition in zebrafish (*Danio rerio*). *Journal of Chemical Ecology*, **33**: 2090–6.

Farré, M., Gajda-Schrantz, K., Kantiani, L., and Barceló, D. (2009) Ecotoxicity and analysis of nanomaterials in the aquatic environment. *Analytical and Bioanalytical Chemistry*, **393**: 81–95.

Fent, K., Weston, A.A., and Caminada, D. (2006) Ecotoxicology of human pharmaceuticals. *Aquatic Toxicology*, **76**: 122–59.

Finnegan, M.C., Pittman, S., and DeLorenzo, M.E. (2009) Lethal and sublethal toxicity of the antifoulant compound Irgarol 1051 to the mud snail *Ilyanassa obsolete*. *Archives of Environmental Contamination and Toxicology*, **56**: 85–95.

Fisher, H.S., Wong, B.B.W., and Rosenthal, G.G. (2006) Alteration of the chemical environment disrupts communication in a freshwater fish. *Proceedings of the Royal Society London B*, **273**: 1187–93.

French, J.B., Nisbett Jr, I.C.T., and Schwabl, H. (2001). Maternal steroids and contaminants in common tern eggs: a mechanism of endocrine disruption? *Comparative Biochemistry and Physiology C*, **128**: 91–8.

Fox, J.E. (2004) Chemical communication threatened by endocrine-disrupting chemicals. *Environmental Health Perspectives*, **112**: 648–53.

Fox, J.E., Starcevic, M., Kow, K.Y., Burow, M.E., and McLachlan, J.A. (2001) Endocrine disrupters and flavonoid signalling. *Nature*, **413**: 128–9.

Fox, J.E., Starcevic, M., Jones, P.E., Burow, M.E., and McLachlan, J.A. (2004) Phytoestrogen signaling and symbiotic gene activation are disrupted by endocrine-disrupting chemicals. *Environmental Health Perspectives*, **112**: 672–7.

Fujishima, A., Rao, T.N., and Tryk, D.A. (2000) Titanium dioxide photocatalysis. *Journal of Photochemistry and Photobiology C-Photochemistry Reviews*, **1**: 1–21.

Griffiths, R.A. (1993) The effect of pH on feeding behaviour in newt larvae (*Triturus*: Amphibia). *Journal of Zoology*, **231**: 285–90.

Hanazato, T. (1991) Pesticides as chemical agents inducing helmet formation in *Daphnia ambigua. Freshwater Biology*, **26**: 419–24.

Hanazato, T. (1999) Anthropogenic chemicals (insecticides) disturb natural organic chemical communication in the plankton community. *Environmental Pollution*, **105**: 137–42.

Hanazato, T. (2001) Pesticide effects on freshwater zooplankton: An ecological perspective. *Environmental Pollution*, **112**: 1–10.

Hanazato, T. and Dodson, S.I. (1992) Complex effects of a kairomone of *Chaoborus* and an insecticide on *Daphnia pulex. Journal of Plankton Research*, **14**: 1743–55.

Hanazato, T. and Dodson, S.I. (1993) Morphological responses of four species of cyclomorphic *Daphnia* to a short-term exposure to the insecticide carbaryl. *Journal of Plankton Research*, **15**: 1087–95.

Hanazato, T. and Dodson, S.I. (1995) Synergistic effects of low oxygen concentration, predator kairomone, and a pesticide on the cladoceran *Daphnia pulex. Limnology and Oceanography*, **40**: 700–9.

Hansson, L-A. (1996) Behavioural response in plants: adjustment in algal recruitment induced by herbivores. *Proceedings of the Royal Society London B*, **263**: 1241–4.

Hara, T.J., Brown, S.B., and Evans, R.E. (1983) Pollutants and chemoreception in aquatic organisms. *Advances in Environmental Science and Technology*, **13**: 247–306.

Heuschele, J. and Candolin, U. (2007) An increase in pH boosts olfactory communication in sticklebacks. *Biology Letters*, **3**: 411–13.

Haubruge, E. and Toutant, J.-P. (1997) Acetylcholinesterase, in the olfactory organ of the common carp *Cyprinus carpio* (Teleost, Cyprinidae): Characterization of molecular forms in vitro and in vivo inhibition by carbofuran. *Belgian Journal of Zoology*, **127**: 63–73.

Hayes, T.B., Collins, A., Lee, M., Mendoza, M., Noriega, N., Stuart, A.A., and Vonk, A. (2002) Hermaphroditic, demasculinized frogs after exposure to the herbicide atrazine at low ecologically relevant doses. *Proceedings of the National Academy of Sciences USA*, **99**: 5476–80.

Holmstrup M, Bindesbøl, A.-M., Oostingh, G.J., Duschl, A., Scheil, V., Köhler, H-R., Loureiro, S., Soares, A.M.V.M., Ferreira, A.L.G., Kienle, G., Gerhardt, A., Laskowski, R., Kramarz, P.E., Bayley, M., Svendsen, C., and Spurgeon, D.J. (2009) Interactions between effects of environmental chemicals and natural stressors: A review. *Science of the Total Environment*, doi:10.1016/j.scitotenv.2009.10.067.

Hubbard, P.C., Barata, E.N., and Canario, A.V.M. (2002) Possible disruption of pheromonal communication by humic acid in the goldfish, *Carassius auratus*. *Aquatic Toxicology*, **60**: 169–83.

Hunter, K. and Pyle, G. (2004) Morphological responses of *Daphnia pulex* to *Chaoborus americanus* kairomone in the presence and absence of metals. *Environmental Toxicology and Chemistry*, **23**: 1311–16.

Jacobson, S.M. and Boylan, D.B. (1973) Effect of seawater soluble fraction of kerosene on chemotaxis in a marine snail, *Nassarius obsoletus*. *Nature*, **241**: 213–15.

Jarvis, P. (2001) Intracellular signalling: The chloroplast talks! *Current Biology*, **11**: 307–10.

Jarvis, P. (2003) Intracellular signalling: The language of the chloroplast. *Current Biology*, **13**: 314–16.

Kasumyan, A.O. and Pashchenko, N.I. (1985) Olfactory way of alarm kairomone perception by fishes. *Vestnik Moskovskii Universitet Biologiya*, **40**: 50–4.

Kats, L.B. and Dill, L.M. (1998) The scent of death: chemosensory assessment of predation risk by prey animals. *Ecoscience*, **5**: 361–94.

Keller, L. and Surette, M.G. (2006) Communication in bacteria: An ecological and evolutionary perspective. *Nature Reviews Microbiology*, **4**: 249–58.

Kieu, N.D., Michels, E., and De Meester, L. (2001) Phototactic behaviour of *Daphnia* and the continuous monitoring of water quality: interference of fish kairomones and food quality. *Environmental Toxicology and Chemistry*, **20**: 1098–103.

Klaine, S.J., Alvarez, P.J.J., Batley, G.E., Fernandes, T.F., Handy, R.D., Lyon, D.Y., Mahendra, S., Mclaughlin, M.J., and Lead, J.R. (2008) Nanomaterials in the environment: behavior, fate, bioavailability, and effects. *Environmental Toxicology and Chemistry*, **27**: 1825–51.

Klaschka, U. (2008) The infochemical effect—A new chapter in ecotoxicology. *Environmental Science and Pollution Research*, **15**: 452–62.

Klaschka, U. and Kolossa-Gehring, M. (2007) Fragrances in the environment: Pleasant odours for nature? *Environmental Science and Pollution Research*, **14**: 44–52.

Kochhann, D., Benaduce, A.P.S., Copatti, C.E., Lorenzatto, K.R., Mesko, M.F., Flores, E.M.M., Dressler, V.L., and Baldisserotto, B. (2009) Protective effect of high alkalinity against the deleterious effects of chronic waterborne cadmium exposure on the detection of alarm cues by juvenile silver catfish (*Rhamdia quelen*). *Archives of Environmental Contamination and Toxicology*, **56**: 770–5.

Koelmans, A.A., Jonker, M.T.O., Cornelissen, G., Bucheli, T.D., Van Noort, P.C.M., and Gustafsson, Ö. (2006) Black carbon: The reverse of its dark side. *Chemosphere*, **63**: 365–77.

Krång, A.-S. (2007) Naphthalene disrupts pheromone induced mate search in the amphipod *Corophium volutator* (Pallas). *Aquatic Toxicology*, **85**: 9–18.

Krång, A.-S. and Dahlström, M. (2006) Effects of a candidate antifouling compound (medetomidine) on pheromone induced mate search in the amphipod *Corophium volutator*. *Marine Pollution Bulletin*, **52**: 1776–83.

Krång, A.-S. and Ekerholm, M. (2006) Copper reduced mating behaviour in male shore crabs (*Carcinus maenas* (L.)). *Aquatic Toxicology*, **80**: 60–9.

Krång, A-S. and Rosenqvist, G. (2006) Effects of manganese on chemically induced food search behaviour of the Norway lobster, *Nephrops norvegicus* (L.) *Aquatic Toxicology*, **78**: 284–91.

Kümmerer, K. (2009) The presence of pharmaceuticals in the environment due to human use—present knowledge and future challenges. *Journal of Environmental Management*, **90**: 2354–66.

Kusch, J. (1993) Induction of defensive morphological changes in ciliates. *Oecologia*, **94**: 571–5.

Lamb, C.F. and Finger, T.E. (1995) Gustatory control of feeding behavior in goldfish. *Physiology and Behavior*, **57**: 483–8.

Lass, S. and Spaak, P. (2003) Chemically induced antipredator defences in plankton: a review. *Hydrobiologia*, **491**: 221–39.

Leduc, A.O.H.C., Noseworthy, M.K., Adrian Jr, J.C., and Brown, G.E. (2003) Detection of conspecific and heterospecific alarm signals by juvenile pumpkinseed under weak acidic conditions. *Journal of Fish Biology*, **63**: 1331–6.

Leduc, A.O.H.C., Kelly, J.M., and Brown, G.E. (2004) Detection of conspecific alarm cues by juvenile salmonids under neutral and weakly acidic conditions: laboratory and field tests. *Oecologia*, **139**: 318–24.

Leduc, A.O.H.C., Roh, E., Harvey, M.C., and Brown, G.E. (2006) Impaired detection of chemical alarm cues by juvenile wild Atlantic salmon (*Salmo salar*) in a weakly acidic environment. *Canadian Journal of Fisheries and Aquatic Sciences*, **63**: 2356–63.

Leduc, A.O.H.C., Lamaze, F.C., McGraw, L., and Brown, G.E. (2008) Response to chemical alarm cues under weakly acidic conditions: a graded loss of antipredator behaviour in juvenile rainbow trout. *Water Air Soil Pollution*, **189**: 179–87.

Lee, S.-W., Kim, S.-M., and Choi, J. (2009) Genotoxicity and ecotoxicity assays using the freshwater crustacean *Daphnia magna* and the larva of the aquatic midge *Chironomus riparius* to screen the ecological risks of nanoparticle exposure. *Environmental Toxicology and Pharmacology*, **28**: 86–91.

Lefcort, H., Meguire, R.A., Wilson, L.A., and Ettinger, W.F. (1998) Heavy metals alter the survival, growth, metamorphosis, and antipredator behavior of Columbia spotted frog (*Rana luteiventris*) tadpoles. *Archives of Environmental Contamination and Toxicology*, **35**: 447–56.

Lefcort, H., Thomson, S.M., Cowles, E.E., Harowicz, H.L., Livaudais, L.M., Roberts, W.E., and Ettinger, W.F. (1999) Ramifications of predator avoidance: predator and heavy metal mediated competition between tadpoles and snails. *Ecological Applications*, **9**: 1477–89.

Lefcort, H., Ammann, E., and Eiger S.M. (2000) Antipredator behavior as an index of heavy metal pollution? A test using snails and caddiesflies. *Archives of Environmental Contamination and Toxicology*, **38**: 311–16.

Lower, N. and Moore, A. (2007) The impact of a brominated flame retardant on smoltification and olfactory function in Atlantic salmon (*Salmo salar* L.) smolts. *Marine and Freshwater Behaviour and Physiology*, **40**: 267–84.

Lürling, M. (2003) Phenotypic plasticity in the green algae *Desmodesmus* and *Scenedesmus* with special reference to the induction of defensive morphology. *Annales de Limnologie – International Journal of Limnology*, **39**: 85–101.

Lürling, M. (2006) Effects of a surfactant (FFD-6) on *Scenedesmus* morphology and growth under different nutrient conditions. *Chemosphere*, **62**: 1351–8.

Lürling, M. and Beekman, W. (2002) Extractable substances (anionic surfactants) from membrane-filters induce morphological changes in the green alga *Scenedesmus obliquus* (Chlorophyceae). *Environmental Toxicology and Chemistry*, **21**: 1213–18.

Lürling, M. and Scheffer, M. (2007) Info-disruption: pollution and the transfer of chemical information between organisms. *Trends in Ecology and Evolution*, **22**: 374–9.

Maul, J.D., Farris, J.L., and Lydy, M.J. (2006) Interaction of chemical cues from fish tissues and organophosphorous pesticides on *Ceriodaphnia dubia* survival. *Environmental Pollution*, **141**: 90–7.

Mandrillon, A-L. and Saglio, P. (2007) Herbicide exposure affects the chemical recognition of a non native predator in common toad tadpoles (*Bufo bufo*). *Chemoecology*, **17**: 31–6.

McLachlan, J.A. (2001) Environmental signaling: What embryos and evolution teach us about endocrine disrupting chemicals. *Endocrine Reviews*, **22**: 319–41.

McPherson, T.D., Mirza, R.S., and Pyle, G.G. (2004) Responses of wild fishes to alarm chemicals in pristine and metal-contaminated lakes. *Canadian Journal of Zoology*, **82**: 694–700.

Meinelt, T., Schreckenbach, K., Knopf, K., Wienke, A., Stüber, A., and Steinberg, C.E.W. (2004) Humic substances increase the constitution of swordtail (*Xiphophorus helleri*). *Aquatic Sciences*, **66**: 239–45.

Mesquita, R.M.R.S., Canário, A.V.M., and Melo, E. (2003) Partition of fish pheromonesbetween water and aggregates of humic acids. Consequences for sexual signaling. *Environmental Science and Technology*, **37**: 742–6.

Michels, E., Leynen, M., Cousyn, C., De Meester, L., and Ollevier, F. (1999) Phototactic behavior of *Daphnia* as a tool in the continuous monitoring of water quality: experiments with positively phototactic *Daphnia magna* clone. *Water Research*, **33**: 401–8.

Mirza, R.S. and Pyle, G.G. (2009) Waterborne metals impair inducible defences in *Daphnia pulex*: morphology, life-history traits and encounters with predators. *Freshwater Biology*, **54**: 1016–27.

Mirza, R.S., Green, W.W., Connor, S., Weeks, A.C.W., Wood, C.M., and Pyle, G.G. (2009) Do you smell what I smell? Olfactory impairment in wild yellow perch from metal-contaminated waters. *Ecotoxicology and Environmental Safety*, **72**: 677–83.

Mitchell, R., Fogel, S., and Chet, I. (1972) Bacterial chemoreception: an important ecological phenomenon inhibited by hydrocarbons. *Water Research*, **6**: 1137–40.

Moore, A. and Waring, C.P. (1996) Sublethal effects of the pesticide diazinon on olfactory function in mature male Atlantic salmon parr. *Journal of Fish Biology*, **48**: 758–75.

Moore, A. and Waring, C.P. (1998) Mechanistic effects of a triazine pesticide on reproductive endocrine function in mature male Atlantic salmon (*Salmo salar* L.) parr. *Pesticide Biochemistry and Physiology*, **62**: 41–50.

Moore, A. and Waring, C.P. (2001) The effects of a synthetic pyrethroid pesticide on some aspects of reproduction in Atlantic salmon (*Salmo salar* L.). *Aquatic Toxicology*, **52**: 1–12.

Moore, A. and Lower, N. (2001) The impact of two pesticides on olfactory-mediated endocrine function in mature male Atlantic salmon (*Salmo salar* L.) parr. *Comparative Biochemistry and Physiology Part B*, **129**: 269–76.

Moore, M.N. (2006) Do nanoparticles present ecotoxicological risks for the health of the aquatic environment? *Environment International*, **32**: 967–76.

Müller, C.T., Priesnitz, F.M., and Beckmann, M. (2005) Pheromonal communication in nereids and the likely intervention by petroleum derived pollutants. *Integrative and Comparative Biology*, **45**: 189–93.

Munday, P.L., Dixson, D.L., Donelson, J.M., Jones, G.P., Pratchetta, M.S., Devitsina, G.V., and Døving, K.B. (2009) Ocean acidification impairs olfactory discrimination and homing ability of a marine fish. *Proceedings of the National Academy of Sciences USA*, **106**: 1848–52.

Newman, M.C. (2001) *Population Ecotoxicology*. Hierarchical Ecotoxicology Series, John Wiley & Sons, Ltd. Chichester, England.

Park, D., Hempleman, S.C., and Propper, C.R. (2001) Endosulfan exposure disrupts pheromonal systems in the red-spotted newt: a mechanism for subtle effects of environmental chemicals. *Environmental Health Perspectives*, **109**: 669–703.

Park, D. and Propper, C.R. (2002) Endosulfan affects pheromonal detection and glands in the male red-spotted newt, *Notophthalmus viridescens*. *Bulletin of Environmental Contamination and Toxicology*, **69**: 609–16.

Pearson, W.H., Sugarman, P.C., Woodruff, D.L., and Olla, B.L. (1981) Impairment of the chemosensory antennular flicking response in the Dungeness crab, *Cancer magister*, by petroleum hydrocarbons. *Fishery Bulletin*, **79**: 641–7.

Pestana, J.L.T., Loureiro, S., Baird, D.J., and Soares, A.M.V.M. (2010) Pesticide exposure and inducible antipredator responses in the zooplankton grazer, *Daphnia magna* Straus. *Chemosphere*, **78**: 241–8.

Preston, B.L., Cecchine, G., and Snell, T.W. (1999) Effects of pentachlorophenol on predator avoidance behavior of the rotifer *Brachionus calyciflorus*. *Aquatic Toxicology*, **44**: 201–12.

Pyle, G.G. and Mirza, R.S. (2007) Copper-impaired chemosensory function and behavior in aquatic animals. *Human and Ecological Risk Assessment*, **13**: 492–505.

Pyle, G., Hunter, K. and Mirza, R. (2006) Effects of copper on kairomone perception in *Daphnia pulex*. *Society of Environmental Toxicology and Chemistry Europe 16th Annual Meeting Abstract Book*, p. 264.

Raimondo, S.M., Rowe, C.L., and Congdon, J.D. (1998) Exposure to coal ash impacts swimming performance and predator avoidance in larval bullfrogs (*Rana catesbeiana*). *Journal of Herpetology*, **32**: 289–92.

Relyea, R.A. and Mills, N. (2001) Predator-induced stress makes the pesticide carbaryl more deadly to gray treefrog tadpoles (*Hyla versicolor*). *Proceedings of the National Academy of Sciences USA*, **98**: 2491–6.

Relyea, R.A. (2005) The lethal impacts of roundup and predatory stress on six species of North American tadpoles. *Archives of Environmental Contamination and Toxicology*, **48**: 351–7.

Rengefors, K., Karlsson, I., and Hansson, L.-A. (1998) Algal cyst dormancy: a temporal escape from herbivory. *Proceedings of the Royal Society London B*, **265**:1353–8.

Roberts, R.O. and Berk, S.G. (1993) Effect of copper, herbicides, and a mixed effluent on chemoattraction of *Tetrahymena pyriformis*. *Environmental Toxicology and Water Quality*, **8**: 73–85.

Ryan, R.P. and Dow, J.M. (2008) Diffusible signals and interspecies communication in bacteria. *Microbiology*, **154**: 1845–58.

Saglio, P. and Trijasse, S. (1998) Behavioral responses to atrazine and diuron in goldfish. *Archives of Environmental Contamination and Toxicology*, **35**: 484–91.

Saglio, P., Trijasse, S., and Azam, D. (1996) Behavioral effects of waterborne carbofuran in goldfish. *Archives of Environmental Contamination and Toxicology*, **31**: 232–8.

Sakamoto, M., Chang, K.H., and Hanazato, T. (2006) Inhibition of development of anti-predator morphology in the small cladoceran *Bosmina* by an insecticide: impact of an anthropogenic chemical on prey-predator interactions. *Freshwater Biology*, **51**: 1974–83.

Sakamoto, M., Hanazato, T., and Y. Tanaka (2009) Impact of an insecticide on persistence of inherent antipredator morphology of a small cladoceran, *Bosmina*. *Archives of Environmental Contamination and Toxicology*, **57**: 68–76.

Santos, L.H.M.L.M., Araujo, A.N., Fachini, A., Pena, A., Delerue-Matos, C., and Montenegro, M.C.B.S.M. (2010) Ecotoxicological aspects related to the presence of pharmaceuticals in the aquatic environment. *Journal of Hazardous Materials*, **175**: 45–95.

Schoeppner, N.M. and Relyea, R.A. (2005) Damage, digestion and defence: the roles of alarm cues and kairomones for inducing prey defences. *Ecology Letters*, **8**: 505–12.

Scholz, N.L., Truelove, N.K., French, B.L., Berejikian, B.A., Quinn, T.P., Casillas, E., and Collier, T.K. (2000) Diazinon disrupts antipredator and homing behaviors in chinook salmon (*Oncorhynchus tshaytscha*). *Canadian Journal of Fisheries and Aquatic Sciences*, **57**: 1911–18.

Scott, G.R., Sloman, K.A., Rouleau, C., and Wood, C.M. (2003) Cadmium disrupts behavioural and physiological responses to alarm substance in juvenile rainbow trout (*Oncorhynchys mykiss*). *The Journal of Experimental Biology*, **206**: 1779–90.

Secondi, J., Hinot, E., Djalout, Z., Sourice, S., and Jadas-Hécart, A. (2009) Realistic nitrate concentration alters the expression of sexual traits and olfactory male attractiveness in newts. *Functional Ecology*, **23**: 800–8.

Sherba, M., Dunham, D.W., and Harvey, H.H. (2000) Sublethal copper toxicity and food response in the freshwater crayfish *Cambarus bartonii* (Cambaridae, Decapoda, Crustacea). *Ecotoxicology and Environmental Safety*, **46**: 329–33.

Singh, N., Manshian, B., Jenkins, G.J.S., Griffiths, S.M., Williams, P.M., Maffeis, T.G.G., Wright, C.J., and Doak, S.H. (2009) NanoGenotoxicology: The DNA damaging potential of engineered nanomaterials. *Biomaterials*, **30**: 3891–914.

Sloman, K.A. (2007) Effects of trace metals on salmonid fish: the role of social hierarchies. *Applied Animal Behaviour*, **104**: 326–45.

Smith, V.H. and Schindler, D.W. (2009) Eutrophication science: where do we go from here? *Trends in Ecology and Evolution*, **24**: 201–7.

Steele, C.W., Strickler-Shaw, S., and Taylor, D.H. (1992) Attraction of crayfishes *Procambarus clarkii*, *Orconectes rusticus* and *Cambarus bartoni* to a feeding stimulant and its suppression by a blend of metals. *Environmental Toxicology and Chemistry*, **11**: 1323–9.

Steinberg, C.E.W., Meinelt, T., Timofeyev, M.A., Bittner, M., and Menzel, R. (2008) Humic Substances (review series). Part 2: Interactions with Organisms. *Environmental Science and Pollution Research*, **15**: 128–35.

Straw, J. and Rittschof, D. (2004) Response of mud snails from low and high imposex sites to sex pheromones. *Marine Pollution Bulletin*, **48**: 1048–54.

Sutterlin, A.M. (1974) Pollutants and the chemical senses of aquatic animals – perspective and review. *Chemical Senses and Flavor*, **1**: 167–74.

Tabb, M. and Blumberg, B. (2006) New modes of action for endocrine-disrupting chemicals. *Molecular Endocrinology*, **20**: 475–82.

Takken, W. and Knols, B.G.J. (1999) Odor-mediated behavior of Afrotropical malaria mosquitoes. *Annual Review of Entomology*, **44**: 131–57.

Tembo, R.N. (2009) The sublethal effects of low-pH exposure on the chemoreception of *Poecilia sphenops*. *Archives of Environmental Contamination and Toxicology*, **57**: 157–63.

Thomas, J. D. (1997) The role of dissolved organic matter, particularly free amino acids and humic substances, in freshwater ecosystems. *Freshwater Biology*, **38**: 1–36.

Thompson, H.E. (2003) Behavioral effects of pesticides in bees - their potential for use in risk assessment. *Ecotoxicology*, **12**: 317–30.

Tierney, K.B., Baldwin, D.H., Hara, T.J., Ross, P.S., Scholz, N.L., and Kennedy, C.J. (2010) Olfactory toxicity in fishes. *Aquatic Toxicology*, **96**: 2–26.

Tollrian, R. and Harvell, C.D. (1999) *The Ecology and Evolution of Inducible Defenses*. Princeton University Press, New Jersey.

Tollrian, R. and von Elert, E. (1994) Enrichment and purification of *Chaoborus* kairomone from water: Further steps toward its chemical characterization. *Limnology and Oceanography*, **39**: 788–96.

Toth, G.B. and Pavia, H. (2000) Water-borne cues induce chemical defense in a marine alga (*Ascophyllum nodosum*). *Proceedings of the National Academy of Sciences USA*, **97**: 14418–20.

Turner, A.M. and Chislock, M.F. (2010) Blinded by the stink: Nutrient enrichment impairs the perception of predation risk by freshwater snails. *Ecological Applications* **20**(8): 2089–95.

Uchida, H., Yasumoto, K., Nishigami, A., Zweigenbaum, J.A., Kusumi, T., and Ooi, T. (2008) Time-of-flight LC/MS identification and confirmation of a kairomone in *Daphnia magna* cultured medium. *Bulletin of the Chemical Society of Japan*, **81**: 298–300.

Van Donk, E. (2006) Food web interactions in lake. What is the impact of chemical information conveyance. In M. Dicke and W. Takken, (eds.) *Chemical Ecology. From Gene to Ecosystem*. Wageningen UR Frontis Series, Vol. 16, Springer, The Netherlands, pp. 145–60.

Van Rijssel, M. and Gieskes, W.W.C. (2002) Temperature, light, and the dimethylsulfoniopropionate (DMSP) content of *Emiliania huxleyi* (Prymnesiophyceae). *Journal of Sea Research*, **48**: 17–27.

Verschoor, A.M., Vos, M., and van der Stap, I. (2004) Inducible defences prevent strong population fluctuations in bi- and tritrophic food chains. *Ecology Letters*, **7**: 1143–8.

Visser, J.H. (1986) Host odor perception in phytophagous insects. *Annual Review of Entomology*, **31**: 121–44.

Vom Saal, F.S., Timms, B.G., Montano, M.M., Palanza, P., Thayer, K.A., Nagel, S.C., Dhar, M.D., Ganjam, V.K., Parmigiani S., and Wekshins, W.V. (1997) Prostate enlargement in mice due to fetal exposure to low doses of estradiol or diethylstilbestrol and opposite effects at high doses. *Proceedings of the National Academy of Sciences USA*, **94**: 2056–61.

Ward, D.T. (2004) Calcium receptor-mediated intracellular signaling. *Cell Calcium*, **35**: 217–28.

Ward, A.J.W., Duff, A.J., Horsfall, J.S., and Currie, S. (2008) Scents and scents-ability: Pollution disrupts chemical social recognition and shoaling in fish. *Proceedings of the Royal Society B: Biological Sciences*, **275**: 101–5.

Waring, C.P. and Moore A. (1997) Sublethal effects of a carbamate pesticide on pheromonal mediated endocrine function in mature male Atlantic salmon (*Salmo salar* L.) parr. *Fish Physiology and Biochemistry*, **17**: 203–11.

Waters, C.M. and Bassler, B.L. (2005) Quorum sensing: Cell-to-cell communication in bacteria. *Annual Review of Cell and Developmental Biology*, **21**: 319–46.

Wauchope, R.D. (1978) The pesticide content of surface water draining from agricultural fields - A review. *Journal of Environmental Quality*, **7**: 459–72.

Webber, H.M. and Haines, T.A. (2003) Mercury effects on predator avoidance behavior of a forage fish, golden shiner (*Notemigonus crysoleucas*). *Environmental Toxicology and Chemistry*, **22**: 1556–61.

Weis, P. and Weis, J.S. (1974) DDT causes changes in activity and schooling behavior in Goldfish. *Environmental Research,* **7**: 68–74.

Wolf, M.C. and Moore, P.A. (2002) Effects of the herbicide metalochlor on the perception of chemical stimuli by *Orconenctes rusticus. Journal of the North American Benthological Society,* **21**: 457–67.

Yasumoto K., Nishigami A., Yasumoto M., Kasai F., Okada Y., Kusumi T., and Ooi, T. (2005) Aliphatic sulfates released from *Daphnia* induce morphological defense of phytoplankton: isolation and synthesis of kairomones. *Tetrahedron Letters,* **46**: 4765–7.

Yasumoto, K., Nishigami, A., Kasai, F., Kusumi, T., and Ooi, T. (2006) Isolation and absolute configuration determination of aliphatic sulfates as the *Daphnia* kairomones inducing morphological defense of a phytoplankton. *Chemical and Pharmaceutical Bulletin,* **54**: 271–4.

Yasumoto, K., Nishigami, A., Aoi, H., Tsuchihashi, C., Kasai, F., Kusumi, T., and Ooi, T. (2008) Isolation and absolute configuration determination of aliphatic sulfates as the *Daphnia* kairomones inducing morphological defense of a phytoplankton—Part 2. *Chemical and Pharmaceutical Bulletin,* **56**: 129–32.

Yasumoto, K., Nishigami, A., Aoi, H., Tsuchihashi, C., Kasai, F., Kusumi, T., and Ooi, T. (2008) Isolation of new aliphatic sulfates and sulfamate as the *Daphnia* kairomones inducing morphological change of a phytoplankton *Scenedesmus gutwinskii. Chemical and Pharmaceutical Bulletin,* **56**: 133–6.

CHAPTER 18

Aquatic chemical ecology: new directions and challenges for the future

Christer Brönmark and Lars-Anders Hansson

Chemical ecology is a vibrant research field and, as you will have gathered when reading the chapters in this volume, we are now getting a more and more complete understanding of just how important chemical defences and information networks are for the structure and function of aquatic ecosystems, freshwater as well as marine. As a base for this volume we organized a workshop at Häckeberga Castle, southern Sweden, in June 2010 and the discussions there and the written contributions to this book have been extremely inspirational to us by providing new insights and food for thought about the importance of all these intricate chemical interactions. For example, we must realize that all these organisms, some of them with brains smaller than sand grains, communicate, navigate, and receive information about resources and threats using a language of which we humans have a very limited understanding. While we, viewing ourselves as the crown of creation, mainly use visual and auditory information, we make relatively little use of our chemical senses; mainly for simple tasks such as to distinguish between sufficiently old wines and to recognize over-ripe cheese. Most other organisms, however, live in a chemical information network, where they exchange information into which we have very limited insight. Although the present generation of humans has developed the huge information network we call the Internet, which has indeed influenced our way of living and, in some cases, even our governmental institutions, similar networks, albeit chemically

based, have been utilized by many organisms to disseminate and share information over evolutionary time. Even such small and evolutionarily old organisms as bacteria use so called 'quorum sensing' to communicate within groups and take unified decisions (see e.g. Chapter 2). It is intriguing to realize that even recent human communication platforms, such as the Internet, are, in principle, just mirroring something that bacteria evolved millions or even billions of years ago. Hence, studies of the 'chemical network', mainly operating beyond our own senses, may give us a sobering perspective on our lives! However, studies of chemical ecology are not only of philosophical interest or for the development of our basic understanding, breakthroughs in this area may also be of economic importance (e.g. for the pharmaceutical industry; see Chapter 13) or for our understanding of how anthropogenic disturbances affect our ecosystems (Chapter 17).

The main aim of this book is to illustrate the importance of chemicals in mediating interactions in aquatic systems, to assess the present status of the field, and to identify areas where more research is urgently needed. Based on the discussions during our workshop and the text received from the authors, we have tried to identify some important challenges for the future, including: (1) *Which are the chemicals involved?* (2) *What is the importance of the chemical network in natural systems?* and (3) *What effects will anthropogenic disturbances have on the chemical interactions in aquatic systems?*

18.1   Which chemicals are involved?

So far, chemists have identified a large number of substances involved in interactions among organisms in aquatic systems (see Chapters 2 and 13). Unfortunately, this knowledge is not evenly distributed among subfields. The chemical identity of many metabolites involved in chemical defences in marine organisms are well known, whereas we still know very little about the chemicals involved in, for example, assessment of predator threats, migration, kin recognition, and reproduction. This may be due to economic incentives—marine algal defence chemistry has been very interesting to the pharmaceutical industry for example (e.g. Sennett 2001). Economic interests as a driver may also account for the discrepancy in chemical understanding of pheromone communication systems in aquatic versus terrestrial systems, where for example the field of insect pheromone communication and host plant finding are highly developed. One potential reason for this difference is that commercial interests, such as forestry and pest control, have driven research in terrestrial systems, whereas there is no such incentive for pheromone research in aquatic ecosystems. Further, there have also been fundamental differences in the scientific approach of different subfields. For example, much of the research on the chemical recognition of predation threat has focused on ecological and evolutionary questions; how has this evolved and what consequences does it have for individuals, populations, and communities? Less focus has been on actually identifying the chemistry behind the change in behaviour or morphology.

However, to be fair, this is not only due to different interests in different fields, but also due to differences in the chemical characteristics of the chemical cue and the present state of analytical tools in chemistry. Pohnert (Chapter 13), for example, suggests that the chemical structure of the majority of metabolites used as defences by organisms in aquatic systems can be determined using analytical approaches available today, even if it is by no means an easy task. To identify infochemicals that are dissolved in water, occur at low concentrations, and are rapidly degraded by bacteria, is significantly more problematic than identifying airborne infochemicals (Chapter 2). Nevertheless, recent improvements in sensitivity and resolution of analytical methods have resulted in progress in this field as well. For example, Yasumotu et al. (2005) extracted a kairomone from a sample (10 kg!) of the herbivore Daphnia that induced colony formation in an alga, Silberbush et al. (2010) identified two hydrocarbons from predatory backswimmers that repel ovipositing in a mosquito, and, further, Ferland-Raymond et al. (2010) linked a single ion from a dragonfly predator feeding on tadpoles to changes in tadpole behaviour. With regards to pheromones, few have been identified from aquatic systems and the high chemical diversity of the ones that have been identified suggests that almost all compounds that can be detected by a receiver against the background noise are potential chemical signals, if the receiver can associate the compound with a specific behavioural context (see e.g. Chapter 3). More chemical identification is required with specific sets of related organisms to develop evolutionary concepts and phylogenetic trees of signalling cues. Actually, pheromones are ideal for studies of signal production, metabolic costs, and evolutionary honesty as they can be measured using analytical techniques; but to date little progress has been made in aquatic systems, at least when compared to the enormous development of the field of pheromone communication in terrestrial insects.

18.2   New techniques: opportunities and pitfalls

In recent years, we have seen the development of a whole new '-omics' tool box that holds great promise for the future—metabolomics, proteomics, and transcriptomics (see also Chapters 2, 13, and 15). The metabolomics approach uses high-resolution chemical analyses to compare the full set of metabolites from control individuals to individuals that have been exposed to, for example, grazers, whereas in proteomics the focus is on comparing proteins. In transcriptomics the focus is on gene expression, especially genes that are up- or downregulated in response to different environments, such as absence/presence of predators. However, the full application of proteomics and transcriptomics to chemical ecology has so far been constrained by the lack of full genome sequence data for most aquatic organisms. One of the possibilities to more rapidly advance the research field, which was intensively discussed during the workshop and brought up in several chapters, is to identify model organisms, that is 'aquatic *Arabidopsis*'. These might be important in generating unifying concepts for studies with different tools, using different approaches. Such organisms should fulfil several criteria, including the following: its genome should be sequenced, the organism should be ecologically important and exhibit diverse interactions with other species, it should be possible to rear in laboratory culture, easy to handle, and have viable natural populations available for study. Actually, relatively few species fulfil a majority of these criteria and during the workshop only zebrafish (*Danio rerio*), the water flea (*Daphnia* spp.), and the mollusc *Aplysia* could be agreed upon. Although there are probably more suggestions for model organisms, we believe that using such models will enhance the possibility of retrieving general information about how chemical senses, information, defences, and communication function and what effects they have at community and ecosystem levels.

There is no doubt that new techniques and methods have been crucial for the recent advancement of chemical ecology and that new methods, instruments, and tools are now available. These include chemical tools, as well as instruments to study chemicals in aquatic environments, but also molecular and genetic tools, such as transcriptomics and gene expression (see e.g. Chapters 2, 8, and 15). We should, of course, embrace these new opportunities in order to improve our understanding of chemical ecology, and there is no question that the development of new technologies is now the main driver of research within many scientific areas, including chemical ecology. These new technologies give us the opportunity to address old questions that were previously out of reach, but they also generate new questions of great interest for the advancement of science. However, new instruments and technologies also provide the possibility of addressing questions that have never been, and never will be, asked, in other words providing answers that lack questions! Here we see a potential risk in becoming too carried away by glimmering chrome, and soft,

blipping sounds from the latest million dollar baby, generating too much information that has not been asked for. However, researchers are all individuals: some like field work, some like theoretical challenges, and some like new technology. Recently, the latter category has indeed got wind in their sails, whereas other approaches are now lagging behind despite being no less useful and, it may be argued, essential for advancing a solid research field! This is absolutely not to recommend that the technology-driven developments should be restricted, but to acknowledge that more ingredients than technology are needed and that we have to encourage theory development, studies in natural systems and in experiments in order to get a true picture of natural interactions, communications, defences, and ecosystems. It may also be of utmost importance to create arenas to communicate for researchers who are using different approaches, for example workshops, specific symposia, and the writing of joint review papers to tie together reductionist and holistic approaches and new technology with hypothesis testing and natural history. Indeed, this mission has driven the process of producing this book!

18.3 Scale and context: chemical interactions in nature

An obvious, albeit often neglected, statement is that ecological questions require data from natural environments, that is at a relevant scale. Many excellent studies in chemical ecology are performed in laboratory environments using abundances, concentrations, or even constellations of organisms far from what is found under natural conditions. For example, the sensational finding that the common substance trimethylamine, which occurs in fish mucus, induces diel vertical migrations in *Daphnia*, was published with fanfare in the journal *Nature* (Boriss *et al.* 1999). However, the concentration of trimethylamine used in the laboratory study was, according to a follow-up study, several orders of magnitude higher than ever recorded in natural systems, and it was later concluded that there was no ecological relevance of trimethylamine in the fish–*Daphnia* interaction (Pohnert and von Elert 2000). This particular study is, unfortunately, not alone, but clearly illustrates that we have to combine studies at different scales to forward the research field. In cases when laboratory experiments were extended to field conditions they delivered unexpectedly different and contradictory results (Johnson and Li 2010; Chapter 9).

Moreover, the environmental context has proved to be very important for the response to chemical substances, and although studies in laboratory settings are highly efficient and allow for controlled conditions and replication, they cannot mimic the physical, physiological, and social contexts associated with natural environments (Johnson and Li 2010). Hence, complementary experiments performed under field conditions may improve the generality and enhance the understanding of laboratory discoveries. However, different aquatic systems (e.g. marine vs. freshwater, plankton vs. benthos, micro- vs. macroorganisms) present inherent differences in terms of experimental possibilities such as manipulative field experiments, experiments over several generations, and so on. Another approach is to let questions regarding chemical ecology originate from field observations of behaviour, growth patterns, and organism interactions. Such observations can then inspire laboratory studies where experiments are comparative, reproducible, relatively inexpensive, easy to set up, and can be conducted even during winter.

In the real world, organisms operate in a landscape in which there are chemical plumes at a wide variety of scales emanating simultaneously from different sources and interacting with each other (see Chapter 1), providing information about both potential risks and rewards. In most laboratory simplifications we may be ignoring some potentially significant ecological effects from these interactions. Thus, in nature, the fluid environment will interact with levels of aversive cues to govern navigation to food and possibly other attractive sources. This constitutes another opportunity to investigate the coupling between biological and physical processes in determining critical ecological activities.

An example of how the physical environment may be involved in cue formation is presented by Gerlach and Atema (Chapter 5). They suggest that the choice of favourable currents by settling marine fish larvae is based on the chemical composition of the current or 'home reef'. Moreover, it is proposed that this mechanism limits dispersal, based on the demonstrated ability of settling larvae to recognize odour differences between reefs. However, very little is known about the composition of such environmental compounds and reef odour differences can either originate from the environment and/or from conspecifics. Hence, the role of the physical environment in modulating chemical sensations will likely be an important area of research in the future. Turner and Peacor (Chapter 10), suggest that integrating studies of perceptual processes with approaches from population and community ecology is a powerful and rich synthesis that is both necessary and desirable.

Another scale issue is that understanding the ecological importance of chemosensory behaviour, such as navigation, implies understanding of the transduction and integration of these chemical signals by neural signal processing, that is responses and consequences at the ecological scale are tightly connected to the physiology of the organisms involved (Chapter 11).

18.4 Responding to multiple cues in a complex environment

Most chemical signals act at very low concentrations, degrade rapidly, and often occur in complex blends instead of being pure compounds, in addition there are multiple cues with different compounds simultaneously carrying information about threats, food, mates, and settling opportunities. Moreover, organisms of course use multiple sensory inputs including vision, hearing and mechanoreception; chemical cues are but one of many sensory inputs that allows organisms to modify their behaviour/physiology/morphology. The information from the different cues and sensory modalities may interact and be synergistic, complementary, or even conflicting. For example, in turbid water, where the use of vision is impaired, fish may compensate by increasing their reliance on chemical cues when determining the local risk from predators (Hartman and Abrahams 2000). Most organisms can, at least partly, separate these multiple cues by multiple sensory modalities and also distinguish between different concentrations or intensities of the cues. For example, when exposed to high concentrations of alarm cues, a prey organism may learn that the predator is highly dangerous, but when concentrations are low they learn the predator as less threatening (Chapters 9 and 10). Organisms may also react to a chemically-induced threat with multiple defences and, for example, phytoplankton may combine different kinds of inducible defences, or even a combination of inducible and constitutive defences. The exact combination may be tuned to a variety of cues for risk of herbivory, including feeding-specific, herbivore-specific, or cell damage-related cues (Van Donk *et al.* 2010).

18.5   Anthropogenic disturbances

Finally, stress on aquatic species will continue to increase, due to impacts stemming from disturbances such as habitat degradation, invasive species, pollution, increased temperature, and ocean acidification. As should be clear from the chapters in this volume, chemical signals play a central role in aquatic systems and may have large effects on the patterns we see in ecosystems through effects on population and community processes. Many of the chemicals involved in the different interactions occur at very low concentrations and over evolutionary time the chemosensory systems have evolved to be very finely tuned to detect and react to these compounds. Even very modest alterations of biogeochemical cycles may be sufficient to alter the chemical background over which the critical cues operate, and thus upset ecological interactions. For example, Dixson *et al.* (2010) showed that newly hatched larvae of a marine fish were able to discriminate between cues from predators versus non-predators, but when exposed to a situation simulating ocean acidification they were instead attracted to predator cues and lost the ability to discriminate between predators and non-predators. Further, an increasing number of studies suggest that different pollutants may severely affect the chemosensory system of aquatic organisms (Chapter 17). Changes in the chemical environment may affect the structure of the chemical

substance so that organisms no longer recognize it as an infochemical, or it may affect the neuro-sensory processes involved in chemical detection. Thus, our ability to understand how community structure and function may be affected by these disturbances is currently limited by our understanding of how stress may influence chemical defences. This is likely an expanding future research field.

18.6   A base for take-off

A major aim of this book has been to gather researchers active within the field of aquatic chemical ecology and have them identify the state of the art and potential future directions. Hence, we would like this book to be a base for take-off for future studies aiming at understanding the role of the chemical interactions in aquatic systems. During the process of producing this volume, the group of authors identified the following constraints and possibilities, which may be used as challenges towards new insights:

- It is necessary to encourage people to take their experiments out in a real setting, that is to determine if the laboratory study provides relevant effects also in *natural systems*.
- Design hypotheses based upon field or behavioural observations, start a study with simple laboratory experiments, then *scale up* to mesocosms and let the results feed modelling. Let the model make predictions that can be tested in the field.
- More *interdisciplinary* work is needed to overcome experimental difficulties and reach general conclusions. Despite the difficulties in getting funding for these network projects, we should continue to ask for programmes that are interdisciplinary. If we are to make progress, researchers from different areas need to work around *common questions*.
- We need aquatic *model organisms*, such as *Arabidopsis* in terrestrial systems. There are several candidates!
- *Libraries for chemical compounds* are missing. If we create such libraries, specific effects of

infochemicals can be analysed much more easily and mechanistic understanding will increase. Moreover, it will improve the possibility of finding specific receptors, identifying the genes, and understanding the evolution of the chemical signal.

- At the cellular level it is important to identify receptors for infochemicals, which calls for *genomic approaches* and studies of receptor expression.
- Take advantage of the *new technology* in the search for the identity of the chemicals involved!

In this book we have demonstrated that chemical information is involved in competition, predation, mate choice, resource finding, navigation, and also in defence and protection. Hence, there is no doubt that chemical sensing and communication is a major phenomenon in natural systems and in the ecology, evolution, and behaviour of the organisms we study. Surprisingly, however, few studies take this chemical information network into account, which is obvious when glancing at recent issues of general ecological and, especially, aquatic ecological journals. This lack of recognition of one of the major forces acting in aquatic systems is probably the result of the difficulties in analysing and in understanding the 'chemical language' of aquatic organisms. Realizing that this is the case was a major driving force behind our decision to invite all these authors to write this volume. During our long scientific careers within aquatic ecology and limnology we have repeatedly had the feeling that organisms at all levels are using cues other than visible ones and that direct confrontation between predator and prey, male and female, or host plant and epiphyte, is not the only interaction, but that there is a chemical network of which we as yet have only a vague idea. Hence, it is with great respect and curiosity that we approach this dimension of organismal interactions and we can only hope that this book can serve as an inspiration and starting point for future understanding of the chemically-based ecological network in aquatic ecosystems.

References

Boriss, H., Boersma, M., and Wiltshire, K. H. (1999) Trimethylamine induces migration of waterfleas. *Nature,* **398**: 382.

Dixson, D. L., Munday, P. L., and Jones, G. P. (2010) Ocean acidification disrupts the innate ability of fish to detect predator olfactory cues. *Ecology Letters,* **13**: 68–75.

Ferland-Raymond, F., March, R. E., Metcalfe, C. D., and Murray, D. L. (2010) Prey detection of aquatic predators: assessing the identity of chemical cues eliciting prey behavioural platsticity. *Biochemical Systematics and Ecology,* **38**: 169–77.

Hartman, E. J. and Abrahams, M. V. (2000) Sensory compensation and the detection of predators: the interaction between chemical and visual information. *Proceedings of the Royal Society of London Series B-Biological Sciences,* **267**: 571–5.

Johnson, N. S. and Li, W. (2010) Understanding behavioural responses of fish to pheromones in natural freshwater environments. *Journal of Comparative Physiology. A.,* **196**: 701–11.

Pohnert and Van Elert (2000) No ecological relevance of trimethylamine in fish - *Daphnia* interactions. *Limnol. Oceanogr.,* **45**: 1153–1156.

Sennett, S. H. (2001) Chemical ecology: Applications in marine biomedical prospecting. In McClintock, J. B. and Baker, B. J. (eds) *Marine Chemical Ecology.* CRC Press, Boca Raton, Florida.

Silberbush, A., Markman, S., Lewinsohn, E., Bar, E., Coeh, J. E., and Blaustein, L. (2010) Predator-released hydrocarbons repel oviposition by a mosquito. *Ecology Letters,* **13**: 1129–38.

Van Donk, E., Ianora, A., and Vos, M. (2010) Induced defences in marine and freshwater phytoplankton: a review. *Hydrobiologia,* **668**: 3–19.

Yasumoto, K., Nishigami, A., Yasumoto, M., Kasai, F., Okada, Y., Kusimi, T., and Ooi, T. (2005) Aliphatic sulfates released from *Daphnia* induce morphological defense of phytoplankton: isolation and synthesis of kairomones. *Tetrahedron Letters,* **46**: 4765–7.

Index

Note: page numbers in *italics* refer to figures, tables, and boxes.